AF783287

V. S. Butylkin A. E. Kaplan
Yu. G. Khronopulo E. I. Yakubovich

Resonant Nonlinear Interactions of Light with Matter

V. S. Butylkin A. E. Kaplan
Yu. G. Khronopulo E. I. Yakubovich

Resonant Nonlinear Interactions of Light with Matter

Translated by O. A. Germogenova

With 70 Figures

Springer-Verlag Berlin Heidelberg New York
London Paris Tokyo

Professor Dr. Valerii S. Butylkin
Institute of Radioengineering and Electronics
Academy of Sciences of the USSR
Marx Prospect, 18, SU-103907 Moscow, USSR

Professor Dr. Alexander E. Kaplan
Department of Electrical Engineering
Barton Hall, John Hopkins University
Baltimore, MD 21218, USA

Professor Dr. Yury G. Khronopulo
701 Empire Blvd., Apt. 1D
Brooklyn, New York 11213, USA

Professor Dr. Evsei I. Yakubovich
Institute of Applied Physics
Academy of Sciences of the USSR
Ul'yanova Street, 46
SU-603600 Gorkii, USSR

Translator:
Dr. O. A. Germogenova
Prospect Vernadskogo, d.95, korp. 2, kv. 37
SU-117526 Moscow, USSR

Title of the original Russian edition: Rezonansnuie vzaimodeĭstviya sveta s veshchestvom.
© Nauka, Moscow 1977

ISBN-13: 978-3-642-68893-5 e-ISBN-13: 978-3-642-68891-1
DOI: 10.1007/978-3-642-68891-1

Library of Congress Cataloging-in-Publication Data. Rezonansnuie vzaimodeĭstviya sveta s veshchestvom. English. Resonant nonlinear interactions of light with matter. Translation of: Rezonansnuie vzaimodeĭstviya sveta s veshchestvom. 1. Nonlinear optics. 2. Resonance. 3. Quantum optics. 4. Multiphoton processes. I. Butylkin, Valeriĭ Semenovich. QC 446.2.R4913 1987 535 86-24856

This work is subject to copyright. All rights are reserved, whether the whole or part of the material is concerned, specifically the rights of translation, reprinting, reuse of illustrations, recitation, broadcasting, reproduction on microfilms or in other ways, and storage in data banks. Duplication of this publication or parts thereof is only permitted under the provisions of the German Copyright Law of September 9, 1965, in its version of June 24, 1985, and a copyright fee must always be paid. Violations fall under the prosecution act of the German Copyright Law.

© Springer-Verlag Berlin Heidelberg 1989

Softcover reprint of the hardcover 1st edition 1989

The use of registered names, trademarks, etc. in this publication does not imply, even in the absence of a specific statement, that such names are exempt from the relevant protective laws and regulations and therefore free for general use.

2157/3150-543210 – Printed on acid-free paper

Preface to the English Edition

This book is devoted primarily to the various kinds of resonant nonlinear interactions of light with two-level (or, in many cases, multilevel) systems. The interactions can involve one-photon as well as multiphoton processes in which some combinations of frequencies of participating photons are close to transitions of atoms or molecules (e.g., we consider stimulated Raman scattering (SRS) as a resonant interaction). This approach involves a broad spectrum of problems. Discussion of some of the basic phenomena as well as the pertinent theory could be found, for instance, in such well-known books as the ones due to N. Bloembergen; S.A. Akhmanov and R.V. Khokhlov; L. Allen and J.H. Eberly, and to V.M. Fain and Ya.I. Khanin. The book "Quantum Electronics" by A. Yariv could serve as an introductory guide to the subject. Thus, some of the basic material in the present book will already be well known to the reader who is an expert in the field. There are, for instance, general density matrix equations; two-level model and basic effects associated with this model, such as saturation of one-photon absorption and Raby oscillations; some basic multiphoton processes such as two-photon absorption, SRS, etc.

However, a large portion of this book is devoted to more recent new results which are not very well known. Among them, the reader will find:

a) the generalized two-level model (Chap. 1), which allows one to retain the main features of the two-level approach when dealing with high-order nonlinear resonant interactions (which involve many photons and/or many levels). The equations of the generalized two-level model (later on used in Chaps. 2, 4, 6–8) allow for a drastic simplification in the description of multiphoton resonance interactions of light waves, accounting at the same time for saturation, dynamical Stark shift of the levels, phase relations, the existence of parametric interactions of waves as well as other physical factors,

b) the theory of two-level nonlinear susceptibility which takes into consideration the entire set of off-resonant levels; this leads to such effects as nonlinear shift of resonant levels and "repolarization" (Chap. 2);

c) the analytic theory of relaxation and Rabi oscillations in the two-level system under action of quasi-resonant radiation with time-dependent amplitude and frequency (Chap. 3); this theory allows one to describe interaction of the system with laser pulses having virtually any envelope shape and duration;

d) the theory of dispersion relations (the generalized Kramers-Kronig theorem for nonlinear media); the polarization properties of the nonlinear response of the system (Chap. 4), and spatial structure of one-dimensional waves in nonlinear resonant media (Chap. 5);

e) the theory of three-photon (Chap. 6) and four-photon (Chap. 7) parametric interactions which can describe such processes as anti-Stokes SRS, generation of high-order frequency combinations under resonant conditions, which are by now widely used in IR, visible and UV ranges to produce coherent radiation; and

f) the theory of self-action of light (i.e., self-focusing, self-defocusing, and self-bending of light beams), which is based on resonant nonlinear interactions (Chap. 8), in particular, self-action which is due to SRS.

This edition is not just an English translation of the Russian original. The material has been substantially updated and revised. The main change, however, is that new results, based on most recent research progress are added. For instance, the use of a non-uniform electrostatic field to increase gain in SRS is discussed (Sect. 6.4.3); both theory and experiment are presented.

One of the most recent and exciting new results is concerned with novel effects that are due to self-action of light at nonlinear interfaces (Sect. 8.6) and in cross-self-focusing beams of light in nonlinear media (Sect. 8.7). The particularly interesting result of these effects is cavityless optical bistability. Optical bistability (OB) has become a rapidly growing field in nonlinear optics. This is related to the potential of OB systems to perform logic and memory functions as elements of optical computer and optical signal processing systems with extremely high operational speed. The use of cavities in the known OB effects imposes certain restrictions. First cavityless OB effects without these restrictions have been proposed and experimentally verified in recent years.

Acknowledgements related to my own contribution to this book can be found at the ends of Chaps. 3 and 8 written by me while the ones concerning the entire book are contained in "From the authors".

Baltimore, February 1989 *Alexander E. Kaplan*

Preface to the Russian Edition

The present book is devoted to resonant nonlinear optical processes. Such processes are, perhaps, the most complex and diverse phenomena in nonlinear optics. Due to this complexity, the abundance of concrete experimental and theoretical papers did not yet find sufficient consideration in the monographical literature.

Their unified approach towards the determination of the resonant response of matter allowed the authors of the present book to describe consistently and with maximum clarity, all prominent features of the behavior of molecular polarization that arise in complex multiphoton interactions with intense coherent fields. As far as the application of the results to the solution of wave problems is concerned, we would like to make the following remark. Because of the specific form of nonlinear polarization associated with resonances of the medium, the behavior of interacting waves differs so sharply from that of the nonresonant cases that it would possibly make sense to introduce a new branch of electrodynamics (by analogy with, for instance, magnetohydrodynamics). The realization of such a program is beyond the scope of the present book, although it does mark the beginning of such a project.

The authors are actively engaged into research in the field of resonant nonlinear optics and they obtained important results, hence, the reader gets first-hand information on this topic. The book should be equally useful for specialists and for those wishing to become familiar with this rapidly developing and interesting branch of physics.

Academician R. V. Khokhlov

From the Authors

The authors were greatly inspired and encouraged to write this book by the late Professor Rem V. Khokhlov, the former President of Moscow State University and one of the Soviet pioneers in the field of nonlinear optics. He will always be remembered by the authors as an outstanding scientist and a considerate man.

The various fragments of this book were discussed with many of our colleagues and researchers at other research institutions. We greatly appreciate their attention, suggestions, and criticism.

The material of the book is distributed between the authors in the following way: Chaps. 1 (excluding Sect. 1.2.3), 2, 6, and 7 (excluding Sect. 7.2,4) are written by V.S. Butylkin and Yu. G. Khronopulo; Sect. 1.2.4 by V.S. Butylkin, Yu. G. Khronopulo, and E.I. Yakubovich; Sects. 4.1 and 7.2,4 by Yu. G. Khronopulo; Chaps. 4 (excluding Sect. 4.1) and 5 by E.I. Yakubovich; Chaps. 3 and 8 are written by A.E. Kaplan.

V.S. Butylkin · A.E. Kaplan
Yu. G. Khronopulo · E.I. Yakubovich

Contents

1. Resonant Multiphoton Interactions and the Generalized Two-Level System

This chapter is devoted to the theory of resonant multi-photon interactions of light with molecules of matter. It presents a classification of resonant interactions, as well as a heuristic approach that demonstrates a possible way of simplifying the equations that govern the variation of the density matrix. The Bogolyubov method for deriving truncated equations and evaluating the polarization of matter is treated in detail. The effective operators, namely, the averaged Hamiltonian and the generalized dipole moment determine the behavior of the density matrix and also the molecular polarization. An obvious pictorial relation between the operators is established. The theory thus constructed allows one to reduce the task of analyzing complex multi-photon interactions of fields with matter to a much simpler problem that involves single-photon resonant interactions.

The above approach makes it possible to take into account accurately all factors that specify the molecular response not only for incoherent, but also for coherent (parametric) multi-photon processes, including saturation of populations, the dynamic Stark effect, different values of linear polarizability of levels, and interference of the probability amplitudes characterizing transitions between them that may be due to various kinds of resonances. The theory developed here also describes nonstationary interactions.

In quantum electronics or radiophysics the word "resonance" means that the field frequency coincides with one of the eigenfrequencies of the system ω_{mn}. They are determined by differences between the energy levels of the particles of the substance. In the case of nonlinear interactions in which several quasimonochromatic field components $\boldsymbol{E}$[1] with the frequencies ω_j

$$\boldsymbol{E} = \sum_j \vec{\mathcal{E}}(\omega_j, \boldsymbol{r}, t)\exp(\mathrm{i}\omega_j t) \ , \tag{1.1}$$

participate, the resonance condition has a more general form:

$$\sum_j n_j\omega_j = \omega_{mn} + \nu \ , \tag{1.2}$$

[1] For technical reasons vectors will either be denoted by symbols like $\boldsymbol{E}$ or $\vec{\mathcal{E}}$. In view of the well specified context this is not expected to give rise to any confusion.

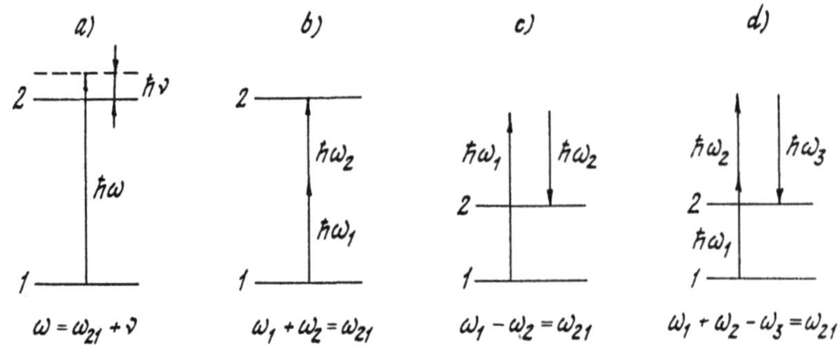

Fig. 1.1a–d. The diagrams of incoherent resonant processes: **(a)** single-photon absorption; **(b)** two-photon absorption; **(c)** Raman scattering; **(d)** three-photon Raman scattering

where ν is a small detuning. The positive integers n_j give the degeneracy of the frequency ω_j, and their sum

$$q = \sum_j n_j$$

indicates the order of the resonance. For instance, when $q = 1$, a first-order resonance is observed which is accompanied by the single-photon absorption or emission of light, depending on the sign of the difference between level populations (the single-photon resonance).

Note that the first phenomenon of nonlinear optics detected experimentally was the self-induced transparency of the medium caused by saturation of the absorption during single-photon resonance [1.1]. The second-order resonances ($q = 2$) are realized for two-photon absorption (emission) [1.2] and Raman scattering of light [1.3]. The third-order resonances correspond, for instance, to three-photon absorption [1.4] and three-photon Raman scattering [1.5] (Fig. 1.1).

In the above processes only one resonance condition is fulfilled, i.e. (1.2). In such cases the spatial and temporal variations of energies of interacting fields do not depend on the relation between their phases. In the literature these kinds of interactions are called incoherent or multiphoton ones (see [1.6] and Akhmanov and Khokhlov's introductory article to the Russian edition of the book by *Bloembergen* [1.7]). Among them we are going to distinguish between the multiphoton absorption (emission) with its elementary acts involving one or more photons, and the Raman processes, in which some photons are emitted and some are absorbed. In the latter case the frequencies of the absorbed and the emitted photons in condition (1.2) have different signs.

In many nonlinear phenomena several resonance relations of different orders (or of the same order but with the participation of different frequencies) are fulfilled. Below it will be shown that an important role in such processes is played by the phase relations between the interacting fields. These sorts

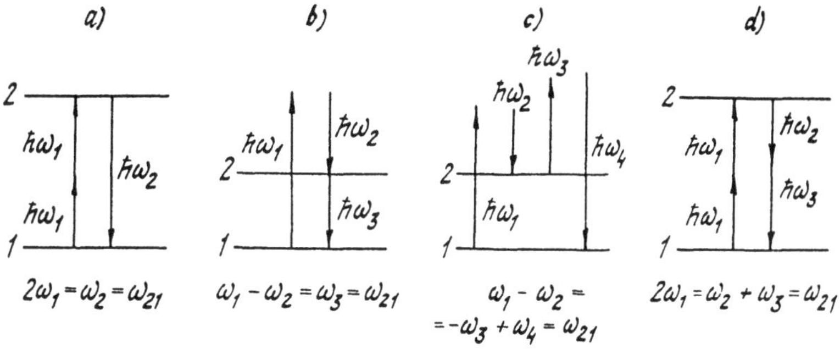

Fig. 1.2a–d. The diagrams of resonant parametric processes: (**a**) resonant generation of the second harmonic; (**b**) generation of the difference frequency during stimulated Raman scattering; (**c**) parametric generation of the anti-Stokes component in stimulated Raman scattering; (**d**) generation of the difference frequency during two-photon absorption

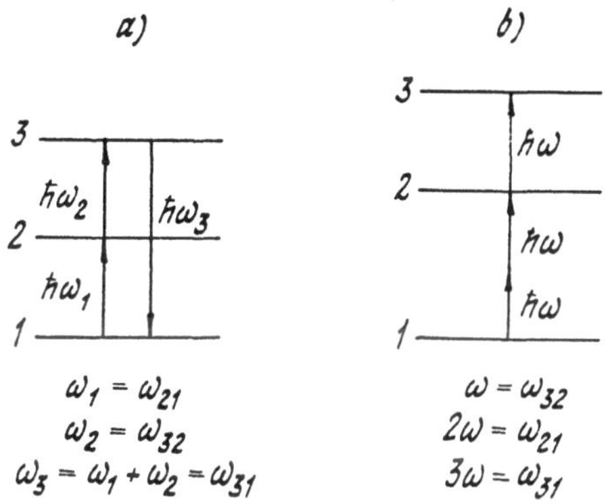

Fig. 1.3a,b. Examples of resonant multiphoton interactions involving three molecular levels

of interactions are classified as coherent or parametric [1,6,8]. To distinguish between these and the similar nonresonant processes, we shall call them resonant parametric interactions. Some of them are depicted in Fig. 1.2. Obviously, transitions involving several pairs of levels, including those having one level in common (see Fig. 1.3), may turn out to be resonant. In the following, the superscript "s" will be employed to denote the resonances that exist simultaneously.

A unified theory which would be applicable for the description of the whole manifold of resonant interactions can be based neither on ordinary perturbation theory, which forms the foundation of the traditional theory of radiation, nor on phenomenological balance-type equations. This is associated with the following circumstances that frequently occur simultaneously: (1) The laser field strengths are so large that a considerable variation in the populations of the excited states is possible, even for higher-order resonances. (2) The response of matter to strong light fields may depend essentially on the energy level shift produced by the dynamic Stark-effect. (3) Radiation may act during a very short time, which requires the nonstationary response to be taken into account. (4) The coherence of the laser radiation leads to a complex dependence of the response of matter during resonant parametric interactions. This response depends not only on the amplitudes but also on the phases of interacting fields.

The description of the single-photon resonance was essentially achieved by the use of the two-level model of the molecule [1.9, 10]. In particular, most results in the theory of quantum generators have been obtained on the basis of two-level models.

The authors of [1.11–16] demonstrated that the form of the equations determining the response in the case of complex resonances in which one of the eigenfrequencies of matter participates is also similar to that of a two-level system. The former differ from the latter in that the coefficients of equations contain as parameters the characteristics of all states of a molecule and the amplitudes and phases of the fields acting on it. Later it was established that for resonant interactions of any order that involve an arbitrary number of levels, the truncated equation for the density matrix has the form of a Neumann equation in which, instead of the interaction energy, a certain effective Hamiltonian appears whose nondiagonal matrix elements differ from zero only for resonant transitions [1.17]. This motivated the introduction of the concept of generalized n-level systems (n being the number of levels participating in resonant interactions) which simplifies very much the description of complex resonant processes.

The present chapter deals with the equations of generalized n-level systems presenting also a calculation of their polarization. The expressions obtained in the first chapter will subsequently be used in the description of multiphoton resonant processes.

1.1 The Basic Equations Describing the Evolution of Radiation Interacting with Matter

Resonant interaction of light with a molecule (atom) leads to the changes of the respective states. The changes in electric and magnetic fields ($\boldsymbol{E}$ and $\boldsymbol{H}$, respectively) of the light wave will be described by classical electrodynamics as is usually done in quantum radiophysics. In the most general case the wave equation

$$[\nabla[\nabla \boldsymbol{E}]] + \frac{1}{c^2}\frac{\partial^2 \boldsymbol{E}}{\partial t^2} + \frac{4\pi}{c^2}\frac{\partial^2 \boldsymbol{P}}{\partial t^2} = 0 \qquad (1.1.1)$$

is applied. In the above formula the polarization $\boldsymbol{P}$ contains a component which is nonlinear with respect to the field. Under general conditions the exact solution of this equation is impossible. However, there is a method that enables one to obtain an approximate solution for most problems of practical interest [1.8, 18]. This method is based on the fact that the characteristic distance and time of variation in the light wave amplitude are much greater than its wavelength and period. This is a consequence of the nonlinear part of the polarization and of the linear losses being small. Hence, the exact solution to (1.1.1) represents to a first approximation a superposition of waves

$$\boldsymbol{E} = \sum \boldsymbol{e}_j C_j(r,t)\exp[\mathrm{i}(\omega_j t - \boldsymbol{k}_j \cdot \boldsymbol{r})] \ .$$

Its slowly varying complex amplitudes C_j satisfy the first-order equations [1.18]

$$\begin{aligned}&[\boldsymbol{e}_j \times [\boldsymbol{k}_j \times \boldsymbol{e}_j]]\boldsymbol{s}_j \frac{\partial C_j}{\partial t} + [\boldsymbol{e}_j \times [\boldsymbol{k}_j \times \boldsymbol{e}_j]]\nabla C_j + \boldsymbol{e}_j \hat{\alpha} \boldsymbol{e}_j C_j \\ &\quad = -\mathrm{i}\frac{2\pi}{c^2}\omega_j^2 \boldsymbol{e}_j \cdot \vec{\mathcal{P}}^{\mathrm{nl}}(\omega_j, \boldsymbol{k}_j)\end{aligned} \tag{1.1.2}$$

(that do not take into account the dispersion and diffraction which would lead to a spreading of wave packets). In these formulas, $\boldsymbol{e}_j$ is the unit vector of the j-th wave polarization, $\boldsymbol{k}_j$ its wave vector, $\boldsymbol{s}_j$ the ray vector (whose modulus is equal to the inverse group velocity), $\boldsymbol{e}_j \hat{\alpha} \boldsymbol{e}_j$ is the linear attenuation coefficient and $\vec{\mathcal{P}}^{\mathrm{nl}}(\omega_j, \boldsymbol{k}_j)$ the slowly varying complex amplitude of the nonlinear polarization wave with the frequency ω_j and the wave vector $\boldsymbol{k}_j$.

It often happens in experiments that the characteristic time in which the field amplitude changes (the pulse duration, etc.) is much greater than the relaxation time in matter, and also much greater than the time in which the light passes through the nonlinear medium $t = L/c$, L being the thickness of the medium and c the velocity of light. In this case the interaction is quasistationary, and one can discard the term with $\partial C_j/\partial t$ in (1.1.2). If the real amplitude $\mathcal{A}_j$ and phase φ_j are introduced in such a way that $C_j = \mathcal{A}_j \exp(-\mathrm{i}\varphi_j)$, then (1.1.2) for quasistationary interaction assumes the form

$$\begin{aligned}&[\boldsymbol{e}_j \times [\boldsymbol{k}_j \times \boldsymbol{e}_j]]\nabla \mathcal{A}_j - \mathrm{i}\mathcal{A}_j[\boldsymbol{e}_j \times [\boldsymbol{k}_j \times \boldsymbol{e}_j]]\nabla\varphi_j \\ &\quad = -\boldsymbol{e}_j \hat{\alpha} \boldsymbol{e}_j \mathcal{A}_j - \mathrm{i}\frac{2\pi}{c^2}\omega_j^2 \boldsymbol{e}_j \cdot \vec{\mathcal{P}}^{\mathrm{nl}}(\omega_j, \boldsymbol{k}_j)\exp(\mathrm{i}\varphi_j) \ .\end{aligned} \tag{1.1.3}$$

Whence the equations for the real field amplitudes and phases

$$\begin{aligned}&[\boldsymbol{e}_j \times [\boldsymbol{k}_j \times \boldsymbol{e}_j]]\nabla \mathcal{A}_j + \boldsymbol{e}_j \hat{\alpha} \boldsymbol{e}_j \mathcal{A}_j \\ &\quad = \frac{2\pi}{c^2}\omega_j^2 \mathrm{Im}\{\boldsymbol{e}_j \cdot \vec{\mathcal{P}}^{\mathrm{nl}}(\omega_j, \boldsymbol{k}_j)\exp(\mathrm{i}\varphi_j)\}\end{aligned} \tag{1.1.4}$$

$$[\boldsymbol{e}_j \times [\boldsymbol{k}_j \times \boldsymbol{e}_j]]\nabla\varphi_j = \frac{2\pi\omega_j^2}{c^2 \mathcal{A}_j}\mathrm{Re}\{\boldsymbol{e}_j \cdot \vec{\mathcal{P}}^{\mathrm{nl}}(\omega_j, \boldsymbol{k}_j)\exp(\mathrm{i}\varphi_j)\} \tag{1.1.5}$$

are derived. The system of equations describing the interaction between light and the molecules of the medium becomes closed if the electrodynamic equations are supplemented by an equation that determines the evolution of the molecular states. According to quantum mechanics, the latter is regarded as specified if the density matrix $\hat{\varrho}$ of the molecule is given (see [1.6], Sects. 1, 2, 6, 7). The density matrix is a Hermit' operator, i.e., its matrix elements ϱ_{mn} satisfy the condition

$$\varrho^*_{mn} = \varrho_{nm} \ . \tag{1.1.6}$$

The diagonal elements of the density matrix ϱ_{nn} correspond to the probabilities to observe the molecule in the n-th state. They are subject to the normalization condition

$$\mathrm{Tr}\{\hat{\varrho}\} = \sum_n \varrho_{nn} = 1 \ . \tag{1.1.7}$$

By making use of the density matrix and the electric dipole moment operator $\hat{\boldsymbol{d}}$ with the matrix elements $\boldsymbol{d}_{mn}$, one can find the average polarization of the molecule which is the source of fields in the electrodynamics equations:

$$\boldsymbol{P} = \mathrm{Tr}\{\hat{\boldsymbol{d}}\hat{\varrho}\} = \sum_{m,n} \boldsymbol{d}_{mn}\varrho_{nm} \ . \tag{1.1.8}$$

When the operators that describe the molecule do not depend on time, then the Schrödinger representation leads to

$$\frac{d\hat{\varrho}}{dt} + \hat{\Gamma}\hat{\varrho} = -\frac{\mathrm{i}}{\hbar}(\hat{\mathcal{H}}\hat{\varrho} - \hat{\varrho}\hat{\mathcal{H}}) = -\frac{\mathrm{i}}{\hbar}[\hat{\mathcal{H}}, \hat{\varrho}] \ , \tag{1.1.9}$$

characterizing the evolution of $\hat{\varrho}$. In this formula $\hat{\mathcal{H}}$ is the Hamiltonian of a molecule in the radiation field, $\hat{\Gamma}$ is the operator of the molecular interaction with the dissipative system which is responsible for the relaxation processes in the molecule[2]. The square brackets denote as usual the commutator of the respective operators.

The Hamiltonian $\hat{\mathcal{H}}$ consists of $\hat{\mathcal{H}}^0$ depicting the free molecule and the operator $\hat{V}$ (with the matrix elements V_{mn}) describing its interaction with the electromagnetic field:

$$\hat{\mathcal{H}} = \hat{\mathcal{H}}^0 + \hat{V} \ . \tag{1.1.10}$$

The eigenvalues of the free molecular Hamiltonian $\hat{\mathcal{H}}^0_m$ correspond to the eigenfunctions ψ_m. In the following it is assumed that the matrix elements of all operators are determined by means of the wavefunctions ψ_m. To put it differently, we shall employ a $\hat{\mathcal{H}}^0$ representation.

[2] Examples of dissipative interactions are the molecular collisions in a gas, interaction of impurity particles with the crystalline lattice and spontaneous emission in free space.

Both the operator of the interaction energy $\hat{V}$ and the operator $\hat{\boldsymbol{d}}$ correspond to real physical quantities and are therefore Hermitian. From the Hermitian property of these operators one obtains

$$\boldsymbol{d}^*_{mn} = \boldsymbol{d}_{nm} \ , \quad V^*_{mn} = V_{nm} \ . \tag{1.1.11}$$

Since the field $\boldsymbol{E}$ is supposed to be represented as a set of quasi-monochromatic waves, see (1.1), the matrix element of the energy of the molecule interacting with this field can be written as a sum:

$$V_{mn} = \sum_j V^{(j)}_{mn}(t)\exp(\mathrm{i}\omega_j t) \ . \tag{1.1.12}$$

The amplitudes $V^{(j)}_{mn}(t)$ are slowly varying functions of time. Because of the hermiticity of the operator $\hat{V}$ they satisfy the condition

$$(V^{(j)}_{mn})^* = V^{(-j)}_{nm} \ . \tag{1.1.13}$$

In the next sections, we shall specify the concrete form of interaction to be an electric dipole. In this case the interaction energy can be expressed through the dipole moment operator (see [1.6], Sect. 5):

$$\hat{V} = -\hat{\boldsymbol{d}} \cdot \boldsymbol{E} \ , \quad V^{(j)}_{mn} = -\boldsymbol{d}_{mn} \cdot \vec{\mathcal{E}}(\omega_j) \ . \tag{1.1.14}$$

For magnetic dipole interactions one has to perform the following substitution in all formulas: $\boldsymbol{d} \to \boldsymbol{\mu}$, $\boldsymbol{E} \to \boldsymbol{H}$, $\boldsymbol{P} \to \boldsymbol{M}$, $\boldsymbol{d}_{mn} \to \boldsymbol{\mu}_{mn}$ where $\boldsymbol{\mu}$ is the magnetic dipole moment operator and $\boldsymbol{M}$ the vector of magnetic polarization of matter.

Consider the relaxation term $\hat{\Gamma}\hat{\varrho}$ of the kinetic equation. Generally speaking, its dependence on $\hat{\varrho}$ can be quite complex (see, for instance, (7.18) in [1.6]). As is the usual practice in quantum radiophysics, we shall utilize the following relaxation operator [1.6, 7, 19]:

$$(\hat{\Gamma}\hat{\varrho})_{mn} = \begin{cases} T^{-1}_{mn}\varrho_{mn} & \text{for} \quad n \neq m \\ \sum_k (W_{mk}\varrho_{mm} - W_{km}\varrho_{kk}) & \text{for} \quad n = m \ . \end{cases} \tag{1.1.15}$$

The transition probability per unit time for a molecule to go over from the state m into the state k, W_{mk}, results from its interaction with only the dissipative system (i.e., for $\hat{V} \equiv 0$). $T_{mn} = T_{nm}$ is the characteristic relaxation time of the non-diagonal elements ϱ_{mn} and ϱ_{nm} of the density matrix.

For the sake of convenience we shall use the interaction representation instead of the Schrödinger representation. The corresponding transformation is performed by means of the unitary operator

$$\hat{U}_0 = \exp\left(-\frac{\mathrm{i}}{\hbar}\hat{\mathcal{H}}^0 t\right) \ .$$

It preserves the operator form of all equations and the operators themselves are transformed according to the rule

$$\hat{A}_{\mathrm{int}} = \hat{U}_0^{-1} \hat{A} \hat{U}_0 \ . \tag{1.1.16}$$

The density matrix, in particular, is transformed in the same way:

$$\hat{\varrho}_{\mathrm{int}} = \hat{\sigma} = \hat{U}_0^{-1} \hat{\varrho} \hat{U}_0 \ . \tag{1.1.17}$$

The kinetic equation in the interaction representation has the form:

$$\frac{d\hat{\sigma}}{dt} + \hat{\Gamma}\hat{\sigma} = -\frac{\mathrm{i}}{\hbar}(\hat{V}_{\mathrm{int}}\hat{\sigma} - \hat{\sigma}\hat{V}_{\mathrm{int}}) = -\frac{\mathrm{i}}{\hbar}[\hat{V}_{\mathrm{int}}, \hat{\sigma}] \ . \tag{1.1.18}$$

In terms of the matrix elements the transition to the interaction representation corresponds to the substitution

$$\varrho_{mn} = \sigma_{mn} \mathrm{e}^{-\mathrm{i}\omega_{mn} t} \ , \quad V_{mn} = (V_{\mathrm{int}})_{mn} \mathrm{e}^{-\mathrm{i}\omega_{mn} t} \ , \tag{1.1.19}$$

where ω_{mn} are the frequencies of transitions between the levels m and n :

$$\omega_{mn} = \hbar^{-1}(\mathcal{H}_m^0 - \mathcal{H}_n^0) \ . \tag{1.1.20}$$

The matrix element $(\hat{\Gamma}\hat{\sigma})_{mn}$ is obtained if in the right-hand and the left-hand sides of (1.1.15), ϱ is replaced by σ.

1.2 The Truncated Equations for the Density Matrix

Let us write the basic equation (1.1.18) in matrix form and make use of (1.1.12, 19), to obtain explicitly the harmonic dependence on time of the coefficients near the matrix elements σ_{mn} :

$$\begin{aligned} &\left(\frac{d}{dt} + T_{mn}^{-1}\right)\sigma_{mn} \\ &\quad = -\frac{\mathrm{i}}{\hbar}\sum_{k,j}\left\{V_{mk}^{(j)}\sigma_{kn}\mathrm{e}^{\mathrm{i}(\omega_{mk}+\omega_j)t} - \sigma_{mk}V_{kn}^{(j)}\mathrm{e}^{\mathrm{i}(\omega_{kn}+\omega_j)t}\right\} \ , \\ &\frac{d}{dt}\sigma_{mm} + \sum_k(\sigma_{mm}W_{mk} - W_{km}\sigma_{kk}) \\ &\quad = -\frac{\mathrm{i}}{\hbar}\sum_{k,j}\left\{V_{mk}^{(j)}\sigma_{km}\mathrm{e}^{\mathrm{i}(\omega_{mk}+\omega_j)t} - \mathrm{c.c.}\right\} \ . \end{aligned} \tag{1.2.1a}$$

The existence of small parameters in the system (1.2.1a) makes it possible to simplify it. In order to single out these parameters, it is sufficient to introduce the dimensionless time $\omega_{mn}t$. Then in the left-hand sides of (1.2.1), parameters of the type $(\omega_{mn}T_{kr})^{-1}$ and $W_{kr}\omega_{mn}^{-1}$ occur. They are small since the period of optical oscillations $2\pi/\omega_{mn}$ is much less than the life time of the molecule in the excited states and the relaxation times T_{kr} that describe the width of the line of the single-photon absorption between the levels k and r [1.6].

On the right-hand side of (1.2.1a), ratios of the amplitudes of transition energy $V_{kr}^{(j)}/\hbar\omega_{mn}$ appear which are also usually much less than unity. Let us estimate this ratio, for instance, for the case when the allowed electric dipole transition ($|d_{kr}| = 10^{-18}$ cgse) is subject to the action of the field due to the focussing of the radiation of a laser with modulated quality. It can also be obtained via picosecond pulses of lasers operating under the conditions of mode synchronization with the radiation power flux density being of the order of 10^{11} W/cm^2. For the optical range frequencies ($\omega_{mn} = 10^{15}$ s^{-1}) even in such intense fields the inequality $|V_{kr}^{(j)}/\hbar\omega_{mn}| \approx 1.5 \times 10^{-2} \ll 1$ holds. Most observed nonlinear optical phenomena occur at much lower radiation intensities.

Equations (1.2.1a) can be reduced to the standard form [1.20]

$$\frac{d}{dt}\sigma_{mn} = \varepsilon_j F_j(\sigma_{kr}, t, \varepsilon_j) \quad , \qquad (1.2.1b)$$

where ε_j are small parameters. Hence, it is possible to say that in addition to the rapid oscillations with the characteristic periods of the order of $2\pi/\omega_j$ and $2\pi/\omega_{mn}$, there should also be slow variations in the density matrix. The consequence of this fact is that the temporal behavior of the average values of physical quantities determined by the density matrix should also contain rapidly and slowly varying parts. From physical considerations it is obvious that the slow variations are those of the molecular response characteristics (for example, changes in the polarization amplitude or in the level populations) during times of the order of T_{mn} and W_{mn}^{-1}, and also those with characteristic times determined by the energy of the molecular interaction with the field $t_{\text{int}} = \hbar/|V_{kr}^{(j)}|$.

It is well known, that the separation of rapid and slow variations in equations of the type (1.2.1b) can lead to their simplification. One of the mathematical techniques usually employed for this purpose is the method of averaging [1.20–22]. We shall apply it directly to the system (1.2.1a) without reducing it to the standard form (1.2.1b). In order to describe the q-th order resonance, it is necessary to use at least the q-th approximation of the averaging method employed.

1.2.1 The Two-Level Model and the First Approximation of the Averaging Method

As the first example, consider the simplest case when a certain frequency of the quasimonochromatic field $\mathcal{E}(\omega_\alpha)$ is close to the transition frequency ω_{21}, i.e., the first-order resonance condition

$$\omega_\alpha = \omega_{21} + \nu_1 \ , \quad |\nu_1| \ll \omega_{21}, \omega_\alpha \ , \tag{1.2.2}$$

is satisfied, and there are no other resonances.

Following the principles of the averaging method, we shall seek the solution of the system (1.2.1a) in the form of the sum of the slowly varying ($\overline{\sigma}_{mn}$) and the rapidly varying ($\tilde{\sigma}_{mn}$) parts:

$$\sigma_{mn} = \overline{\sigma}_{mn} + \tilde{\sigma}_{mn} \ . \tag{1.2.3}$$

Let us substitute (1.2.3) into (1.2.1a). Assuming that at least some of the $\overline{\sigma}_{mn}$ are much greater than all $\tilde{\sigma}_{mn}$[3], only the quantities $\overline{\sigma}_{mn}$ will be retained on the right-hand sides of (1.2.1a). The resulting equations will be averaged by integrating them over a time interval which is much greater than the characteristic periods of the rapid variations $2\pi/\omega_{mn}$ and $2\pi/\omega_j$, but much less than ν_1^{-1}, W_{mn}^{-1}, $\hbar/V_{mn}^{(j)}$ and T_{mn}. All slowly varying functions of time $[V_{mn}^{(\alpha)}, \overline{\sigma}_{mn}, \exp(-t/T_{mn}), \exp(\mathrm{i}\nu_1 t)]$ will be taken out of the integration sign.

This procedure leads to the truncated equations of the first approximation:

$$\frac{d}{dt}\overline{\sigma}_{12} + \frac{\overline{\sigma}_{12}}{T_{12}} = \frac{\mathrm{i}}{\hbar} V_{12}^{(\alpha)} \mathrm{e}^{\mathrm{i}\nu_1 t} (\overline{\sigma}_{11} - \overline{\sigma}_{22}) \ , \tag{1.2.4a}$$

$$\frac{d}{dt}\overline{\sigma}_{11} + \sum_k (\overline{\sigma}_{11} W_{1k} - \overline{\sigma}_{kk} W_{k1}) = -\frac{2}{\hbar} \mathrm{Im}\left\{ \overline{\sigma}_{12} V_{21}^{(-\alpha)} \mathrm{e}^{-\mathrm{i}\nu_1 t} \right\} \ , \tag{1.2.4b}$$

$$\frac{d}{dt}\overline{\sigma}_{22} + \sum_k (\overline{\sigma}_{22} W_{2k} - \overline{\sigma}_{kk} W_{k2}) = \frac{2}{\hbar} \mathrm{Im}\left\{ \overline{\sigma}_{12} V_{21}^{(-\alpha)} \mathrm{e}^{-\mathrm{i}\nu_1 t} \right\} \ , \tag{1.2.4c}$$

$$\frac{d}{dt}\overline{\sigma}_{mm} + \sum_k (\overline{\sigma}_{mm} W_{mk} - \overline{\sigma}_{kk} W_{km}) = 0 \ , \quad m \neq 1, 2 \ , \tag{1.2.4d}$$

$$\frac{d}{dt}\overline{\sigma}_{mn} + \frac{\overline{\sigma}_{mn}}{T_{mn}} = 0 \ , \quad (m,n) \neq (1,2), (2,1) \ . \tag{1.2.4e}$$

Note that in *Landau* and *Lifshitz* [1.23], techniques close to the averaging method were applied for obtaining the equations for the wave functions that describe the resonance interaction of molecules with the monochromatic radi-

[3] One can be sure that at least the populations of some levels of the system turn out to be large because of the condition $\mathrm{Tr}\{\overline{\sigma}\} = 1$.

ation field in the absence of relaxation ($T_{mn}^{-1} = W_{mn} = 0$). Formulas (1.2.4) represent a generalization of the equations given in [1.23] for the system with relaxation.

When levels 1 and 2 are the lower levels of the molecule, and the populations of all other states can be neglected, (1.2.4a–c) describe the behavior of the well-studied two-level system ([1.6]. Sect. 17), the equations of which were introduced in a phenomenological way. In this case it is more convenient to use the difference $\eta = \overline{\sigma}_{11} - \overline{\sigma}_{22}$ instead of the populations $\overline{\sigma}_{11}$ and $\overline{\sigma}_{22}$ themselves. An equation for η can be easily obtained from (1.2.4b, c) if one uses the normalization condition $\overline{\sigma}_{11} + \overline{\sigma}_{22} = 1$:

$$\frac{d}{dt}\eta + (\eta - \eta_0)\tau^{-1} = -\frac{4}{\hbar}\mathrm{Im}\left\{\overline{\sigma}_{12}V_{21}^{(-\alpha)}\mathrm{e}^{-\mathrm{i}\nu_1 t}\right\} . \tag{1.2.5}$$

Here η_0 is the equilibrium difference of populations $\sigma_{11}^{(0)} - \sigma_{22}^{(0)}$ in the absence of the electromagnetic field, and τ is the lifetime of the molecule in the excited state 2. These quantities are expressed through the transition probabilities W_{12} and W_{21}[4]:

$$\eta_0 = \frac{W_{21} - W_{12}}{W_{21} + W_{12}} , \quad \tau = (W_{12} + W_{21})^{-1} . \tag{1.2.6}$$

Thus, we have obtained equations describing the slowly varying part of the density matrix for the first order resonance. Their solutions determine the response of the molecule to the field $\mathcal{E}(\omega_\alpha)$. If one also computes the rapidly varying part $\tilde{\sigma}_{mn}$ of the density matrix, it is possible to obtain within the same approximation a more accurate solution. To this end, we shall subtract from (1.2.1a), where $\sigma_{mn} = \overline{\sigma}_{mn}$ is assumed to hold for the right-hand side, the truncated equations of the first approximation (1.2.4). This will lead to the equations for $\tilde{\sigma}_{mn}^{(1)}$ and $\tilde{\sigma}_{mm}^{(1)}$, the rapidly varying part of the density matrix in the first approximation of the averaging method:

$$\frac{d}{dt}\tilde{\sigma}_{mn}^{(1)} + \frac{\tilde{\sigma}_{mn}^{(1)}}{T_{mn}} = -\frac{\mathrm{i}}{\hbar}\widetilde{\sum_{k,j}}\left\{V_{mk}^{(j)}\overline{\sigma}_{kn}\mathrm{e}^{\mathrm{i}(\omega_{mk}+\omega_j)t} - \overline{\sigma}_{mk}V_{kn}^{(j)}\mathrm{e}^{\mathrm{i}(\omega_{kn}+\omega_j)t}\right\} ,$$

$$\frac{d}{dt}\tilde{\sigma}_{mm}^{(1)} + \sum_k [W_{mk}\tilde{\sigma}_{mm}^{(1)} - W_{km}\tilde{\sigma}_{kk}^{(1)}]$$
$$= -\frac{\mathrm{i}}{\hbar}\widetilde{\sum_{k,j}}\left\{V_{mk}^{(j)}\overline{\sigma}_{km}\mathrm{e}^{\mathrm{i}(\omega_{mk}+\omega_j)t} - \mathrm{c.c.}\right\} . \tag{1.2.7}$$

---

[4] In thermodynamic equilibrium, all transition probabilities W_{mn} are related to the equilibrium populations σ_{nn}^0 through equations $W_{mn} = \sigma_{nn}^0/\tau_{mn}$ where $\tau_{mn} = \tau_{nm}$, therefore, $\tau = \tau_{21}(\sigma_{11}^0 + \sigma_{22}^0)^{-1} = \tau_{21}$.

The sign $\sim$ above the sum symbol means that on the right-hand sides of (1.2.7) only the rapidly oscillating terms are retained.

The solution of these equations allows to establish a connection relation between rapidly and slowly varying parts of the density matrix. As usual, one can assume in its derivation that at the initial moment of time t_0, the field of the frequency ω_j was zero implying all $V_{mk}^{(j)}(t_0) = 0$. When a stationary or any other stationary-state regime is studied, it is possible to set t_0 equal to $-\infty$. Up to the accuracy of terms of the order $(V/\hbar\omega)(1/\omega T)\overline{\sigma}$, the wanted solution is given by

$$\tilde{\sigma}_{mn}^{(1)} = \frac{1}{\hbar} \widetilde{\sum_{k,j}} \left[\frac{\overline{\sigma}_{mk} V_{kn}^{(j)} \mathrm{e}^{\mathrm{i}(\omega_{kn}+\omega_j)t}}{\omega_{kn} + \omega_j} - \frac{V_{mk}^{(j)} \overline{\sigma}_{kn} \mathrm{e}^{\mathrm{i}(\omega_{mk}+\omega_j)t}}{\omega_{mk} + \omega_j} \right] . \tag{1.2.8}$$

It will be shown in Sect. 1.3 that the rapidly varying part of the density matrix $\tilde{\sigma}_{mn}^{(1)}$ determines the correction $\mathrm{Tr}\{\hat{\boldsymbol{d}}\hat{\tilde{\sigma}}^{(1)}\}$ to the polarization of the two-level system $\mathrm{Tr}\{\hat{d}\hat{\overline{\sigma}}\}$. This enables one to take into account the contribution from the nonresonance levels of the molecule in the dielectric constant at the frequency ω_α.

Below, $\tilde{\sigma}_{mn}^{(1)}$ will be used for the construction of the higher approximation equations that describe the slow variation of the density matrix under the action of fields, the frequencies of which satisfy the resonance conditions of order higher than (1.2.2).

1.2.2 Second-Order Resonances and an Example of the Simultaneous Realization of Two Resonance Conditions

Frequently nonlinear optical phenomena are accompanied by interactions of fields. Between the frequencies of the fields and the frequency of matter, there exist several resonances of different order. For instance, during stimulated Raman light scattering the anti-Stokes component is generated in addition to the Stokes one [1.7], i.e., two second-order resonances are observed (Fig. 1.2c). Another process of this type is the generation of the sum (difference) frequency in a medium that absorbs this frequency (Fig. 1.2a,b).

The second example contains all the most typical features of multiphoton interactions of different order that occur simultaneously. Therefore, at the beginning of the derivation of the truncated equations for multiphoton processes, we consider such a case.

Assume that, in addition to the first-order resonance condition (1.2.2), the same molecular levels 1 and 2 and some field frequencies ω_β and ω_γ are subject to the condition of the second-order resonance:

$$\omega_\beta + \omega_\gamma = \omega_{21} + \nu_2 \ , \quad |\nu_2| \ll \omega_\beta, \omega_\gamma, \omega_{21} \ . \tag{1.2.9}$$

If $\omega_\beta>0$ and $\omega_\gamma>0$, this condition means that a transition with absorption (emission) of two photons (see Fig. 1.1b) is possible between levels 1 and 2. When $\omega_\beta<0$ or $\omega_\gamma<0$, condition (1.2.9) corresponds to Raman light scattering involving the transition between these levels (see Fig. 1.1c). We shall also see below that more complex interactions of fields $\mathcal{E}(\omega_\alpha)$, $\mathcal{E}(\omega_\beta)$ and $\mathcal{E}(\omega_\gamma)$ with matter that depend on phase relations between these fields, can occur in the presence of the field $\mathcal{E}(\omega_\alpha)$.

The second-order resonance (1.2.9) cannot be taken into account in the first approximation of the averaging method, and therefore it is necessary to obtain the second approximation equations for the density matrix. An outline of its derivation follows below. Let us substitute the quantity $\overline{\sigma}_{mn}+\tilde{\sigma}_{mn}^{(1)}$ (and not just $\overline{\sigma}_{mn}$ as in the derivation of the first approximation equations) in the place of σ_{mn} in the right-hand side of (1.2.1a) and average the resulting expressions taking, as before, all slowly varying functions of time out of the integral sign. Now not only the functions proportional to $\exp[\mathrm{i}(\omega_{21}-\omega_\alpha)t]=\exp(\mathrm{i}\nu_1 t)$, will be slowly varying, but so will the functions on the right-hand sides. The latter emerge after the substitution of $\tilde{\sigma}_{mn}^{(1)}$ given by (1.2.8) and are proportional to $\exp[\mathrm{i}(\omega_{21}-\omega_\beta-\omega_\gamma)t]=\exp(\mathrm{i}\nu_2 t)$ [see condition (1.29)] and also to $\exp[\mathrm{i}(\omega_{mk}-\omega_j+\omega_j)t]=1$ (if $m=k$), respectively. This procedure leads to the second approximation truncated equations for $\overline{\sigma}_{mn}$:

$$\left(\frac{d}{dt}+\frac{1}{T_{12}}-\mathrm{i}\Omega_{12}\right)\overline{\sigma}_{12}=\frac{\mathrm{i}}{\hbar}(\overline{\sigma}_{11}-\overline{\sigma}_{22})\sum_{s=1,2}v_{12}^{(s)}\mathrm{e}^{\mathrm{i}\nu_s t} , \tag{1.2.10a}$$

$$\frac{d}{dt}\overline{\sigma}_{11}+\sum_k(W_{1k}\overline{\sigma}_{11}-W_{k1}\overline{\sigma}_{kk})=-\frac{2}{\hbar}\times\mathrm{Im}\left\{\overline{\sigma}_{12}\left(\sum_{s=1,2}v_{12}^{(s)}\mathrm{e}^{\mathrm{i}\nu_s t}\right)^*\right\} , \tag{1.2.10b}$$

$$\frac{d}{dt}\overline{\sigma}_{22}+\sum_k(W_{2k}\overline{\sigma}_{22}-W_{k2}\overline{\sigma}_{kk})=\frac{2}{\hbar}\times\mathrm{Im}\left\{\overline{\sigma}_{12}\left(\sum_{s=1,2}v_{12}^{(s)}\mathrm{e}^{\mathrm{i}\nu_s t}\right)^*\right\} , \tag{1.2.10c}$$

$$\frac{d}{dt}\overline{\sigma}_{mm}+\sum_k(W_{mk}\overline{\sigma}_{mm}-W_{km}\overline{\sigma}_{kk})=0 \quad (m\neq 1,\, 2) , \tag{1.2.10d}$$

$$\left(\frac{d}{dt}+\frac{1}{T_{mn}}-\mathrm{i}\Omega_{mn}\right)\overline{\sigma}_{mn}=0 , \quad (m,\,n)\neq(1,\,2),\,(2,\,1) . \tag{1.2.10e}$$

The quantity Ω_{mn} represents the difference between the Stark shifts of the levels m and n produced by the action of all optical fields $\mathcal{E}(\omega_j)$:

$$\Omega_{mn} = \hbar^{-1} \sum_{\substack{a,b \\ j>0}} [\kappa_{ab}^{mm}(\omega_j) - \kappa_{ab}^{nn}(\omega_j)]\mathcal{E}_a(\omega_j)\mathcal{E}_b^*(\omega_j) \ . \tag{1.2.11}$$

In this equation, $\kappa_{ab}^{mm}(\omega_j)$ are the linear polarizability tensor components for a molecule which is in the m-th energy state at the frequency ω_j[5]. In the electric dipole approximation, the expressions for all polarizabilities except $\kappa_{ab}^{11}(\omega_\alpha)$ and $\kappa_{ab}^{22}(\omega_\alpha)$, coincide with the well-known formulas of the perturbation theory that determine the polarizability of matter in the transparency region [1.6, 24, 25]:

$$\kappa_{ab}^{nn}(\omega_j) = \frac{2}{\hbar}\sum_k \frac{(d_{nk})_a(d_{kn})_b\omega_{kn}}{\omega_{kn}^2 - \omega_j^2} \ , \qquad n{\neq}1, 2 \quad \text{for } j = \alpha \ . \tag{1.2.12a}$$

For the resonant field $\mathcal{E}(\omega_\alpha)$ the sum of the type (1.2.12a) in the expression for the polarizability of the levels 1 and 2 contains no terms with resonant denominators:

$$\kappa_{ab}^{11}(\omega_\alpha) = \frac{2}{\hbar}\sum_{k\neq 2} \frac{(d_{1k})_a(d_{k1})_b\omega_{k1}}{\omega_{k1}^2 - \omega_\alpha^2} - \frac{(d_{12})_a(d_{21})_b}{\hbar(\omega_{21}+\omega_\alpha)} \ ,$$

$$\kappa_{ab}^{22}(\omega_\alpha) = \frac{2}{\hbar}\sum_{k\neq 1} \frac{(d_{2k})_a(d_{k2})_b\omega_{k2}}{\omega_{k2}^2 - \omega_\alpha^2} + \frac{(d_{12})_a(d_{21})_b}{\hbar(\omega_{21}+\omega_\alpha)} \ . \tag{1.2.12b}$$

It follows from (1.2.11) that there is a relationship between the change in the transition frequency produced by the electromagnetic field and the macroscopic quantities – the refractive indices of a medium consisting of molecules in the m-th energy state: $n_{ab} = \sqrt{1 + 4\pi N\overline{\sigma}_{mm}\kappa_{ab}^{mm}}$, where $N\overline{\sigma}_{mm}$ is the density of the number of molecules in this state. We would like to emphasize the universal nature of the manner in which Ω_{mn} is determined by polarizabilities; it characterizes both resonant and nonresonant actions of the field on the molecule. In the limiting case when $\omega_j \to 0$, the formula for Ω_{mn} gives the difference between the quadratic Stark shifts of the levels m and n produced by a constant field [1.24].

Consider the right-hand sides of (1.2.10a–c) determining the behavior of the matrix elements $\overline{\sigma}_{12}$, $\overline{\sigma}_{11}$ and $\overline{\sigma}_{22}$ of the transition 1–2 that interacts resonantly with the fields $\mathcal{E}(\omega_\alpha)$, $\mathcal{E}(\omega_\beta)$ and $\mathcal{E}(\omega_\gamma)$. The quantities $v_{12}^{(s)}$ appearing here have the dimension of energy and are equal to

$$v_{12}^{(1)} = V_{12}^{(\alpha)} \ , \quad \text{and} \tag{1.2.13}$$

[5] Here and in the following, the subscripts a, b, and a_i denote the projection of vectors onto the corresponding axes in the Cartesian reference frame.

$$v_{12}^{(2)} = \hbar^{-1} \sum_k \left(\frac{V_{1k}^{(\beta)} V_{k2}^{(\gamma)}}{\omega_{1k} + \omega_\beta} + \frac{V_{1k}^{(\gamma)} V_{k2}^{(\beta)}}{\omega_{1k} + \omega_\gamma} \right) . \tag{1.2.14}$$

The first of them is the energy of interaction of the field $\mathcal{E}(\omega_\alpha)$ with the resonance transition; it occurred already in the first approximation equations (1.2.4). The second one, $v_{12}^{(2)}$, is caused by the second-order resonance. Note that in the absence of such resonances $v_{12}^{(2)} = 0$, and the only difference between the second approximation equations (1.2.10) and the first approximation ones (1.2.4) is that in (1.2.10), the Stark shifts Ω_{mn} of the transition frequencies are taken into account. If the first-order resonance (1.2.2) is absent, the right-hand sides of the second approximation equations contain only the term proportional to $v_{12}^{(2)} \exp(\mathrm{i}\nu_2 t)$ which describes two-photon processes with the participation of the fields $\boldsymbol{E}(\omega_\beta)$ and $\boldsymbol{E}(\omega_\gamma)$.

It can easily be shown that when several second-order resonance conditions similar to (1.2.9) are fulfilled, on the right-hand sides of the second approximation equations (1.2.10a–c), the quantities of the type (1.2.14) will appear where β and γ will be replaced by the subscripts of fields, the frequencies of which satisfy the above resonance conditions.

1.2.3 The Hamiltonian of the Averaged Motion

Before considering more complex resonant interactions, let us demonstrate that (1.2.10) for the slowly varying part of the density matrix can be written in the form resembling that of the initial equation (1.1.18) for the whole density matrix. To this end we shall introduce the Hermitian matrix $\hat{\mathcal{V}}$:

$$\hat{\mathcal{V}} = \begin{bmatrix} \mathcal{V}_{11} & \mathcal{V}_{21} & 0 & \cdots & 0 & \cdots \\ \mathcal{V}_{12} & \mathcal{V}_{22} & 0 & \cdots & 0 & \cdots \\ 0 & 0 & \mathcal{V}_{33} & \cdots & 0 & \cdots \\ \cdot & \cdot & \cdot & \cdot & \cdot & \cdots \\ 0 & 0 & 0 & \cdots & \mathcal{V}_{mm} & \cdots \\ \cdot & \cdot & \cdot & \cdot & \cdot & \cdots \end{bmatrix} , \tag{1.2.15}$$

whose nondiagonal element $\mathcal{V}_{12}$ equals the sum appearing on the right-hand sides of (1.2.10):

$$\mathcal{V}_{12} = \sum_s v_{12}^{(s)} \mathrm{e}^{\mathrm{i}\nu_s t} \tag{1.2.16}$$

where s is the ordinal number in the list of resonance conditions of the type (1.2.2, 9). The coefficient $\mathcal{V}_{21}$ in the equation for $\overline{\sigma}_{21}$ cannot be obtained through simple permutation of the subscripts 1 and 2 in the expressions for $v_{12}^{(s)}$, see (1.2.13, 14); besides such a permutation it is necessary to make in these expressions the substitution $\omega_j \to \omega_{-j} = -\omega_j$. The matrix element $\mathcal{V}_{21}$ can be determined from the Hermitian property of the matrix $\hat{\mathcal{V}}$, namely, $\mathcal{V}_{21} = \mathcal{V}_{12}^*$. The diagonal elements in the matrix $\mathcal{V}$ represent the energies of the Stark shift:

$$\mathcal{V}_{mm} = v_{mm} = \sum_{\substack{j>0 \\ a,b}} \kappa_{ab}^{mm}(\omega_j)\mathcal{E}_a(\omega_j)\mathcal{E}_b^*(\omega_j) \ . \tag{1.2.17}$$

By making use of the matrix $\hat{\mathcal{V}}$ one can write (1.2.10) in the form that coincides with that of (1.1.18):

$$\frac{d}{dt}\hat{\bar{\sigma}} + \hat{\Gamma}\hat{\bar{\sigma}} = -\frac{\mathrm{i}}{\hbar}[\hat{\mathcal{V}}, \hat{\bar{\sigma}}] \ . \tag{1.2.18}$$

The operator $\hat{\mathcal{V}}$ in (1.2.18) describing the averaged motion plays the same role as the operator giving the energy of interaction of matter with the field $\hat{V}$ in (1.1.18). Therefore, $\hat{\mathcal{V}}$ will be called the operator of interaction energy for the averaged motion, or, for the sake of brevity, the *Hamiltonian of the averaged motion.* We would like to emphasize the fact that among the nondiagonal elements of this Hamiltonian, only those are different from zero which correspond to the transition for which the resonance conditions (1.2.2, 9), etc. are valid.

Below we shall give a rigorous proof that the slow variation of the density matrix is for any number of resonances of arbitrary orders described by (1.2.18). In the nondiagonal element (1.2.16) of the averaged Hamiltonian, there is a term $v_{12}^{(q)}\exp(\mathrm{i}\nu_q t)$, where q is the order of the resonance, and

$$v_{12}^{(q)} = \hbar^{(1-q)} \sum_{k',k'',\dots,k^{(q-1)}} \hat{C}_j V_{1k'}^{(j_1)} V_{k'k''}^{(j_2)} \dots V_{k^{(q-1)}2}^{(j_q)} \{(\omega_{1k'} + \omega_{j_1}) \times (\omega_{1k''} + \omega_{j_1} + \omega_{j_2}) \dots (\omega_{1k^{(q-1)}} + \omega_{j_1} + \dots + \omega_{j_{q-1}})\}^{-1} \tag{1.2.19}$$

for each resonance.

Note that $v_{12}^{(q)}$ coincides with the composite matrix element of the order q, which is obtained when a q-photon process is described by the methods of ordinary perturbation theory ([1.26] Sect. 15 and also [1.27]). When there are also resonances of the order less than q, the terms with resonance denominators should be excluded from (1.2.19). The presence of the operator $\hat{C}_j$ means that the terms in (1.2.19) that result from all possible permutations of the subscripts of the frequencies ω_{j_i}, for which the resonance condition of the q-th order (1.2) is valid, should be summed up. If there are several resonances of the q-th order, it is necessary to include in (1.2.16) all terms of the type (1.2.19) that are associated with these resonances. In this case in (1.2.19), q should be replaced by q_s, the order of the s-th resonance.

In the next chapters we shall need an expression for $v_{12}^{(q)}$ which contains the explicit form of its dependence on the amplitudes. To obtain such a form, let us introduce the q-th order polarizability:

$$
\begin{aligned}
\kappa^{12}_{a_1 a_2 \dots a_q} &= \left(\kappa^{21}_{a_1 a_2 \dots a_q}\right)^* \\
&= \hbar^{1-q} \sum_{k', k'', \dots, k^{(q-1)}} \hat{C}(j_i, a_i)(d_{1k'})_{a_1}(d_{k'k''})_{a_2} \dots (d_{k^{(q-1)}2})_{a_q} \\
&\times \{(\omega_{k'1} - \omega_{j_1})(\omega_{k''1} - \omega_{j_1} - \omega_{j_2}) \dots \\
&\times (\omega_{k^{(q-1)}1} - \omega_{j_1} - \dots - \omega_{j_{q-1}})\}^{-1} \ . \qquad (1.2.20)
\end{aligned}
$$

Then

$$
v^{(q)}_{12} = - \sum_{a_1, a_2, \dots, a_q} \kappa^{12}_{a_1, a_2 \dots a_q} \mathcal{E}_{a_1}(\omega_1) \mathcal{E}_{a_2}(\omega_2) \dots \mathcal{E}_{a_q}(\omega_{j_q}) \ . \qquad (1.2.21a)
$$

In (1.2.20), the quantities $(d_{mn})_{a_j}$ represent the projections of the matrix elements $\boldsymbol{d}_{mn}$ of the dipole moment on the respective axis' $a_i = x, y, z$ of the Cartesian reference frame on which in (1.2.21a), the field $\mathcal{E}(\omega_{j_i})$ of the frequency ω_{j_i} is projected. $\hat{C}(j_i, a_i)$ means that one has to add up the terms appearing in (1.2.20) due to all possible simultaneous permutations of the subscripts of the frequencies j_i and coordinates a_i. In order to avoid errors, it is necessary to keep in mind the following: the frequencies ω_j occur in the expression for polarizability and in (1.2.19) with the same sign as in the resonance condition (1.2).

In the case of linearly polarized fields, it is convenient to choose as a_i the directions that coincide with those of the field vectors $\vec{\mathcal{E}}(\omega_j)$. Then the expression for $v^{(q)}_{12}$ can be simplified:

$$
v^{(q)}_{12} = -\kappa^{12}_{a_1 a_2 \dots a_q} \mathcal{E}(\omega_1) \mathcal{E}(\omega_2) \dots \mathcal{E}(\omega_q) = -\kappa^{12}_{(q)} \prod_j \mathcal{E}(\omega_j) \ . \qquad (1.2.21b)
$$

It is obvious that the second-order polarizability κ^{12}_{ab} (or $\kappa^{12}_{(2)}$) is responsible for the two-photon transitions between levels 1 and 2, the quantity κ^{12}_{abc} (or $\kappa^{12}_{(3)}$) is responsible for various three-photon transitions, etc.

1.2.4 The Truncated Equations for Resonances of Arbitrary Order Involving Many Levels

Let us demonstrate the applicability of (1.2.18) in the general case. We shall start with (1.1.18) in the absence of relaxation; the question of its introduction into the truncated equations will be discussed later. Taking $(p+1)$ to be no less than $\max\{q_s\}$, i.e., no less than the highest order of the resonances (1.2), we apply $(p+1)$ times the procedure of the averaging method to obtain for $\bar{\hat{\sigma}}$ the averaged equation of the $(p+1)$-th approximation:

$$
\frac{d}{dt}\bar{\hat{\sigma}} = -\frac{i}{\hbar}\overline{[\hat{V}, (\bar{\hat{\sigma}} + \tilde{\hat{\sigma}}^{(1)} + \dots + \tilde{\hat{\sigma}}^{(p)})]} \ , \qquad (1.2.22)
$$

where $\hat{V} \equiv \hat{V}_{\text{int}}$. The bar above the commutator symbol implies averaging with respect to time, and the q-th term of the rapidly varying part of σ is

$$\hat{\tilde{\sigma}}^{(q)} = \left(-\frac{i}{\hbar}\right)^q \underbrace{\tilde{\int}\left[\hat{V}, \tilde{\int}\left[\dots\tilde{\int}[\hat{V},\hat{\bar{\sigma}}]dt\dots\right]dt\right]}_{q\text{ integrals}}dt \ . \tag{1.2.23}$$

The sign $\sim$ above the integral symbols means that the integration is performed over the "rapidly varying" time (i.e., all slowly varying functions of time are regarded as constants), the rapidly varying part of the result of integration being retained.

Consider the q-th term on the right-hand side of (1.2.22). It will be shown below that this term differs from zero if there are resonances of the q-th order. Substituting the expression for $\hat{\tilde{\sigma}}^{(q-1)}$ from (1.2.23) into (1.2.22), one can see that it consists of 2^q terms. They can be divided into groups in which $\hat{\bar{\sigma}}$ occurs at the first, second, ..., m-th place. The first and the last groups contain just one term and they are equal, respectively, to

$$\begin{aligned}&\hat{\bar{\sigma}}\left(-\frac{i}{\hbar}\right)^q\overline{\left(\tilde{\int}\left(\dots\left(\tilde{\int}\left(\tilde{\int}\hat{V}\,dt\right)\hat{V}\,dt\right)\dots\right)\hat{V}\,dt\right)\hat{V}}\\ &\quad= \left(-\frac{i}{\hbar}\right)^q\hat{\bar{\sigma}}\hat{\bar{u}}_q^{(1)}\end{aligned} \tag{1.2.24}$$

and

$$\left(-\frac{i}{\hbar}\right)^q\overline{\hat{V}\tilde{\int}\hat{V}\left(\dots\left(\tilde{\int}\hat{V}\left(\tilde{\int}\hat{V}\,dt\right)dt\right)\dots\right)dt}\cdot\hat{\bar{\sigma}} = \left(-\frac{i}{\hbar}\right)^q\hat{\bar{u}}_q^{(2)}\hat{\bar{\sigma}} \ . \tag{1.2.25}$$

If one uses the obvious property of the averaged quantity $\overline{\tilde{d}/dt(\dots)} = 0$, where $\tilde{d}/dt$ denotes differentiation with respect to the "rapidly varying" time, one can easily show that

$$\hat{\bar{u}}_q^{(2)} = (-1)^{q-1}\hat{\bar{u}}_q^{(1)} \ .$$

Thus, the sum (1.2.24, 25) can be presented as

$$-\frac{i}{\hbar}[\hat{\mathcal{V}}^{(q)}\hat{\bar{\sigma}}] \ , \quad \text{where}$$

$$\hat{\mathcal{V}}^{(q)} = \left(-\frac{i}{\hbar}\right)^{(q-1)}\hat{\bar{u}}_q^{(2)} = \left(\frac{i}{\hbar}\right)^{(q-1)}\hat{\bar{u}}_q^{(1)} \ .$$

It can be demonstrated that all other groups of terms of the commutator $[\hat{V}, \hat{\tilde{\sigma}}^{(q-1)}]$ in which $\hat{\bar{\sigma}}$ appears at the m-th place ($m \neq 1$, $q+1$) vanish. To this end, it is sufficient to note that they can be written in the form of the averaged derivative with respect to the rapidly varying time:

$$(-1)^{q-m+1}\overline{\frac{\tilde{d}}{dt}(\hat{\mu}^{(2)}_{m-1}\hat{\tilde{\sigma}}\hat{\mu}^{(1)}_{q-m+1})} = 0 \ , \quad \text{where} \tag{1.2.26}$$

$$\hat{\mu}^{(1)}_j = \widetilde{\int}\hat{u}^{(1)}_j \, dt \ , \quad \hat{\mu}^{(2)}_j = \widetilde{\int}\hat{u}^{(2)}_j \, dt \ . \tag{1.2.27}$$

The quantities $\hat{u}^{(1)}_j, \hat{u}^{(2)}_j$ are determined by (1.2.24, 25). The validity of condition (1.2.26) for small values of q follows directly from (1.2.23, 24). Then the induction method may be used to prove it for the $(q+1)$-th term on the right-hand side of (1.2.22).

Since the (1.2.26) holds for an arbitrary term from (1.2.22), the latter finally assumes the form

$$\frac{d}{dt}\hat{\tilde{\sigma}} = -\frac{\mathrm{i}}{\hbar}[\hat{\mathcal{V}}, \hat{\tilde{\sigma}}] \ , \quad \text{where} \tag{1.2.28}$$

$$\hat{\mathcal{V}} = \sum_{q=1}^{p} \hat{\mathcal{V}}^{(q)} = \overline{\hat{V}} + \frac{\mathrm{i}}{\hbar}\overline{\left(\widetilde{\int}\hat{V}\,dt\right)\hat{V}}$$

$$+ \ldots + \left(\frac{\mathrm{i}}{\hbar}\right)^{p-1}\overline{\left(\widetilde{\int}\left(\ldots\left(\widetilde{\int}\left(\widetilde{\int}\hat{V}\,dt\right)\hat{V}\,dt\right)\ldots\right)\hat{V}\,dt\right)\hat{V}} \ . \tag{1.2.29}$$

Thus, the slow variation of the density matrix for any number of resonances of arbitrary order is described by the canonical equation (1.2.28) with the averaged Hamiltonian (1.2.29). The relationships existing between $u^{(2)}_q$ and $u^{(1)}_q$ also make it possible to employ (1.2.25) to express the Hamiltonian through $\hat{u}^{(2)}_q$.

Consider the matrix elements of the averaged Hamiltonian. It follows from (1.1.12, 19) and (1.2.29) that the nondiagonal matrix elements of the term $\hat{\mathcal{V}}^{(q)}$ of the Hamiltonian $\hat{\mathcal{V}}$ differ from zero only for those transitions for which the resonance conditions of the q-th order are fulfilled. For example, the quantities $\mathcal{V}^{(q)}_{12}$ for levels 1 and 2 between which such resonance occurs, are expressed through the slowly varying amplitudes of interaction energy by means of (1.2.19). When the fields interact resonantly with several pairs of molecular levels, (1.2.16, 21) should be used for determining those matrix elements of $\hat{\mathcal{V}}$ that differ from zero; in these equations subscripts 1 and 2 should be replaced by the subscripts of levels of the corresponding resonant transitions. In the case of an interaction as depicted in Fig. 1.3a, the nonzero nondiagonal elements of the Hamiltonian of the averaged motion are in the first approximation [1.28]:

$$\mathcal{V}_{12} = V^{(\omega_1)}_{12}\mathrm{e}^{\mathrm{i}\nu_1 t} \ , \quad \mathcal{V}_{23} = V^{(\omega_2)}_{23}\mathrm{e}^{\mathrm{i}\nu_2 t} \ , \quad \mathcal{V}_{13} = V^{(\omega_3)}_{13}\mathrm{e}^{\mathrm{i}\nu_3 t} \ .$$

In the second approximation they will have additional terms. For instance, [1.29]

$$\mathcal{V}_{13} = \hbar^{-1} \sum_{m\neq 2} \left(\frac{V_{1m}^{(\omega_1)} V_{m3}^{(\omega_2)}}{\omega_{1m} + \omega_1} + \frac{V_{1m}^{(\omega_2)} V_{m2}^{(\omega_1)}}{\omega_{1m} + \omega_2} \right) e^{i(\nu_1+\nu_2)t} + V_{13}^{(\omega_3)} e^{i\nu_3 t} \ .$$

In contrast to the nondiagonal elements, the diagonal ones $\mathcal{V}_{mm}^{(q)} = v_{mm}^{(q)}$ for even q differ from zero, not only for those levels that participate in resonant interactions but for all levels of the system. For $q>2$, the quantities $v_{mm}^{(q)}$ represent corrections to the dynamic Stark shift of the order of $v_{mm}^{(2)}(V/\hbar\omega)^{q-2} \ll v_{mm}^{(2)}$. In the case of coherent multiphoton processes, small corrections to the quadratic Stark shift can also appear for odd q. Thus, the quantity $\mathcal{V}_{mm}^{(3)}$ may contain a term associated with the parametric addition of frequencies and proportional to the corresponding susceptibility χ_{abc}.

A remark on the limits of applicability of (1.2.28) with the averaged Hamiltonian (1.2.29) would be in order.

The results obtained above by the averaging method are valid when no new resonances appear because of the change of the frequencies of the matter subject to the action of the external fields [1.30]. In other words, not only the existing detunings but also the detunings that result from interactions with the field and are nonlinear should be small in comparison with all linear combinations of the field frequencies and the eigenfrequencies of the molecule [except the combinations that have been taken into account by the resonance conditions (1.2)[6]]. The nonlinear detunings are, obviously, the frequencies of Stark shifts (1.2.11). Thus, (1.2.28) holds if, besides the condition

$$V_{kr}^{(j)}/\hbar\omega_{mn} \ll 1 \ , \tag{1.2.30}$$

the inequalities

$$\Omega_{mn} \ll \sum_j \ell_j \omega_j - \omega_{sr} \quad (\ell_j = 1, 2, \ldots, q_{\max}) \ , \tag{1.2.31}$$

are valid. In the above formula $\ell_j \neq n_j$ for $(s,r) = (m,n)$. Note that when $\sum \ell_j\omega_j - \omega_{sr} \sim \omega_{mn}$, condition (1.2.31) actually coincides with (1.2.30).

To conclude this section, let us discuss the relaxation terms in the averaged equation.

Note that when the truncated equations of the second approximation were derived, we took the slowly varying functions $\exp(t/T_{mn})$, $\exp(t/\tau_{mn})$ out of the integral. As has been demonstrated in [1.11, 12], in this case in (1.2.10a), small corrections to the line widths T_{mn}^{-1} will appear:

$$\frac{2}{\hbar^2} \sum_{k,j} \left\{ \frac{\omega_{km}^2 + \omega_j^2}{(\omega_{km}^2 - \omega_j^2)^2 T_{kn}} |V_{km}|^2 + \frac{\omega_{nk}^2 + \omega_j^2}{(\omega_{nk}^2 - \omega_j^2) T_{km}} |V_{nk}|^2 \right\} \ .$$

[6] Certainly, it is sufficient that this condition is fulfilled for combinations of not more than $q_{\max}$ frequencies, where $q_{\max}$ is the highest order of resonances (1.2) that occur in the linear approximation.

Since the relative magnitude of these corrections is of the order of $(V_{mk}/\hbar\omega_{mn})^2 \ll 1$, they can be neglected. The inter-level transition probabilities W_{mk} will also contain corrections caused by the interaction between field and matter at the tails of the absorption lines:

$$\frac{2}{\hbar^2}\sum_j \frac{\omega_{mk}^2+\omega_j^2}{T_{mk}(\omega_{mk}^2-\omega_j^2)^2}|V_{mk}^{(j)}|^2 \ .$$

These corrections are negligible if the detuning of frequencies from all transitions that do not satisfy the first-order resonance conditions is much greater than the corresponding line widths.

In such situations, relaxation can obviously be taken into account by introducing the relaxation operator (1.1.15) directly into (1.2.28). Whence the applicability of (1.2.18) for the description of arbitrary resonance processes follows.

It remains to add a last note. The whole previous discussion, like the papers [1.11–17, 31], deals with the resonant processes with the participation of transitions only in the discrete energy spectrum. The method of averaging, however, can be successfully applied also to the problems of interaction of radiation with matter in which the transitions into the continuous spectrum are essential. As an example, we can cite [1.32, 33] where the averaging method is used for the analysis of resonant multiphoton ionization of atoms.

1.3 Polarization of Matter and the Generalized Dipole Moment

As is well known, the response of matter that characterizes its interaction with an electromagnetic field is given by polarization

$$\boldsymbol{P}=\sum_j \vec{\mathcal{P}}_j \exp(\mathrm{i}\omega_j t) \ . \qquad (1.3.1)$$

Its spectral components with the frequencies ω_j are the sources of fields in the Maxwell equations and are therefore responsible for the absorption or the emission of fields with these frequencies.

Polarization produced by resonant interactions of the molecule with light is equal to $\mathrm{Tr}\{\hat{\boldsymbol{d}},\ \bar{\hat{\sigma}}+\tilde{\hat{\sigma}}\}$, where the slowly varying part of the density matrix $\bar{\hat{\sigma}}$ is determined by (1.2.18). One can easily see that the term $\mathrm{Tr}\{\hat{\boldsymbol{d}}\bar{\hat{\sigma}}\}$ respresents the contribution of the polarization involving the frequencies $\omega_{mn}=\omega_\alpha-\nu_\alpha$ of the resonant transitions. The spectral components $\boldsymbol{P}(\omega_j)$ at the frequencies ω_j participating in the multiphoton resonance processes are caused by the second term which is proportional to $\tilde{\hat{\sigma}}$. Since there is a dependence between the rapidly varying part of the density matrix and its slowly varying part $\bar{\hat{\sigma}}$, see (1.2.23), it is obvious that the polarization as a whole can be expressed through $\bar{\hat{\sigma}}$. The proportionality coefficient (the dimensions of which are the some ones as those of the dipole moment) will depend on fields that participate in the multiphoton process.

We shall obtain the operator of this generalized dipole moment and demonstrate that, as in the case of the Hamiltonian of the averaged motion (1.2.29), its nondiagonal matrix elements differ only for resonant transitions from zero, and the existence of the diagonal ones is associated with the condition $\sum \ell_j\omega_j = 0$. To this end, let us determine the amplitude of a spectral component of polarization for one of the frequencies ω_j that occur in the q-th order resonance condition ($q>1$; first the nondegenerate case will be considered):

$$\vec{\mathcal{P}}^{(q)}(\omega_j) = \overline{\mathrm{Tr}\{\boldsymbol{d}e^{-\mathrm{i}\omega_j t}\hat{\tilde{\sigma}}^{(q-1)}\}} \ . \tag{1.3.2}$$

Multiplying (1.3.2) by $-\vec{\mathcal{E}}^*(\omega_j)$ and adding the result to its complex conjugate, we obtain

$$\vec{\mathcal{P}}^{(q)}(\omega_j)\vec{\mathcal{E}}^*(\omega_j) + \text{c.c.} = \overline{\boldsymbol{P}^{(q)}(\omega_j)\boldsymbol{E}(\omega_j)} = -\overline{\mathrm{Tr}\{\hat{V}^{(j)}\hat{\tilde{\sigma}}^{(q-1)}\}} \ . \tag{1.3.3a}$$

Let us now substitute the expression for $\hat{\tilde{\sigma}}^{(q-1)}$ from (1.2.23) into (1.3.3); taking into acount (1.2.24, 25) and the fact that the commutator under the sign of the external integral (1.2.23) can be presented as $\tilde{d}/dt(\dots)$ [see (1.2.26)], we can write (1.3.3) as

$$\begin{aligned}\overline{\boldsymbol{P}^{(q)}(\omega_j)\boldsymbol{E}(\omega_j)} \\ = -\left(-\frac{\mathrm{i}}{\hbar}\right)^{q-1} \mathrm{Tr}\left\{\sum_{m=1}^{q}(-1)^{q-m}\overline{\hat{\mu}^{(1)}_{q-m}\hat{V}^{(j)}\hat{\mu}^{(2)}_{m-1}}\hat{\bar{\sigma}}\right\} \ ,\end{aligned} \tag{1.3.3b}$$

where $\hat{\mu}_j$ ($j>1$) is determined in correspondence with (1.2.27) and the quantities $\hat{\mu}^{(1)}_0 = \hat{\mu}^{(2)}_0 = 1$ are introduced. Further, by making use of the equation

$$\overline{\frac{\tilde{d}}{dt}\left(\tilde{\int}\hat{\mu}^{(1)}_{q-m}\hat{V}^{(j)}dt\right)\hat{\mu}^{(2)}_{m-1}} = 0$$

and the definition of $\hat{u}^{(2)}$ and $\hat{\mu}^{(2)}$ [see (1.2.25, 27)], it can be shown that

$$\overline{\hat{\mu}^{(1)}_{q-m}\hat{V}^{(j)}\hat{\mu}^{(2)}_{m-1}} = -\overline{\left(\tilde{\int}\hat{\mu}^{(1)}_{q-m}\hat{V}^{(j)}dt\right)\hat{V}\hat{\mu}^{(2)}_{m-1}} \ . \tag{1.3.4}$$

Applying this procedure $(m-1)$ times, we find that the m-th term in the sum in (1.3.3b) together with the factor $(-\mathrm{i}/h)^{q-1}$ is equal to $\hat{\mathcal{V}}^{(q)}$ from (1.2.29) in which at the m-th place, $\hat{V}$ is replaced by $\hat{V}^{(j)}$. It is obvious that the sum of all terms in (1.3.3b) is $\hat{\mathcal{V}}^{(q)}$. Therefore, polarization at the frequency of any of the fields participating in the q-photon process is related to $\hat{\mathcal{V}}^{(q)}$:

$$\overline{\boldsymbol{P}^{(q)}(\omega_j)\boldsymbol{E}(\omega_j)} = -\mathrm{Tr}\{\hat{\mathcal{V}}^{(q)}\hat{\bar{\sigma}}\} \ . \tag{1.3.5}$$

Accordingly, the total polarization $\boldsymbol{P}^{(q)}$ is given by

$$\overline{\boldsymbol{P}^{(q)}\boldsymbol{E}} = \sum_{j=-q}^{q} \vec{\mathcal{P}}^{(q)}(\omega_j)\vec{\mathcal{E}}^*(\omega_j) = -q\mathrm{Tr}\{\hat{\mathcal{V}}^{(q)}\hat{\bar{\sigma}}\} \ . \tag{1.3.6}$$

Let us now present (1.3.5) in the form

$$\vec{\mathcal{P}}^{(q)}(\omega_j)\vec{\mathcal{E}}^*(\omega_j) + \text{c.c.}$$
$$= -\left(\sum_{m>n} \mathcal{V}_{mn}^{(q)}\bar{\sigma}_{nm} + \text{c.c.} + \sum_{m} \mathcal{V}_{mm}^{(q)}\bar{\sigma}_{mm}\right) \ . \tag{1.3.7}$$

Obviously, an operator of the generalized dipole moment

$$\hat{\boldsymbol{D}}^{(q)} = \sum_j \hat{\mathcal{D}}^{(q)}(\omega_j)\mathrm{e}^{\mathrm{i}\omega_j t} \ , \tag{1.3.8}$$

can be introduced, the amplitude matrix of which determines the polarization amplitudes

$$\vec{\mathcal{P}}^{(q)}(\omega_j) = \mathrm{Tr}\{\hat{\mathcal{D}}^{(q)}(\omega_j)\hat{\bar{\sigma}}\} \tag{1.3.9}$$

and is expressed through the averaged Hamiltonian in the following way:

$$\mathcal{V}_{mn}^{(q)} = -\vec{\mathcal{D}}_{mn}^{(q)}(\omega_j)\vec{\mathcal{E}}^*(\omega_j) \ . \tag{1.3.10}$$

Since $\mathcal{V}_{nm}^{(q)} = \{\mathcal{V}_{mn}^{(q)}\}^*$, the condition

$$\vec{\mathcal{D}}_{nm}^{(q)}(\omega_j) = \{\vec{\mathcal{D}}_{mn}^{(q)}(-\omega_j)\}^* \tag{1.3.11}$$

is fulfilled.

It must be noted that (1.3.10) determines $\vec{\mathcal{D}}_{mn}^{(q)}(\omega_j)$ at a frequency ω_j which has the same sign as in the resonance condition

$$\sum_{j=1}^{q} \omega_j = \omega_{mn} \ .$$

Thus, if $\omega_1 - \omega_2 = \omega_{21}$ then $\vec{\mathcal{D}}_{21}^{(2)}(\omega_1)$ is the amplitude near $\exp(\mathrm{i}\omega_1 t)$ and $\vec{\mathcal{D}}_{21}^{(2)}(-\omega_2)$ the one near $\exp(-\mathrm{i}\omega_2 t)$.

Expression (1.3.10) permits a simple physical interpretation, namely, that the averaged interaction energy for any of the fields participating in a q-photon interaction with the transition $m - n$ is similar to the energy of interaction of the field with a two-level system having the dipole moment $\vec{\mathcal{D}}_{mn}^{(q)}$, the quantity $\mathcal{V}_{mn}^{(q)}$ being the same for all fields.

It can easily be demonstrated on the basis of (1.2.29, 15 and 1.3.10) that, because of the additivity of $\mathcal{V}$, both for resonances of different orders and for several resonances of the same order, the generalized dipole moment for an arbitrary number of resonances can be written as

$$\boldsymbol{D}_{mn} = \sum_{q_s,j} \vec{\mathcal{D}}_{mn}^{(q_s)}(\omega_j)\mathrm{e}^{\mathrm{i}\omega_j t} = \sum_{j} \vec{\mathcal{D}}_{mn}(\omega_j)\mathrm{e}^{\mathrm{i}\omega_j t} \ , \tag{1.3.12}$$

where the presence of the subscript s means that there may be several resonances of the q-th order. Consequently, the total polarization is

$$\boldsymbol{P} = \sum_{q} \boldsymbol{P}^{(q)} = \mathrm{Tr}\{\hat{\boldsymbol{D}}\hat{\overline{\sigma}}\} \ . \tag{1.3.13}$$

It can be readily established that all amplitudes $\vec{\mathcal{P}}^{(q)}(\omega_j)$ (and, therefore, the total polarization $\boldsymbol{P}$) consist of two parts: $\vec{\mathcal{P}}^{(q)\mathrm{res}}(\omega_j)$ associated with the resonance condition of the q-th order and $\vec{\mathcal{P}}^{(q)\mathrm{nonres}}(\omega_j)$ which may differ from zero also in the absence of resonances. Indeed, in Sect. 1.2 it has been shown that the nondiagonal elements $\mathcal{V}_{mn}^{(q)}$ are only in the presence of the q-th order resonances different from zero. For this reason the first term in (1.3.7) determines the resonance part of the polarization:

$$\vec{\mathcal{P}}^{(q)\mathrm{res}}(\omega_j) = \sum_{mn} \vec{\mathcal{D}}_{mn}^{(q)}(\omega_j)\overline{\sigma}_{nm} \ . \tag{1.3.14}$$

Equation (1.2.21) makes it possible to express the quantities $\{\vec{\mathcal{D}}_{mn}^{(q)}(\omega_j)\}_a$, i.e., the projections of $\vec{\mathcal{D}}_{mn}^{(q)}(\omega_j)$, through polarizability tensors of the q-th order:

$$\begin{aligned}\underset{(m>n)}{\{\mathcal{D}_{mn}^{(q)}(\omega_j)\}_{a_j}} &= \sum_{\substack{a_1,\dots,a_{j-1},\\ a_{j+1},\dots,a_q}} \kappa_{a_1\dots a_q}^{mn}\mathcal{E}_{a_1}^*(\omega_1) \\ &\quad \dots\mathcal{E}_{a_{j-1}}^*(\omega_{j-1})\mathcal{E}_{a_{j+1}}^*(\omega_{j+1})\dots\mathcal{E}_{a_q}^*(\omega_q)\mathrm{e}^{-\mathrm{i}\nu_q t} \ .\end{aligned} \tag{1.3.15}$$

All the above formulas hold for the nondegenerate case. For a resonance with n_j-fold degeneracy with respect to the frequency ω_j, (1.3.6) yields

$$\vec{\mathcal{P}}^{(q)}(\omega_j)\vec{\mathcal{E}}^*(\omega_j) = -n_j\mathrm{Tr}\{\hat{\mathcal{V}}^{(q)}\hat{\overline{\sigma}}\} \tag{1.3.16}$$

and, therefore,

$$\underset{(m>n)}{\mathcal{V}_{mn}^{(q)}} = -n_j^{-1}\vec{\mathcal{D}}_{mn}^{(q)}(\omega_j)\vec{\mathcal{E}}^*(\omega_j) \ . \tag{1.3.17}$$

The last term in (1.3.7), which is proportional to $\mathcal{V}_{mm}^{(q)}$, differs from zero if the field spectrum contains frequencies that satisfy the condition $\sum_j \ell_j \omega_j = 0$. Because of this, the respective part of polarization is nonresonant; its amplitudes can be written as

$$\begin{aligned}&\vec{\mathcal{P}}^{(q)\mathrm{nonres}}(\omega_j = \omega_1 + \ldots + \omega_{j-1} + \omega_{j+1} + \ldots + \omega_q)\\&= \sum_n \vec{\mathcal{D}}_{nn}^{(q)}(\omega_j)\overline{\sigma}_{nn} \ ,\end{aligned} \tag{1.3.18}$$

where the projections of $\mathcal{D}_{nn}^{(q)}$ are expressed through the diagonal components of the polarizability tensors $\kappa_{a_1,\ldots,a_q}^{nn}$ by means of (1.3.15) in which it must be assumed that $m = n$ and $\nu_q = 0$.

In the first order the nonresonant polarization is obviously absent. For $q = 2$ it is determined by the linear susceptibility

$$\{\mathcal{P}^{(2)\mathrm{nonres}}(\omega_j)\}_a = \sum_b \chi_{ab}\mathcal{E}_b(\omega_j) = \sum_{b,n} \kappa_{ab}^{nn}\mathcal{E}_b(\omega_j)\overline{\sigma}_{nn} \ .$$

The part of polarization which is quadratic with respect to the field $\boldsymbol{P}^{(3)\mathrm{nonres}}$ coincides with the corresponding component of polarization calculated in second order perturbation theory (see [1.6], Sect. 13), provided the terms with resonance denominators are excluded from the susceptibility tensor

$$\begin{aligned}\{\mathcal{P}^{(3)\mathrm{nonres}}(\omega_k = \omega_s + \omega_\ell)\}_a &= \sum_n \{\mathcal{D}_{nn}^{(3)}(\omega_k)\}_a \overline{\sigma}_{nn}\\&= \sum_{n,b,c} \overline{\sigma}_{nn}\kappa_{abc}^{nn}(\omega_\ell, \omega_s)\mathcal{E}_b(\omega_\ell)\mathcal{E}_c(\omega_s)\\&= \sum_{b,c} \chi_{abc}(\omega_\ell, \omega_s)\mathcal{E}_b(\omega_\ell)\mathcal{E}_c(\omega_s) \ .\end{aligned} \tag{1.3.19}$$

As is well known, the polarization (1.3.19) is responsible for nonresonant three-photon parametric interactions, such as addition and doubling of frequencies in a transparent nonlinear medium.

Note that the nonresonant part of the polarization depends on the populations $\overline{\sigma}_{nn}$ and can change rapidly in intense fields as the field frequency passes through the resonance. Therefore, the terms "resonant" and "nonresonant" applied to the polarization components (1.3.14, 18) are conditional – in general in nonlinear problems.

1.4 The Generalized Two-Level System

Interactions in which resonances of different orders occur only for one transition are important both for clarification of the main regularities of the resonant interactions of radiation with matter and for practice. If those populations of the levels that are not related to the resonant transition can be neglected, (1.2.18) [accounting for relaxation and quadratic Stark shift] can be written in the form of matrix equations for two quantities: the nondiagonal matrix element $\overline{\sigma}_{12} = \sigma$ and the difference of resonant level populations $\eta = \overline{\sigma}_{11} - \overline{\sigma}_{22}$:

$$\begin{aligned}
&\frac{d\overline{\sigma}}{dt} + (T^{-1} - \mathrm{i}\Omega)\overline{\sigma} = \frac{\mathrm{i}}{\hbar}\mathcal{V}\eta \ , \\
&\frac{d\eta}{dt} + \frac{\eta - \eta_0}{\tau} = -\frac{4}{\hbar}\mathrm{Im}\{\overline{\sigma}\mathcal{V}^*\} \ .
\end{aligned} \tag{1.4.1}$$

In the above expression, subscripts 1 and 2 of the quantities T_{12}, Ω_{12}, $\mathcal{V}_{12}$ are omitted, τ is the relaxation time for the population difference to reach its equilibrium value.

Thus, it turns out to be possible to generalize the equations of the two-level system for resonant multiphoton interactions. Therefore, (1.4.1) will be called the equations of the generalized two-level system [1.34, 35][7].

As in the case of the ordinary two-level system, the populations $\overline{\sigma}_{11}$ and $\overline{\sigma}_{22}$ satisfy the normalization condition

$$\mathrm{Tr}\{\sigma\} = \mathrm{Tr}\{\overline{\sigma}\} = \overline{\sigma}_{11} + \overline{\sigma}_{22} = 1 \ . \tag{1.4.2}$$

This follows from the form of the density matrix $\sigma = \overline{\sigma} + \tilde{\sigma}$ since all terms in the rapidly varying part of the density matrix $\tilde{\sigma} = \sum \tilde{\sigma}^{(q)}$ satisfy the equation $\mathrm{Tr}\{\tilde{\sigma}^{(q)}\} = 0$. The latter equation can easily be obtained from (1.2.23): the trace of the commutator is identically zero.

When the conditions for the applicability of the generalized two-level system are fulfilled, it is more convenient to use, instead of (1.3.14, 18), an expression for the polarization amplitudes containing the variables $\overline{\sigma}$ and η:

$$\begin{aligned}
\vec{\mathcal{P}}(\omega_j) &= \vec{\mathcal{P}}^{\mathrm{nonres}}(\omega_j) + \vec{\mathcal{P}}^{\mathrm{res}}(\omega_j) \\
&= \tfrac{1}{2}\{\vec{\mathcal{D}}_{11}(\omega_j) + \vec{\mathcal{D}}_{22}(\omega_j) + [\vec{\mathcal{D}}_{11}(\omega_j) - \vec{\mathcal{D}}_{22}(\omega_j)]\eta\} + \vec{\mathcal{D}}_{21}(\omega_j)\overline{\sigma} \ .
\end{aligned} \tag{1.4.3}$$

On the basis of (1.4.1, 3), as well as of the more general expressions for the density matrix and polarization (1.2.18 and 1.3.13), it is possible to take into account all physical factors that determine the behavior of the molecular response both, in the case of incoherent and coherent (parametric) multiphoton processes. For example, the saturation of populations, the dynamic Stark shift and the difference of the linear polarizabilities of levels, and also the interference of the probability amplitudes for transitions between them as caused by different resonances.

[7] When several transitions are resonant, (1.2.18) can be used to derive equations for the generalized n-level system, where n is the number of levels participating in resonant interactions with the field.

2. The Molecular Response to the Resonant Effects of Quasimonochromatic Fields

In this chapter the reader will find a detailed analysis of the behavior of populations and the real and imaginary parts of molecular nonlinear resonant polarization. There is a discussion of the relation between contributions produced by the difference in the values of polarizability of resonant levels, the saturation of populations, the dynamic Stark effect, into nonlinear dielectric polarizability, with special attention paid to single- and two-photon resonances.

We consider here parametric self-induced transparency of matter and the possibility of studying the polarizability of excited states by measuring the nonlinear refractive index. The estimates obtained enable us to arrive at the conclusion that the resonant mechanisms responsible for the nonlinearity of the refractive index of matter may affect substantially the nature of the self-action of light (e.g., its self-focussing). The results derived in this chapter form the foundation of the theory of resonant interactions to be developed in Chaps. 6–8.

2.1 The Change of Populations of the Generalized Two-Level System in Quasimonochromatic Fields

It is of considerable interest to study the behavior of populations in light fields, particularly because the population of excited levels is an important stage in the development of chemical reactions induced by laser radiation [2.1], and also of the laser spark and multiphoton ionization [2.2–6]. The authors of [2.7] applied the two-level model to the description of populations in single-photon interaction when the Stark shift can be neglected. The saturation of populations may also be accomplished via multiphoton processes produced by the radiation of powerful lasers. In the preceding chapter it has been shown that the populations for resonances of arbitrary orders are described by (1.4.1). Hence, the population dynamics in multiphoton interactions of fields with matter can be reduced to the study of populations of the generalized two-level system.

To illustrate the main features of their behavior we discuss in here two extremely simple examples. First of all, we consider the saturation of populations accompanying the q-photon stationary interaction of the field with a molecule and find the conditions under which the effect of the Stark shift is essential. Then we shall analyze the quasi-stationary variation of populations whenever it can be described by balance equations, and when the concept of the transition probability in the presence of fields can be introduced. In this case the equations of the generalized two-level system also turn out to be quite useful since they enable one to obtain easily the probability of the molecular transition into an excited state even in the presence of several resonances of different orders; this probability proves to depend on the phases of the interacting fields.

2.1.1 Saturation of Populations of Resonant Levels and the Effect Which the Level Shift Under the Influence of Light Has on Saturation

Our argument will be based on (1.4.1). Consider a situation when the values of detuning ν_s [see (1.2)] are the same for all resonances and are equal to ν, and the fields determining the quantitites $v^{(s)}$ have constant amplitudes and phases. Then a stationary solution of (1.4.1) exists:

$$\sigma_{\mathrm{st}} = \mathrm{i}\hbar^{-1}\mathcal{V}T\eta_{\mathrm{st}}[1+\mathrm{i}(\nu-\Omega)T]^{-1} \ , \tag{2.1.1}$$

$$\eta_{\mathrm{st}} = \eta_0\left\{1+\frac{|\mathcal{V}|^2}{U^2}[1+(\nu-\Omega)^2T^2]^{-1}\right\}^{-1} \ , \quad \text{where} \tag{2.1.2}$$

$$U = \hbar/2\sqrt{\tau T} \ . \tag{2.1.3}$$

The physical meaning of this quantity will be clarified below. Let us first compare (2.1.2) with (17.68) of [2.8] which describes the saturation of populations in the ordinary two-level system. There are only two differences between them. (i) the dependence of the quantity $|\mathcal{V}|^2$ occurring in (2.1.2) on the intensity of light fields may be more complex than a direct proportionality; (ii) in (2.1.2) the Stark shift under the effect of light appears in addition to detuning. In the case of the single-photon resonance this shift represents the only difference between the generalized and the ordinary two-level systems.

As before, the dependence of the population difference on detuning is depicted by a downwards facing Lorentz curve the asymptotion of which merge into the straight line $\eta = \eta_0$. Its center is at the point $\nu = \Omega$ and its half-width equals $T^{-1}(1+|\mathcal{V}|^2/U^2)^{1/2}$.

Let us study the dependence of η_{st} on the fields acting on the molecule. First we consider the case with no Stark shift. This may occur if the polarizabilities of the ground and excited states are equal: $\kappa_{ab}^{11} = \kappa_{ab}^{22}$ [see (1.2.11)]. Then the dependence of populations on the field is determined only by the nondiagonal matrix element $\mathcal{V}$ of the energy of the averaged motion.

As usual, it is possible to introduce the field energy saturating the transition, for which $\eta = 0.5\eta_0$:

$$v_{\text{sat}}(\nu) = \frac{\hbar}{2}\left(\frac{1+\nu^2T^2}{\tau T}\right)^{1/2} . \tag{2.1.4}$$

The quantity U appearing in (2.1.2) is v_{sat} for vanishing detuning; it will be called the saturation energy. It is a constant which characterizes a given transition.

Let us compare the saturation of populations in the case of multiphoton absorption of different orders. Such a comparison is only meaningful for incoherent multiphoton processes when the transition frequency is a multiple of the frequency of the field:

$$q\omega = \omega_{21} + \nu \ , \quad q = 1, 2, 3, \ldots \ . \tag{2.1.5}$$

At the same time

$$|\mathcal{V}|^2 = |\kappa^{12}_{(q)}(\omega)|^2 \mathcal{J}^q \ ,$$

holds where

$$\mathcal{J} = |\mathcal{E}(\omega)|^2 \ .$$

The q-th order polarizability $\kappa^{12}_{(q)}$ [1] is determined by (1.2.20).

As follows from (2.1.2), for low intensity fields the function $\eta_{\text{st}}(\mathcal{J})$ has the form of a parabola of power q. For the saturating field

$$\mathcal{E}_{\text{sat}}(\nu) = \left|\frac{v_{\text{sat}}(\nu)}{\kappa^{12}_{(q)}(\omega)}\right|^{1/q} , \tag{2.1.6}$$

the population difference is $\eta_{\text{st}} = 0.5\eta_0$. As the field increases further, the curve $\eta_{\text{st}}(\mathcal{J})$ turns into a hyperbola of the same power q. The common parameter of both, parabola and hyperbola, is the quantity $\mathcal{E}_{\text{sat}}^{-2q}$.

One can roughly estimate the magnitude of $\mathcal{E}_{\text{sat}}$ of the allowed q-photon transition if it is assumed that [see (1.2.20)]

$$\kappa^{12}_{(q)} \approx d^q/(\hbar\omega)^{q-1} , \tag{2.1.7}$$

where d is a quantity of the order of magnitude coinciding with that of the dipole moment of the allowed single-photon transition, and the order of mag-

[1] The subscript (q) will denote the order of polarizability in the degenerate case [when (2.1.5) is valid].

nitude of ω is the frequency in the optical range. Then (2.1.4, 6, 7) yield[2]

$$\mathcal{E}_{\rm sat}(0) \approx \frac{\hbar\omega}{d(4\omega^2\tau T)^{1/2q}} \approx E_{\rm at}(2\omega\sqrt{\tau T})^{-1/q} \ . \tag{2.1.8}$$

Obviously, in the general case it is meaningless to compare the saturating fields in multiphoton absorption of different orders with each other since the quantity $\kappa^{12}_{(q)}$ depends on the selection rules for the q-photon transition between levels 1 and 2, and it may turn out that $\kappa^{12}_{(q_2)} \gg \kappa^{12}_{(q_1)}$, although $q_2 > q_1$. For allowed q_2 and q_1 photon acts of absorption, the magnitude of the field required for saturation of the same transition correlates with that of the difference between q_2 and q_1.

Let us consider the effect which the Stark shift of levels in the fields involved in a multiphoton process has on the saturation of populations induced by this process.

From (2.1.2) it is seen that the observation of the peculiarities in the behavior of the population difference $\eta_{\rm st}$ associated with this shift requires fields that would displace the center of the transition line by a distance of the order of its half-width.

Let us discuss these peculiarities in detail. As follows from (2.1.2), there can only be two essentially different situations: (i) ν and Ω have different signs ($\nu/\Omega<0$), and (ii) ν and Ω have the same sign ($\nu/\Omega>0$).

In the first case the increase of the field energy leads to an increase of the resultant detuning $|\nu-\Omega|$. The interaction of the molecule with the field becomes weaker, and the saturation is less pronounced than in the absence of the Stark effect. In the limit $|\Omega| \gg T^{-1}$, ν and $|\mathcal{V}| \gg v_{\rm sat}$ the function $\eta_{\rm st}(\mathcal{J})$ tends in the case of a single-photon (Fig. 2.1a) process to η_0, for a two-photon (Fig. 2.1b, solid curves) process to

$$\eta_0\left\{1+\frac{\hbar|\kappa^{12}_{(2)}|^2}{v^2_{\rm sat}|\kappa^{22}_{ab}-\kappa^{11}_{ab}|T^2}\right\}^{-1} ,$$

and for $q\geq 3$ to a hyperbola of the power $q-2$ (Fig. 2.1b, dashed lines). For the single-photon absorption the curve $\eta_{\rm st}(\mathcal{J})$ acquires a minimum at $\Omega T = -(1+\nu^2T^2)^{1/2}{\rm sign}\nu$ for which

$$\eta^{(1)}_{\rm min} = \eta_0\left\{1+\frac{\hbar|d|^2}{2|\kappa^{11}_{ab}-\kappa^{22}_{ab}|TU^2(|\nu T|+\sqrt{1+\nu^2T^2})}\right\}^{-1} .$$

If $\nu/\Omega>0$, the increase of $|\mathcal{E}(\omega)|^2$ induces first a decrease of the resulting detuning and then a new increase. Accordingly, the interaction of the field with

[2] $E_{\rm at} \sim 10^6 - 10^7$cgse is the strength of "intraatomic" fields for the optical electron.

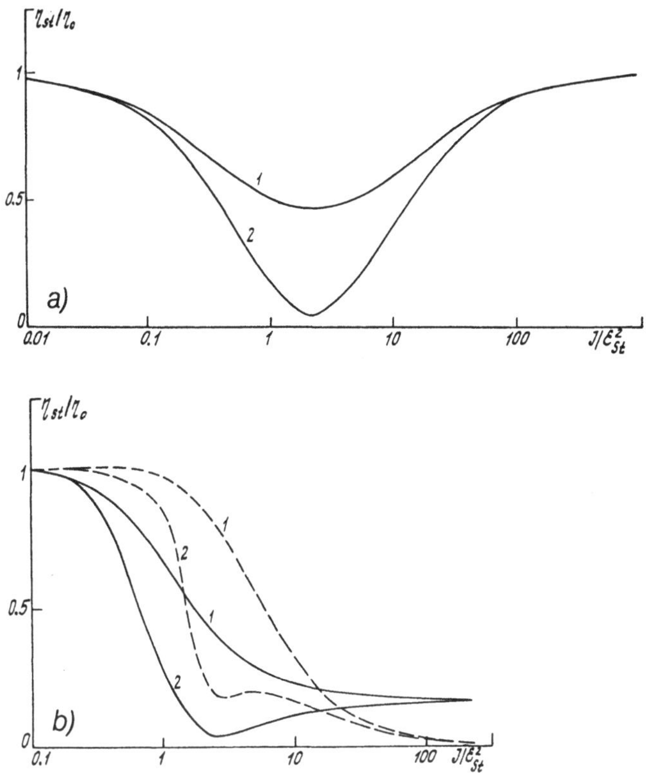

Fig. 2.1a,b. The stationary population difference as a function of the radiation intensity. **(a)** Single-photon resonance, $|\nu T| = 2$, $\mathcal{E}^2_{\rm sat}(0) = 0.1\mathcal{E}^2_{\rm St}$; **(b)** (—) two-photon resonance, $|\nu T| = 2$, $\mathcal{E}^2_{\rm sat}(0) = 0.45\mathcal{E}^2_{\rm St}$; (- - -) three-photon resonance, $|\nu T| = 2$, $\mathcal{E}^2_{\rm sat}(0) = 1.45\mathcal{E}^2_{\rm St}$, $\mathcal{E}_{\rm St}$ being the Stark field, see (2.1.9). Curves *1* correspond to $\nu/\Omega<0$, curves *2* to $\nu/\Omega>0$

the molecule first becomes stronger and is then diminished. In correspondence with this the saturation accompanying the increase of $\mathcal{J} = |\mathcal{E}(\omega)|^2$ will at low intensities occur more rapidly than in the absence of the Stark effect. If the case $|\Omega| \gg T^{-1}$, ν the asymptotic behavior of $\eta_{\rm st}(\mathcal{J})$ coincides with that for $\nu/\Omega<0$. For $\nu/\Omega>0$ and $q>1$, this function does not necessarily decrease monotonically as in the case $\nu/\Omega<0$. It can acquire extrema whose positions are determined from the conditions

$$\Omega T = \nu T + \frac{1}{\nu T} \quad \text{for} \quad q = 2 \ ;$$

$$\Omega T = \frac{\nu T(q-1) \pm \sqrt{\nu^2T^2 - q(q-2)}}{q-2} \quad \text{for} \quad q>2 \ .$$

When $q = 1$, the function $\eta(\mathcal{J})$ again has a minimum (Fig. 2.1a) whose position is specified by the condition $\Omega T = \operatorname{sign}\nu\sqrt{1+\nu^2T^2}$. The value of the population difference at this point is

$$\eta_{\min}^{(2)} = \eta_0\left\{1 + \frac{\hbar|d|^2}{2U^2|\kappa^{11} - \kappa^{22}|T(\sqrt{1+\nu^2T^2} - |\nu T|)}\right\}^{-1} .$$

For $q>2$ and $\nu^2T^2>q(q-2)$, the population difference as a function of the energy of the field has both a minimum and a maximum (Fig. 2.1b, dashed line). The field intensity corresponding to the minima of the curves on Fig. 2.1 is smallest for the value of detuning

$$\nu_{\text{opt}} = (q-1)T^{-1}\operatorname{sign}\Omega .$$

At the same time the absorption line is shifted by $|\Omega_{\min}| = qT^{-1}$. The maximum of $\eta(\mathcal{J})$ occurs at higher values of the field than its minimum. For example, in order to observe the maximum of η_{st} for $\nu = \nu_{\text{opt}}$, $q\geq 3$, a field is required which is sufficient for a shift of the line by

$$|\Omega_{\max}| = (q^2+2q+2)/(q-2)T .$$

In this case the difference $|\Omega_{\max} - \Omega_{\min}|$ equals $(4q+2)/(q-2)T$ and varies from $14T^{-1}$ (for $q=3$) to $4T^{-1}$ (for $q\gg 1$).

The field typical for such effects is determined by the condition $|\Omega T| = 1$ and will be called a Stark field:

$$\mathcal{E}_{\text{St}} = \left(\frac{\hbar}{|\kappa_{ab}^{11} - \kappa_{ab}^{22}|T}\right)^{1/2} . \tag{2.1.9}$$

The magnitude of the Stark field can be roughly estimated in along the same lines as in the case of the saturating field. Setting $|\kappa_{ab}^{11} - \kappa_{ab}^{22}| \approx d^2/\hbar\omega$ one obtains

$$\mathcal{E}_{\text{St}} \approx \hbar\omega/d(\omega T)^{1/2} \approx E_{\text{at}}(\omega T)^{-1/2} . \tag{2.1.10}$$

By comparing $\mathcal{E}_{\text{St}}$ and $\mathcal{E}_{\text{sat}}(0)$ [see (2.1.8)], one can easily see that for allowed transitions in the single-photon absorption, $\mathcal{E}_{\text{sat}}(0)\ll\mathcal{E}_{\text{St}}$ since $\omega\tau\gg 1$. This inequality is also valid for two-photon processes if the life time of the molecule in the excited state is much greater than the inverse transition line width (or, more precisely, if $4\tau\gg T$)[3]. For higher order multiphoton interactions $(q>2)\mathcal{E}_{\text{sat}}\gtrsim\mathcal{E}_{\text{St}}$, and the Stark shift will be essential in population saturation.

[3] For electron transitions in vapors and gases the condition $2\tau \approx T$ can be satisfied. In this case the saturating field for the two-photon processes will be of the order of the Stark field.

It can also considerably affect the redistribution of populations in the case of the single- and two-photon acts of absorption on weakly allowed transitions.

Let us give a numerical evaluation of the quantities $\mathcal{E}_{\mathrm{St}}$ and $\mathcal{E}_{\mathrm{sat}}$. This evaluation can only be tentative since the range of values of constants κ_{ab}^{nn}, $\kappa_{(q)}^{12}$, T and τ is very broad. For the values $d \approx 10^{-18}$ cgse, $\omega = 10^{15}\,\mathrm{s}^{-1}$, the inverse line width typical for gases $T \approx 10^{-8}$ s and $\tau \approx 10^{-8}$ s (the excited state is not metastable), (2.1.10) yields $\mathcal{E}_{\mathrm{St}}^2 \approx 10^5$ cgse (which corresponds to the power flux density $S = 50\,\mathrm{MW/cm^2}$). For the single-photon absorption $\mathcal{E}_{\mathrm{sat}}^2 \approx 2.5 \times 10^{-3}$ cgse ($S{\sim}1\,\mathrm{W/cm^2}$) is arrived at, for the two-photon absorption $\mathcal{E}_{\mathrm{sat}}^2 \approx 0.5 \times 10^5$ cgse ($S \approx 25\,\mathrm{MW/cm^2}$), and for the three-photon absorption $\mathcal{E}_{\mathrm{sat}}^2 = 1.3 \times 10^7$ cgse ($S = c\sqrt{\varepsilon}/\mathcal{E}(\omega)/^2/2\pi \approx 6.5\,\mathrm{GW/cm^2}$). For metastable states the magnitudes of the saturating fields are much less. For instance, when $\tau = 10^{-4}$ s for the three-photon absorption, $\mathcal{E}_{\mathrm{sat}}^2 \approx 6.5 \times 10^5$ cgse ($S = 300\,\mathrm{MW/cm^2}$).

We have considered the saturation of populations resulting from the multiphoton absorption which is degenerate with respect to frequency. We shall now discuss the case when the resonance condition does not degenerate into (2.1.5) but preserves its general from (1.2). For linearly polarized fields, the matrix element $\mathcal{V}$ can be written as

$$\mathcal{V} = -\kappa_{(q)}^{12} \prod_j \mathcal{E}^{n_j}(\omega_j) \tag{2.1.11}$$

see (1.2.21b). Substituting (2.1.11) into (2.1.2), introducing the normalized detuning $\Delta = \nu T$ and constants

$$\mathcal{K}_j = \hbar^{-1}[\kappa_{jj}^{22}(\omega_j) - \kappa_{jj}^{11}(\omega_j)]T \;, \quad \mathcal{J}_{\mathrm{sat}} = \mathcal{E}_{\mathrm{sat}}^2(0) \;, \tag{2.1.12}$$

we obtain an expression for the stationary population difference

$$\eta_{\mathrm{st}} = \eta_0 \left\{ 1 + \mathcal{J}_{\mathrm{sat}}^{-q} \prod_{j>0} \mathcal{J}_j^{n_j} \left[1 + \left(\Delta + \sum_{j>0} \mathcal{K}_j \mathcal{J}_j \right)^2 \right]^{-1} \right\}^{-1} \;, \tag{2.1.13}$$

where

$$\mathcal{J}_j = |\mathcal{E}(\omega_j)|^2 \;. \tag{2.1.14}$$

As follows from (2.1.13), the dependence of η_{st} on the energy of one of the fields (whose frequency appears n_j times in the resonance condition) behaves in the same way as in the n_j-fold degenerate case considered above, provided the magnitudes of the other fields are fixed.

We would like to emphasize the fact that effective saturation of populations under nondegenerate conditions is only possible if the intensity of all fields participating in the multiphoton process is sufficiently high.

For resonant parametric processes, when several resonance conditions are fulfilled simultaneously, the stationary population difference turns out to depend not only on the intensities of the fields acting on the molecule but also on their pases. At the end of the Sect. 2.1.2 we shall return to the discussion of such cases.

The question of the value of the field that saturates the population difference for transitions which are forbidden for single- or two-photon absorption should be discussed separately. Whereas for forbidden electron transitions $\mathcal{E}_{\text{sat}} \sim \mathcal{E}_{\text{at}}$, the saturation of forbidden vibrational-rotational transitions in polyatomic molecules can occur at intensities of the order of several tens of MW/cm^2 because of the removal of selection rules for these transitions that actually represents a manifestation of higher order multiphoton processes [2.24]. For instance, excitation (up to the state of saturation) of molecules of the type SF_6, BCl_3 during the first order resonance with forbidden vibrational-rotational transitions $|v,j\rangle \to |v+1, j\pm 3\rangle$, where v and j are vibrational and rotational quantum numbers, respectively, can be due to a three-photon process associated with resonance condition of the form $\omega - \omega + \omega = \omega_{21}$.

2.1.2 Balance Equations and Interference of Transition Probability Amplitudes in Resonant Parametric Interactions

Consider the variation of populations in time. Frequently the characteristic time for the building-up of stationary populations is much greater than the one required to generate the nondiagonal element σ of the density matrix. The respective necessary condition is that $T \ll \tau$. If one is interested in the development of the process over times $t \gg T$, the derivative in the first of the equations in (1.4.1) may be neglected, and one can assume that the quantities σ and η are again related through the algebraic equation (2.1.1), but now with η_{st} replaced by η, the actual value of the population difference. When (2.1.1) is substituted into the second equation of (1.4.1), the balance equation for η results in

$$\frac{d\eta}{dt} + \frac{\eta_0}{\tau}\left(\frac{\eta}{\eta_{\text{st}}} - 1\right) = 0 \ . \tag{2.1.15}$$

The solution of this equation with the initial condition $\eta(t=0) = \eta_{\text{in}}$ has the form

$$\eta = \eta_{\text{st}} + (\eta_{\text{in}} - \eta_{\text{st}})\exp\left(-\frac{t}{\tau}\frac{\eta_0}{\eta_{\text{st}}}\right) \ . \tag{2.1.16}$$

The characteristic time of the variation of the population difference under the influence of fields (the induced relaxation time) is

$$\tau_{\text{ind}} = \tau\eta_{\text{st}}/\eta_0 \ . \tag{2.1.17}$$

In strong fields, relaxation is accelerated [until the Stark effect becomes noticeable, see (2.1.2)].

The behavior of the populations σ_{11} and σ_{22} themselves is also described by balance equations:

$$\frac{d}{dt}\sigma_{nn} + w_{mn}\sigma_{nn} - w_{nm}\sigma_{mm} = 0 \ , \quad (m,n) = (1,2) \ , \tag{2.1.18}$$

in which the probability w_{12} for a transition from state 1 to state 2 includes the probability W_{12} of this transition in the absence of fields (when $\mathcal{V} = 0$) and the probability of the induced transition

$$W_{12}^{\mathrm{ind}} = W_{21}^{\mathrm{ind}} = 2|\mathcal{V}|^2 T\hbar^{-2}[1 + (\nu - \Omega)^2 T^2]^{-1} \ . \tag{2.1.19}$$

If all fields are weaker than the Stark field (2.1.10), we have

$$W_{12}^{\mathrm{ind}} \sim \prod_{j>0} |\mathcal{E}(\omega_j)|^{2n_j} \tag{2.1.20}$$

for any multiphoton and Raman interactions. When the other fields become comparable with the Stark ones, the probability of induced transitions increases faster than in low intensity fields if Ω approaches ν. It increases slower or even decreases if the growth of the field causes Ω to move away from ν.

Consider the field-dependent peculiarities of the behavior of the induced transition probability which are typical for resonant parametric processes. The latter always appear when there are several resonant processes that occur simultaneously (several resonance conditions of the type (1.2) are fullfilled). In such cases the matrix element $\mathcal{V}$ of the Hamiltonian of the averaged motion consists of terms associated with different resonances [see (1.2.16)]. Each of the terms is proportional to the product of complex field amplitudes that participate in the corresponding process, see (1.2.21). Therefore, we observe an interference of the probability amplitudes of individual processes (instead of the addition of the probabilities themselves), and the probability of the induced transition W_{12}^{ind}, which is proportional to $|\mathcal{V}|^2$, will depend on the phase relations between the fields.

Consider as an example the interference of the probability amplitudes for the single-photon and the degenerate two-photon processes [conditions (1.2.2, 9) are satisfied with $\omega_\beta = \omega_\gamma = \frac{1}{2}\omega_\alpha = \omega$, see Fig. 1.2a]. It will be assumed for the sake of simplicity that the fields $\vec{\mathcal{E}}(\omega)$ and $\vec{\mathcal{E}}(2\omega)$ are polarized, respectively, in the directions of Cartesian axes a and b. Then (1.2.20, 21) and (1.2.16) yield:

$$\mathcal{V} = \mathcal{V}^{(1)} + \mathcal{V}^{(2)} = -(d_{12})_b \mathcal{E}(2\omega) - \kappa_{aa}^{12}\mathcal{E}^2(\omega) \ . \tag{2.1.21}$$

By substituting (2.1.21) into (2.1.19), we obtain a term in $W_{12}^{\rm ind}$ proportional to $\mathcal{E}(2\omega)\mathcal{E}^2(\omega)$ which depends on the phases of fields corresponding to the main and the doubled frequencies. As the quantity $\arg \kappa_{aa}^{12}\mathcal{E}^2(\omega)/\mathcal{E}(2\omega)d_{12b}$ varies from 0 to π, the probability of the induced transition 1–2 varies from its maximal value associated with

$$|\mathcal{V}|^2 = [d_{12b}|\mathcal{E}(2\omega)| + \kappa_{aa}^{12}|\mathcal{E}(\omega)|^2]^2 \ ,$$

to its minimal value

$$|\mathcal{V}|^2 = [d_{12b}|\mathcal{E}(2\omega)| - \kappa_{aa}^{12}|\mathcal{E}(\omega)|^2]^2 \ .$$

Obviously, amplitudes and phases of fields with $\mathcal{V} = 0$ may be related to each other. In our example $\mathcal{V} = 0$ holds when

$$\arg\frac{\mathcal{E}^2(\omega)}{\mathcal{E}(2\omega)} = \pi \ , \quad \frac{|\mathcal{E}(\omega)|^2}{|\mathcal{E}(2\omega)|} = \frac{d_{12b}}{\kappa_{aa}^{12}} \ .$$

In this case the probability of induced transitions is also equal to zero. When a molecule is subject to such a field, the populations of its levels are at any moment of time in equilibrium and therefore no field energy is absorbed.

This phenomenon which may be called the parametric self-induced transparency, represents a general property of resonant parametric interactions. It will be shown below that under certain conditions the parametric self-induced transparency of macroscopic volumes of matter is possible. The parametric self-induced transparency was observed by the authors [2.25] for the case of a resonant interaction where $\omega_{21} = \omega_1 + \omega_2 = \omega_4 - \omega_3$.

2.2 Susceptibility in Incoherent Multiphoton Processes

2.2.1 Expressions for Susceptibility

The nonlinear susceptibility of matter can be defined in a way similar to that of the linear one provided: the evolution of the field for the slowly varying amplitudes $\mathcal{E}(\omega_j, t)$ can be described by (1.1.2) and the evolution of the state of matter for the slowly varying components of the density matrix by (1.2.18). That is, it can be defined as the coefficient of the complex amplitude $\mathcal{E}(\omega_j)$ in a Fourier-transform of the nonlinear polarization [2.9]:

$$\mathcal{P}_a[\omega_j, \mathcal{E}(\omega_l), t] = \chi_{ab}[\omega_j, \mathcal{E}(\omega_l), t]\mathcal{E}_b(\omega_j, t) \ .$$

This can easily be shown if one takes into account the fact that when the Fourier integrals of the nonlinear polarization (1.3.13) are computed, the slowly varying functions of time should be taken out of the integration sign (in the same manner as done in Sects. 1.2, 3 for the calculation of the polarization and the components of the density matrix).

Let us consider the resonant and nonresonant polarizations of a molecule separately. The nonresonant part of the polarization (1.3.13) expressed in terms of the population difference η has the form

$$\boldsymbol{P}^{\text{nonres}} = \tfrac{1}{2}[\boldsymbol{D}_{11} + \boldsymbol{D}_{22} + (\boldsymbol{D}_{11} - \boldsymbol{D}_{22})\eta] \ . \tag{2.2.1}$$

In this section we assume that only one resonance condition is fulfilled. This means that no resonant parametric phenomena occur, and only multiphoton absorption (emission) and Raman transitions are possible between levels 1 and 2. It will also be supposed that neither of the fields participating in a given multiphoton process takes part in nonresonant parametric phenomena. Then the vectors $\boldsymbol{D}_{11}$ and $\boldsymbol{D}_{22}$ are determined by the polarizability tensors of the ground (κ_{ab}^{11}) and excited (κ_{ab}^{22}) states of the molecule (1.2.12):

$$(\boldsymbol{D}_{nn})_a = \sum_j \kappa_{ab}^{nn}(\omega_j)\mathcal{E}_b(\omega_j)\exp(i\omega_j t) \ .$$

Substituting this equation into (2.2.1) and isolating the amplitude of the polarization spectral component at the frequency ω_j, one easily finds the nonresonant susceptibility of the molecule at this frequency:

$$\chi_{ab}^{\text{nonres}} = \tfrac{1}{2}[\kappa_{ab}^{11} + \kappa_{ab}^{22} + (\kappa_{ab}^{11} - \kappa_{ab}^{22})\eta] \ . \tag{2.2.2}$$

Since the κ^{nn} are real quantities, the nonresonant susceptibility (2.2.2) is also real. As the energy of the light fields increases, the nonresonant susceptibility in the stationary case becomes saturated in the same way as the stationary population difference. The spectral component of the resonant part of polarization $\chi^{\text{res}}(\omega_j)$ (the frequency ω_j having the same sign as it has in the resonance condition) can be found by means of (1.4.3) and (1.2.16) and (1.3.17) for the matrix element of the Hamiltonian of the averaged motion $\mathcal{V}_{12}$ and the spectral component of the generalized dipole moment $\vec{\mathcal{D}}_{21}(\omega_j)$. In the case of an incoherent multiphoton process, $\vec{\mathcal{D}}_{21}$ and $\mathcal{V}_{12}$ contain one term each which will be denoted by $\vec{\mathcal{D}}^*(\omega_j)$ and $\mathcal{V}$, respectively. We have chosen the quantity $\vec{\mathcal{D}}_{12}(\omega_j)$ for $\vec{\mathcal{D}}(\omega_j)$, therefore (1.3.17) can now be written as

$$\mathcal{V} = -n_j^{-1}\vec{\mathcal{D}}(\omega_j)\vec{\mathcal{E}}(\omega_j) \ . \tag{2.2.3}$$

Substitution of (2.1.1) into (1.4.3) results in an expression for quasistationary polarization which can always be used when the population balance equation (2.1.15) is valid:

$$\vec{\mathcal{P}}^{\text{res}}(\omega_j) = \hbar^{-1}T\mathcal{V}\vec{\mathcal{D}}^*(\omega_j)\eta\frac{\mathrm{i} + (\nu - \Omega)T}{1 + (\nu - \Omega)^2T^2} = [\vec{\mathcal{P}}^{\text{res}}(\omega_{-j})]^* \ . \tag{2.2.4}$$

The resonant part of susceptibility corresponding to it can be obtained from this equation if one makes use of (2.2.3):

$$\chi_{ab}^{\mathrm{res}}(\omega_j) = -\hbar^{-1} n_j^{-1} T \mathcal{D}_a^*(\omega_j)\mathcal{D}_b(\omega_j)\eta \frac{\mathrm{i} + (\nu - \Omega)T}{1 + (\nu - \Omega)^2 T^2}$$
$$= [\chi_{ab}^{\mathrm{res}}(\omega_{-j})]^* \ . \qquad (2.2.5)$$

The total quasistationary susceptibility of a particle is the sum of (2.2.2, 5):

$$\chi_{ab}(\omega_j) = \chi_{ab}^{\mathrm{res}}(\omega_j) + \chi_{ab}^{\mathrm{nonres}}(\omega_j) \ .$$

If the balance equations are not applicable, it is no longer possible to express polarization and susceptibility just through the population difference. In this case the general relation (1.4.3) and the common solution of (1.4.1) for η and σ should be used.

In this section we shall be concerned mainly with the stationary susceptibility. The last part of the section, however, deals with the real part of susceptibility generated at the ends of a rectangular light pulses.

To study the stationary susceptibility we consider as an example the diagonal component $\chi_{jj}(\omega_j)$ that determines the projection of polarization $\vec{\mathcal{P}}_j(\omega_j)$ onto the direction of the field $\vec{\mathcal{E}}(\omega_j)$. This component contains, instead of $\mathcal{D}_a$ and $\mathcal{D}_b$, the projection of the generalized dipole moment $\mathcal{D}_j(\omega_j)$ on the direction of the field $\vec{\mathcal{E}}(\omega_j)$. From (2.2.3) and (2.1.11),

$$\mathcal{D}_j(\omega_j) = n_j \kappa_{(q)} \prod_{l \neq j} \mathcal{E}^{n_l}(\omega_l) \mathcal{E}^{n_j - 1}(\omega_j) \qquad (2.2.6)$$

is had. Making use of this expression together with (2.1.13), (1.2.11) and adding (2.2.2, 5), we obtain the diagonal component of the stationary susceptibility for the q-th order resonance:

$$\chi_{jj}(\omega_j) = \frac{\eta_0 T}{\hbar} |\kappa_{(q)}|^2 \frac{\mathcal{K}_j \mathcal{J}_j \vartheta - \left(\Delta + \sum_l \mathcal{K}_l \mathcal{J}_l\right) n_j - \mathrm{i} n_j}{1 + \left(\Delta + \sum_l \mathcal{K}_l \mathcal{J}_l\right)^2 + \mathcal{J}_{\mathrm{sat}}^{-q} \prod_l \mathcal{J}_l^{n_l}} \mathcal{J}_j^{n_j - 1} \prod_{l \neq j} \mathcal{J}_l^{n_l}$$
$$+ \frac{1}{2}\{\kappa_{jj}^{11}(\omega_j) + \kappa_{jj}^{22}(\omega_j) + [\kappa_{jj}^{11}(\omega_j) - \kappa_{jj}^{22}(\omega_j)]\eta_0\}$$
$$= \chi_{jj}^*(\omega_{-j}) \ ; \qquad (2.2.7)$$

where $\vartheta = 2\tau/T$ and the quantities $\mathcal{K}_j$, $\mathcal{J}_j$, $\mathcal{J}_{\mathrm{sat}}$ are defined by (2.1.12, 14).

Each of the physical factors responsible for the behavior of the susceptibility during the variation of fields corresponds to specific terms in (2.2.7). The term $\mathcal{J}_j \mathcal{K}_j \vartheta$ in the numerator of the fraction that appears in (2.2.7), describes the effect of redistribution of populations between levels 1 and 2 with different linear polarizabilities $\kappa_{jj}^{11}(\omega_j)$ and $\kappa_{jj}^{22}(\omega_j)$ on the diagonal component of susceptibility. The presence of products $\mathcal{J}_j \mathcal{K}_j$ that are summed up with Δ leads to the change of susceptibility caused by the Stark shift of the resonant transition line center. The quantity $\mathcal{J}_{\text{sat}}^{-q}\prod_l \mathcal{J}_l^{n_l}$ in the denominator is responsible for the saturation of $\chi_{jj}(\omega_j)$ associated with the population saturation.

The imaginary part $\mathrm{Im}\{\chi_{jj}(\omega_j)\}$ of the diagonal component of susceptibility describes the absorption (emission) of the energy of the field $\vec{\mathcal{E}}(\omega_j)$ by matter. Its real part $\mathrm{Re}\{\chi_{jj}(\omega_j)\}$ determines, in the case of sufficiently weak absorption (i.e., when $4\pi N\ \mathrm{Im}\{\chi_{jj}\}\ll 1$ as usual), the refractive index of matter for this field,

$$n(\omega_j)\approx\sqrt{1+4\pi N\ \mathrm{Re}\{\chi_{jj}(\omega_j)\}}\ .$$

2.2.2 The Imaginary Part of Susceptibility as a Function of Fields and the Energy Absorbed by Matter

Consider, first of all, a q-fold degenerate resonance. From (2.2.7) we obtain the imaginary part of susceptibility in the form

$$\mathrm{Im}\{\chi(\omega)\}=-\frac{q\eta_0 T|\kappa_{(q)}|^2}{\hbar}\frac{\mathcal{J}^{q-1}}{1+(\Delta+\mathcal{K}\mathcal{J})^2+(\mathcal{J}/\mathcal{J}_{\text{sat}})^q}\ . \tag{2.2.8}$$

where the subscripts of $\mathcal{K}_j$, $\mathcal{J}_j$, χ_{jj} and ω_j have been omitted. One can easily see from this expression that in a noninverted medium ($\eta_0>0$), the quantity $\mathrm{Im}\{\chi(\omega)\}$ is negative and the q-photon absorption will take place [see (2.2.11)]. In an active medium ($\eta_0<0$), we have $\mathrm{Im}\{\chi(\omega)\}>0$ and one observes q-photon emission of light. To make matters simpler, we shall speak of absorption, although everything discussed below applies equally to absorbing as well as to active media.

Consider the behavior of the quantity $\mathrm{Im}\{\chi(\omega)\}$ as a function of the radiation frequency and intensity.

If the relatively weak dependence of the quantities $\mathcal{K}$ and $\mathcal{J}_{\text{sat}}$ occurring in (2.2.8) on ω can be neglected, it is possible to assume that the frequency dependence of $\mathrm{Im}\{\chi(\omega)\}$ can be depicted by a Lorentz curve with the frequency at its center

$$\omega=(\omega_{21}+T^{-1}\mathcal{K}\mathcal{J})q^{-1}=(\omega_{21}-\Omega)/q\ , \tag{2.2.9}$$

and the half-width

$$(qT)^{-1}\sqrt{1+(\mathcal{J}/\mathcal{J}_{\text{sat}})^q}\ . \tag{2.2.10}$$

Figure 2.2 gives a qualitative illustration of $|\mathrm{Im}\{\chi(\omega)\}|$ as a function of the field intensity. In low intensity fields ($\mathcal{J}\ll\mathcal{J}_{\text{sat}}$; $\mathcal{J}\ll\mathcal{J}_{\text{St}}=|\mathcal{K}|^{-1}=\mathcal{E}_{\text{St}}^2$), the imaginary part of the susceptibility is proportional to $\mathcal{J}^{q-1}$. In strong fields ($\mathcal{J}\gg J_{\text{sat}},\mathcal{J}_{\text{St}}$) increasing $\mathcal{J}$ make the quantity $\mathrm{Im}\{\chi(\omega)\}$ to approach zero as $\mathcal{J}^{-2}$ for single-photon absorption and as $\mathcal{J}^{-1}$ for $q>1$.

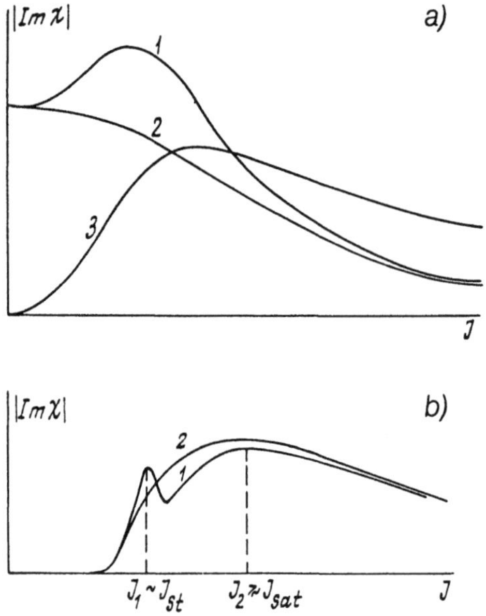

Fig. 2.2a,b. The dependence of the imaginary part of the susceptibility on the radiation intensity. **(a)** *1* single-photon resonance, $\mathcal{J}_{\text{sat}} > (-2\mathcal{K}\Delta)^{-1} > 0$; *2* single-photon resonance, one of these conditions is violated; *3* two- and three-photon resonances; **(b)** the higher order resonance (greater than three). *1* for $\mathcal{K}\Delta < 0$, $q^2 - 4q + 3 \lesssim \Delta^2(2q+1)$; *2* at least one of these conditions is violated

The presence of a maximum (minimum) in the curves of Fig. 2.2 implies that the relative variation of the radiation intensity resulting from the absorption over a distance of unit length is maximal (minimal) for the corresponding value of $\mathcal{J}$. The position of the extrema of the function $\text{Im}\{\chi(\omega)\}$ is determined by

$$(\mathcal{J}/\mathcal{J}_{\text{sat}})^q + \mathcal{K}^2\mathcal{J}^2(3-q) + 2\mathcal{K}\mathcal{J}\Delta(2-q) + (1+\Delta^2)(1-q) = 0 \ .$$

Let us compute the energy Q which is absorbed per second by a unit volume of matter containing N molecules of a given type. Substituting (2.2.8) into the expression

$$Q = -2N\omega\mathcal{J}\text{Im}\{\chi(\omega)\} \ , \tag{2.2.11}$$

(see [2.8], Sect. 9)[4] and comparing the result with (2.1.2), we obtain:

$$Q = \frac{Nq\hbar\omega}{2\tau}(\eta_0 - \eta_{\text{st}}) \ . \tag{2.2.12}$$

From this formula it follows that the behavior of the absorbed energy Q accompanying the variation of the light wave intensity is completely determined by the behavior of the stationary population difference (see Sect. 2.1).

[4] The difference between the signs of the right-hand sides of (2.2.11) and (9.9) of [2.8] is associated with the fact we define the susceptibility via $P = \chi\mathcal{E}\exp(\text{i}\omega t)$ + complex conjugate, while *Fain* uses $P = \chi\mathcal{E}\exp(-\text{i}\omega t)$ + complex conjugate.

Consider now the incoherent multiphoton interactions that include fields with different frequencies. One can show on the basis of the general expression (2.2.7), that the dependence of the imaginary part of susceptibility at the frequency ω_j on the energy of the field at the same frequency has a form similar to that of (2.2.8) for the n_j fold degenerate process ($q = n_j$). The quantity $\mathrm{Im}\{\chi_{jj}(\omega_j)\}$ as a function of the energies of fields with other frequencies ($\omega_k \neq \omega_j$), is in weak fields proportional to $\mathcal{J}_k^{n_k}$. In the limit of strong fields ($\mathcal{J}_k \to \infty$), $\mathrm{Im}\{\chi_{jj}(\omega_j)\}$ approaches a constant value for $n_k \geq 2$, and zero if $n_k = 1$. In the latter case there may be a broad range of the field energy variation, the field frequency being ω_k, in which $\mathrm{Im}\{\chi_{jj}(\omega_j)\}$ can be regarded as approximately constant.

For instance, if the frequency ω_k is close to the center of the transition line which is shifted by all other fields, and

$$|\mathcal{K}_k| \mathcal{J}_{\mathrm{sat}}^{n_k} \prod_{l \neq k} \mathcal{J}_l^{-n_l} \ll 1 \ ,$$

then $\mathrm{Im}\{\chi_{jj}(\omega_j)\} \approx \mathrm{const}$ for values of $\mathcal{J}_k$ satisfying the inequalities

$$\mathcal{J}_{\mathrm{sat}}^{n_k} \prod_{l \neq k} \mathcal{J}_l^{-n_l} \ll \mathcal{J}_k \ll \mathcal{K}_k^{-2} \mathcal{J}_{\mathrm{sat}}^{-n_j} \prod_{l \neq k} \mathcal{J}_l^{n_l} \ .$$

The weak dependence of $\mathrm{Im}\{\chi_{jj}(\omega_j)\}$ on $\mathcal{J}_k$ in the above energy range can be explained by the fact that here the main role is played by the saturation effect, while the influence of the Stark effect on the imaginary part of susceptibility is negligible. The asymptotic behavior of $\mathrm{Im}\{\chi_{jj}(\omega_j)\}$ for $\mathcal{J}_k \to +\infty$ and $n_k \geq 2$ is due to similar causes.

The energy absorbed per unit of time at each of the frequencies ω_j in a multiphoton process (we shall denote it by Q_j) is determined, as before, by means of (2.2.7, 11), where ω should be replaced by ω_j. The total power scattered in matter is equal to

$$Q = \sum_j Q_j = \frac{N\hbar}{2\tau} \sum_j n_j \omega_j (\eta_0 - \eta_{\mathrm{st}}) \ . \tag{2.2.13}$$

The summation here includes the frequencies that occur in condition (1.2).

Let us compute, for example, the power absorbed by matter during stimulated Raman scattering (SRS). The imaginary part of susceptibilities at the frequencies ω_1 and ω_2 can be found from (2.2.7):

$$\begin{aligned} \mathrm{Im}\{\chi_{11}(\omega_1)\} &= -\mathrm{Im}\{\chi_{11}(-\omega_1)\} \\ &= -\frac{\eta_0 T}{\hbar} |\kappa|^2 \\ &\quad \times \frac{\mathcal{J}_2}{1 + (\Delta + \mathcal{K}_1 \mathcal{J}_1 + \mathcal{K}_2 \mathcal{J}_2)^2 + (4\tau T/\hbar^2)|\kappa|^2 \mathcal{J}_1 \mathcal{J}_2} \ , \end{aligned}$$

$$
\begin{aligned}
\mathrm{Im}\{\chi_{22}(-\omega_2)\} &= -\mathrm{Im}\{\chi_{22}(\omega_2)\} \\
&= -\frac{\eta_0 T}{\hbar}|\kappa|^2 \\
&\quad\times \frac{\mathcal{J}_1}{1+(\Delta+\mathcal{K}_1\mathcal{J}_1+\mathcal{K}_2\mathcal{J}_2)^2+(4\tau T/\hbar^2)|\kappa|^2\mathcal{J}_1\mathcal{J}_2} \quad .
\end{aligned}
\tag{2.2.14}
$$

In these expressions κ is the second order polarizability [(1.2.20)]:

$$
\begin{aligned}
\kappa &= \kappa^{12}_{(2)} = \kappa^{12}_{12} \\
&= \frac{1}{\hbar}\sum_m \left[\frac{(\boldsymbol{d}_{1m}\boldsymbol{e}_1)(\boldsymbol{d}_{m2}\boldsymbol{e}_2)}{\omega_{m1}-\omega_1} + \frac{(\boldsymbol{d}_{1m}\boldsymbol{e}_2)(\boldsymbol{d}_{m2}\boldsymbol{e}_1)}{\omega_{m1}+\omega_2}\right] \quad ,
\end{aligned}
\tag{2.2.15}
$$

where $\boldsymbol{e}_1$ and $\boldsymbol{e}_2$ are unit vectors whose directions coincide with those of the electric fields at the frequencies ω_1 and ω_2.

In a noninverted medium ($\eta_0 > 0$), the power

$$
Q_1 = 2\omega_1 N \mathcal{J}_1 |\mathrm{Im}\{\chi_{11}(\omega_1)\}|
$$

is positive, and

$$
Q_2 = -2N\omega_2 \mathcal{J}_2 |\mathrm{Im}\{\chi_{22}(\omega_2)\}|
$$

is negative. Therefore, in such a medium one observes the absorption of the field at the frequency ω_1 and its emission at ω_2. The power equal to

$$
Q = \frac{N\hbar\omega_{21}}{2\tau}(\eta_0 - \eta_{\mathrm{st}})
$$

is transferred to the matter (where $\omega_1 - \omega_2 = \omega_{21}$).

The energy transfer from the frequency ω_1 to the frequency ω_2 can occur also on transitions with $\kappa = 0$; this is due to the higher order processes, such as $\omega_1 - \omega_1 + \omega_1 - \omega_2 = \omega_{21}$ [2.26]. The contributions into Q_1 and Q_2 associated with this resonance are proportional, respectively, to $\mathcal{J}_1^3$ and $\mathcal{J}_1^2\mathcal{J}_2$.

2.2.3 The Real Part of Susceptibility for the Single-Photon Resonance

The dependence of the real part of susceptibility $\mathrm{Re}\{\chi_{kk}(\omega_k)\}$ on interacting fields is more complex than that of $\mathrm{Im}\{\chi_{kk}(\omega_k)\}$ since the former is not only affected by saturation and Stark shift, but also by redistribution of populations between levels with different κ^{11}_{kk} and κ^{22}_{kk}. For this reason we shall study the behavior of $\mathrm{Re}\{\chi_{kk}(\omega_k)\}$ in the simplest and most frequently observed resonant processes, namely, in the single-photon absorption, and also in two-photon absorption and stimulated Raman scattering. (Sections 2.2.3–5 are based on the results of [2.9].)

In the case of a single-photon resonance the real part of susceptibility (2.2.7) consists of two terms:

$$\mathrm{Re}\{\chi_{kk}(\omega_k)\} = \chi_0(\omega) + \chi_1(\omega) \quad , \quad \text{where}$$

$$\chi_0(\omega) = \tfrac{1}{2}\{\kappa^{11}(\omega) + \kappa^{22}(\omega) + [\kappa^{11}(\omega) - \kappa^{22}(\omega)]\eta_0\} \quad ,$$
$$\chi_1(\omega) = \frac{\eta_0 T}{\hbar}|d|^2 \frac{\mathcal{J}\mathcal{K}\vartheta - (\Delta + \mathcal{J}\mathcal{K})}{1 + (\Delta + \mathcal{J}\mathcal{K})^2 + \mathcal{J}/\mathcal{J}_{\mathrm{sat}}} \quad . \tag{2.2.16}$$

The first of these terms comes from the nonresonant susceptibility and does not depend on the field. The second term describes in the limit of low intensity fields the anomalous dispersion of the linear susceptibility in the resonance region. For large detunings ($\nu{\sim}\omega$) both terms are transformed into a formula which characterizes the normal dispersion of susceptibility, if one takes (1.2.12) into account.

In high intensity fields the term χ_1 depends on the detuning Δ and the square of the field amplitude in a more complex way. This dependence is simpler if $\kappa^{11} = \kappa^{22} (\mathcal{K} = 0)$. Then the formula for χ_1 coincides with the well-known expression for the susceptibility of a two-level system (see, for instance, [2.10]), and the behavior of $\mathrm{Re}\{\chi\}$ during the variation of the field is determined only by the saturation of populations of the resonant transition. In this case the plot of $\chi_1(\Delta)$ represents the curve of anomalous dispersion, which corresponds to the Lorentz curve of the line with the central frequency $\omega = \omega_{21} (\Delta = 0)$ and the half-width obtained from (2.2.10) when $q = 1$. As $\mathcal{J}$ increases, this curve becomes broader, and its amplitude approaches zero. If the nonzero value of $\mathcal{K}$ had only resulted in the appearance of the Stark shift (the term $\mathcal{J}\mathcal{K}\vartheta$ being absent), the only change of the curve $\chi_1(\Delta)$ would have been a displacement of the central frequency, i.e., $\omega = \omega_{21} - \Omega$. However, when $\mathcal{K}{\neq}0$, because of the redistribution of populations between levels 1 and 2, χ is increased by a quantity depicted by the Lorentz curve with the same central frequency and half-width. As $\mathcal{J}{\to}\infty$, the amplitude of this curve increases up to the value $\eta_0(\kappa^{22} - \kappa^{11})/2$ determined by saturation.

Of interest is the dependence of χ_1 on Δ in the medium with $\vartheta = 1 (\tau = T/2)$. In the case of substances such as rarefied gases, it is possible that the life time of the upper level equals the inverse width of the homogeneous absorption line. For media of this kind the change of the nominator of χ_1 produced by the term $\mathcal{J}\mathcal{K}\vartheta$ is compensated for by the change due to the Stark effect. As long as $\mathcal{J}{\ll}\mathcal{J}_{\mathrm{St}}$, the quantity χ_1 practically coincides with the real part of susceptibility of a two-level system.

In the general case $\vartheta{\neq}1$, and such compensation is absent.

Figure 2.3 depicts the dependence of χ_1 on Δ for $\vartheta = 10$, and several values of $\mathcal{J}$ with the assumption that η_0 and $\mathcal{K}$ are positive. The corresponding curves for $\eta_0{<}0$, $\mathcal{K}{<}0$, $\eta_0{>}0$, $\mathcal{K}{<}0$; $\eta_0{<}0$, and $\mathcal{K}{>}0$ can be obtained from this plot if one takes into account the fact that for the same value of $\mathcal{J}$,

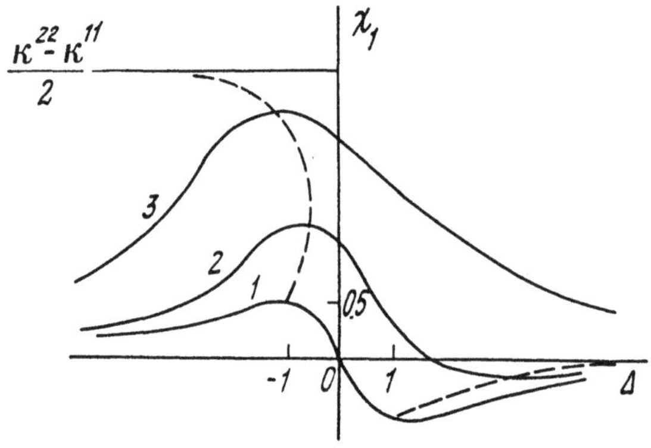

Fig. 2.3. The quantity χ_1 as a function of detuning Δ. The values of paramters: $\mathcal{K} = 5|d|^2 T\hbar^{-1}$, $\vartheta = 10$. (*1*) $\mathcal{J} = 0$; (*2*) $\mathcal{J} = 0.0037\,\hbar^2|d|^{-2}T^{-2}$; (*3*) $\mathcal{J} = 0.2\,\hbar^2|d|^{-2}T^{-2}$. (---) represent the curves along which the maximum and the minimum of the function $\chi_1(\Delta)$ move as the field increases

$$\chi_1(\eta_0, -\mathcal{K}, -\Delta) = -\chi_1(\eta_0, \mathcal{K}, \Delta) \quad ;$$
$$\chi_1(-\eta_0, \mathcal{K}, \Delta) = -\chi_1(\eta_0, \mathcal{K}, \Delta) \quad . \tag{2.2.17}$$

Let us discuss the extremal and the zero points of these plots in detail. The extremum of $\chi_1(\Delta)$ is associated with the greatest deviations of refractive index and phase velocity of the light wave from the values assumed by them far from the resonance. The zero on the curve $\chi_1(\Delta)$ means that these quantities coincide with their nonresonance values. The position of the extremal points on the curves on Fig. 2.3 is determined by

$$\Delta_{1,2} = \mathcal{J}\mathcal{K}(\vartheta - 1) \mp \sqrt{1 + \mathcal{J}\mathcal{J}_{\text{sat}}^{-1} + (\mathcal{J}\mathcal{K}\vartheta)^2} \quad , \tag{2.2.18a}$$

and the quantity χ_1 at these points is given by

$$\chi_1(\Delta_{1,2}) = \pm\frac{|d|^2 T\eta_0}{2\hbar}\left[\sqrt{1 + \mathcal{J}\mathcal{J}_{\text{sat}}^{-1} + (\mathcal{J}\mathcal{K}\vartheta)^2} \mp \mathcal{J}\mathcal{K}\vartheta\right]^{-1} \quad . \tag{2.2.18b}$$

One can easily see that for $2\mathcal{K}\mathcal{J}_{\text{sat}}\vartheta > \vartheta/(\vartheta - 1)$ when the field increases, the quantity Δ_1 first increases. Then, after the value

$$\Delta_1 = -(2\mathcal{K}\mathcal{J}_{\text{sat}}\vartheta^2)^{-1}\{\vartheta - 1 + \sqrt{[(2\mathcal{K}\mathcal{J}_{\text{sat}}\vartheta)^2 - 1](2\vartheta - 1)\}} \quad ,$$

it begins to diminish, yet still approaching, the center of the line[5] which is subject to a shift due to the high-frequency Stark effect. In the case of high intensity fields ($2\mathcal{J}\mathcal{J}_{\text{sat}}\mathcal{K}^2\vartheta^2 \gg 1$)

[5] The quantity Δ_1 never turns out to be exactly at the center of the line, although it is sufficiently close to it when $\mathcal{K} \gg 1$.

$$\Delta_1 \approx -\left(\mathcal{J}\mathcal{K} + \frac{1}{2\mathcal{K}\mathcal{J}_{\text{sat}}\vartheta}\right) \quad ,$$

and for this value of detuning $\text{Re}\{\chi\}$ is

$$\chi_0 + \chi_1(\Delta_1) = (\chi^{11} + \chi^{22})/2 \quad .$$

Note that the real part of the molecular susceptibility takes on the same value if the populations of states 1 and 2 coincide, not because of the resonant field, but due to a different reason. If in the description of the nonlinear resonant susceptibility one limits oneself to the two-level approximation, this obvious result cannot be obtained. Indeed, in this approximation

$$\lim_{\mathcal{J}\to\infty} \chi_1 = 0 \quad \text{and} \quad \lim_{\mathcal{J}\to\infty} \text{Re}\{\chi\} = \chi_0 = \kappa^{22}\sigma_{22}^0 + \kappa^{11}\sigma_{11}^0 \approx \kappa^{11} \quad .$$

When $2\mathcal{K}\mathcal{J}_{\text{sat}}\vartheta<\vartheta/(\vartheta-1)$, the increase of the field is accompanied by the monotonic decrease of Δ_1. The asymptotic behavior in strong fields of the quantity Δ_1 and the limiting value of $\text{Re}\{\chi(\Delta_1)\}$ are the same as in the previous case.

The second extremum moves away from the center of the line for all $\mathcal{K}>0$ as the field increases. When $2\mathcal{J}\mathcal{J}_{\text{sat}}\mathcal{K}^2\vartheta^2\gg 1$, we have

$$\Delta_2 \approx \mathcal{J}\mathcal{K}(2\vartheta-1) + (2\mathcal{J}_{\text{sat}}\mathcal{K}\vartheta)^{-1} \quad \text{and} \quad \lim_{\mathcal{J}\to\infty} \chi(\Delta_2) = \chi_0 \quad .$$

The behavior of extrema of $\chi_1(\Delta)$ for $\mathcal{K}<0$ can be easily described on the basis of the above pattern for $\mathcal{K}>0$ and (2.2.17).

The zeros of the plots in Fig. 2.3, $\Delta_0 = \mathcal{J}\mathcal{K}(\vartheta-1)$, move towards one side as $\mathcal{J}$ increases. When $\mathcal{J}\mathcal{K}\vartheta = 1$, which corresponds to the field amplitude

$$\mathcal{E}_\kappa = \sqrt{\hbar/2\tau|\kappa^{22}-\kappa^{11}|} \quad , \tag{2.2.19}$$

the zero point turns out to be at the edge of the absorption line (with its Stark shift taken into account). If $|\mathcal{E}|>\mathcal{E}_\kappa$, the quantity χ_1 has the same sign over the whole absorption band. This happens because the redistribution of populations between levels with different linear polarizabilities influences the susceptibility. Hence, the field $\mathcal{E}_\kappa$ plays a characteristic role in determining this factor.

In some applications, particularly in the analysis of the self-action of electromagnetic waves, the nonlinear part of susceptibility is essential. That is, that part of it which depends on the field and vanishes if it is absent

$$\chi_{\text{nl}} = \frac{|d|^2T\eta_0}{\hbar}\mathcal{J}\frac{\mathcal{K}[\vartheta(1+\Delta^2)-(1-\Delta^2)] + \Delta\mathcal{J}_{\text{sat}}^{-1} + \Delta\mathcal{K}^2\mathcal{J}}{(1+\Delta^2)[1+(\Delta+\mathcal{K}\mathcal{J})^2 + \mathcal{J}\mathcal{J}_{\text{sat}}^{-1}]} \quad . \tag{2.2.20}$$

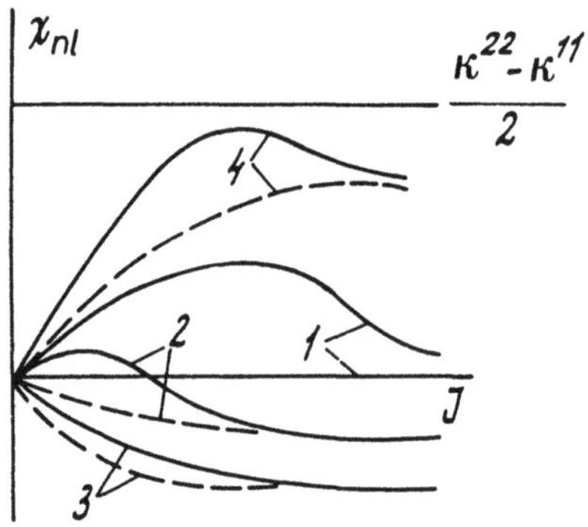

Fig. 2.4. The dependence of $\chi_{\rm nl}$ on the field intensity in single-photon resonance, see (2.2.20). (*1*) $\Delta_1 = 0$; (*2*) $\Delta_2<0$, $R = (\vartheta+1)\Delta_2^2+\Delta_2(\mathcal{K}\mathcal{J}_{\rm sat})^{-1}+\vartheta-1>0$; (*3*) $\Delta_3<0$, $R(\Delta_3)<0$; (*4*) $\Delta_4>0$. (---) depict the functions $\chi_{\rm nl}(\mathcal{J})$ in the two-level model for the same values of detuning Δ

A qualitative representation of this part of the susceptibility as a function of $\mathcal{J}$ for positive $\mathcal{K}$ and η_0 is given in Fig. 2.4 for four typical values of detuning Δ. The plots for the cases when $\mathcal{K}$ and/or η_0 are negative can be obtained from Fig. 2.4 by applying relations (2.2.17). If

$$4\mathcal{K}^2\mathcal{J}_{\rm sat}^2\vartheta^2 = \frac{\hbar^2(\kappa^{22}-\kappa^{11})^2}{|d|^4T^2} > \frac{\vartheta^2}{\vartheta^2-1} \qquad (2.2.21a)$$

holds for any values of detuning, the curves on Fig. 2.4 have an extremum which is caused by the Stark effect. If the lhs of (2.21a) is smaller than $\vartheta^2/(\vartheta^2-1)$, this extremum is absent for detunings satisfying the condition

$$(\vartheta+1)\Delta^2+\frac{\Delta}{\mathcal{K}\mathcal{J}_{\rm sat}}+\vartheta-1<0$$

(see curve 3 in Fig. 2.4).

If $\vartheta \gg 1$, the inequality (2.2.21a) can be written as

$$|\kappa^{22}-\kappa^{11}|>\hbar^{-1}|d|^2T \quad . \qquad (2.2.21b)$$

This implies that the difference between the polarizabilities of the ground and excited states of a molecule should be greater than the amplitude of the anomalous dispersion curve for the transition between these two levels. If at the same time condition $\mathcal{J}_{\rm sat} \ll \mathcal{J}_{\rm St} = |\mathcal{K}|^{-1}$ is satisfied, the peaks of the curves in Fig. 2.4 become flat for all $|\Delta| \lesssim 1$, and the value of the extremum is almost equal to the limiting (for $\mathcal{J}\to\infty$) value of the nonlinear susceptibility in the absence of the Stark effect (denoted by $\tilde{\chi}_{\rm nl}$). For instance, in the above case for $\Delta = 0$, we have

$$\lim_{\mathcal{J}\to\infty}(\tilde{\chi}_{\rm nl}/\chi_{\rm nl}^{\rm max}) = \vartheta(1+2|\mathcal{K}|\mathcal{J}_{\rm sat})(\vartheta-1)^{-1} \approx 1 \quad .$$

The difference in the behavior of $\tilde{\chi}_{\text{nl}}$ and of the nonlinear susceptibility of a two-level system is caused by the term $\mathcal{K}\vartheta(1+\Delta^2)$ in the nominator of (2.2.20). This term is also responsible for the fact that $\lim_{\mathcal{J}\to\infty} \tilde{\chi}_{\text{nl}}$ is by $(\kappa^{22} - \kappa^{11}/2)\eta_0$ greater than the corresponding quantity in the two-level approximation, which equals to

$$\frac{|d|^2 T\eta_0}{\hbar}\frac{\Delta}{1+\Delta^2} \quad .$$

Thus, when $\mathcal{K}\neq 0$ and $\vartheta\neq 1$, the susceptibility of matter as a function of the field frequency and intensity may differ essentially from the behavior of the susceptibility of a two-level system. In correspondence with this, the nature of the self-action of electromagnetic waves in such substances should also be different from that of the self-action in a medium consisting of two-level particles. Thus, if (2.2.21a) is valid, the self-focussing of radiation of any frequency within the limits of the width of the active transition line is possible provided $\mathcal{K}>0$ for an absorbing medium and $\mathcal{K}<0$ for an amplifying medium. Indeed, the necessary condition for self-focussing is the positive sign of the derivative $\partial\chi_{\text{nl}}(\Delta,\mathcal{J})/\partial\mathcal{J}$. It follows from (2.2.20) that in the above cases the quantity

$$\left.\frac{\partial\chi_{\text{nl}}(\Delta,\mathcal{J})}{\partial\mathcal{J}}\right|_{J=0} = \frac{|d|^2 T\eta_0\mathcal{K}}{\hbar(1+\Delta^2)}\left[\vartheta - \frac{1-\Delta^2}{1+\Delta^2}(\mathcal{K}\mathcal{J}_{\text{sat}})^{-1}\right] \tag{2.2.22}$$

is positive, regardless of the value of Δ. In the two-level approximation the only waves which experience self-focussing are those with $\Delta>0$ for an absorbing medium and with $\Delta<0$ for an amplifying one [2.10].

A more detailed treatment of self-action during single-photon absorption will be given in Chap. 8. In here we shall just discuss the possibility of conditions under which (2.2.21a) is valid. In solutions and liquids the characteristic values of T and τ are $T{\sim}10^{-12}-10^{-13}$ s and $\tau{\sim}10^{-7}-10^{-9}$ s; in substances suitable for lasers $T{\sim}10^{-11}$ s and $\tau{\sim}10^{-3}-10^{-4}$ s. In both cases $\vartheta\gg 1$, and (2.2.21a) can be reduced to (2.2.21b) which is satisfied if $|\kappa^{22}-\kappa^{11}|>(10^{16}-10^{14})\cdot|d|^2$. For example, in ruby $|d|\approx 3\times 10^{-21}$ cgse, and the difference of polarizabilities should be greater than 10^{-25} cm^3.

The rough estimate of this quantity given in the previous section resulted in $\kappa^{22}-\kappa^{11}{\sim}10^{-24}$ cm^3. The difference of polarizabilities, however, can vary within broad limits. For instance, if the quantity $\kappa^{22}-\kappa^{11}$ is determined from the value of the quadratic Stark shift [2.11] [see (1.2.11)], we obtain $\kappa^{22}-\kappa^{11} = 4\times 10^{-21}$ cm^3 for the transition $2S-2P$ in lithium. The same number results from the data on the shift of argon lines with $\lambda = 4475$ Å and $\lambda = 4538$ Å [2.12]. Generally speaking, it is for weakly allowed transitions, also possible that $\kappa^{22}-\kappa^{11}{\sim}10^{-21}$ cm^3. In such cases

$$\frac{|\kappa^{22}-\kappa^{11}|\hbar}{|d|^2 T} = 10^4 \gg 1 \quad .$$

The authors of [2.13] describe the measurements of induced anisotropy. The results of that work enable one to derive an estimate for the difference between polarizabilities of the ground (singlet) and the excited (triplet) states of deuterated naphthalene. For the wave length $\lambda = 4200$, this difference is no less than $4 \times 10^{-24}\,\mathrm{cm}^3$.

Section 2.3 contains an evaluation of the difference between the polarizabilities of ground and excited vibrational states. For a hydrogen molecule it turns out that $\kappa^{22} - \kappa^{11} \approx 1.55 \times 10^{-25}\,\mathrm{cm}^3$.

In the case of substances with a short life time in the excited state ($\tau = 10^{-8}\,\mathrm{s}$, $N = 0.5 \times 10^{22}\,\mathrm{cm}^{-3}$, $|d| = 10^{-21}$ cgse), the saturating field is difficult to attain (it corresponds to the flux $\sim$10 Gw/cm^2). When $\mathcal{J} \ll \mathcal{J}_{\mathrm{sat}}$ and $\kappa^{22} - \kappa^{11} = 10^{-25}\,\mathrm{cm}^3$, we have $\varepsilon_{\mathrm{nl}} \approx 2 \times 10^{-10}\,\mathcal{J}$, i.e., the coefficient near $\mathcal{J}$ can be greater than the respective coefficient associated with the Kerr effect (for the sake of comparison we give the latter for nitrobenzene $\varepsilon_{\mathrm{nl}}^{\mathrm{kerr}} \approx 10^{-11}\,\mathcal{J}$).

2.2.4 The Real Part of Susceptibility for Two-Photon Absorption (TPA) and Stimulated Raman Scattering (SRS)

As in the case of single-photon absorption, the real part of the stationary susceptibility at the frequency ω_k of any of the fields participating in TPA or SRS can be presented as a sum [see (2.2.7)]

$$\mathrm{Re}\,\{\chi_{kk}(\omega_k)\} = \chi_0(\omega_k) + \chi_{\mathrm{nl}}(\omega_k) \quad . \tag{2.2.23}$$

The linear susceptibility $\chi_0(\omega_k)$ is defined by the first of equations (2.2.16) and the nonlinear susceptibility $\chi_{\mathrm{nl}}(\omega)$ by

$$\chi_{\mathrm{nl}}(\omega_k) = \frac{\eta_0 T}{\hbar}|\kappa|^2 \mathcal{J}_l \frac{\mathcal{K}_k \mathcal{J}_k \vartheta - (\Delta + \mathcal{J}_1\mathcal{K}_1 + \mathcal{J}_2\mathcal{K}_2)}{1 + (\Delta + \mathcal{J}_1\mathcal{K}_1 + \mathcal{J}_2\mathcal{K}_2)^2 + \mathcal{J}_1\mathcal{J}_2/\mathcal{J}_{\mathrm{sat}}^2} \quad ,$$
$$(k,l) = (1,2); \quad l \neq k \quad . \tag{2.2.24a}$$

The second order polarizability $\kappa = \kappa_{(2)}^{12} = \kappa_{12}^{12}$ appearing here is given for SRS by (2.2.15). In order to describe susceptibility in the case of two-photon absorption, the quantity $+\omega_2$ in (2.2.15) should be replaced by $-\omega_2$.

In TPA experiments, one frequently comes across a situation when the frequencies are degenerate $\omega_1 = \omega_2 = \omega$. The nonlinear susceptibility associated with the two-photon absorption of the field characterized by a degenerate frequency ω differs somewhat from (2.2.24):

$$\chi_{\mathrm{nl}}(\omega) = \frac{\eta_0 T}{\hbar}|\kappa|^2 \mathcal{J} \frac{\mathcal{K}\mathcal{J}\vartheta - 2(\Delta + \mathcal{K}\mathcal{J})}{1 + (\Delta + \mathcal{K}\mathcal{J})^2 + \mathcal{J}^2/\mathcal{J}_{\mathrm{sat}}^2} \quad . \tag{2.2.25a}$$

In this expression $\mathcal{J} = |\mathcal{E}(\omega)|^2$ and the quantity κ is no longer determined via (2.2.15), but by

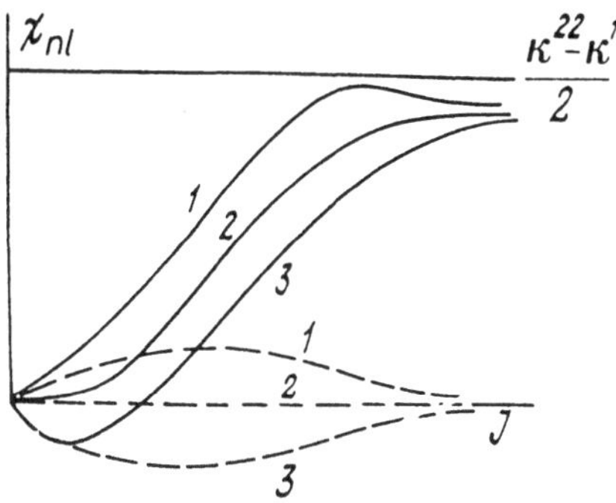

Fig. 2.5. The dependence of the real part of the nonlinear susceptibility on the field intensity in degenerate two-photon absorption for $\mathcal{K}>0$, $\eta_0>0$. (1) $\Delta = \tilde{\Delta}<0$, (2) $\Delta = 0$, (3) $\Delta = -\tilde{\Delta}>0$. (---) depict $\chi_{\mathrm{nl}}(\mathcal{J})$ for $\kappa^{11} = \kappa^{22} (\mathcal{K} = 0)$ and the same values of detuning

$$\kappa = \hbar^{-1} \sum_m \frac{(d_{1m} e)(d_{m2} e)}{\omega_{m1} - \omega} \quad . \tag{2.2.26}$$

Let us consider the nonlinear susceptibility (2.2.25a) in detail. Its dependence on $\mathcal{J}$ is depicted in Fig. 2.5 for several values of detuning and $\mathcal{K}>0$, $\eta_0>0$. The case of negative $\mathcal{K}$ or η_0 can be obtained by means of

$$\begin{aligned} &\chi_{\mathrm{nl}}(\eta_0, \,-\mathcal{K}, \,-\Delta) = -\chi_{\mathrm{nl}}(\eta_0, \mathcal{K}, \Delta) \;; \\ &\chi_{\mathrm{nl}}(-\eta_0, \mathcal{K}, \Delta) = -\chi_{\mathrm{nl}}(\eta_0, \mathcal{K}, \Delta) \quad . \end{aligned}$$

One can see from Fig. 2.5 that

$$\lim_{\mathcal{J}\to\infty} \chi_{\mathrm{nl}}(\mathcal{J}, \Delta) = \eta_0 |\kappa| \frac{\tau - T}{\sqrt{\tau T}} \frac{\mathcal{K}\mathcal{J}_{\mathrm{sat}}}{1 + \mathcal{K}^2 \mathcal{J}_{\mathrm{sat}}^2} \quad ,$$

in contrast to the susceptibility for the single-photon resonance, does not depend on detuning. If $\tau \gg T$ and $\mathcal{J}_{\mathrm{St}} \gg \mathcal{J}_{\mathrm{sat}}$, i.e., if

$$T^{1/2} |\kappa^{22} - \kappa^{11}| \ll 2\tau^{1/2} |\kappa| \quad , \tag{2.2.27}$$

then the limiting value of χ_{nl} is close to $\eta_0(\kappa^{22} - \kappa^{11})/2$. For (2.2.27) with $\gg$ instead of $\ll$, the quantity

$$\lim_{\mathcal{J}\to\infty} \chi_{\mathrm{nl}} = \frac{2\eta_0 |\kappa|^2}{\kappa^{22} - \kappa^{11}} \left(\frac{\tau}{T} - 1 \right)$$

is always much less than $\eta_0(\kappa^{22} - \kappa^{11})/2$.

The curves in Fig. 2.5 corresponding to $\Delta>0$, have minima, and those corresponding to $\Delta<0$, maxima. The position of the extrema is determined by the positive roots of the equation

$$\Delta \left[\mathcal{K}^2 (\vartheta - 1) + \mathcal{J}_{\mathrm{sat}}^{-2} \right] \mathcal{J}^2 + \mathcal{K}(\vartheta - 2)(1 + \Delta^2) \mathcal{J} - \Delta(1 + \Delta^2) = 0 \quad .$$

It can be easily shown that if $\mathcal{J}_{\rm sat}$ and $\mathcal{J}_{\rm St}$ differ substantially, the value $\chi_{\rm nl}^{\rm max}$ is attained at $\mathcal{J}\sim\min\{\mathcal{J}_{\rm sat},\mathcal{J}_{\rm St}\}$. When $\mathcal{J}\ll\min\{\mathcal{J}_{\rm sat},\mathcal{J}_{\rm St}\}$, the nonlinear susceptibility (2.2.25a) can be computed from an approximate expression

$$\chi_{\rm nl}\approx\frac{\eta_0 T}{\hbar}\frac{|\kappa|^2}{1+\Delta^2}\mathcal{J}\left[-2\Delta+\mathcal{K}\left(\vartheta-2\frac{1-\Delta^2}{1+\Delta^2}\right)\mathcal{J}\right] \quad . \tag{2.2.25b}$$

The first term in (2.2.25b) coincides with the usual expression for the susceptibility caused by two-photon processes (see [2.14], pp. 85–87). It is obvious that in general the remaining terms cannot be smaller than the first one. For example, when $\mathcal{J}\gg2\mathcal{E}_\kappa^2$ and $\vartheta\gg2$, the contribution of the second term is greater than that of the first over the whole band $|\nu|\leq T^{-1}$. When $\vartheta\gg2$, the effect of the Stark shift on the susceptibility is small, and only the quantity $\mathcal{J}\mathcal{K}\vartheta$ should be retained among those terms in the square brackets of (2.2.25b) which depend on the field. Hence, we may in this case for $\mathcal{J}$ with $2\mathcal{E}_\kappa^2\ll\mathcal{J}\ll\mathcal{J}_{\rm sat}$ use the following expression for the susceptibility

$$\chi_{\rm nl}\approx\frac{\eta_0 T|\kappa|^2\mathcal{K}\vartheta}{\hbar(1+\Delta^2)}\mathcal{J}^2=\frac{2\eta_0|\kappa|^2\tau T(\kappa^{22}-\kappa^{11})|\mathcal{E}|^4}{\hbar^2(1+\Delta^2)} \quad . \tag{2.2.25c}$$

The range of applicability of (2.2.25c) can be quite wide, since often

$$\frac{\mathcal{J}_{\rm sat}}{2\mathcal{E}_\kappa^2}=\frac{|\kappa^{22}-\kappa^{11}|}{2|\kappa|}\sqrt{\frac{\tau}{T}}\gg1 \quad .$$

For instance, when TPA takes place in vapors and gases ($\tau=10^{-7}$ s, $T=10^{-9}$ s, $2|\kappa|=10^{-24}\,{\rm cm}^3$) we have $|\kappa^{22}-\kappa^{11}|\sim10^{-24}-10^{-21}\,{\rm cm}^3$ and for the ratio $\mathcal{J}_{\rm sat}/2\mathcal{E}_\kappa^2\sim10-10^4$, so that the quantity $\mathcal{E}_\kappa^2$ corresponds in this case to the energy flux $5\times10^3-5\,{\rm KW/cm}^2$.

Let us estimate $\varepsilon_{\rm nl}=4\pi N\chi_{\rm nl}$ for $\chi_{\rm nl}$ being given by (2.2.25c). For a pressure of 1 atm ($N\approx10^{19}\,{\rm cm}^{-3}$) in gases with $\Delta=0$ and the above values of $\kappa^{22}-\kappa^{11}$, the variation of the dielectric constant $\varepsilon_{\rm nl}\approx0.25\times(10^{-6}-10^{-3})$ is in a flux of $4\,{\rm MW/cm}^2$. In condensed media, TPA on electron transitions ($\tau=10^{-9}$ s, $T=10^{-11}$ s, $N=10^{22}\,{\rm cm}^{-3}$) may even for $\kappa^{22}-\kappa^{11}=10^{-25}\,{\rm cm}^3$ result in $\varepsilon_{\rm nl}\sim10^{-6}$ in a flux of $100\,{\rm MW/cm}^2$.

Note that the above formulas can be applied for describing the susceptibility produced by pulsed radiation of lasers with modulated quality ($t_{\rm pulse}=2\times10^{-8}$ s). Thus, $\varepsilon_{\rm nl}$ due to TPA is comparable with the nonlinear dielectric constant in liquids, associated with the Kerr effect. Therefore, it is most likely that the presence of TPA will substantially affect the self-action of light, in gases as well as in condensed media.

Let us now discuss the nonlinear susceptibility produced by SRS. For fields that are small in comparison with the saturating ($\mathcal{J}_1\mathcal{J}_2\ll\mathcal{J}_{\rm sat}^2$) and Stark ones, one can easily obtain from (2.2.24) that

$$\chi_{\rm nl}(\omega_k) \approx \frac{\eta_0 T|\kappa|^2}{\hbar(1+\Delta^2)}\mathcal{J}_l$$
$$\times \left[-\Delta + \mathcal{J}_k\mathcal{K}_k\left(\vartheta - \frac{1-\Delta^2)}{1+\Delta^2)}\right) - \mathcal{J}_l\mathcal{K}_l\frac{1-\Delta^2}{1+\Delta^2}\right] \quad . \qquad (2.2.24b)$$

Generation of the Stokes component occurs most effectively near the center of the line (with its Stark shift taken into account), i.e.,

$$\Delta_{\rm opt} = -\sum_l \mathcal{J}_l\mathcal{K}_l \quad .$$

When $|\mathcal{K}_l\mathcal{J}_l| \ll 1$, the inequality $|\Delta_{\rm opt}| \ll 1$ holds. Substituting $\Delta_{\rm opt}$ into (2.2.24b), it can easily be shown that if $\vartheta \gg 1$ then the essential term in $\chi_{\rm nl}(\omega_k)$ is that proportional to $\mathcal{J}_1\mathcal{J}_2$ and

$$\chi_{\rm nl}(\omega_k) \approx \frac{\eta_0 T|\kappa|^2\mathcal{K}_k\vartheta}{\hbar}\mathcal{J}_k\mathcal{J}_l \quad . \qquad (2.2.24c)$$

When $\vartheta \approx 1$, the terms associated with the Stark effect provide essential contributions for $\chi_{\rm nl}(\omega_k)$ and hence (2.2.24b) should be used.

Although formulas (2.2.24a–c) describe stationary susceptibility, they can also be applied to the calculation of $\chi_{\rm nl}$ caused by SRS on vibrational levels in liquids for nanosecond, and sometimes for picosecond, pulses under the condition that $t_{\rm pulse} \gg \tau$ (in liquids $\tau = 10^{-10} - 10^{-12}$ s). The quantity $\varepsilon_{\rm nl}$ can be substantial in thin filaments that are formed during self-focussing (the energy flux density $\sim 2\,{\rm GW/cm^2}$ [2.15]). For $\kappa^{22} - \kappa^{11} = 10^{-25}\,{\rm cm^3}$, $\tau = 10^{-10}$ s, $T = 10^{-11}$ s, $\mathcal{J}_2/\mathcal{J}_1 = 0.5$ and $|\kappa| = 10^{-24}\,{\rm cm^3}$, we obtain $\varepsilon_{\rm nl} = 5 \times 10^{-5}$. In the case of SRS of a picosecond pulse (the energy flux density $\sim 10^3\,{\rm GW/cm^2}$, $t_{\rm pulse} = 3 \times 10^{-12}$ s [2.16]), one has $\varepsilon_{\rm nl} = 5 \times 10^{-4}$ even for very wide lines ($T = 10^{-13}$ s) with a small life time $\tau = 10^{-12}$ s. Therefore, SRS can also influence the self-action of light.

2.2.5 The Real Part of Susceptibility Generated by Light Pulses

Concluding our study of the susceptibility of matter that arises during incoherent multiphoton interactions, let us consider its real part as produced by light pulses. This analysis is topical because of the observation of self-focussing of nanosecond ($t_{\rm pulse} \approx 10^{-9} - 10^{-8}$ s) and picosecond ($t_{\rm pulse} \approx 10^{-11} - 10^{-12}$ s) light pulses in air [2.16, 17], where the contribution of the Kerr effect and electrostriction into the nonlinear dielectric constant is negligible: $\varepsilon_{\rm nl}^{\rm kerr} \sim (10^{-15} - 10^{-16})\mathcal{J}$; $\varepsilon_{\rm nl}^{\rm str} \approx 5 \times 10^{-15}\,\mathcal{J}$.

The relaxation times in gases under the conditions of atmospheric pressure are $\tau \approx 10^{-3}$ s for vibrational transitions and $\tau \sim 10^{-7} - 10^{-8}$ s for electron transitions. The line width is in both cases of the same order of magnitude and corresponds to $T \approx 10^{-10}$ s. Therefore, it can always be assumed that $\vartheta = 2\tau/T \gg 1$.

The duration of nanosecond pulses satisfies the condition $\tau \gtrsim t_{\text{pulse}} \gg T$; for picosecond pulses we have $T \gg t_{\text{pulse}}$. Consider both these cases separately. For the sake of simplicity the shape of the pulse will be supposed to be rectangular.

1) A short pulse with $\tau \gtrsim t_{\text{pulse}} \gg T$. In this case both the resonant and nonresonant susceptibilities can be expressed through the population difference η, see (2.2.2, 1.4.3, 2.1.1). Its behavior is described by (2.1.16), the solution of the balance equation. Before the arrival of the light pulse the populations were in equilibrium, so the quantity η_{in} in (2.1.16) should be assumed to equal η_0. Making use of the solution obtained, we can determine the real part of the susceptibility for any moment of time t within $t_{\text{pulse}} \gtrsim t \gg T$:

$$\begin{aligned}\operatorname{Re}\{\chi_{jj}(\omega_j)\} - \chi_0(\omega_j) &= \frac{\eta_0 T}{\hbar}\frac{|\mathcal{V}|^2}{\mathcal{J}_j}\left[1 + (\Delta + \sum \mathcal{J}_l\mathcal{K}_l)^2 + \frac{|\mathcal{V}|^2}{U^2}\right]^{-1} \\ &\quad \times \Bigg\{\mathcal{K}_j\mathcal{J}_j\vartheta(1 - \mathrm{e}^{-t/\tau_{\text{ind}}}) - n_j(\Delta + \sum \mathcal{J}_l\mathcal{K}_l) \\ &\quad \times \left[1 + \frac{|\mathcal{V}|^2/U^2}{1 + (\Delta + \sum \mathcal{J}_l\mathcal{K}_l)^2}\mathrm{e}^{-t/\tau_{\text{ind}}}\right]\Bigg\}\,. \qquad (2.2.28)\end{aligned}$$

In this formula we have used the induced relaxation time τ_{ind} introduced previously (2.1.17). Note that when $t \gg \tau_{\text{ind}}$ (2.2.28) is transformed into the equation for stationary susceptibility, see (2.2.7).

On the rhs of (2.2.28) there are two terms in curly brackets. The first of them, proportional to $\mathcal{K}_j\mathcal{J}_j\vartheta$, describes the effect on the susceptibility of population redistribution between levels with different nonlinear polarizabilities. The second, proportional to n_j, corresponds to the susceptibility of a two-level system with its time-dependent saturation being taken into account; it also contains the contribution of the Stark effect.

The behavior of susceptibility after the termination of the pulse (at $t - t_{\text{pulse}} \gg T$), is determined by the population relaxation and is described by

$$\begin{aligned}&\operatorname{Re}\{\chi_{jj}(\omega_j) - \chi_0(\omega_j)\} \\ &= \frac{\eta_0 T}{\hbar}\frac{|\mathcal{V}|^2\mathcal{K}_j\vartheta[1 - \exp(-t_{\text{pl}}/\tau_{\text{ind}})]\exp[(t_{\text{pl}} - t)/\tau]}{1 + (\Delta + \sum \mathcal{J}_l\mathcal{K}_l)^2 + |\mathcal{V}|^2/U^2}\quad . \qquad (2.2.29)\end{aligned}$$

This result can be easily obtained if one takes into account that (i) the variation of population is governed by (2.1.16), where $\eta_{\text{st}} = \eta_0$, the time is measured from $t = t_{\text{pulse}}$ onwards and η_{in} denotes the population difference produced by light at the end of the pulse, and (ii) the nondiagonal matrix element σ approaches zero over a time of the order of T. The susceptibility associated with the two-level system and the Stark shift disappears simultaneously with σ. As follows from (2.2.28), the contribution of this susceptibility is maximal at the beginning of the pulse. However, as will be shown below it is even for very short time intervals practically never the dominant contribution.

2) A supershort pulse characterized by

$$t_{\rm pl} \ll T \;; \quad \tau \;; \quad T|\Delta + \sum \mathcal{J}_l \mathcal{K}_l|^{-1} \quad . \tag{2.2.30}$$

The third of these conditions means that the difference between the field frequency and the line center shifted by the Stark effect is much less than the pulse's spectral width.

In this case the equations for σ and η can be solved via iteration. In the first approximation $\eta = \eta_0$. Substituting this value into the first equation of (1.4.1), and its solution into the second, and taking into account (1.4.3) and (2.2.30), we obtain

$$\begin{aligned} &\mathrm{Re}\,\{\chi_{jj}(\omega_j) - \chi_0(\omega_j)\} \\ &= \frac{\eta_0 T}{\hbar} \frac{|\mathcal{V}|^2}{\mathcal{J}_j} \left\{ \mathcal{J}_j \mathcal{K}_j - \frac{n_j}{2} \left(\Delta + \sum \mathcal{J}_l \mathcal{K}_l \right) \right\} \frac{t^2}{T^2} \quad . \end{aligned} \tag{2.2.31}$$

The condition of applicability of the iteration method $|\eta - \eta_0| \ll \eta_0$ has the form

$$2\hbar^{-2} |\mathcal{V}|^2 t^2 \ll 1 \quad . \tag{2.2.32}$$

One can see from (2.2.31) that for single-photon absorption ($n_j = 1$, $l = j$) the main contribution to $\chi_{\rm nl}$ comes not from the Stark effect [the term $(n_j/2) \sum_l \mathcal{J}_l \mathcal{K}_l$], but from the excitation of matter [the term $\mathcal{K}_j J_j$]. For SRS ($n_j = 1$), the sign of $\chi_{\rm nl}$ at the pumping frequency is also determined by the excitation of matter if the transformation of pumping radiation into the Stokes component does not exceed 50 %. Then the sign of $\chi_{\rm nl}$ at the frequency of the Stokes component is determined by the Stark effect. It would be of interest to note that in the other two-photon process, namely the degenerate TPA ($n_j = 2$), both factors compensate each other exactly and the sign of $\chi_{\rm nl}$ only depends on the frequency detuning Δ. For absorption in which three or more photons participate the main contribution into $\chi_{\rm nl}$ comes from the Stark effect.

As the pulse duration or the magnitude of detuning with respect to the center of the line (which is shifted by the action of fields) increases, the susceptibility exhibits oscillations with the frequency

$$\nu + \sum_l [\kappa^{22}(\omega_l) - \kappa^{11}(\omega_l)] \mathcal{J}_l \hbar^{-1} \quad ,$$

which decays as $\exp(-t/T)$.

Figure 2.6 schematically depicts the evolution of the real part of matter susceptibility which accompanies a rectangular light pulse.

Let us estimate the quantity $\varepsilon_{\rm nl}$ which is formed at the end of a pulse with a duration of 10^{-11} s and the energy flux density of $100\,\mathrm{GW/cm^2}$ for a single-photon resonance with a weakly allowed transition ($|d| = 2 \times 10^{-21}$ cgse) in a

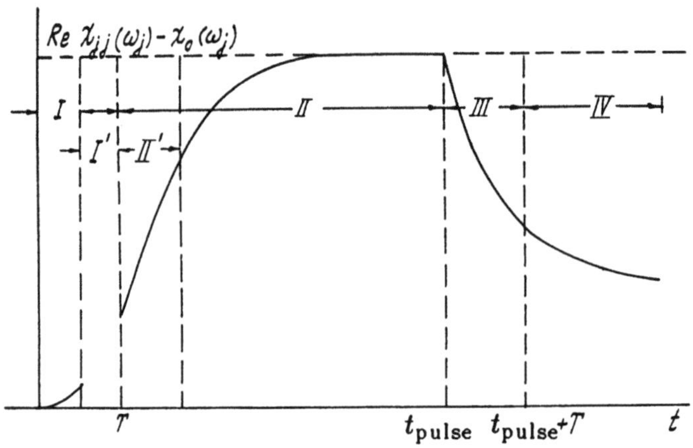

Fig. 2.6. The evolution of the real part of the susceptibility of matter produced by the light pulse. I in the region of parabolic increase of susceptibility, see (2.2.31), II in the region of exponential build-up (with the characteristic times τ_{ind}) of the stationary susceptibility, see (2.2.28), II' in the linear part of this region, III in the region of damping of the contribution from the Stark effect and the "two-level" system, IV in the region of exponential damping (with the characteristic time τ) of the "residual" susceptibility, see (2.2.29). In the region I' where oscillations of susceptibility may occur no curve is plotted

gas subject to a pressure of the order of 1 atm ($N = 10^{19}\,\text{cm}^{-3}$, $\tau = 10^{-9}\,\text{s}$, $T = 10^{-10}\,\text{s}$). One can easily see that in this case the first two conditions of (2.2.30) and the condition (2.2.32) are satisfied. The third condition in (2.2.30) is fulfilled for $\nu = 0$ only if $|\kappa^{22} - \kappa^{11}| \leq 10^{-25}\,\text{cm}^3$. Then ε_{nl} can be expressed through susceptibility (2.2.31), and its value is equal to 10^{-7}. If the quantity $|\kappa^{22} - \kappa^{11}|$ is greater than $10^{-25}\,\text{cm}^3$, then, in order to satisfy condition (2.2.30), it is necessary to demand $\nu = \hbar^{-1}(\kappa^{11} - \kappa^{22})\mathcal{J}$. In this situation ε_{nl} can reach a value of the order of 10^{-3} (for $\kappa^{22} - \kappa^{11} \sim 10^{-21}\,\text{cm}^3$). This means that self-focussing caused by resonance mechanisms is possible even for ultrashort pulses.

2.3 Spectroscopy of Polarizabilities of Excited States

Here we shall discuss one of the spectroscopic aspects in the study of the nonlinear dielectric constant which arises in resonant processes. That concerns the possibility of the direct measurement of the linear polarizability of the excited state of matter by recording ε_{nl} via the interference method.

The polarizability κ^{22}_{ab} determines the Stark shift of the energy of the excited state 2 in a light wave (see Sect. 1.2). The quantity $(\kappa^{22}_{ab} - \kappa^{11}_{ab})$ may also be associated with the main contribution to ε_{nl}, which is responsible for the self-action of waves participating in resonant interactions [see (2.2.24c, 25c)]. For this reason the measurements of κ^{22} are certainly of interest.

As follows from (2.2.20 and 25c), in the case of single- and two-photon absorption, the quantity κ^{22} can be found from the measurements of ε_{nl} associated with the redistribution of populations between levels with different polarizabilities. For when condition (2.2.21b) is satisfied (for the single-photon absorption) and $\vartheta \gg 1$ (for TPA), ε_{nl} is proportional to the difference between polarizabilities of the ground and the excited states in fields with intensities $\mathcal{J} \ll \mathcal{J}_{\text{sat}}$.

Let us estimate the nonlinear increment of the optical path

$$\Delta l = 0.5 \int_0^L \varepsilon_{\text{nl}} dz \quad ,$$

which is traversed by the light wave as it passes a layer of the absorbing substance with a thickness L. If $\mathcal{J} \ll \mathcal{J}_{\text{sat}}$ and the resonance is exact, i.e.,

$$\varepsilon_{\text{nl}} = 8\pi\hbar^{-2} N \eta_0 (\kappa^{22} - \kappa^{11}) |d_{12}|^2 \tau T \mathcal{J} \quad ,$$

the intensity distribution is described by the Bouguer law

$$\mathcal{J} = \mathcal{J}_0 \exp(-z/L_{\text{a}})$$

with the attenuation length

$$L_{\text{a}} = \lambda\hbar\sqrt{\varepsilon}/8\pi^2 N \eta_0 T |d_{12}|^2 \quad .$$

For the optical path increment we obtain

$$\Delta l = \frac{1}{2\pi\hbar} \lambda (\kappa^{22} - \kappa^{11}) \tau \sqrt{\varepsilon} \mathcal{J}_0 \left[1 - \exp\left(-\frac{L}{L_{\text{a}}} \right) \right] \quad .$$

For the values of parameters of matter given at the end of Sect. 2.2.3, under the assumption that $\sqrt{\varepsilon} = 1.6$ and $L \gg L_{\text{a}}$, the optical path increment is $\Delta l = 2 \times 10^{-11} \mathcal{J}_0$. It is equal to half the wave length of the neodymium laser second harmonic for $\mathcal{J}_0 = 1.3 \times 10^6$ cgse (note that $\mathcal{J}_0 \ll \mathcal{J}_{\text{sat}} = 2.5 \times 10^7$ cgse).

If $\mathcal{J} \ll \mathcal{J}_{\text{sat}}$ and $\nu = 0$, the two-photon absorption of light is described by an equation

$$L_{\text{TPA}} \frac{d}{dz} \left(\frac{\mathcal{J}}{\mathcal{J}_0} \right) = - \left(\frac{\mathcal{J}}{\mathcal{J}_0} \right)^2$$

where the two-photon absorption length is given by

$$L_{\text{TPA}} = \lambda\hbar\sqrt{\varepsilon}/8\pi^2 N \eta_0 T |\kappa^{21}|^2 \mathcal{J}_0 \quad .$$

Hence

$$\mathcal{J} = \mathcal{J}_0(1 + Z/L_{\mathrm{TPA}})^{-1} \quad .$$

By making use of this expression and (2.2.25c) for the estimate of the nonlinear optical path increment arising during TPA, we obtain

$$\Delta l = \frac{1}{\pi\hbar}\lambda(\kappa^{22} - \kappa^{11})\tau\sqrt{\varepsilon}\mathcal{J}_0\left(1 - \frac{1}{1 + L/L_{\mathrm{TPA}}}\right) \quad .$$

For a value of L such that $L \gg L_{\mathrm{TAP}}$, it is possible that $\Delta l = 0.5\,\lambda$ with the intensity at the input $\mathcal{J}_0 = \pi\hbar/2\tau\sqrt{\varepsilon}(\kappa^{22} - \kappa^{11})$, which is equal to half the corresponding quantity in the case of single-photon absorption.

Thus, the nonlinear dielectric constant $\varepsilon_{\mathrm{nl}}$, which is formed in matter subject to the action of light under the conditions of single- and two-photon resonance, can certainly be measured by interference methods. Therefore, it appears possible to determine the quantity $\kappa^{22} - \kappa^{11}$ and, consequently, also the Stark shift constant [see (1.2.11)] for weakly allowed and much broadened transitions.

Such measurements can be performed even for vibrational states, the differences between their polarizabilities is in the general case less than that of the electron states. Let us estimate this difference for the excited and the ground vibrational states.

If the contribution of the dipole moments of transitions between vibrational states is neglected, and the dipole moments of transitions from these levels to the excited electron state are assumed to be equal, one can easily arrive at the following formula:

$$\kappa^{22}(\omega) - \kappa^{11}(\omega) \approx \kappa^{11}(\omega)\frac{\omega_{\mathrm{v}}}{\omega_{\mathrm{e}}}\frac{\omega_{\mathrm{e}}^2 + \omega^2}{\omega_{\mathrm{e}}^2 - \omega^2} \quad , \tag{2.3.1}$$

where ω_{v} is the vibrational transition frequency, ω_{e} is the minimal frequency of the allowed electron transitions and ω is the radiation frequency. For a hydrogen molecule $\omega_{\mathrm{v}}/\omega_{\mathrm{e}} \approx 0.05$ holds and the polarizability of the ground state at a wavelength 5890 Å determined on the basis of the hydrogen refractive index (for a pressure of 1 atm $n \approx 1.000138$) is $\kappa^{11} = 0.9 \times 10^{-24}\,\mathrm{cm}^3$. Neglecting ω^2 in comparison with ω_{e}^2 we have $\kappa^{22} - \kappa^{11} = 0.5 \times 10^{-25}\,\mathrm{cm}^3$.

The approximate expression (2.3.1) obviously gives a value for $\kappa^{22} - \kappa^{11}$ which is too low. As follows from a more accurate quantum-mechanical computation, for hydrogen, $\kappa^{22} - \kappa^{11} = 1.55 \times 10^{-25}\,\mathrm{cm}^3$ [2.18].

The vibrational states which are forbidden in infrared-transitions can be excited by SRS. If pumping pulses with a duration of the order of serveral tens of nanoseconds are used, the variation of ε produced by thermal effects does not exceed 10 % of the value of $\varepsilon_{\mathrm{nl}}$ caused by SRS for $\kappa^{22} - \kappa^{11} \geq 10^{-25}\,\mathrm{cm}^3$ [2.19].

Let us calculate the optical path increment Δl arising because of such nonlinear polarizability of the molecule. Taking into account the fact that $4\pi N\chi_{\mathrm{nl}}$

is much less than unity, one can easily obtain by means of (2.2.2) that

$$\Delta l = \frac{2\pi N}{n(\omega_p)} \int_0^L \chi_{nl} dz = \frac{\pi N(\langle\kappa^{22}\rangle - \langle\kappa^{11}\rangle)}{n(\omega)} \int_0^L (\eta_0 - \eta) dz \quad , \tag{2.3.2}$$

where $\langle\kappa^{22}\rangle$ and $\langle\kappa^{11}\rangle$ are mean polarizabilities of the ground and the excited vibrational molecular states.

The life time of the excited vibrational states of diatomic gases is sufficiently large, i.e., for H_2 under a pressure of 50 atm, $\tau \approx 5 \times 10^{-6}\,\text{s} \gg \tau_{\text{pulse}}$.

Then, if the intensities of the pumping field $\mathcal{J}_p$ and the Stokes component $\mathcal{J}_s$ satisfy the condition

$$|\kappa^{21}|\mathcal{J}_p\mathcal{J}_s \ll \frac{\hbar}{2\sqrt{T\tau_{\text{pulse}}}} \quad , \tag{2.3.3}$$

it follows from (2.1.16 and 1.2.14) that

$$\eta_0 - \eta = 4\hbar^{-2} T\eta_0 |\kappa^{21}|^2 \int_0^{t_{\text{pulse}}} \mathcal{J}_p\mathcal{J}_s \, dt \quad . \tag{2.3.4}$$

The quantity on the right-hand side of (2.3.3) is the saturating energy under nonstationary conditions and it is greater by a factor of $\sqrt{\tau/t_{\text{pulse}}}$ than the stationary saturating energy (2.1.3). For hydrogen under a pressure of 50 atm, $(\mathcal{J}_p\mathcal{J}_s)^{1/2}_{\text{sat}}$ corresponds to a flux of 10 GW/cm^2. If inequality (2.3.3) is valid, the spatial variation of the Stokes component for plane waves is described by

$$\frac{d\mathcal{J}_s}{dz} = \frac{8\pi^2 N|\kappa^{21}|^2 T\eta_0}{\hbar n_0(\omega_s)\lambda_s} \mathcal{J}_p\mathcal{J}_s \quad . \tag{2.3.5}$$

We shall assume that the Stokes component and the pumping beams have a Gaussian transverse distribution of intensities:

$$\mathcal{J}_{p,s} \sim \exp\left[-\frac{x^2}{a_{p,s}^2} - \frac{y^2}{b_{p,s}^2}\right] \quad .$$

Here, $a_{p,s}$, $b_{p,s}$ are the half-widths of the pumping field and the Stokes component beams in the direction of coordinates x and y, respectively. Substituting (2.3.5 and 4) into (2.3.2) and expressing $\mathcal{J}_s$ at the center of the beam through the total energy of the pulse W_s, we obtain

$$\langle\kappa^{22}\rangle - \langle\kappa^{11}\rangle = \frac{\Delta l}{\lambda_p} \frac{\lambda_p}{\lambda_s} \frac{n(\omega_p)}{n(\omega_s)} \frac{\hbar\pi c a_s b_s}{W_s - W_{s0}} \quad , \tag{2.3.6}$$

where W_{s0} is the energy of the Stokes component as it enters the medium.

Thus, the difference between polarizabilities of the excited and the ground vibrational states of a molecule is determined by the shift of interference fringes $\Delta l/\lambda_p$ and the energy of the Stokes component; both these quantities can be measured.

For hydrogen, such measurements have been performed in [2.20]. The role of pumping was played by the second harmonic of a single-mode neodymium laser ($\lambda_p = 0.53\,\mu$m). The main portion of pumping radiation was directed into an interferometer of a Mach-Zender type, one arm of which contained a chamber of length $L = 4$ cm filled with compressed hydrogen (under a pressure of 60 atm). The vibrations of the H_2 molecules were excited in the process of amplification of SRS Stokes component ($\lambda_s = 0.68\,\mu$m) produced in a separate chamber. Counterbeams were used in the amplification which made it possible to avoid the appearance of a SRS anti-Stokes component (including the axial one, see Chap. 7) thereby destroying the excited state.

The experiment was performed under the following conditions: (i) inequality (2.3.3) was valid, (ii) the spatial lengths of the pulses were much greater than the length L of the chamber, (iii) the transverse distribution of beam intensity was close to a Gaussian and (iv) during the beams' passage through the hydrogen chamber, its dimensions remained practically constant. The measurements were conducted at the pumping frequency; since $t_{\text{pulse}} \ll \tau \approx 5 \times 10^{-6}$ s, the interference pattern that was established at the end of the pumping pulse was recorded.

For the pumping energy $W_p = 0.15$ J and $W_{s0} = 5 \times 10^{-3}$ J for $a_c b_c \approx 1.5 \times 10^{-8}$ m^2, the energy $W_s = 15 \times 10^{-3}$ J was obtained, and the relative deviation $\Delta l/\lambda_p = 0.75$ at the center of the beam was observed.

When the error in the determination of energy, the beam cross section area and the quantity $\Delta l/\lambda_p$ are taken into account, from (2.3.6) the difference between polarizabilities of the ground ($v = 0$) and the excited ($v = 1$) vibrational states can be obtained at a pumping frequency equal to $(1.2-1.8) \times 10^{-25}$ cm^3, which is close to the computed value of 1.55×10^{-25} cm^3 [2.18].

Let us formulate the results presented in Sects. 2.2 and 2.3.

1) The real part of nonlinear susceptibility is due to the interplay of three factors:

a) Light-induced redistribution of populations of the excited and ground states with different polarizabilities κ^{11} and κ^{22}.

b) Saturation caused by the population difference approaching zero as the field intensity increases.

c) The Stark shift of the anomalous dispersion curve and the radiation absorption line center.

Factors (a) and (b) are associated with the change of populations which also depends on the value of the Stark shift of levels. The latter, in its turn, is directly related to the difference $\kappa^{22} - \kappa^{11}$. Therefore, except for the case when $\kappa^{11} = \kappa^{22}$, there are no situations in which the above effects could act separately. If one uses the equations for the generalized two-level system then it is possible to obtain the expressions for nonlinear susceptibilities with the combined action of these causes taken into account [for instance, in the quasi-stationary case χ is described by (2.2.7)]. The range of applicability of these expressions extends up to fields that exceed the saturating and Stark ones.

2) As the light pulse duration decreases, the relative role of the Stark effect in the formation of $\chi_{\rm nl}$ becomes more important since it is inertialess while changes in the populations do require finite time $\tau_{\rm ind}$. The Stark shift, however, participates in the formation of $\chi_{\rm nl}$ through the nondiagonal element of the density matrix, which is not inertialess. Therefore, the contribution of the Stark effect will be of the same order of magnitude as that associated with the redistribution of populations, even for very short pulses. In the case of pulses of large duration, the influence of the Stark effect on $\chi_{\rm nl}$ is negligible if $2\tau \gg T$, and $\mathcal{J} \ll \mathcal{J}_{\rm St}$.

3) The quantity $\chi_{\rm nl}$ that appears in the above processes is sufficiently large and can be measured, for example, by interference methods. The measurements of $\chi_{\rm nl}$ make it possible to determine the polarizability of the excited state, as well as the constant of the resonant transition Stark shift produced by light. Such measurements can frequently be conducted with light fields of lower intensities than those of the fields which are necessary for the measurement of the Stark shift by conventional methods [2.11].

Note that the quantity $\kappa^{22} - \kappa^{11}$ can also be determined by measuring the increment of the matter dielectric constant formed in resonant processes by a test nonresonant field. In this case, the susceptibility for the test field is χ_0 [see (2.2.16)]; in the formula for χ_0, the quantity η_0 should be replaced by η which appears during the corresponding resonant process[6].

4) On the basis of the estimates of the quantity $\chi_{\rm nl}$ associated with the effects considered here, it is possible to say that the resonant mechanisms for the formation of $\chi_{\rm nl}$ can substantially affect the nature of the self-action of light (even in the presence of the Kerr effect and electrostriction). Sometimes these mechanisms constitute the main factor responsible for self-focussing. As an example, we can cite the self-action of pico and nanosecond pulses in gases [2.16, 17], since for gases the quantity $\chi_{\rm nl}$ associated with the Kerr effect and electrostriction is small, and there is not enough time for thermal self-focussing to develop during the action of such pulses.

Under these circumstances, the self-action of light should possess a number of specific features resulting from the fact that the formation of $\chi_{\rm nl}$ is accompanied by the decrease (increase) of the light beam total energy, and in

---

[6] At present the methods for measuring population differences are developed sufficiently well (see, for instance, [2.21–23]).

the case of stimulated Raman processes, by the transfer of energy from some frequencies to others. In Chap. 8 we shall discuss the self-action of light caused by resonance mechanisms, taking into account the variation of the energy of waves participating in the process.

Concluding this section, let us note that in [2.27, 28] the nonlinear refractive indices due to single- and two-photon light absorption by Nd^{3+} ions in glasses were measured. In the context of radiation self-focusing and self-phase-modulation in laser media (see also [2.29–31]) such investigations are of particular interest. The self-focusing caused by the population of the excited levels under the laser action in molecular gases (atmospheric gases and the ones used in optical frequency standards as nonlinear absorbers) are also under investigation [2.32–36]. The formation of a light-induced resonant nonlinear refractive index leads also to polarization effects [2.37–39].

2.4 Molecular Response for Resonant Parametric Interactions

A typical feature of resonant parametric interactions is the simultaneous realization of resonance conditions of the type (1.2). Let us assume, as in Sect. 2.1, that the detunings appearing in these conditions are equal ($\nu_s = \nu$). We shall also suppose that the interaction is quasi-stationary, i.e., that the behavior of populations is described by the balance equation (2.1.15), and the nondiagonal matrix element σ is defined by (2.1.1). In this case, it is obvious that the amplitude of resonance polarization at the frequency of the field that participates in a resonant parametric process is expressed through the population difference by means of (2.2.4), where

$$\mathcal{V} = \mathrm{e}^{\mathrm{i}\nu t} \sum_s v^{(s)} \quad , \tag{2.4.1}$$

$$\vec{\mathcal{D}}(\omega_l) = \mathrm{e}^{\mathrm{i}\nu t} \sum_{s'} \vec{\mathcal{D}}^{(s')}(\omega_l) \quad . \tag{2.4.2}$$

[The summation in (2.4.1) is carried out over all resonances, and in (2.4.2) – only over those in which the frequency ω_l participates].

It follows from the above that saturation and Stark shift affect molecular polarization during resonant parametric interactions as well as during incoherent multiphoton processes. Because of this we shall not dwell on these effects. Instead, we shall consider phenomena that are specific for resonant parametric interactions.

First of all, note the following: if the amplitudes and phases of fields participating in parametric interaction are such that $\mathcal{V} = 0$, all components of polarization at the frequencies of all these fields vanish. It must also be recalled that under these kinds of circumstances the populations of molecular levels assume equilibrium values, and therefore the nonresonant part of polarization turns out to be the same as in the absence of any resonances (parametric self-induced transparency, see Sect. 2.1). Because of the existence of a nonresonant part of polarization, however, the energy can also be transferred from the fields of some frequencies into the fields of other frequencies when $\mathcal{V} = 0$. The resonant polarization at a given frequency ω_l may also vanish for $\mathcal{V} \neq 0$ provided the sum $\sum_{s'} \vec{\mathcal{D}}^{(s')}(\omega_l)$ is zero.

Strictly speaking, for the parametric processes that are nondegenerate with respect to frequency, the introduction of susceptibility is not always possible. This is associated with the absence of the component $\vec{\mathcal{P}}(\omega_l)$ of the field of the same frequency in parametric terms and, therefore, with the fact that one cannot define susceptibility unambiguously by means of an expression $\vec{\mathcal{P}}(\omega_l) = \hat{\chi} : \vec{\mathcal{E}}(\omega_l)$. Nevertheless, it should be noted, that the quantity $\chi_{ll}^{\text{res}} = e_l \cdot \vec{\mathcal{P}}^{\text{res}}(\omega_l)\exp(\mathrm{i}\varphi_l)\mathcal{A}_l^{-1}$ appears in (1.1.5) for the field phase in exactly the same way as the susceptibility does in the case of incoherent processes. Another typical feature of resonant parametric interactions is associated with the behavior of this quantity at small amplitudes of the field with frequency ω_l. Of practical interest is the situation in which this amplitude is small. This is due to the fact that the experiments are usually set up as follows: a substance is exposed to pumping radiation of one or several frequencies; at the initial moment of time (or at the boundary of the substance) the fields of other frequencies are absent; the energy transfer is studied. The development of the latter depends essentially on the relationship between the phases of interacting fields.

Substituting $\mathcal{V}$ and $\vec{\mathcal{D}}$ expressed through $\mathcal{A}_l$ and φ_l [see Chap. 1, Sect. 1.1] into (2.2.4), one easily finds

$$\begin{aligned}\chi_{ll}^{\text{res}} = e_l \cdot \vec{\mathcal{P}}_l \mathcal{A}_l^{-1} \exp(\mathrm{i}\varphi_l) = &-\frac{\eta}{\hbar} T \frac{\mathrm{i} + (\nu - \Omega)}{1 + (\nu - \Omega)^2 T^2} \sum_{s,s'} \kappa_{(s)} \kappa_{(s')}^* n_l^{(s')} \\ &\times \exp\left\{\mathrm{i} \sum_m (n_m^{(s)} - n_m^{(s')}) \Phi_m\right\} \\ &\times \mathcal{A}_l^{n_l^{(s)} + n_l^{(s')} - 2} \prod_{m \neq l} \mathcal{A}_m^{n_m^{(s)} + n_m^{(s')}} \quad . \end{aligned} \tag{2.4.3}$$

In this formula $n_m^{(s)}$ is the multiplicity of degeneration with respect to the frequency ω_m of a resonance with a number s. The terms with $s = s'$ do not depend on the field phases and correspond to the incoherent part of the interaction. The remaining terms describe parametric interaction.

As follows from (2.4.3) for the field of the frequency ω_l, all resonant parametric processes can be divided into three groups.

The first group includes resonant parametric processes with resonance relations of the type (1.2) which contain at least one term with $n_l^{(s)} = 0$ and the resonance involving the frequency ω_l contain at least one resonance which is not degenerate with respect to this frequency [i.e., $n_l^{(s')} = 1$]. Then, for a sufficiently small amplitude $\mathcal{A}_l$, the quantity $\chi_{ll}^{\mathrm{res}}(\omega_l)$ turns out to be proportional to $\mathcal{A}_l^{-1}$ and tends to infinity if $\mathcal{A}_l$ approaches zero, while the amplitudes of the other fields remain finite. Because of this, the phase φ_l of the field with frequency ω_l at the entrance surface of the layer should immediately assume a value for which the derivative φ_l' [being proportional to $\mathrm{Re}\{\chi_{ll}(\omega_l)\}$] vanishes. In other words, the phase locking of the weak field occurs. It is obvious that this phase will remain locked near the above value until the field intensity ($\sim\mathcal{A}_l^2$) becomes sufficiently large.

Therefore, the spatial (or temporal) scale of variation of phase φ_l will be determined not by the detuning of wave vectors of the interacting waves, but by the characteristic scale of the increase of $\mathcal{A}_l$.

In the resonant parametric processes of the second group, the minimal value of the sum $n_l^{(s)} + n_l^{(s')}$ is equal to two. In this case, for small amplitudes $\mathcal{A}_l$ the quantity $\chi_{ll}^{\mathrm{res}}(\omega_l)$ does not depend on $\mathcal{A}_l$. The motion of the phase is determined both by the resonant and nonresonant parts of polarization [see (2.4.3)]. If there is a resonance without the participation of the frequency ω_l [$n_l^{(s)} = 0$], and if the fields of other frequencies have a sufficiently high intensity, the phase may turn out to be locked in the region where $\mathrm{Re}\{\chi_{ll}(\omega_l)\} \approx 0$.

In the third group the minimal value of $n_l^{(s)} + n_l^{(s')}$ is greater than 2. Since at the same time

$$\lim_{\mathcal{A}_l \to 0} \chi_{ll}^{\mathrm{res}}(\omega_l) = 0 \quad ,$$

no phase locking will occur and its behavior is determined mainly by the nonresonant part of polarization.

In Chaps. 6, 7, certain resonant parametric processes that are of practical interest will be studied. Nearly all of them are included in the first group, and their development is strongly affected by the phenomenon of phase locking. Resonant frequency division is the only process among the ones considered by us in which there is no locking of the weak field phase. This process belongs to the second group.

3. The Dynamics of Quantum Systems for Resonant Interactions with Strong Nonstationary Fields *

One of the simplest and yet fundamental models used in quantum electronics to describe the resonant interaction of light with matter is the two-level system. The reason for this is that, in spite of its simplicity, the two-level model is capable of describing most of the effects observed in real systems, in particular, most of the nonlinear effects. One of the main manifestations of the quantum nature of the two-level system is its temporal dynamics under the action of a resonant (or quasi-resonant) driving force. Probably the most interesting effect is related to the so-called Rabi oscillations with their frequencies determined by the intensity and frequency of the driving field. When both these parameters of the driving field change rapidly, the Rabi oscillations as well as the entire dynamics become very complicated.

Stationary (i.e., ideally monochromatic) fields of constant amplitude and frequency do by no means represent the normal case to be dealt with. The problem of the dynamics of a quantum system in strong nonstationary (non-monochromatic) fields is of interest in almost all real situations. This holds for laser radiation sources as well, especially in the case of high intensity radiation. At the same time, the study of this problem makes it possible to discover a number of new properties of quantum systems in the resonance region. This chapter deals with the interaction of two-level systems with modulated (or short-pulsed) driving fields, in particular, with the dynamics of the Rabi oscillations.

3.1 The Equation of Motion and Its Properties

3.1.1 The Specific Features of the Relaxation of the System in a Strong Quasi-Resonant Field

Under the resonance conditions, the relaxation times of a quantum transition strongly affects the dynamics of the system in a quasi-monochromatic field. The most interesting example is the case in which the driving field is strong (i.e., when perturbation theory is inapplicable) and varies sufficiently fast (i.e. the characteristic time of variation of its amplitude and/or frequency is of the order of, or less than, the relaxation times of the transitions).

The relaxation of the system is usually described using two characteristic times introduced phenomenologically: the transverse relaxation time T which equals the inverse half-width of a resonance line in a weak field, and the longitudinal relaxtion time τ, i.e., the life-time of a particle in an excited state. In general these times are different. They may be the same, $T = \tau$, for a low energy transition (for instance, in the microwave range), and also for the optical frequencies in some gases [3.1]. From the mathematical point of view, the case $T = \tau$ is advantageous since it considerably simplifies the analysis of the dynamic system. Hence, this case will be singled out below. In the optical range the usual condition for gases is $\tau \gg T$, which is due to the so-called optical collisions of atoms being responsible for the "shaking" of the phase of polarization without affecting the level populations (see, for example, [3.2]).

In the limiting case of purely radiative mechanism of relaxation, (e.g. in rarified gases), these relaxation times are related to each other as $\tau = T/2$, since in this case the energy of excited system (which is proportional to the population of the excited level) decays twice as fast as its polarization. The radiation (spontaneous) lifetime (see e.g. [3.3]) is

$$\tau = \hbar\lambda^3/8\pi^2 d^2 n^3$$

where d is a dipole moment of the transition, and n is a refractive index at the wavelength of the transition λ.

Consider first the case in which there is no relaxation, but the system is subject to the action of a stationary quasi-resonant field (with the amplitude $\boldsymbol{R}$ and the frequency $\omega = \omega_{21} + \nu$, where ω_{21} is the transition frequency and $\nu^2 \ll \omega_{21}^2$). It is well known (see, for instance [3.4, 5]) that in such a case the probability for a two-level system to be in one of the states undergoes continuous harmonic oscillations with frequency

$$\Omega_{\text{os}} = (r^2 + \nu^2)^{1/2} \quad ,$$

where $r = \boldsymbol{d} \cdot \boldsymbol{R}/\hbar$ is the magnitude of the energy of interaction. The first theoretical description of these oscillations was due to *Rabi* [3.6] (see also [3.5]), and the magnitude r is often called the Rabi frequency. On the other hand, it is well known [3.5–9] that in the presence of relaxation a system in a stationary field reaches the stationary regime associated with saturation of absorption, in which the level populations do not vary in time and tend to become equalized as the field intensity increases.

These two descriptions can be accepted only in the limiting case of either (i) short intense field pulses (shorter than the relaxation times), when there is not enough time for the population oscillations to decay, or (ii) a quasi-stationary (very slowly varying) field, when the population at each moment may be regarded as quasi-stationary with no Rabi oscillations.

Besides, when relaxation times are different, the excitation of Rabi oscillations is characterized by the intensity threshold, which is associated with the difference between the relaxation times. In this connection the concepts of weak or strong fields can be introduced. For the weak field case, perturbation theory is applicable, which implies that the field is much less than a certain scale r_{cr} (when the longitudinal τ and the transverse T relaxation times are far apart, it can differ significantly from the "saturating" amplitude). In this approximation, there are no population oscillations and the change in the nature of relaxation can be interpreted by using varying relaxation times that tend to equalize as the field intensity increases. In the case of a strong field, when the field amplitude is much greater than the r_{cr}, oscillations can be excited. As will be shown, the oscillation component of relaxation can be determined by assuming that the instantaneous oscillation frequency is close to $\Omega_{os} \approx (r^2 + \nu^2)^{1/2}$, and their logarithmic decrement is close to the averaged inversed time $(\tau^{-1} + T^{-1})/2$.

In a relaxation-free system ($T = \tau = \infty$) with an exact frequency resonance ($\nu \equiv 0$) and an arbitrary temporal variation of the amplitude $r(t)$, the level populations undergo undamped oscillations with the instantaneous Rabi frequency $r(t)$, i.e.,

$$\eta \propto \cos\left(\int r\, dt\right)$$

[3.9, 10, 11]. In the presence of relaxation with equal times ($T = \tau$),

$$\eta \propto e^{-t/T} \cos\left(\int r\, dt\right)$$

[3.12–14]. The authors of [3.12] studied approximate solutions in a nonmonochromatic field in two limiting situations: (i) in the weak field approximation (i.e., essentially the expansion into the perturbation theory series) and (ii) in a strong field approximation (see also [3.15–19]). In a number of papers the solution is constructed on the basis of perturbation theory [3.20–22] or its modifications. If the radiation intensity is greater than a certain critical value, the series of the perturbation theory diverges. Their approximate summation for certain special cases of noise fields was carried out in [3.23, 24]. The authors of [3.25] obtained a solution for the probability amplitudes in the special case of a field amplitude that increases for a fixed value of frequency detuning (ν = const) like an exponential function. A number of papers ([3.26–28] and the references in [3.27]) contain calculations using models for the temporal variation of the noise field in the form of oscillation intervals with constant parameters. References [3.24, 29] also deal with the effect of a strong noise field for negligible intrinsic relaxation times.

The authors of a considerable number of studies, both theoretical (see [3.12, 17, 18, 23, 30–35] and references in [3.12, 31]) and experimental [3.36, 37] analyzed the systems response to a weak signal in the presence of a strong monochromatic field. A modified perturbation theory is sufficient to describe the phenomena observed here but we shall not concentrate on them in this chapter [except in Sect. 3.4; (3)]. The change of the spectrum can be interpreted as the Stark splitting of levels in the atom + field system (in particular, the removal of degeneracy in the case $\omega = \omega_{21}$ by the interaction) as treated in a large number of papers (reviews [3.38, 39] and references therein). Also note the experimental measurement of the effect (for example, of self-action) in fields characterized by a broad spectrum [3.38, 40]. The authors of [3.1] observed polarization oscillations in a strong monochromatic field in the optical range.

In our work [3.14, 41, 42] we used the method of reducing the kinetic equations for the density matrix to an equation for the population to study the resonant interaction of a strong nonstationary field with a two-level system. The exact solutions to the latter were obtained for a good number of cases in which the solutions were not known previously. Approximate solutions to it were arrived at for practically all problems involving nonstationary fields that are interesting from the physical point of view. This made it possible to apply a unified approach to the whole range of possible relationships between the parameters of the field and those of the system. This means in particular that in this method, there are no stipulations as to intensity and duration of the driving field. The present chapter is based mainly on the results of these studies.

Although the exact solutions can only be found for certain types of relationships between amplitude $r(t)$ and frequency detuning $\nu(t)$, they enable one to cover a broad class of physically interesting dependences of r and ν on time. This class includes, in particular, fields with exponentially increasing or decreasing amplitudes, allowing one to describe the dynamics of the system at the leading or trailing edge of practically any pulse; the transient change of the field frequency when its amplitude is constant; bell-shaped pulses; cases in which the field goes through zero (especially in a periodic manner); etc. All solutions obtained have one basic property, that is, they are valid for an arbitrary maximal amplitude of the field. This enables one to trace the evolution of the dynamics of the system as the amplitude varies "from zero" onwards. This is of particular importance when $\nu \neq 0$ and $T \neq \tau$, i.e., when the relaxation pattern undergoes a qualitative change as the amplitude r exceeds the value

$$r_{\mathrm{cr}} \propto [\nu^2 + \tfrac{1}{4}(T^{-1} - \tau^{-1})]^{1/2} \quad ,$$

which separates the weak field region where the perturbation theory is still applicable, from the strong field region, where the characteristic Rabi oscillations can be observed. Finally, having a large number of exact solutions, it is possible to construct a sufficiently accurate approximate theory without the shortcomings of perturbation theory whose series diverge for $r > r_{\mathrm{cr}}$.

These new results may be used to measure the characteristics of a material in experiments by recording the behaviour of the population or the polarization by means of any of the known methods that use a pulse of a given shape resembling one of those considered here. It is also possible to calculate the value of the dipole moment of the resonant transition directly, as well as relaxation times from the known intensity of the incident wave, if the frequency of population or polarization oscillations and also their amplitude and the damping time are measured. The observation of polarization can be conducted directly by using the scattered field, and that of population by detecting the variation of the passage of a weak test field which is resonant to a transition from one of the lasing levels to some third, unperturbed level of the system. In order to perform observations for a nondepleted driving field, the thickness of the layer of matter must not be greater than

$$L_{\rm cr} \sim c \min(t_{\rm pl}, r_{\rm max}^{-1})/n \quad .$$

Here c is the velocity of light, $t_{\rm pl}$ is the pulse duration and $r_{\rm max}$ is the maximal amplitude of the field. If $t_{\rm pl} \sim t$, and $r_{\rm max}$ is several times greater than T^{-1} (under these conditions pronounced oscillations should appear) then $T \approx 10^{-9} - 10^{-8}$ s corresponds to $L_{\rm cr} \sim 10 - 10^2$ cm. For $L > L_{\rm cr}$ one has to deal with the problem of nonlinear nonstationary propagation (the exceptional special case being 2π pulses [3.43]), in which case the equations for polarization should be solved in combination with the Maxwell equations. The problem of propagating 2π pulses is well studied [3.4, 43, 44, 45] and we shall not concentrate on it at this point.

The theory of transition processes in lasers represents a natural field of application of these results where, apparently, one can make effective use of the approximate solutions for the strong field. Another natural application of this theory is a selective multiphoton excitation of atoms from the ground state to very high excitation levels using resonant intermediate states.

3.1.2 The Equation of Population Motion

The kinetic equations for the elements ϱ_{ik} of the density matrix of a "pure" two-level system whose molecules are characterized by a dipole moment $\boldsymbol{d}$, which interacts with the electric field $\boldsymbol{E}(t)$, can be written as (see, for instance, [3.8])

$$\begin{aligned} &\dot{\eta} + \tau^{-1}(\eta - \eta_0) = 2\mathrm{i}\hbar^{-1}\boldsymbol{d}\cdot\boldsymbol{E}(t)(\varrho_{21} - \varrho_{12}) \ , \\ &\dot{\varrho}_{21} + (T^{-1} + \mathrm{i}\omega_{21})\varrho_{21} = \mathrm{i}\hbar^{-1}\boldsymbol{d}\cdot\boldsymbol{E}(t)\eta \ , \\ &\eta = \varrho_{11} - \varrho_{22} \ ; \quad \varrho_{12} = \varrho_{21}^{*} \ ; \quad \varrho_{11} + \varrho_{22} = 1 \quad , \end{aligned} \tag{3.1.1}$$

where η is the difference between populations of the ground and the excited levels (per molecule), η_0 is the equilibrium population difference for a given temperature in the absence of the field, T and τ are the transverse and the longitudinal relaxation times, respectively, and ω_{21} is the resonant transition frequency.

As usual, the interaction energy $V(t)$ is assumed to be small:

$$|V(t)| = |\boldsymbol{d}\cdot\boldsymbol{E}(t)| \ll \hbar\omega_{21} \quad . \tag{3.1.2}$$

The field is supposed to be linearly polarized and all dipole moments of molecules are assumed to have the same orientation.

We are only interested in the quasi-resonant case, i.e., the following conditions

$$|\omega-\omega_{21}| \ll \omega_{21} \ ; \quad \Delta\omega \ll \omega_{21} \ , \quad T^{-1}, \tau^{-1} \ll \omega_{21} \tag{3.1.3}$$

will be postulated for the field frequency ω and the spectral width $\Delta\omega$. No stipulations, however, are imposed on the quantity $\Delta\omega T$. In the following we shall assume the amplitude of the field $\boldsymbol{R}(t)$ and its instantaneous frequency $\omega(t) = \omega_{21} + \nu(t)$ to be given functions of time. That is, it will be assumed that

$$\boldsymbol{E}(t) = \boldsymbol{R}(t)\cos[\omega_{21}t + \varphi(t)] \ , \quad \dot{\varphi}(t) = \nu(t) = \omega(t) - \omega_{21} \quad , \tag{3.1.4}$$

where, due to (3.1.3), the amplitude $R(t)$ and the instantaneous detuning $\nu(t)$ are functions of time that vary slowly in comparison with $\cos\omega_{21}t$ (for quasi-monochromatic oscillations such a representation is unique [3.46]).

Let us renormalize (3.1.1) by introducing variables

$$X(t) = \eta(t)/\eta_0 \ ; \quad A(t) = \hbar^{-1}\boldsymbol{d}\cdot\boldsymbol{E} \ ; \quad r(t) = \hbar^{-1}\boldsymbol{d}\cdot\boldsymbol{R}(t) \quad . \tag{3.1.5}$$

Now the reduced interaction amplitude $r(t)$ has the dimensions of frequency and is, in fact, a Rabi frequency. Instead of the nondiagonal element $\varrho_{21} = \varrho_{12}^*$, we introduce a variable

$$Y(t) = \frac{2i}{\eta_0}\varrho_{21}\,\mathrm{e}^{\mathrm{i}\omega_{21}t} \quad . \tag{3.1.6}$$

We write now $A(t)$ as

$$A(t) = \tfrac{1}{2}[a(t)\,\mathrm{e}^{\mathrm{i}\omega_{21}t} + a^*(t)\,\mathrm{e}^{-\mathrm{i}\omega_{21}t}] \quad , \tag{3.1.7}$$

where

$$a(t) = r(t)\,\mathrm{e}^{\mathrm{i}\varphi(t)} \tag{3.1.8}$$

is a slowly varying function. Next we use condition (3.1.3), substituting (3.1.5–7) into (3.1.1), and perform the truncation procedure usual for slow envelope approximation, i.e., we retain only slowly varying terms discarding all rapidly varying terms having frequencies $2\omega_{21}$, $4\omega_{21}$, etc. [this is justified because of conditions (3.1.2, 3)]. As a result we arrive at the simple system

$$\hat{D}_\tau(X-1) = \tfrac{1}{2}(aY + a^*Y^*) \tag{3.1.9}$$

$$\hat{D}_T(Y) = -Xa^* \quad , \tag{3.1.10}$$

where $\hat{D}_\tau$ and $\hat{D}_T$ are linear differential operators:

$$\hat{D}_\tau = \frac{d}{dt} + \tau^{-1} \;\; ; \quad \hat{D}_T = \frac{d}{dt} + T^{-1} \quad .$$

Application of operator $\hat{D}_T$ to both sides of (3.1.9) and modification of the rhs of the resulting equation, taking into account (3.1.8) and (3.1.10), leads to

$$\hat{D}_T[\hat{D}_\tau(X-1)] = \tfrac{1}{2}(\dot{a}Y + \dot{a}^*Y^*) - r^2X \quad . \tag{3.1.11}$$

Subjecting (3.1.11) to a similar procedure once again, we obtain

$$\hat{D}_T^2[\hat{D}_\tau(X-1)] = \tfrac{1}{2}(\ddot{a}Y + \ddot{a}^*Y^*) - r\dot{r}X - \hat{D}_\tau(r^2X) \quad . \tag{3.1.12}$$

Now, eliminating Y and Y^* from (3.1.9, 11, 12), we arrive at a single equation for the population $X(t)$:

$$\begin{aligned}\hat{D}_T^2[\hat{D}_\tau(X-1)] = {} & (\dot{a}a^* - a\dot{a}^*)^{-1}\{(\ddot{a}a^* - \ddot{a}^*a)[\hat{D}_T\hat{D}_\tau(X-1) \\ & + r^2X] + (\dot{a}\ddot{a}^* - \dot{a}^*\ddot{a})\hat{D}_\tau(X-1) - r\dot{r}X - \hat{D}_T(r^2X) \quad .\end{aligned}$$

Bearing in mind (3.1.8) further transformations yield a sought equation

$$\frac{r}{\nu}\hat{D}_T\left\{\frac{r}{\nu}\left[\frac{1}{r}\hat{D}_\tau(\frac{1}{r}\hat{D}_\tau(X-1)) + X\right]\right\} + \hat{D}_\tau(X-1) = 0 \quad . \tag{3.1.13}$$

Equation (3.1.13) is inhomogeneous; its free term

$$\tau^{-1}\left\{\frac{r}{\nu}\hat{D}_T\left[\frac{r}{\nu}\hat{D}_T(\frac{1}{r})\right] + 1\right\}$$

and its coefficients, are determined by the varying parameters of the field. Thus, the field exciting the system seems to act on it in two completely different ways. Firstly, through the coefficients of (3.1.13), it sets the instantaneous values of the resonance eigenfrequencies (i.e., the parameters of level splitting in the atom + field system), and thereby the frequency and the logarithmic decrement of Rabi oscillations. Secondly, it forms the varying signal (the free term of the equation) which, as a matter of fact, is the very cause for oscillations in the system (unlike the decaying motion due to the initial conditions). Hence it follows that in order to create considerable oscillations, the field should not only be of sufficient intensity, but also vary rapidly enough so that the frequency spectrum of its amplitude and frequency deviation contains frequencies of the order of Ω_{os}. In other words, its temporal-variation scale should be sufficiently short in comparison with relaxation times. Below it will be shown that no oscillations appear, even in a strong splitting-producing field if it varies too slowly.

The initial conditions for (3.1.13) are determined by the initial values of all elements of the density matrix. Usually at the moment $t = t_0$ when the field is turned on, the polarization of the system and the nondiagonal elements of the density matrix ϱ_{21} and ϱ_{12}, are equal to zero, i.e., $Y(t_0) = 0$, and the population difference equals its equilibrium value in the absence of the field, $\eta(t_0) = \eta_0$. Using this fact and (3.1.9, 10), we obtain the initial conditions for the population and its derivatives:

$$X(t_0) = 1 \ ; \quad \dot{X}(t_0) = 0 \ ; \quad \ddot{X}(t_0) = -r^2(t_0) \quad . \tag{3.1.14}$$

In the general case, (3.1.13) is a third-order equation with the coefficients varying in time but sometimes its order may be reduced. For instance, when $r \equiv 0$, (3.1.13) degenerates into a first-order equation

$$\hat{D}_\tau(X - 1) = 0 \quad , \tag{3.1.15}$$

which describes population relaxation in the absence of the field.

An arbitrary modulation of amplitude in the case of the exact resonance ($\nu \equiv 0$) represents an important case which permits the most detailed study. Formula (3.1.13) degenerates then into a second-order equation

$$\frac{1}{r}\hat{D}_T\left[\frac{1}{r}\hat{D}_\tau(X - 1)\right] + X = 0 \quad . \tag{3.1.16}$$

The initial conditions for this equation, which correspond to $\eta(t_0) = \eta_0$, $\varrho_{12} = 0$, are

$$X(t_0) = 1 \ , \quad \dot{X}(t_0) = 0 \quad . \tag{3.1.17}$$

3.1.3 Equation of Population Dynamics for Two-Photon Processes

Most of the direct calculations presented in this chapter pertain to single-photon absorption (or emission, for $\eta_0 < 0$) in an ideal two-level system. The same method, however, can be applied to multi-level systems and multiphoton resonances by reducing the problem to a certain generalized two-level system. It is only necessary to relate the parameters of such a system to the characteristics of the original multi-level system. Two-photon absorption (TPA) and stimulated Raman scattering (SRS) processes are described by two-photon interactions. Procedures similar to that performed here make it possible to obtain equations corresponding to the generalized two-level system from the equations for the density matrix of an arbitrary multi-level system. The equations for the dynamics of population of lasing levels can then be of the type (3.1.9, 10) and therefore also (3.1.13). However, instead of r and ν they will contain now the quantities $r_{2\text{ph}}$ and $\nu_{2\text{ph}}$ defined as

$$r_{2\text{ph}} = -\frac{8\kappa^{12}_{(2)}}{\hbar} R_1 R_2 \ ,$$

$$\nu_{2\text{ph}} = \omega_1(t) \pm \omega_2(t) - \omega_{21} - \sum_{j=1}^{2} \frac{4T^{-1}}{\mathcal{E}^2_{j\text{St}}} R_j^2 \ , \qquad (3.1.18)$$

where $\kappa^{12}_{(2)}$ is the polarizability for the transition for TPA or SRS (Chap. 1), the R_j are the amplitudes of interacting fields, and the ω_j are their frequencies [in the second equation of (3.1.18) the upper sign corresponds to TPA and the lower to SRS],

$$\sum 4R_j^2 T^{-1} \mathcal{E}^{-2}_{j\text{St}}$$

is the Stark shift of the transition frequency due to the presence of other, nonresonant levels, and

$$\mathcal{E}^2_{j\text{St}} = \frac{\hbar}{T(\kappa_j^{22} - \kappa_j^{11})} \ ,$$

i.e., (2.1.9). In Sect. 3.3.1, (1), below, the exact solution for the population dynamics during resonant two-photon absorption will be given. This takes into account the temporal variation of the Stark shifts of levels in the case of arbitrary dependences of the field amplitude on time.

A great deal of work ([3.47, 48] and references therein) was devoted to multiphoton processes in strong stationary fields. Among the recent articles dealing with the behavior of a system in nonstationary fields, it is worth noting [3.49], where the transition regime of a two-level system was studied in the experiment on two-photon absorption in the microwave range, and the change of the relaxation time produced by the field and predicted by the theory was observed.

3.2 Amplitude Modulation for Exact Frequency Resonance, $\omega \equiv 0$ (Exact Solutions)

The relationship between relaxation times is very important in finding exact solutions for the arbitrary modulation. When $T = \tau$, (3.1.16) can be solved for all $r(t)$, i.e., for arbitrary amplitude modulation of the field.

3.2.1 Equal Relaxation Times ($T = \tau$)

Introducing new variables v and ξ

$$v = (X - 1)\mathrm{e}^{t/T} \quad , \qquad \xi = \int r\,dt \quad , \tag{3.2.1}$$

we can transform (3.1.16) into

$$\frac{d^2 v}{d\xi^2} + v = -\exp[t(\xi)/T] \tag{3.2.2}$$

with the solution

$$X = \frac{1}{u}\int \cos\left(\int_{t'}^{t} r\,dt\right) du' \quad ; \qquad u = \mathrm{e}^{t/T} \quad . \tag{3.2.3}$$

The initial conditions (3.1.17) specify a solution

$$X(t) = T^{-1}\int_{t_0}^{t} \mathrm{e}^{(t'-t)/T} \cos\left(\int_{t'}^{t} r\,dt\right) dt'$$

$$+ \mathrm{e}^{(t_0-t)/T} \cos\left(\int_{t_0}^{t} r\,dt\right) \quad . \tag{3.2.4}$$

Below we consider several particular amplitude shapes that are of physical significance.

1) In the absence of damping ($T^{-1} = \tau^{-1} = 0$) [3.10, 11], it follows from (3.2.4) that

$$X = \cos\left(\int_{t_0}^{t} r\,dt\right) \quad , \tag{3.2.5}$$

i.e., there are continuous Rabi oscillations of population with the instantaneous frequency being equal to the field amplitude $r(t)$.

2) An abrupt "step", i.e., the field is turned on discontinuously ($r = 0$ for $t<0$ and $r = r_0$ for $t \geq 0$). Equation (3.2.4) yields, for ($t_0 = 0$),

$$X = (1 + r_0^2 T^2)^{-1}[1 + r_0 T\,\mathrm{e}^{-t/T}(\sin r_0 t + r_0 T \cos r_0 t)] \quad . \tag{3.2.6}$$

In other words, $X(t)$ describes relaxation of oscillating population to a new equilibrium value $X_{\mathrm{sat}} = (1 + r_0^2 T^2)^{-1}$. The quantity $r_0^2 T^2$ determines not only the saturation of the system, but also the intensity of oscillations; when $r_0^2T^2 \gg 1$, their amplitude at the beginning of relaxation is close to 1.

3) A smooth, sloping step

$$r(t) = \frac{r_0}{2}\left(1 + \tanh\frac{1}{qT}\right) \quad , \qquad (q \text{ is an integer}) \quad . \tag{3.2.7}$$

When $q = 1$ and $t_0 \rightarrow -\infty$ (the field increases from zero), we obtain

$$X = (1 + r_0^2T^2)^{-1}[1 + \mathrm{e}^{-t/T}(1 + r_0T \sin\xi - \cos\xi)] \quad , \tag{3.2.8}$$

where

$$\xi = \int_{-\infty}^{t} r\,dt = r_0T \ln(1 + \mathrm{e}^{t/T}) \quad .$$

When this equation is compared with (3.2.6), one can note that for the same stationary field, the oscillation amplitude in this case is smaller which is associated with the finite switch-on time of the field ($\sim T$). For $r_0^2T^2 \gg 1$, the maximum oscillation amplitude is $\approx 2/(3\pi)$.

For $q = 2$ (i.e., for a less steep step) and $t_0 \rightarrow -\infty$, it follows from (3.2.4) that

$$X = (1 + B^2)^{-1}\{1 + u^{-1}(1 + 4B^2)^{-1} \times [3B(2Bu^{1/2} - \sin\xi) + (2B^2 - 1)(1 - \cos\xi)]\}, \qquad \text{where}$$

$$B = r_0T \; ; \quad u = \mathrm{e}^{t/T} \; ; \quad \xi = 2B \ln(1 + u^{1/2}) \quad .$$

Here the maximum amplitude of oscillations is yet smaller, and for $r_0^2T^2 \gg 1$, it is $\approx 2/(3\pi)^2$. Naturally, as the switch-on time increases, the amplitude of oscillations in the relaxation process diminishes. Pronounced oscillations can still be observed at rather long switch-on times of the field, (2–5)T.

4) A radiation pulse of arbitrary duration and amplitude

$$r(t) = \frac{r_0\,\mathrm{e}^{t/T}}{\cosh^2[q(\mathrm{e}^{t/T} - 1)]} \quad ; \qquad q = \text{ const} \quad . \tag{3.2.9}$$

The pulse duration is given by the quantity q. In particular, when $q \gg 1$,

$$r \approx \frac{r_0}{\cosh^2(qt/T)} \quad ,$$

which describes a supershort pulse with $t_{\text{pulse}} \sim 2T/q$, where t_{pulse} is the pulse duration at the level of approximately half the intensity. For $t_0 \sim -\infty$, the exact solution (3.2.4) corresponding to the pulse (3.2.9) is

$$X(t) = (2q\,\mathrm{e}^{t/T})^{-1}\{\sin(\xi+b_1)[\mathrm{si}(\xi+b_1)-\mathrm{si}(b_1)] + \cos(\xi+b_1)[\mathrm{ci}(\xi+b_1)-\mathrm{ci}(b_1)] - \sin(b_2-\xi) \times [\mathrm{si}(b_2-\xi)-\mathrm{si}(b_2)] - \cos(b_2-\xi)[\mathrm{ci}(b_2-\xi)-\mathrm{ci}(b_2)]\}, \quad (3.2.10)$$

where

$$\xi = \frac{r_0 t}{q}[\tanh q(\mathrm{e}^{t/T}-1)+\tanh q] \;;$$
$$b_1 = \frac{r_0 T}{q}(1-\tanh q) \;; \quad b_2 = \frac{r_0 T}{q}(r+\tanh q) \;;$$
$$\mathrm{si}(\xi) = \int_{\infty}^{\xi} \frac{\sin x}{x}dx \;; \quad \mathrm{ci}(\xi) = \int_{\infty}^{\xi} \frac{\cos x}{x}dx \quad . \quad (3.2.11)$$

Of special interest is the case in which strong oscillations can be observed in an intense short pulse ($q \gg 1$). The number of oscillations during the pulse duration is determined by

$$\phi = \frac{1}{2\pi}\int_{-\infty}^{\infty} r\,dt \quad , \quad (3.2.12)$$

therefore, their observation requires that the condition $b_2 > 2\pi$ or, for $q \gg 1$, $r_0 > \pi q T$ is satisfied. Let us trace, on the basis of (3.2.10), the evolution of the system behavior as the maximum amplitude r_0 of the pulse increases and its duration remains constant, $t_{\mathrm{pulse}} \ll T$ (Fig. 3.1). When the condition $r_0 > \pi q/T$ is satisfied, the first oscillations appear with the motion of population inside the pulse governed approximately by (3.2.5). The system behaves as if there is no damping. After the end of the field pulse, the population relaxes exponentially to the stationary state with $X = 1$ [in agreement with (3.1.15)], the initial value of the exponential function being that population which existed in the system at the end of the pulse, i.e., in this case

$$X \approx 1 - \mathrm{e}^{-t/T}[1-\cos(\xi_\infty - \xi_{-\infty})] \quad .$$

Thus, first, the number of population oscillations excited by a single pulse of the field is determined directly by the integral of the field (3.2.12), and second, for a sufficiently short pulse the oscillation amplitude is large (≈ 1) and the magnitude of the exponential relaxation tail behind them is a periodic function of the total integral of intensity.

As the amplitude r_0 increases further, oscillations become damped during the pulse duration (Fig. 3.1). The condition for this is

$$r_0 T \gtrsim \cosh^2 q \sim \cosh^2(T/t_{\mathrm{pulse}}) \quad .$$

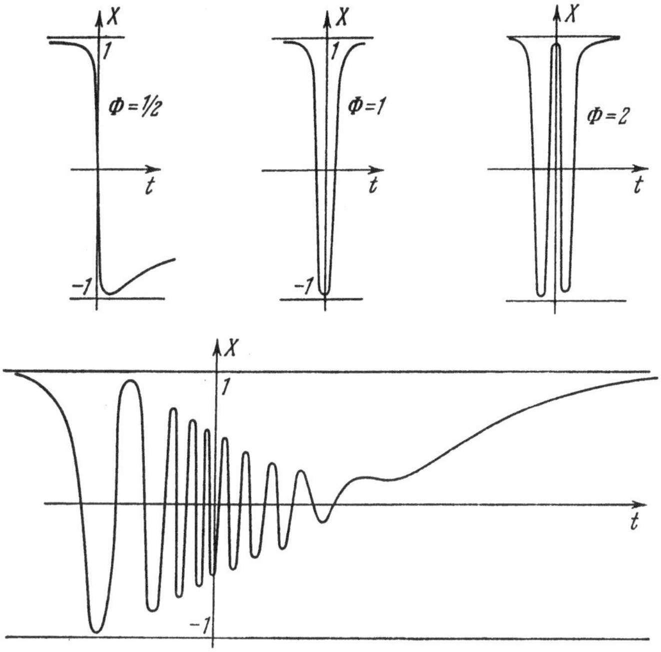

Fig. 3.1. Relaxation of population in a short pulse field ($t_{\text{pulse}} \ll T$)

Whence it follows that in order to observe strong oscillations over the whole pulse, the pulse should be "sufficiently short", i.e.,

$$(t_{\text{pulse}})_{\text{sat}} \equiv \left| t_2\left(r = \frac{1}{T}\right) - t_1\left(r = \frac{1}{T}\right) \right| \ll T \quad , \tag{3.2.13}$$

where $(t_{\text{pulse}})_{\text{sat}}$ is the pulse duration at the level of the saturation intensity, $r(t) = 1/T$, and not of half the intensity. Note that [3.14] also presents the exact solution for a pulse of the form $r = r_0 \cosh^{-2}(t/2T)$.

5) A field pulse that increases or decreases as an exponential function:

$$r(t) = r_0 \exp(t/T_E) \quad , \tag{3.2.14}$$

where T_E is an (arbitrary) time of the field increase. It follows from (3.2.3) that

$$X = 1 - \xi^{-q}[I_q(\xi) + C_1 \sin \xi + C_2 \cos \xi] \quad , \tag{3.2.15}$$

where $q = T_E/T$, $\xi = r_0 T_E \exp(t/T_E)$ and a function

$$I_q(\xi) = \int x^q \sin(\xi - x)dx \tag{3.2.16}$$

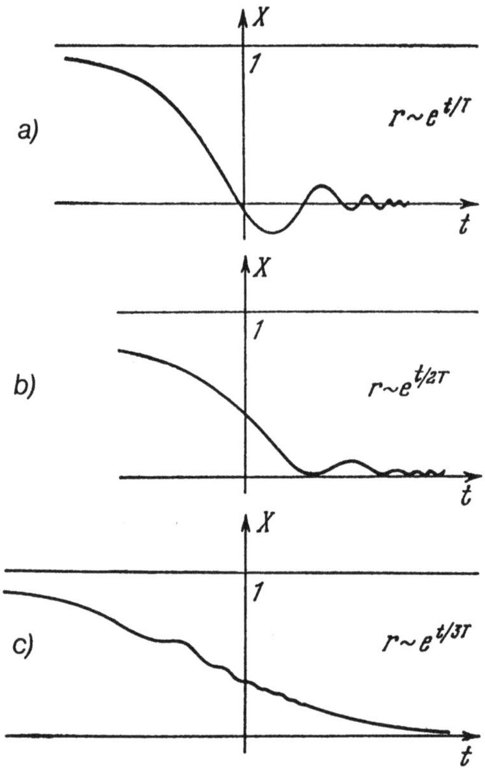

Fig. 3.2. Relaxation of population in an exponentially increasing field

is introduced, for which a recursion relation

$$I_q(\xi) = \xi^q - q(q-1)I_{q-2}$$

is valid. When q is an integer (positive or negative) or a half-integer, the solution (3.2.15) can be expressed as a finite power series and/or integral sine (and cosine) or Fresnel integrals.

In particular, for the increasing pulse ($T_E > 0$) and the initial conditions (3.1.17), when the field is turned on from zero ($t_0 \to -\infty$), we have for $T_E = T\,(q = 1)$ (Fig. 3.2a)

$$X = \frac{1}{\xi}\sin\xi \ ; \quad \xi = r_0 T\,\mathrm{e}^{t/T} \ ; \tag{3.2.17}$$

for $T_E = 2T\,(q = 2)$ (Fig. 3.2b)

$$X = \left(\frac{2}{\xi}\sin\frac{\xi}{2}\right)^2 \ ; \quad \xi = 2r_0 T\,\mathrm{e}^{t/2T} \ ; \tag{3.2.18}$$

for $T_E = 3T\,(q = 3)$ (Fig. 3.2c)

$$X = \frac{6}{\xi^2}\left(1 - \frac{\sin \xi}{\xi}\right) \; ; \quad \xi = 3r_0 T \, e^{t/3T} \; ; \tag{3.2.19}$$

and for $T_E = T/2 \, (q = 1/2)$

$$X = \left(\frac{\pi}{2\xi}\right)^{1/2} [\sin \xi S(\xi) - \cos \xi C(\xi)] \; , \quad \xi = \frac{r_0 T}{2} e^{2t/T} \; , \tag{3.2.20}$$

where $S(\xi)$ and $C(\xi)$ are Fresnel integrals

$$S(\xi) = \int_0^{\xi} (2\pi z)^{-1/2} \sin z \, dz \; ,$$

$$C(\xi) = \int_0^{\xi} (2\pi z)^{-1/2} \cos z \, dz \; .$$

The solutions for the decreasing pulses ($q<0$) can be constructed in a similar way. In particular, for a decreasing pulse turned on at $t = t_0$ for $T_E = -T \, (q = -1)$, one obtains

$$X = 1 - \xi \{ \sin \xi [\mathrm{ci}(\xi) - \mathrm{ci}(\xi_0)] - \cos \xi [\mathrm{si}(\xi) - \mathrm{si}(\xi_0)] \} \; ,$$

where $\xi = -r_0 T \exp(-t/T)$, $\xi_0 = \xi(t_0)$. For $T_E = -2T \, (q = -2)$, one has

$$X = 1 + \xi^2 \{ \sin \xi [\mathrm{si}(\xi) - \mathrm{si}(\xi_0) - \xi_0^{-1} \cos \xi_0] + \cos \xi [\mathrm{ci}(\xi) - \mathrm{ci}(\xi_0) + \xi_0^{-1} \sin \xi_0] \} \; ,$$

and for $T_E = -T/2 \, (q = -1/2)$

$$X = 1 - (2\pi\xi)^{1/2} \{ \sin \xi [C(\xi) - C(\xi_0)] - \cos \xi [S(\xi) - S(\xi_0)] \} \; .$$

For a rapidly decreasing pulse ($|T_E| \ll T$, $T_E<0$, $|q| \ll 1$) which is turned on at $t_0 = 0$ and is sufficiently intense ($r_0 \gg 1/T$), it is possible, using the relation $I_q(\xi) \approx \xi^q$ for $|q| \ll |\xi|$, to obtain an approximate formula that describes population relaxation for the initial conditions (3.1.17)

$$X \approx \begin{cases} e^{-t/T} \cos(\xi - \xi_0) & \text{for} \quad |r|>1/T \\ 1 - e^{-(t-t_1)/T}(1 - \cos \xi_0) & \text{for} \quad |r|<1/T \end{cases} \tag{3.2.21}$$

where

$$\xi = r_0 T_E \, e^{t/T_E} \; ; \quad \xi_0 = \xi(0) \; ; \quad t_1 = |T_E| \ln |r_0 T| = (t_{\mathrm{pulse}})_{\mathrm{sat}}$$

(t_1 being the pulse duration at the saturation level $r = 1/T$).

Similarly, for a rapidly increasing pulse ($T_E \ll T$) we obtain

$$X \approx e^{-(t-t_1)/T} \cos\xi \quad \text{for} \quad |r|>T^{-1} \quad \text{and}$$

$$X \approx \cos\xi \quad \text{for} \quad |r|<T^{-1} \quad . \tag{3.2.22}$$

The number of oscillations is determined by the quantity Φ (3.2.12). One can see that under a condition analogous to (3.2.13) for a pulse of finite duration, the oscillations are also strong.

6) The exact solutions can be calculated when the field amplitude passes through zero [3.14] with an arbitrary value of the derivative at this point either as:

$$r(t) = r_0 \exp(t/T)[\exp(t/T) - 1] \quad \text{or} \tag{3.2.23}$$

$$r(t) = r_0 \sinh(2t/T) \quad . \tag{3.2.24}$$

In the first case, (3.2.23), the solution computed on the basis of (3.2.4) for $t_0 \to -\infty$ (the field increases from zero, then decreases and at $t = 0$ passes through zero) is

$$X(t) = e^{-t/T}\left|\sqrt{\frac{\pi}{r_0 T}}\right|\left\{\sin\xi\left[S\left(\left|\frac{r_0 T}{2}\right)+\text{sign}(t)S(\xi)\right]\right.\right.$$
$$\left.+\cos\xi\left[C\left(\left|\frac{r_0 T}{2}\right|\right)+\text{sign}(t)C(\xi)\right]\right\} \quad , \quad \text{where}$$

$$\xi = \left|\frac{r_0 T}{2}\right|(e^{t/T} - 1)^2 \quad .$$

In the second case, (3.2.24), when the field amplitude becomes infinite for $t \to \pm\infty$, the initial condition, instead of (3.1.17), should have the form $X \to 0$ for $t \to -\infty$. Then the solution (3.2.3) becomes

$$X(t) = e^{-t/T}\left|\frac{\pi}{2B}\right|^{1/2}\left\{\sin\left(B\sinh^2\frac{t}{T}\right)\right.$$
$$\times\left[\frac{1}{2}+\text{sign}(t)S\left(B\sinh^2\frac{t}{T}\right)\right]+\cos\left(B\sinh^2\frac{t}{T}\right)$$
$$\times\left[\frac{1}{2}+\text{sign}(t)C\left(B\sinh^2\frac{t}{T}\right)\right]+\sin\left(B\cosh^2\frac{t}{T}\right)$$
$$\times\left[S\left(B\cosh^2\frac{t}{T}\right)-\frac{1}{2}\right]+\cos\left(B\cosh^2\frac{t}{T}\right)$$
$$\left.\times\left[C\left(B\cosh^2\frac{t}{T}\right)-\frac{1}{2}\right]\right\} \; ; \quad B = |r_0 T| \quad .$$

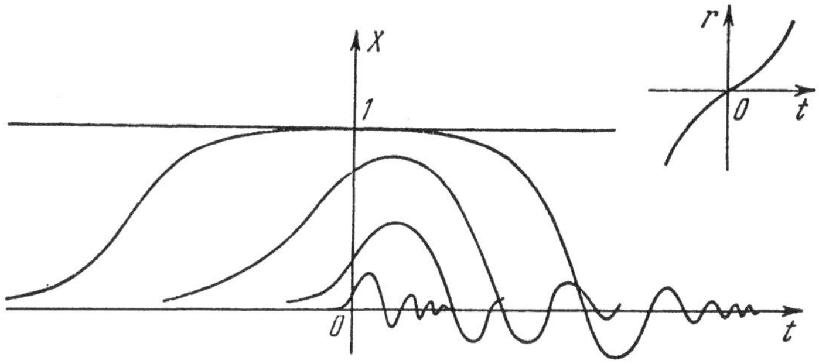

Fig. 3.3. Relaxation of population when the field passes through zero and its evolution accompanying the variation of the slope of the field increase at the zero point

Figure 3.3 depicts the evolution of the population relaxation when (3.2.24) is valid and the rate B of the field amplitude variation at zero increases. When the B's are large, there is no time for the population to approach the equilibrium value as the field passes through zero. In the general case of a field $r(t)$ that changes sign rapidly (i.e., $rT^2 \gg 1$ when $r = 0$) and assumes large values on both sides of zero, the following approximate solution is valid:

$$\begin{aligned} X(t) \approx\; & T^{-1}\,\mathrm{e}^{-(t-t_1)/T}\left|\frac{\pi}{\dot{r}(t_1)}\right|^{1/2} \\ & \times\left\{\sin\xi\left[\frac{1}{2}+\mathrm{sign}(t-t_1)S(\xi)\right]\right. \\ & \left.\times\cos\xi\left[\frac{1}{2}+\mathrm{sign}(t-t_1)C(\xi)\right]\right\}\,, \quad \xi=\left|\int_{t_1}^{t} r\,dt\right| \end{aligned} \qquad (3.2.25)$$

where t_1 is the time at which the amplitude assumes zero value, $r(t_1) = 0$. After field passes through zero, rather intense oscillations are excited, with the maximum amplitude

$$\Delta X_{\max} \sim T^{-1}|2\pi/\dot{r}(t_1)|^{1/2} \quad ,$$

with $X(t_1) \approx \Delta X_{\max}/2\sqrt{2}$, $\dot{X}(t_1) \sim 1/T$, and the oscillations decay as $\exp(-t/T)$. Thus, the passage of the amplitude through zero may serve as an effective means of exciting oscillations. The intense oscillations of population can be obtained just by the increase of the magnitude of the field, without decreasing the pulse duration.

3.2.2 The Case of Unequal Relaxation Times ($T \neq \tau$)

Under the condition $T \neq \tau$, the exact solutions can be found only for a limited class of functions $r(t)$ that are of physical interest.

1) A discontinuous step ($r = 0$ for $t<0$ and $r = \text{const} = r_0$ for $t \geq 0$). The solution of (3.1.16) under the initial conditions (3.1.17) is

$$X = X_{\text{sat}} + (1 - X_{\text{sat}})\exp(-\delta_+ t)\left[\cos \Omega_{\text{os}} t + \frac{\delta_+}{\Omega_{\text{os}}} \sin \Omega_{\text{os}} t\right] \quad , \tag{3.2.26}$$

where

$$X_{\text{sat}} = (1 + r_0^2 T\tau)^{-1} \; ; \quad \Omega_{\text{os}} = (r_0^2 - \delta_-^2)^{1/2} \; ;$$
$$\delta_- = \tfrac{1}{2}(T^{-1} - \tau^{-1}) \; ; \quad \delta_+ = \tfrac{1}{2}(T^{-1} + \tau^{-1}) \quad . \tag{3.2.27}$$

Thus, the solution corresponds to oscillations only if $r_0^2 > \delta_-^2$, which is in contrast to the case $T = \tau$ in which there is a possibility of oscillations for any amplitude of the field. Note that in general the condition of reaching the oscillatory region $r_0^2 > \delta_-^2$ does not coincide with the saturation condition $r_0^2 T\tau > 1$. When $6T \gtrsim \tau$, oscillations occur before saturation (which is valid particularly for the limiting case of radiation dumping, e.g. in a rarefied gas, when $T = 2\tau$ and all the more so for $T \sim \tau$). In the other limiting case, when $\tau \gg T$ (the relaxation time inside the line is much less than the life time of the excited state) it is only under conditions of strong saturation of the system that oscillations can be observed.

2) A relatively "smooth" step, the amplitude of which increases from zero as $\exp(\delta_- t)$ with an arbitrary stationary amplitude ($r_\infty = B\delta_-$) :

$$r(t) = B\delta_-[1 + B\exp(-2t\delta_-)]^{-1/2} \quad . \tag{3.2.28}$$

By using the change of variables

$$\gamma = \int \exp(-2t\delta_-) r(t) dt \; ; \quad v(\gamma) = (X - 1)\mathrm{e}^{t/\tau} \quad , \tag{3.2.29}$$

we can transform (3.1.16) to

$$\frac{d^2 v}{d\gamma^2} + \exp(4t\delta_-) v = -\exp(2t\delta_\perp) \; ;$$
$$\delta_\perp = T^{-1} - \tfrac{1}{2}\tau^{-1} \quad , \tag{3.2.30a}$$

where $\exp(4t\delta_-)$ and $\exp(2t\delta_\perp)$ should be expressed through γ. By substituting (3.2.28) into (3.2.29) we obtain

$$\gamma = [1 + B\exp(-2t\delta_-)]^{1/2} \quad ,$$

and (3.2.30) assumes the form

$$\frac{d^2v}{d\gamma^2} + \frac{B^2}{(\gamma^2-1)^2}v = -\left(\frac{B}{\gamma^2-1}\right)^{\delta_\perp/\delta_-} .$$

The solution of the corresponding homogeneous equation is [3.50]

$$v(\gamma) = (\gamma^2-1)^{1/2}\left[C_1 \sin\left(\frac{b}{2}\ln\left|\frac{1+\gamma}{1-\gamma}\right|\right) + C_2 \cos\left(\frac{b}{2}\ln\left|\frac{1+\gamma}{1-\gamma}\right|\right)\right] ;$$

$$b = (B^2-1)^{1/2} .$$

In this case oscillations also appear only when $|r_\infty|>|\delta_-|$. The solution of the inhomogeneous equation (3.2.30) yields

$$X = 1 + B^2(b \sinh^q \psi)^{-1} \int \sin b(\psi'-\psi) \sinh^q \psi' d\psi' , \qquad (3.2.31)$$

where $b=(B^2-1)^{1/2}$, $\psi = \mathrm{arcsinh}[\exp(t\delta_-)/\sqrt{B}]$ and $q=\delta_+/\delta_-$. In particular, for $\tau \gg T$ ($q \to 1$) and under the initial conditions (3.1.17) when $t_0 \to -\infty$, we obtain from (3.2.31)[1]

$$X \approx \frac{\sin b\psi}{b \sinh \psi} = \left(|B| - \frac{1}{|B|}\right)^{-1/2} \mathrm{e}^{-t/2T} \sin b\psi . \qquad (3.2.32)$$

3) A steeper step for which the field increases at zero as $\exp(6t\delta_-)$:

$$r(t) = B\delta_- \exp(6t\delta_-)[1+\exp(4t\delta_-)]^{-1} \times [1+B^2+\exp(4t\delta_-)]^{-1/2} . \qquad (3.2.33)$$

For such $r(t)$, (3.2.30) becomes

$$\frac{d^2v}{d\gamma^2} + \left(\frac{B^2}{\sinh^2 2\gamma} - 1\right)v = -\left(\frac{B^2}{\sinh^2 2\gamma} - 1\right)^{\delta_\perp/2\delta_-} , \qquad (3.2.34)$$

where

$$2\gamma = \mathrm{arcsinh}[B(1+\exp 4t\delta_-)^{-1/2}] ; \quad \delta_\perp = T^{-1} - \tfrac{1}{2}\tau^{-1} .$$

Like two previous cases, this equation has an oscillatory solution only for $|r_\infty|>|\delta_-|$ ($|B|>1$). The homogeneous equation corresponding to (3.2.34) has a general solution of the form

[1] Note that in (3.2.32), $X \to 0$ when $t \to \infty$, although $|r| < \infty$. This can be explained by the fact that the "oscillatory" condition $|r_\infty| > |\delta_-|$ for $\tau \gg T$ corresponds to very strong saturation, since in this case $\tau_{sat} \ll |\delta_-|$. Therefore, as one computes the approximate formula, the quantity X_{sat} disappears. Its exact value, however, is determined as before, from (3.2.27).

$$v = \sqrt{\sinh 2\gamma}\left[C_2 \sin\left(\frac{b}{2}\ln|\tanh \gamma|\right) C_2 \cos\left(\frac{b}{2}\ln|\tanh \gamma|\right)\right] \quad ;$$
$$b = \sqrt{B^2 - 1} \quad .$$

When $\tau \gg T$, by making τ^{-1} approach zero, we find that the particular solution of (3.2.34) is $v_1 = -1$. It follows from this that for the quantity X[2] under the initial conditions (3.1.17) when $t_0 \to -\infty$,

$$X(t) \approx [1 + e^{2t/T}]^{-1/4} \cos\left(\frac{b}{2}\ln\left|\frac{\tanh \gamma}{\tanh \gamma_0}\right|\right) \quad , \quad \text{where} \tag{3.2.35}$$

$$\gamma = \tfrac{1}{2}\text{arcsinh}[B(1 + e^{2t/T})^{-1/2}] \; ;$$
$$\gamma_0 = \tfrac{1}{2}\text{arcsinh}\, B \quad .$$

The comparison of (3.2.35) and (3.2.32) yields that the more sudden switching on of the field (3.2.33) corresponds to the greater oscillation amplitude; namely, when $B \gg 1$, the maximal oscillation amplitude in (3.2.35) is $(1 + 2\pi)^{-1/2} \approx 0.37$, whereas in (3.2.32) it is equal to $2/3\pi \approx 0.2$. For a discontinuous step, (1) the maximal oscillation amplitude for $|r_\infty| \gg |\delta_-|$ is ≈ 1.

4) Field pulses that increase or decrease exponentially (3.2.14). The change of variables

$$\xi = \int r\, dt = r_0 T_E\, e^{t/T} \quad ; \quad v(\xi) = (1 - X)\, e^{t/\tau} \quad , \tag{3.2.36}$$

in (3.1.16) leads to

$$\frac{d^2 v}{d\xi^2} + \frac{\beta - \alpha}{\xi}\frac{dv}{d\xi} + v = -\left(\frac{\xi}{r_0 T_E}\right)^{\alpha} \quad ; \tag{3.2.37}$$

$$\alpha = \frac{T_E}{\tau} \; ; \quad \beta = \frac{T_E}{T} \quad .$$

The solution of the corresponding homogeneous equation is

$$v = \xi^m [C_1 J_m(\xi) + C_2 Y_m(\xi)] \; ;$$
$$m = \tfrac{1}{2}(1 + \alpha - \beta) = \tfrac{1}{2} - T_E \delta_- \quad ,$$

where $J_m(\xi)$ and $Y_m(\xi)$ are Bessel functions of the first and second kind, respectively. Solving the inhomogeneous equation (3.2.37) and going over to X, we obtain:

$$X = 1 - \xi^{m-\alpha}[s_{m+\beta,m}(\xi) + C_1 J_m(\xi) + C_2 Y_m(\xi)] \quad , \tag{3.2.38}$$

[2] See footnote 1.

where the Lommel function $s_{k,m}(\xi)$ is used. It is defined by [3.51]

$$s_{k,m}(\xi) = \frac{\pi}{2}\left[Y_m(\xi)\int_0^{\xi} \xi^k J_m(\xi)d\xi - J_m(\xi)\int_0^{\xi} \xi^k Y_m(\xi)d\xi\right] \tag{3.2.39}$$

and satisfies the recursion relations

$$s_{k+2,m}(\xi) = \xi^{k+1} - [(k+1)^2 - m^2]s_{k,m}(\xi) \quad . \tag{3.2.40}$$

Exploiting the relation

$$\begin{aligned} s_{k,m}(\xi) = \xi^{k+1} \sum_{h=0}^{\infty} (-1)^n \left(\frac{\xi}{2}\right)^{2n} \\ \times \frac{\gamma(k-m+1/2)\Gamma(k+m+1/2)}{\Gamma(k-m+2n+3/2)\Gamma(k+m+2n+3/2)} \end{aligned} \tag{3.2.41}$$

for $s_{k,m}(\xi)$ [3.51] (i.e., for $\xi \to 0$ the quantity $s_{k,m}\xi^{-m-1}[(k+1)^2 - m^2] \to 1$) and also, the relations for $J_m(\xi)$ and $Y_m(\xi)$ and the initial conditions (3.1.17) for $t_0 \to -\infty$ (the field increases from zero), we arrive at the conclusion that for $T_E > 0$ (the increasing field pulse) the constants C_1 and C_2 in (3.2.38) should be set equal to zero, i.e.,

$$X = 1 - \xi^{m-\alpha} s_{m+\beta,m}(\xi) \quad . \tag{3.2.42}$$

When one of the quantities $k \pm m$ (in our situation it is either $\alpha + 1$ or β) is an odd negative integer, the expansion (3.2.41) becomes meaningless, however, one can use another Lommel function, $S_{k,m}(\xi)$ [3.51] instead of $s_{k,m}(\xi)$. $S_{k,m}(\xi)$ has a finite series:

$$\begin{aligned} S_{m+\beta,m}(\xi) = \xi^{m+\beta-1} - [(m+\beta-1)^2 - m^2]\xi^{m+\beta-3} \\ + [(m+\beta-1)^2 - m^2][(m+\beta-3)^2 - m^2]\xi^{m+\beta-5} - \dots, \end{aligned} \tag{3.2.43}$$

making the results much simpler. Let us consider some typical special cases (it is assumed that for $T_E > 0$ we have $t_0 \to -\infty$).

a) For $T_E = T\,(\beta = 1)$

$$\begin{aligned} X &= \Gamma(1+\vartheta^{-1})\left(\frac{2}{\xi}\right)^{1/\vartheta} J_{1/\vartheta}(\xi) \\ &= \frac{\Gamma(1+\vartheta^{-1})}{(r_0 T/2)^{1/\vartheta}} e^{-t/2\tau} J_{1/\vartheta}(r_0 T\, e^{t/T}) \quad , \end{aligned} \tag{3.2.44}$$

where $\vartheta = 2\tau/T \geq 1$. In particular, when $T = 2\tau\,(\vartheta = 1)$ (the limiting case of a radiation damping, e.g. in rarefied gas)

$$X = (r_0 T)^{-1}\,\mathrm{e}^{-t/T} J_1(r_0 T\,\mathrm{e}^{t/T}) \quad ,$$

and when $\tau \gg T\,(\vartheta \gg 1)$,

$$X \approx J_0(r_0 T\,\mathrm{e}^{t/T}) \quad .$$

b) For $T_E = 3T\,(\beta = 3)$,

$$X = \frac{12}{\vartheta\xi^2}\left[1 - \Gamma\left(\frac{3}{\vartheta}\right)\left(\frac{2}{\xi}\right)^{(3/\vartheta)-1} J_{(3/\vartheta)-1}(\xi)\right] \quad ; \tag{3.2.45}$$

in particular, when $T = (2/3)\tau\,(\vartheta = 3)$ and $T_E = 3T = 2\tau$,

$$X = \frac{4}{\xi^2}[1 - J_0(\xi)] \quad .$$

When $T = \tau/3\,(\vartheta = 6)$ and $T_E = 3T = \tau$, we obtain (3.2.18), in which case it must be assumed that $\xi = r_0 T\,\mathrm{e}^{t/T}$. The solutions for $\beta = 5$, 7, 9, etc. are found in a similar way.

c) The situation when $T_E \ll \tau\,(\alpha \to 0)$ (the condition $T_E \ll T$ is not necessary) is important since it makes it possible to describe the relaxation of a system accompanying the sharp increase of the field amplitude. On the basis of (3.2.43) with $\alpha = 0$ we obtain

$$X \approx \Gamma\left(\frac{\beta+1}{2}\right)\left(\frac{2}{\xi}\right)^{(\beta-1)/2} J_{(\beta-1)/2}(\xi) \quad . \tag{3.2.46}$$

Note that all the solutions given above for the field amplitude increasing as an exponential function ($T_E > 0$) can, naturally, be used for the description of relaxation in the field of a finite duration pulse that terminates at a certain time moment t_1. Namely, after the termination of the pulse, the system is described by (3.1.15) with the initial condition $X(t_1) = X(t_1 - 0)$, or

$$X(t > t_1) = 1 - \exp\left(\frac{t_1 - t}{\tau}\right)[1 - X(t_1)] \quad . \tag{3.2.47}$$

d) The case when $T_E < 0$ and $|T_E| \ll \tau$ [i.e. $|\alpha| \ll 1$], but the condition $|\beta| = |T_E|/T \ll 1$ is not required. This case corresponds to a sharply decreasing pulse which is switched on discontinuously at $t = 0$. Under the assumption that in

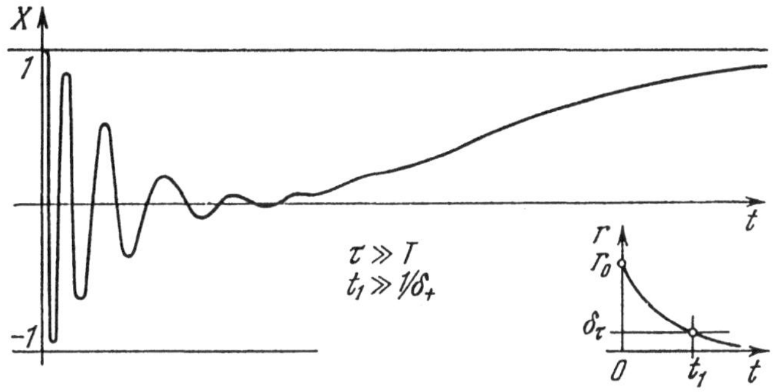

Fig. 3.4. Relaxation of population in the field of an exponentially decaying short pulse ($t_{\text{pulse}} \ll \delta_+^{-1}$) for different relaxation times ($T \neq \tau$)

(3.2.43) $S_{k,m}(\xi) \approx \xi^{(\beta-1)/2}$ and the initial conditions (3.1.17) we obtain for $|r_0| \gg T^{-1}$:

$$X(t) \approx \left(\frac{\xi}{\xi_0}\right)^{m-\alpha} \frac{\pi\xi_0}{2}[J_m(\xi)Y_{m-1}(\xi_0) - Y_m(\xi)J_{m-1}(\xi_0)] \quad , \tag{3.2.48}$$

i.e.,

$$X \approx \exp(-t\delta_+)\cos(\xi - \xi_0) \quad \text{for} \quad |\xi|>m \quad \text{where}$$

$$\xi = |r_0 T|\, e^{t/T_E} \ ; \quad \xi_0 = \xi(0) \ ,$$
$$m = \frac{1-\beta}{2} \ ; \quad \delta_+ = \frac{1}{2}(T^{-1} + \tau^{-1}) \quad .$$

After the time moment t_1 for which $\tau \sim 1/2T$ ($\xi \sim m$), i.e., $t_1 \sim |T_E| \ln |2r_0 T|$ is reached, the population relaxes as an exponential function $\exp(-t/\tau)$ approaching the stationary value $X = 1$ in correspondence with (3.2.47). Figure 3.4 depicts the relaxation of population for $t_1 \gg \delta_+^{-1}$.

e) When $T_E = 2\tau$ (i.e. $\alpha = 2$), by using of (3.2.43) we obtain:

$$X = \frac{2(\vartheta-1)}{\xi^2}\left[1 - \Gamma\left(\frac{\vartheta-1}{2}\right)\left(\frac{2}{\xi}\right)^{(\vartheta-3)/2} J_{(\vartheta-3)/2}(\xi)\right] \quad . \tag{3.2.49}$$

In particular, for $T = \tau/2$ ($\vartheta = 4$), (3.2.19) can be derived in which it must now be assumed that $\xi = 2r_0\tau\, e^{t/2\tau}$. Solutions for $\alpha = 4$, 6, 8, etc. are found in a similar manner.

f) The calculation of integrals in (3.2.39) also becomes easier when m is a half-integer (positive or negative) in which case the result is expressed in terms of trigonometric functions. In particular, for $m = 1/2$ (i.e., $\alpha = \beta$ and, therefore, $T = \tau$) we obtain solutions discussed in Sect. 3.2.1, (v). Let us consider some other cases.

$T_E = \delta_-^{-1} = 2/(T^{-1} - \tau^{-1})$, i.e., $m = -1/2$. When $\tau > T$ (i.e. $\vartheta > 2$), this corresponds to the increasing field pulse, and when $\tau < T$ ($\vartheta < 2$), to the decreasing one. The substitution of $m = -1/2$ in (3.2.39) results in a relation for X of the type (3.2.15) where the quantity $q = T_E/T$ should now be replaced by

$$q = 1 + \alpha = 1 + \frac{T_E}{\tau} = \frac{\vartheta + 2}{\vartheta - 2} \quad .$$

One of the consequences of this is that for $T \ll \tau$ ($\vartheta \gg 1$), the initial conditions (3.1.17) and $t_0 \to -\infty$, the relaxation is described by (3.2.18), where $\xi = 2\tau_0 T \exp(t/2T)$ (as a matter of fact, this is a special case of relation (3.2.46) for $\beta = 2$).

When $T_E = -4\tau$ and $T = 2\tau$ ($\vartheta = 1$) (the pulse is decreasing exponentially, i.e. the field is turned on discontinuously at $t = t_0$ and then diminishes as $r_0\,\mathrm{e}^{-t/2T}$) and one takes into account the recursion formula for I_q under the initial conditions (3.1.17), it follows from (3.2.15) that

$$X = 1 - \frac{1}{2}\xi^2 \left[1 - \frac{\xi}{\xi_0}\cos(\xi - \xi_0) + \frac{\xi}{\xi_0^2}\sin(\xi - \xi_0)\right]$$
$$+ \tfrac{1}{2}\xi^3 \{\sin \xi[\mathrm{ci}(\xi) - \mathrm{ci}(\xi_0)] - \cos \xi[\mathrm{si}(\xi) - \mathrm{si}(\xi_0)]\} \quad ; \quad \text{where}$$
$$\xi = -2r_0 T\,\mathrm{e}^{-t/2T} \quad ; \quad \xi_0 = \xi(t_0) \quad .$$

g) $T_E = -\delta_-^{-1}$. $m = 3/2$ is the pulse which is a mirror reflection in time of the previous one. By substituting $m = 3/2$ into (3.2.29), allowing for the well-known relations for $J_{3/2}(\xi)$ and $Y_{3/2}(\xi)$ and making use of the recursion formula for I_q, we obtain in notations of (3.2.16):

$$X = 1 - (\alpha - 1)^{-1}[\alpha I_{\alpha-1}(\xi) - \xi^{-\alpha} \dot{I_\alpha}(\xi)] \; ;$$
$$\alpha = \frac{T_E}{\tau} \quad .$$

Expressions for $T_E = 2\delta_-^{-1}$ ($m = -3/2$), $T_E = -2\delta_-^{-1}$ ($m = 5/2$), etc. can be calculated in a similar manner.

3.2.3 Relaxation in the Field of a Single Pulse for $T \neq \tau$, and Methods for Exact Solutions

As far as applications (especially laser technology) are concerned, the most important is the case of a pulse limited in time and characterized by an arbitrary amplitude (denoted by $B\delta_-$) and duration (T_{pulse}). Since for $T \neq \tau$ the equations of motion (3.1.16) cannot in general be integrated in quadratures, it is desirable to find the exact solutions for at least some kinds of pulses of special shapes which may be special but which permit, however, arbitrary variation of duration and amplitude. Work [3.42] presents a method for determining the shape of such a pulse and involves solving a certain inverse problem, as well as the calculation of the system dynamics corresponding to it.

Equation (3.1.16) can be reduced to

$$\frac{d^2 v}{d\gamma^2} + G^2(\gamma) v = -G^{\delta_\perp/\delta_-} \quad , \quad \text{where}$$

$$G \equiv \exp[2t(\gamma)\delta_-] \tag{3.2.30b}$$

in a way similar to that used by us before, by means of the change of variables (3.2.29). The function $t(\gamma)$ and, therefore, $G(\gamma)$ are specified by (3.2.29) for γ. If $T \neq \tau$, the function $G(\gamma)$ related to $r(t)$ through (3.2.29) may turn out to be quite complex. As is well known, the second-order equation of the type (3.2.30b) with arbitrary $G^2(\gamma)$ cannot be integrated in quadratures. Consider now a pulse of the form

$$r(t) = B\delta_-[1 + \sinh^2(t\delta_-)/\sinh^2(\tfrac{1}{2}T_{\text{pulse}}\delta_-)] \quad , \tag{3.2.50}$$

which reduces to $r = B\delta_-/\cosh(t\delta_-)$ if $\sinh^2(T_{\text{pulse}}\delta_-) = 1$. By introducing an auxiliary function $g(t)$

$$g(t) = \frac{b_1 G}{b_2 + G} \; ; \quad G = e^{2t\delta_-} \quad , \tag{3.2.51}$$

where constants b_1 and b_2 are defined as

$$b_2 = 4B^2 \sinh^2(\tfrac{1}{2}T_{\text{pulse}}\delta_-) \quad ,$$

$$b_1 = \left[(b_2+1)^2 - \frac{b_2^2}{B^2}\right]^{1/2} \quad ,$$

one can transform (3.2.30b) (without the rhs) into

$$v''_\gamma + \left(\frac{g''_\gamma}{2g} - \frac{3}{4}\frac{g'^2}{g^2} + g^2\right) v = 0 \quad . \tag{3.2.52}$$

It can be readily verified that the solution of (3.2.52) for arbitrary functions $g(\gamma)$ or $g(G)$ is

$$v = \frac{1}{g^{1/2}} \exp\left(\pm \mathrm{i} \int g \, d\gamma \right) \quad , \quad \text{where} \quad \gamma = \int r \, \mathrm{e}^{-2t\delta_-} dt \quad .$$

Determining the solution of the complete inhomogeneous equation (3.2.30b) and changing over to the population value X in correspondence with (3.2.29), one finds

$$X = 1 + \frac{\mathrm{e}^{-t/\tau}}{g^{1/2}} \int_0^{\gamma} \frac{G(\gamma')^{\delta_\perp/\delta_-}}{g(\gamma')^{1/2}} \left[\int_{\gamma}^{\gamma'} g(\gamma) d\gamma \right] d\gamma' \quad , \tag{3.2.53}$$

where the functions $r(t)$ and $g(t)$ for any given case are defined by (3.2.50, 51). Note that in calculating (3.2.53) the fact that $d\gamma = (r(t)/G(t))dt$ can be used, and so the value of the instantaneous oscillation frequency can be obtained

$$\Omega_{\mathrm{os}} = \frac{d}{dt} \left[\int g(\gamma) d\gamma \right] = \frac{r(t)g(t)}{G(t)} \quad . \tag{3.2.54}$$

This expression differs from the quantity Ω_{os} calculated in the absence of relaxation or for $T = \tau$, when it is equal to $r(t)$.

The total number of oscillations during the pulse duration is

$$\Phi = \frac{1}{2\pi} \int_{-\infty}^{\infty} \Omega_{\mathrm{os}} \, dt \quad .$$

Thus, the oscillations may appear only if $|\Phi|>1$. For a pulse of large duration ($T_{\mathrm{pulse}}\delta_- \gg 1$), this condition leads to $|B|>1$, i.e., $|r_{\mathrm{max}}|>|\delta_-|$ which coincides with the condition obtained previously for infinite duration signals. When $\sinh^2((1/2)T_{\mathrm{pulse}}\delta_-) = 1$, we find that $\Phi = (1/2)(B-(1/2))$, i.e., in order that oscillations appear it is necessary that $B>5/2$. If $\sinh^2((1/2)T_{\mathrm{pulse}}\delta_-)<1$, the critical value of the amplitude B is increased.

When $(B-(1/2))^2 < 1 - \sinh^{-2}((1/2)T_{\mathrm{pulse}}\delta_-)$, the quantity $\int g \, d\gamma$ in (3.2.53), in correspondence with the definition (3.2.53), becomes imaginary. This condition, however, always leads to a value of the threshold amplitude of producing oscillations, which is less than that obtained from condition $|\Phi|>1$.

In a similar way, by choosing other functions $g(\gamma)$ or $g(G)$ that differ from (3.2.51), we can select new forms of dependence $r(t)$ that are of interest in some real situation. Relations (3.2.52–54) remain valid for arbitrary $g(\gamma)$ or $g(G)$, and the dependence $r(t)$ for a given function $g(G)$ is determined by

$$r(t) = \delta_- \frac{G^2}{g^{3/2}} \frac{dg}{dG} \left[-\left(g + \frac{G^2}{g} \right) + 2 \int \frac{G}{g} dG \right]^{-1/2} \quad . \tag{3.2.55}$$

This expression can readily be derived by comparing the lhs of equations (3.2.30b) and (3.2.52), whence

$$G^2 = \frac{g''_\gamma}{2g} - \frac{3}{4}\left(\frac{g'_\gamma}{g}\right)^2 + g^2 \quad .$$

This relation represents the first-order differential equation with respect to $r(t)$ (allowing for the fact that $d\gamma = (1/\delta_-)(r/G^2)dG$:

$$2g\frac{G^2}{r}\frac{d}{dG}\left(\frac{G^2}{r}\frac{dg}{dG}\right) - 3\left(\frac{G^2}{r}\frac{dg}{dG}\right)^2 = \frac{g^2(G^2-g^2)}{\delta_-^2} \quad .$$

Its solution yields (3.2.55).

In conclusion, let us describe an effective means for finding new forms of the dependences $r(t)$ for which exact solutions of (3.2.30b) can be obtained and which are expressed in a class of functions whose type is specified beforehand. To this end, consider

$$\frac{d^2v}{d\gamma^2} + \hat{U}_\gamma[u(\gamma)]v = 0 \quad , \tag{3.2.56}$$

which contains an arbitrary function $u(\gamma)$, where the nonlinear differential operator $\hat{U}_\gamma$ is chosen in such a way that (3.2.56) has an exact solution. This solution then can be expressed through the function $u(\gamma)$, regardless of its concrete form. The following examples can be given for the operator $\hat{U}_\gamma$ and the solution $v(\gamma)$ of (3.2.56):

$$\begin{aligned}
&\text{a)}\ \hat{U}_\gamma(u) = -u''_\gamma/u \ ; \quad v = u(\gamma)\left(C_1 + C_2\int\frac{d\gamma}{u^2}\right) \ ;\\
&\text{b)}\ \hat{U}_\gamma(u) = -(u^2 + u'_\gamma) \ ; \quad v = \exp\left(\int u\,d\gamma\right)\\
&\qquad\qquad \times\left(C_1 + C_2\int\exp\left(-2\int u\,d\gamma'\right)d\gamma\right) \ ;\\
&\text{c)}\ \hat{U}_\gamma(u) = \frac{1}{u^4} - \frac{u''_\gamma}{u} \ ; \quad v = u(\gamma)\left(C_1\sin\int\frac{d\gamma}{u^2} + C_2\cos\int\frac{d\gamma}{u^2}\right) \ ;\\
&\text{d)}\ \hat{U}_\gamma(u) = \frac{1}{2}\frac{u'''}{u'} - \frac{3}{4}\left(\frac{u''}{u'}\right)^2 + \left(\frac{1}{4} - m^2\right)\left(\frac{u'}{u}\right)^2 + u'^2 \ ;\\
&\qquad v = (u/u')^{1/2}[C_1 J_m(u) + C_2 Y_m(u)] \quad .
\end{aligned} \tag{3.2.57}$$

If (3.2.56) is considered as a realization of (3.2.30) without the rhs, than

$$\hat{U}_\gamma[u(\gamma)] = G^2 \quad . \tag{3.2.58}$$

For given $\hat{U}_\gamma$ and $u(\gamma)$, (3.2.58) is an ordinary (nonoperator) equation with respect to γ, the solution of which is $\gamma = \gamma(t)$ or $\gamma = \gamma(G)$ whence, by definition (3.2.29), $r(t) = G(t)d\gamma/dt$.

Thus, by choosing various functions $u(\gamma)$ and making use of one of the operators $U_\gamma(u)$ (3.2.57), one can find the form of $r(t)$ in such a way that it most closely approximates the specific physical situation. In this case the solution of (3.2.56) is automatically determined from the chosen operator (3.2.57).

3.3 Amplitude-Frequency Modulation of the Field (Exact Solutions)

Let us consider certain typical cases of simultaneous temporal variations of both the amplitude $r(t)$ and the frequency detuning of the quasi-resonant field $\nu(t)$, when the exact solution of the equation of motion (3.1.13) can be obtained. The exact solutions are useful not only for the direct comparison with the possible experimental measurements when field modulation is close to the above kinds of functions $r(t)$ and $\nu(t)$, but also for establishing the detailed qualitative features of the parameters of the solution in the more general case. The latter is essential particularly for approximate solutions for a strong field ($r^2 \gg \nu^2 + \delta_-^2$), when the dependences $r(t)$ and $\nu(t)$ are assumed to be arbitrary, and for short field pulses ($T_{\text{pulse}} \ll T$) as well.

3.3.1 The Case of Equal Relaxation Times ($T = \tau$)

For $T = \tau$, the introduction of the new function $v = (X - 1)e^{t/T}$ and the new independent variable $\xi = \int r\,dt$ makes it possible to reduce (3.1.13) to the form

$$\hat{F}(v) \equiv \frac{1}{f} \cdot \frac{d}{d\xi}\left[\frac{1}{f}(v''_\xi + v)\right] + v'_\xi = -\frac{1}{f} \cdot \frac{d}{d\xi}\left[\frac{1}{f}e^{[t(\xi)/T)]}\right] \tag{3.3.1}$$

where $f(\xi) \equiv \nu/r$.

The homogeneous equation corresponding to the third-order equation (3.3.1) can be reduced into a second-order equation. The physical aspect is, in this case as well as in Sect. 3.2.1, that for $T = \tau$ (and for $\eta_0 = 0$) the two-level system can be described not only by the density matrix, but also by the wave function ψ with a complex phase, which immediately lowers the order of the equations. Indeed, if for a weak interaction (3.1.2), the function ψ is represented as a linear combination of the initial unperturbed wave functions $\psi = a_1\psi_1 + a_2\psi_2$, it is possible to obtain from the Heisenberg equation two first-order equations for the amplitudes a_1 and a_2. Therein the relaxation terms are usually introduced phenomenologically (in the construction of the density matrix they lead to equal times T and τ). These two equations can be reduced to a second-order equation similar to (3.3.3) (see below). By direct substitution it can be demonstrated that all real solutions of equation $\hat{F}(v) = 0$ can be written as

$$v = zz^* - 4z'_\xi z^{*\prime}_\xi \quad , \tag{3.3.2}$$

where z is the general solution of

$$z''_\xi + \mathrm{i}f(\xi)z'_\xi + \tfrac{1}{4}z = 0 \ , \quad f \equiv \frac{\nu}{r} \quad . \tag{3.3.3}$$

In the following we shall give a number of classes of relations between ν and r when the exact solutions of (3.3.3) and (3.3.1) exist.

1) $f \equiv \nu/r = \text{const} = q$ (i.e. there is a linear relationship between the frequency deviation and the interaction amplitude; however both of them vary arbitrarily with time). Equation (3.3.3) can be solved by elementary methods, e.g., the solution of the inhomogeneous equation (3.3.1) and changing over to X yields:

$$X(t) = (1+q^2)^{-1}\Bigg[q^2 + C\,\mathrm{e}^{-t/T} + \frac{1}{T}\int \mathrm{e}^{[(t'-t)/T)]} \times \cos\left(\int_{t'}^{t} r\sqrt{1+q^2}\,dt\right)dt'\Bigg] \quad . \tag{3.3.4a}$$

The initial conditions (3.1.14) correspond to $C = 0$; and the remaining two constants of solution (3.3.4a) can be obtained from conditions (3.1.17). In that case (3.3.4a) coincides with solution (3.2.3) of Sect. 3.2 for $\nu \equiv 0$, $T = \tau$ and arbitrary $r(t)$. In the latter formula the quantities r and X should be replaced by

$$r_{\mathrm{ef}} = r(t)\sqrt{1+q^2} \ , \qquad X_{\mathrm{ef}} = X(1+q^2) - q^2 \quad . \tag{3.3.4b}$$

In other words, the instantaneous oscillation frequency is $(r^2+\nu^2)^{1/2}$, and the amplitude of the whole process will be less than that under the conditions of the exact frequency resonance by a factor of $1 + \nu^2/r^2$. Naturally, whenever the integral in (3.2.3) can be calculated exactly, this can be done in a given situation with (3.3.4b) taken into account.

It is of interest to note that the above formulas describe, in particular, the behavior of populations for two-photon absorption (TPA) and the arbitrary temporal variation of the field intensity. Assume that the double frequency of the field 2ω is constant and exactly equal to the unperturbed resonant transition frequency ω_{21}. Then the TPA instantaneous frequency detuning is only determined by the Stark shift $\nu_{2\mathrm{p}} = 4R^2/T\mathcal{E}^2_{\mathrm{St}}$ which is quadratic with respect to the field. At the same time, the interaction amplitude $r_{2\mathrm{p}} = -8\kappa^{12}_{(2)}R^2/\hbar$ here is also quadratic with respect to the field, i.e., their ratio is constant.

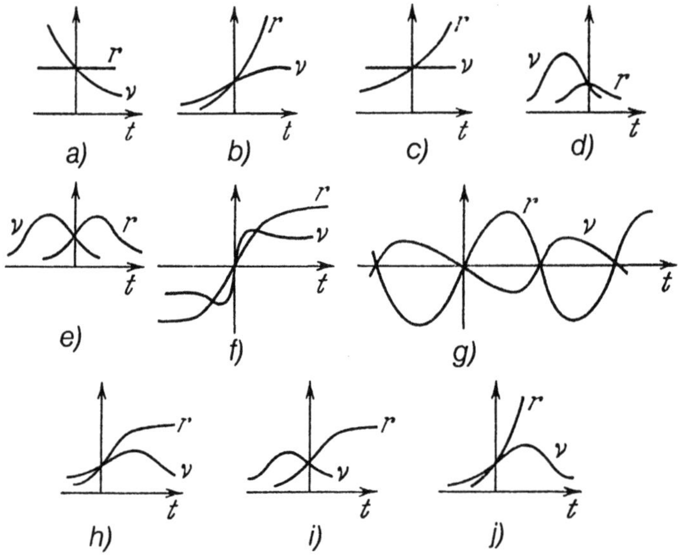

Fig. 3.5. Different time dependency of the field frequency ν and amplitude r that permit exact solutions of the equation for the population dynamics

Therefore, the temporal variation of population is described by (3.3.4a, 4b), where the constant q is given by the ratio of Stark shift constant and TPA effectiveness:

$$q = \frac{\nu_{2\text{ph}}}{r_{2\text{ph}}} = \frac{\kappa^{11} - \kappa^{22}}{2\kappa^{12}_{(2)}} \quad .$$

2) $f \equiv \nu/r = q/\xi$, where $\xi = \int r\,dt$, q being an arbitrary quantity. The solution of (3.3.3) in such a case is

$$z = \xi^m \left[C_1 J_m\left(\frac{\xi}{2}\right) + C_2 J_{-m}\left(\frac{\xi}{2}\right) \right] \quad ; \quad m = \tfrac{1}{2}(1 - \mathrm{i}q) \quad . \tag{3.3.5}$$

The function of the type $f = q/\xi$ can result in most diversified forms of the dependence $r(t)$ and $\nu(t)$. Let us consider a few examples that are interesting from the physical point of view.

a) For a fixed amplitude of the field the frequency detuning decreases or increases as a hyperbolic function (Fig. 3.5a):

$$r = \text{const} \quad , \qquad \nu = q/t \quad . \tag{3.3.6}$$

b) The field amplitude decreases or increases as an exponential function with an arbitrary characteristic time T_E (Fig. 3.5b):

$$r = r_0\,\mathrm{e}^{t/T_E} \quad ; \quad \nu = \nu_0(1 + C\,\mathrm{e}^{-t/T_E})^{-1} \quad ; \quad q = \nu_0 T_E \quad . \tag{3.3.7}$$

In these formulas r_0, ν_0 and C are arbitrary constants [a special case of $\nu =$ const $= \nu_0$ (Fig. 3.5c), i.e., of $C = 0$, which was discussed in [3.245] in terms of probability amplitudes].

c) Bell-shaped field pulses of arbitrary duration T_{pulse} and amplitude r_0 (Figs. 3.5d,e):

$$\begin{aligned} r &= \frac{r_0}{\cosh(t/T_{\text{pulse}})} \quad ; \\ \nu &= \frac{\nu_0}{\cosh(t/T_{\text{pulse}})\left[\operatorname{arctan} h\left(\sinh \frac{t}{T_{\text{pulse}}}\right) + C\right]} \quad . \end{aligned} \tag{3.3.8}$$

d) The amplitude passes through zero (Fig. 3.5f):

$$r = r_0\,\tanh(t/T_E) \;\; ; \quad \nu = \frac{\nu_0\,\tanh(t/T_E)}{\ln|C\,\tanh(t/T_E)|} \quad ; \quad q = \nu_0 T_E \quad . \tag{3.3.9}$$

e) The amplitude varies periodically with an arbitrary frequency Ω_m (Fig. 3.5g):

$$r = r_0\,\sin\Omega_m t \;\; ; \quad \nu = \frac{\nu_0\,\sinh\Omega_m t}{C + \cos\omega_m t} \quad ; \quad q = \frac{\nu_0}{\Omega_m} \quad . \tag{3.3.10}$$

3) Note that (3.3.1) (without the rhs) is invariant with respect to the substitution

$$t \rightarrow -t \;\; ; \quad r(t) \rightarrow r(-t) \;\; ; \quad \nu(t) \rightarrow \nu(-t) \quad .$$

Therefore, if a solution of the homogeneous equation $\hat{F}(v) = 0$ is found for a certain pair of functions $\nu(t)$, $r(t)$, in the case of a field inverted in time (i.e., for $\nu_1(t) = \nu(-t)$, $r_1(t) = r(-t)$) the solution is also determined by the simple inversion of the sign of time (i.e., $v_1(t) = v(-t)$). In particular, for a very short field pulse ($T_{\text{pulse}} \ll T$), when it can be assumed that there is no relaxation, the above applies directly to the behavior of population $X(t)$ itself.

3.3.2 The Non-Equal Relaxation Times ($T \neq \tau$)

When $T \neq \tau$, (3.1.16) can readily be reduced to the second-order equation if the frequency is related to the amplitude through one of the formulas

$$\frac{\nu}{r} = \text{const}\,\exp(\pm 2t\delta_-) = \text{const}\,G^{\pm 1} \quad ; \tag{3.3.11}$$

where $\delta_- = \frac{1}{2}(T^{-1} - \tau^{-1})$. By choosing the time reference point it is possible to set const $= 1$. Then, if the exponential function (3.3.11) has a negative sign, (3.1.13) becomes

$$\left(\hat{D}_\tau - 2\frac{\dot{\nu}}{\nu}\right)\left[\left(\hat{D}_\tau - \frac{\dot{\nu}}{\nu}\right)\hat{D}_\tau(X-1) + (r^2+\nu^2)X\right] = \frac{\nu^2}{\tau} \quad .$$

The first integral of this equation is

$$\left(\hat{D}_\tau - \frac{\dot{\nu}}{\nu}\right)\hat{D}_\tau(X-1) + (r^2+\nu^2)X = \nu^2(1+C\,\mathrm{e}^{-t/\tau}) \quad .$$

Under the initial conditions (3.1.14) $C=0$; by introducing a variable $\varphi = \int \nu\, dt$ and a function $v(\varphi) = (X-1)\exp(t/\tau)$ we obtain

$$v''_\varphi + [1+G^2(\varphi)]v = -G^{\delta_\perp/\delta_-} \; ; \quad G = \mathrm{e}^{2t\delta} \quad . \tag{3.3.12}$$

In the limiting case $T = 2\tau$, $\delta_\perp = 0$, and the rhs in (3.3.11) is equal to -1; for the case for $\tau \gg T$ it is approximately equal to $-G^2$. It is of interest to note that if a certain particular solution $v_1(\varphi)$ of (3.3.11) for $\delta_\perp = 0$ is found, the particular solution for $\tau \gg T$ is $v_2(\varphi) = -(1+v_1)$, where $v_1(\varphi)$ is computed for the corresponding T and τ.

Let us consider certain kinds of functions $\nu(t)$ and $r(t)$ that are of interest from the physical point of view and for which the exact solution of (3.3.11) can be obtained.

1) The frequency detuning is constant, $\nu = \text{const} = \nu_0$, and the amplitude increases as the exponential function $r = \nu_0\exp(2t\delta_-)$ (Fig. 3.5c). This example will be used to illustrate the oscillatory properties of the solution. Here $\varphi = \nu_0(t)$ and (3.3.11) reduces to

$$v''_\varphi + (1+\mathrm{e}^{2\varphi/m})v = -\exp(2\varphi\delta_\perp/v_0) \; ; \quad m = \frac{\nu_0}{2\delta_-} \quad . \tag{3.3.13}$$

The solutions to the corresponding homogeneous equation [3.50] are Bessel functions with the imaginary subscript

$$J_{\pm\mathrm{i}m}(m\,\mathrm{e}^{\varphi/m}) = J_{\pm\mathrm{i}m}(mG) \quad .$$

Using them to solve (3.3.13) and again changing to X, we obtain

$$X = 1 - m^2G^{1-q}[s_{q,\mathrm{i}m}(mG) + CJ_{\mathrm{i}m}(mG) + C^*J_{-\mathrm{i}m}(mG)] \; ;$$
$$q = (2T\delta_-)^{-1} \quad , \tag{3.3.14}$$

where $s_{q,k}(\xi)$ is the Lommel function [3.51] which is defined for a noninteger q as

$$s_{q,k}(\xi) = \frac{\pi}{2\sin k\pi}\left[J_k(\xi)\int_0^{\xi}\xi'^q J_{-k}(\xi')d\xi' - J_{-k}(\xi)\int_0^{\xi}\xi'^k J_k(\xi')d\xi'\right] \quad . \tag{3.3.15}$$

Because of the fact that $s_{q,\mathrm{im}}(\xi)$ and X are real and $J_{\mathrm{im}}(\xi) = J^*_{-\mathrm{im}}(\xi)$, the constants near J_{im} and $J_{-\mathrm{im}}$ in solution (3.3.14) are chosen as complex conjugate. When $\delta_->0\,(\tau>T)$, by decomposing the function $s_{q,\mathrm{im}}(\xi)$ into a power series with respect to the argument

$$\begin{aligned} s_{q,\mathrm{im}}(\xi) = \frac{\xi^{q+1}}{(q+1)^2+m^2}\Bigg\{1 - \frac{\xi^2}{(q+3)^2+m^2} \\ + \frac{\xi^4}{[(q+3)^2+m^2][(q+s)^2+m^2]} - \dots\Bigg\} \quad , \end{aligned}$$

and taking into account the well-known expansion for $J(\xi)$, we realize that conditions (3.1.14) for $t_0\to-\infty$ (the field is switched on "from zero") correspond to $C = C^* = 0$, i.e.,

$$X = 1 - m^2\,\mathrm{e}^{-t/\tau}s_{g,\mathrm{im}}(mG) \;;\quad q = (2T\delta_-)^{-1} \quad ; \tag{3.3.16}$$

$$m = \frac{\nu_0}{2\delta_-} \quad .$$

In the general case ($t_0\neq-\infty$) there is an oscillation component in solution (3.3.14), with the frequency and the logarithmic decrement being different and characterized by different stages of the process. For small $r(t)$, when the argument of the Bessel function is much less than the quantity $\max(1,|m|)$ (i.e., if $r^2\ll\nu^2+\delta_-^2$, since $mG = r/2\delta_-$), one can retain only the first term of the expansion of $J_{\pm\mathrm{im}}(mG)$ with respect to mG, whence

$$\begin{aligned} CJ_{\mathrm{im}} + C^*J_{-\mathrm{im}}(mG) &\approx C\frac{(mG/2)^{\mathrm{im}}}{\Gamma(1+\mathrm{im})} + C^*\frac{(mG/2)^{-\mathrm{im}}}{\Gamma(1-\mathrm{im})} \\ &= B\cos(m\ln mG+\varphi_0) = B\cos(\nu_0 t+\varphi_0) \quad . \end{aligned}$$

In this expression B and φ_0 are determined from the initial conditions. Thus, for $r^2\ll\nu^2+\delta_-^2$, a component $\mathrm{e}^{-t/\tau}\cos\nu_0 t$ may be present in the relaxation of populations, which is caused by the usual linear beats between the weak field frequency ω and the intrinsic polarization at the resonant frequency ω_{21} of the as yet "undistorted" system. The beats' ν logarithmic decrement τ^{-1} represents a sum of the polarization logarithmic decrement T^{-1} and the increment of the field increase (in this case $-2\delta_-$), in other words, $\tau^{-1} = T^{-1} - 2\delta_-$.

For large values of the argument mG (or for $r^2\gg\nu^2+\delta_-^2$), the usual asymptotic expansion is valid for the function $J_{\mathrm{im}}(mG)$ from which it follows that

$$C J_{\mathrm{im}}(mG) + C^* J_{-\mathrm{im}}(mG) \propto G^{-1/2} \cos(mG + \varphi_0) \quad ,$$

i.e., here the population contains oscillations

$$G^{(1/2-q)} \cos mG = \mathrm{e}^{-t\delta_+} \cos\left(\int r\,dt\right) \quad ,$$

where $\delta_+ = (1/2)(T^{-1} + \tau^{-1})$ [it is as if an averaging of the relaxation times occurs, and the new decrement does not depend on the rate of the field increase]. Thus, when $r^2 \gg \nu^2 + \delta_-^2$, oscillations approach the regime of zero detuning ($\nu \equiv 0$, Sect. 3.2) for $r^2 \gg \delta_-^2$, i.e., they already correspond to the specific oscillatory conditions in a strong field.

Note that, like the case $\nu = 0$ (Sect. 3.2.2), the "strong" field requirement for producing such oscillations, $r^2 \gg \nu^2 + \delta_-^2$, does not generally coincide with the condition of saturation. Under a quasi-stationary regime

$$X_{\mathrm{sat}} = \left(1 + \frac{r^2}{r_{\mathrm{sat}}^2}\right)^{-1} \; ; \quad r_{\mathrm{sat}}^2 = \frac{1 + \nu^2 T^2}{\tau T} \quad , \tag{3.3.17}$$

i.e., for $T\,Zd\,\tau$ and ν^2 (the field frequency is within the limits of the resonance line) the critical "oscillatory" amplitude can be lower than the saturating amplitude, whereas when $\tau \gg T$, r_{cr} is substantially greater than r_{sat}.

2) The variation of the frequency detuning and the amplitude is given by

$$\nu = \nu_0 G^2 \frac{2^\sigma}{(1 + G^2)^\sigma} \; ; \quad r = \nu G \; ;$$
$$\nu_0 = \mathrm{const} \; ; \quad \sigma = \mathrm{const} \quad . \tag{3.3.18}$$

By varying σ one can use the functions in (3.3.18) for describing a number of various physical situations, e.g., (3.3.12) can be solved in quadratures for any σ.

a) $\nu = 2\nu_0 G^2 (1 + G^2)^{-1}$, $\sigma = 1$ (Fig. 3.5b). When $\delta_- > 0$, the frequency approaches a constant quantity $2\nu_0$ for $t \to \infty$. Here $\varphi = \nu_0 (2\delta_-)^{-1} \ln(1 + G^2)$, and (3.3.12) assumes the form

$$v''_\varphi + \mathrm{e}^{2\varphi/m} v = -(\mathrm{e}^{2\varphi/m} - 1)^{\delta_\perp/2\delta_-} \; ; \quad m = \nu_0/\delta_- \quad . \tag{3.3.19}$$

The solutions to the corresponding homogeneous equation are the Bessel functions $J_0(\psi)$, $Y_0(\psi)$, where $\psi = m\,\mathrm{e}^{\varphi/m} = m(1 + G^2)^{1/2}$.

b) $\omega = r_0 G^2 (1 + G^2)^{-3/2}$, $r = G\nu$ and $\sigma = 3/2$ (Fig. 3.5c). When $t \to \infty$, the amplitude $r(t)$ tends to a constant quantity r_0, and the frequency $\nu(t)$ has a maximum at $t = (\ln 2)/4\delta_-$ [the reference point for the time is assumed to be the moment when $r(t) = \nu(t)$], with $\omega_{\max} = 2r_0/3^{3/2}$. In this case

$\varphi = -r_0(2\delta_-)^{-1}(1+G^2)^{-1}$ and (3.3.12) can be written as

$$v''_\varphi + \frac{m^2}{4\varphi^2}v = -\left(\frac{m^2}{4\varphi^2} - 1\right)^{\delta_\perp/2\delta_-} \quad ; \quad m = \frac{r_0}{\delta_-} \quad . \tag{3.3.20}$$

The solution of the corresponding homogeneous equation (the Euler equation) is

$$\varphi^{(1\pm\sqrt{1-m^2})/2} \quad ;$$

when $m^2>1(r_0^2>\delta_-^2)$ it becomes oscillatory:

$$\exp\left[\pm\mathrm{i}\frac{\sqrt{m^2-1}}{4}\ln(G^2+1)\right] \quad .$$

The instantaneous oscillation frequency is

$$\delta_-(m^2-1)^{1/2}G^2(1+G^2)^{-1} \quad ,$$

for $\nu\to 0$ and $r\to r_0$ it approaches the quantity $(r_0^2-\delta_-^2)^{1/2}$, in correspondence with Sect. 3.2.2.

The inhomogeneous equation (3.3.20) and changing over to X we have

$$X = 1 - \left(1+\frac{8}{m^2}\right)^{-1} G^2(1+G^2)^{-1} \quad ,$$

[here $r\to$ const when $t\to -\infty$, since $\delta_-<0$. Therefore, instead of the conditions (3.1.14), the condition $X\to$ const >0 for $t_0\to -\infty$ should be chosen].

Under the above initial condition there are no oscillations; if $|t_0|<\infty$, they are excited at the beginning of the process. In the other limiting case, when $|\tau|\gg 1$ and under the initial conditions (3.1.14) we obtain for $t_0\to -\infty$

$$X \approx \left[1+\frac{8}{m^2}(1+G^2)^{3/4}\left(\cos\psi - \frac{m^2}{2(m^2-1)^{1/2}}\sin\psi\right)\right]$$
$$\times\left(1+\frac{8}{m^2}\right)^{-1}(1+G^2)^{-1} \; ;$$
$$\psi = \frac{(m^2-1)^{1/2}}{4}\ln(1+G^2) \quad . \tag{3.3.21}$$

For $m^2\gg 1$, the maximal oscillation amplitude is $\approx 4/|m|$. When $\nu\equiv 0$, their amplitude, on the contrary, increases with the increase of r_0 (see Sect. 3.2.2) and approaches a constant quantity. This difference is due to the fact that when $\omega\equiv 0$, the first oscillations occur when $r^2\ll\nu^2$, therefore they must be weak.

c) If in (3.3.18) $\sigma \neq 1$ and $\sigma \neq 3/2$, then

$$\varphi = \nu_0 (2\delta_-)^{-1} (1-\sigma)^{-1} \left(\frac{1+G^2}{2} \right)^{1-\sigma} \quad ,$$

and (3.3.12) assumes the form

$$\begin{aligned} &v''_\varphi + B^2 \varphi^{(1/1-\sigma)} v = -(B^2 \varphi^{(1/1-\sigma)} - 1)^{\delta_\perp / 2\delta_-} \quad ; \\ &B^2 = \left| \frac{\nu_0}{\delta_-(1-\sigma)} \right|^{(1/\sigma - 1)} \quad . \end{aligned} \tag{3.3.22}$$

The solution of the corresponding homogeneous equation is

$$\varphi^{1/2} [C_1 J_m(\psi) + C_2 Y_m(\psi)] \; ; \quad m = \frac{1-\sigma}{3-2\sigma} \; ; \quad \psi = 2mB\varphi^{1/2m} \quad .$$

Hence one can obtain the solution of (3.3.22):

$$\begin{aligned} v = m\pi B \varphi^{1/2} \int & \varphi'^{1/2} (B^2 \varphi'^{(1/1-\sigma)} - 1)^{\delta_\perp / 2\delta_-} \\ & \times [J_m(\psi) Y_m(\psi') - Y_m(\psi) J_m(\psi')] d\varphi' \quad , \end{aligned} \tag{3.3.23}$$

where $\psi' = \psi(\varphi')$ and $X = 1 + v\,\mathrm{e}^{-t/\tau}$.

d) In particular, when $\nu = \nu_0 / \cosh^2(2\delta_- t)$, $r = \nu G (\sigma = 2, m = 1)$, the frequency and the amplitude variation are described by short pulses that are shifted with respect to each other (Fig. 3.5d). If $\delta_1 = 0$, under conditions (3.1.14) for $t_0 \to -\infty$ we have

$$\begin{aligned} &X = 1 - \frac{G^2}{B^2} \left[\varphi - \frac{\sqrt{\varphi}}{B} J_1(2B\sqrt{\varphi}) \right] \; ; \\ &\varphi = \frac{B^2}{1+G^2} \; ; \quad B = \left| \frac{\nu_0}{\delta_-} \right| \quad , \end{aligned}$$

and in the limiting case $\tau \gg T$

$$X = \frac{\varphi}{B^2} + \frac{\pi\sqrt{\varphi}}{B} [J_1(2B\sqrt{\varphi}) Y_1(2B^2) - Y_1(2B\sqrt{\varphi}) J_a(2B^2)] \quad .$$

e) When $\sigma = 5/4\,(m = -1/2)$ [Fig. 3.5j], it follows from (3.3.23) that

$$v(\varphi) = B\varphi \int x^{-3} \left(\frac{x^4}{B^2} - 1 \right)^{\delta_\perp / 2\delta_-} \sin\left(x - \frac{B}{\varphi} \right) dx \quad .$$

Hence for $\delta_\perp = 0$ under the condition $X \to 0$ for $t_0 \to -\infty$, we obtain for X

$$X = 1 - \frac{B\varphi}{2}\left[\frac{\varphi}{B} - \sin\left(\frac{B}{\varphi}\right)\mathrm{si}\left(\frac{B}{\varphi}\right) + \cos\left(\frac{B}{\varphi}\right)\mathrm{si}\left(\frac{B}{\varphi}\right)\right] \; ;$$
$$B = \sqrt{2}(2\nu_0/\delta_-)^2 \quad .$$

Similarly, one can express the solution through $\mathrm{si}(B/\varphi)$ and $\mathrm{ci}(B/\varphi)$ if the quantity $\delta_\perp/2\delta_-$ is an integer, particularly when $\tau \gg T$, which corresponds to $\delta_\perp/2\delta_- \approx 1$.

f) An interesting special case of (3.3.18) corresponds to a field whose amplitude and frequency detuning represent pulses which are like mirror images with respect to each other at a certain moment of time. This takes place when $\sigma = 5/2$ (Fig. 3.5e):

$$r = \nu_0 G \cosh^{-5/2}(2t\delta_-) \; ; \quad \nu = \nu_0 G^{-1} \cosh^{-5/2}(2t\delta_-) \quad .$$

3) The variation of the amplitude and the frequency detuning is given by

$$\nu = \nu_0 G^2(b^2 + G^2)^{-3/2} \; ; \quad r = \nu G \; ; \quad b = \mathrm{const} \quad . \tag{3.3.24}$$

At the same time $\varphi = r_0(2\delta_-)^{-1}(b^2 + G^2)^{-1/2}$, and (3.3.12) transforms into

$$v''_\varphi + \left[1 - b^2 + \left(\frac{r_0}{2\varphi\delta_-}\right)^2\right]v = -\left[\left(\frac{r_0}{2\varphi\delta_-}\right)^2 - b^2\right]^{\delta_\perp/2\delta_-} \quad . \tag{3.3.25}$$

The solution of the corresponding homogeneous equation is

$$\sqrt{\varphi}J_{\pm m}(\varphi\sqrt{1-b^2}) \; , \quad \text{where}$$
$$m = \tfrac{1}{2}[1 - (r_0/\delta_-)^2]^{1/2} \quad ,$$

hence the solution in quadratures can easily be found. If $b^2 = 0$ or if the quantity $\delta_\perp/2\delta_-$ is an integer, this solution is expressed through the Lommel functions.

a) In particular, when $b^2 = 0$, (3.3.24) describes a field with a constant amplitude $r_0 = \mathrm{const}$ and the frequency detuning $\nu = r_0\,\mathrm{e}^{-2t\delta_-}$ that varies as an exponential function [compare Fig. 3.5a with (1)]. Solving (3.3.25) and changing over to $X = 1 + v\,\mathrm{e}^{-t/\tau}$, we obtain

$$X = 1 - \mathrm{e}^{-t\delta_+} \times \left[B^{(1+q)/2} s_{(1-q)/2,m}\left(\frac{B}{G}\right) + C_1 J_m\left(\frac{B}{G}\right) + C_2 Y_m\left(\frac{B}{G}\right)\right] \quad , \tag{3.3.26}$$

where

$$B = \frac{r_0}{2\delta_-} \quad ; \quad q = \frac{\delta_\perp}{\delta_-} \quad ; \quad m = (\tfrac{1}{4} - B^2)^{1/2} \quad .$$

When $\tau > T$, under the initial conditions (3.1.14) for $t_0 \to -\infty$, the asymptotic form of $J_m(\psi)$, $Y_m(\psi)$ and $s_{(1-q)/2,m}(\psi)$ for $\psi \to \infty$,

$$s_{(1-q)/2,m}(\psi)\psi^{(1+q)/2} \to 1 \quad \text{for} \quad \psi \to \infty$$

yields $C_1 = C_2 = 0$.

Note that the subscript $m = (1/4 - b^2)^{1/2}$ becomes imaginary when $r_0^2 > \delta_-^2$. Because of this, the oscillatory components in (3.3.26) have a frequency close to $\nu(t)$ only as long as $\nu^2 \gg r^2$. When $\nu^2 \ll r_0^2$ ($\psi^2 \ll |m|^2$), the oscillation frequency is already determined by the magnitude of the imaginary subscript [compare with (1)], i.e., it is close to $(r_0^2 - \delta_-^2)^{1/2}$.

b) If in (3.3.24) the quantity $b^2 > 0$, the field amplitude tends to a constant r_0 as $t \to \infty$ (for $\tau > T$), and the frequency detuning has a maximum $\nu_{\max} = b^{-1}(2\tau_0/3^{3/2})$ at the time $t = (\ln 2b^2)/4\delta_-$ [with the time reference point corresponding to $\nu(t) = r(t)$]. When $0 < b^2 < 1$, these curves are similar to those depicted on Fig. 3.5i, and when $b^2 > 1$, they are close to the plots on Fig. 3.5h, the only difference being that now the position of the maximum on the curve $\nu(t)$ depends on b^2. The situation when $b^2 = 1$ is considered in (2b).

Let us return to (3.3.11). In the case of a positive sign in the exponential function of (3.3.11), we again set const $= 1$ thus reducing (3.1.13) to

$$\hat{D}_\tau\left\{\frac{1}{r}\hat{D}_T\left[\frac{1}{r}\hat{D}_\tau(X-1)\right]\right\} + (1+G^2)\hat{D}_\tau(X-1) = \tau^{-1} \quad ;$$
$$G = \exp(2t\delta_-) \quad .$$

The introduction of a new variable $\varphi = \int \nu\, dt$ and a new function $u(\varphi) = r^{-1}\,\mathrm{e}^{t/\tau}\hat{D}_\tau(X-1)$ makes it possible to transform this equation to the second-order one:

$$u''_\varphi + u(1 + G^{-2}) = -(\tau\nu)^{-1}\exp[t(\varphi)/\tau] \quad , \tag{3.3.27}$$

with the motion of population $X(t)$ determined by

$$\begin{aligned} X(t) &= 1 + \mathrm{e}^{-t/\tau}\int \nu(t')u(t')G^{-2}(t')dt' \\ &= 1 + \mathrm{e}^{-t/\tau}\int uG^{-2}d\varphi \quad . \end{aligned}$$

We shall demonstrate that the solutions of (3.3.12, 27) (taken with the zero rhs) are mutually reversible in the sense that a solution $v(t)$ of (3.3.12) corresponding to a certain function $\nu(t)$ (and $r = \nu G$) is at the same time a solution $u(t)$ of (3.3.27), corresponding to the ampltiude and frequency detuning inverted in time, i.e., to $\nu_i(t) = \nu(-t)$ and $r_1(t) = r(-t)$.

Indeed, assuming that in the lhs of (3.3.12)

$$t_1 = -t \ ; \quad \nu(t) = \nu_1(-t) = \nu_1(t_1) \ ;$$
$$\varphi = -\varphi_1(t_1) \ ; \quad G^2(t) = G_1^{-2}(t_1) \quad ,$$

we arrive at the expression $v''_{\varphi_1} + (1 + G_1^{-2})v$, which is equal to the left-hand side of (3.3.27). And since for (3.3.27) $r = \nu \exp(-2t\delta_-)$, the solution also corresponds to the inversion in time of the amplitude $r(t)$. Thus, the exact solutions applied to the field inverted in time can now be easily found.

3.4 Approximate Solutions in Various Limiting Cases

A good approximate solution of (3.1.13) can be found for arbitrary forms of the dependence of r and ν on time, and on the relationship between relaxation times T and τ of the field amplitude and frequency in the limiting cases of the weak field $[r^2 \ll \max(r_{cr}^2, r_{sat}^2)]$ and the strong field $(r^2 \gg r_{cr}^2)$, where

$$r_{cr}^2 = \nu^2 + \delta_-^2 \ ; \quad r_{sat}^2 = \frac{1 + \nu^2 T^2}{\tau T} \quad , \tag{3.4.1}$$

and also for an arbitrary amplitude of a slowly or rapidly varying field.

1) In the weak field $(r^2 \ll r_{sat}^2)$, because of the fact that the system is far from the state of saturation and, therefore, the population is close to its equilibium value (i.e., $1 - X \ll 1$), (3.1.13) can be written as

$$\frac{1}{\nu}\hat{D}_T\left\{\frac{1}{\nu}\hat{D}_T\left[\frac{1}{r}\hat{D}_\tau(X-1)\right]\right\} + \frac{1}{r}\hat{D}_\tau(X-1)$$
$$= -\frac{1}{\nu}\hat{D}_T\left(X\frac{r}{\nu}\right) \approx -\frac{1}{\nu}\hat{D}_T\left(\frac{r}{\nu}\right) \quad . \tag{3.4.2}$$

By setting

$$\varphi = \int \nu\, dt \ ; \quad u(\varphi) = \frac{1}{r}\mathrm{e}^{t/T}\hat{D}_\tau(1-X) \quad ,$$

we reduce it to the form

$$u''_\varphi + u \approx \frac{d}{d\varphi}\left(\frac{r}{\nu}\mathrm{e}^{t/T}\right) \quad . \tag{3.4.3}$$

Hence, by calculating $u(\varphi)$ and changing over to $X(t)$, from the definition of u we obtain

$$1-X \approx \frac{1}{2}e^{-t/\tau}\int^{t}\int^{t'} e^{t'/\tau}\, e^{(t''-t')/T} r(t')r(t'')\cos\left(\int_{t''}^{t'}\nu\, dt\right)dt'\, dt'' \quad . \tag{3.4.4}$$

In particular, in the limiting case $T=2\tau$,

$$1-X \approx \frac{1}{2}e^{-2t/T}\left[\left(\int r\, e^{t/T}\sin\varphi\, dt\right)^2+\left(\int r\, e^{t/T}\cos\varphi\, dt\right)^2\right] \quad . \tag{3.4.5}$$

For an exact frequency resonance ($\nu\equiv 0$, $\varphi=\text{const}$) and under the conditions of (3.1.17),

$$1-X \approx \frac{1}{2}e^{-2t/T}\left(\int r\, e^{t/T}dt\right)^2 \quad . \tag{3.4.6}$$

Let us consider the case, in which $r^2 \ll r_{\rm cr}^2$, and the relationship between r^2 and $r_{\rm sat}^2$ is arbitrary [one can see from (3.4.1) that although for $\tau \gg T$ the system can already be in a state of strong saturation, $r^2 \gg r_{\rm sat}^2$, the condition of "weakness" of the field $r^2 \ll r_{\rm cr}^2$ is still valid, see Sects. 3.2.2 and 3.3]. Then in the right-hand side of (3.4.2) it must be assumed in the first approximation that instead of a solution satisfying the condition $X_0=1$, the quasi-stationary solution $X_{\rm sat}$ (3.3.17) is present. Therefore, the right-hand side of (3.4.3) will have the form

$$\frac{d}{d\varphi}\left[\frac{r}{\nu}e^{t/T}\left(1+\frac{r^2}{r_{\rm sat}^2}\right)^{-1}\right] \quad ,$$

whence one can easily find $u(t)$ and then X.

2) For a strong field $\nu^2/r^2 \ll 1$, and for this reason the solution of (3.1.13) should be close to that of the degenerate equation corresponding to $\nu\equiv 0$ (3.1.16). (Note, that a sufficient condition in the search for an approximate solution is the small difference between the values of oscillation frequencies and logarithmic decrements, and this can be demonstrated by studying the exact solutions of Sects. 3.2.2 and 3.3). In view of this, we shall write (3.1.13) as

$$\frac{1}{r}\hat{D}_T\left[\frac{1}{r}\hat{D}_\tau(X-1)\right]+X=-\frac{\nu}{r}\hat{D}_T^{-1}\left[\frac{\nu}{r}\hat{D}_\tau(X-1)\right]\equiv Q_D(t) \quad , \tag{3.4.7}$$

where $\hat{D}_T^{-1}$ is the inverse operator with respect to $\hat{D}_T$, i.e.,

$$\hat{D}_T^{-1}\hat{D}_T = 1 \ , \quad \text{whence} \quad \hat{D}_T^{-1}(\xi) = \mathrm{e}^{-t/T} \int \xi\, \mathrm{e}^{t/T} dt \quad .$$

We shall look for its solution by considering (3.4.7) as a second-order equation with the additional rhs $Q_D(t)$, for which the unknown function $X(t)$ can be chosen for $\nu^2/r^2 \ll 1$ as a solution of (3.1.16) valid for $\nu \equiv 0$. Section 3.2 presents solutions of (3.3.16) for certain functions $r(t)$ (if $T \neq \tau$) and for an arbitrary $r(t)$ if $T = \tau$. In the general case the approximate solution of a homogeneous equation corresponding to (3.1.16) is

$$X \approx \exp\left[-t\delta_+ \pm \mathrm{i} \int (r^2 - \delta_-^2)^{1/2} dt\right] \quad ,$$

and by making use of it one can readily find the approximate solution of the inhomogeneous equation (3.1.16) itself:

$$X_0 \approx -\frac{\delta_-^2}{\bar{r}^2} + \int^t \mathrm{e}^{(t'-t)\delta_+} \hat{D}_+\left(\frac{r^2}{\bar{r}^2}\right) \cos\left(\int_{t'}^t \bar{r}\, dt\right) dt' \quad \text{where} \tag{3.4.8}$$

$$\bar{r} = (r^2 - \delta_-^2)^{1/2} \ ; \quad \hat{D}_+ = \frac{d}{dt} + \delta_+ \quad ; \quad \delta_+ = \frac{1}{2}(T^{-1} + \tau^{-1}) \quad .$$

If $T = \tau$, this solution is exact for an arbitrary function $r(t)$, and if $T \neq \tau$, it is exact for $r = \text{const}$. It can be shown that it is close to exact solutions in the case $T \neq \tau$ for those functions $r(t)$ for which the exact solutions are found in Sect. 3.2.2. This is especially true as far as the values $r^2 \gg \delta_-^2$ are concerned. Then expression (3.4.8) can be simplified since $D_+(r^2/r^2) \approx \delta_+$. With the aid of (3.4.8), the magnitude of the rhs Q_D in (3.4.7) can be easily calculated. Now we use again the approximate solution of the homogeneous equation corresponding to (3.1.16) to obtain the correction $\Delta X = X - X_0$ to the solution X_0 (3.4.8) of the initial equation (3.1.16). The correction is caused by the rhs $Q_D(t)$ in (3.4.7) for the strong field ($r^2 \gg r_{\text{cr}}^2$):

$$\Delta X = X - X_0 \approx \int^t \mathrm{e}^{(t'-t)\delta_+} r(t') Q_D(t') \sin\left(\int_{t'}^t r\, dt\right) dt' \quad . \tag{3.4.9}$$

Note that the expression for $Q_D(t)$ can be simplified if the amplitude of the field is either constant or varies sufficiently slowly, so that $|\dot{r}/r^2| \ll 1$, since then $\hat{D}_\tau(X-1) \approx -\tau^{-1}$ and, therefore,

$$Q_D \approx \frac{\nu}{r\tau}\mathrm{e}^{-t/\tau} \int \frac{\nu}{r}\mathrm{e}^{t'/T} dt' \quad .$$

The above results make it possible to describe the resonant interaction of the quantum system with the field characterized by an arbitrary spectrum whose width is less or greater than the resonant width of the system lines. The only necessary condition is that the field should be sufficiently strong.

3) Let us consider separately the case of a strong field with an almost constant amplitude $r = r_0 + \Delta r(t)$ where $r_0 = \text{const}$. For such a field, the temporal variation of amplitude and frequency deviations is arbitrary but much less than $r_0(\Delta r^2, \nu^2 \ll r_0^2$, the relationship between ν and T^{-1} being arbitrary). It can be demonstrated that under these conditions the behavior of the correction to the stationary value of the population $\Delta X = X - X_{\text{sat}}$ is governed by an equation with constant coefficients:

$$\Delta\ddot{X} + 2\delta_+\Delta\dot{X} + (r_0^2 + r_{\text{sat}}^2)\Delta X$$
$$= \frac{1}{\tau}\left(1 + \frac{r_{\text{sat}}^2}{r_0^2}\right)^{-1}[q_r(t) + q_\nu(t)] \quad , \quad \text{where} \tag{3.4.10}$$

$$q_r(t) = \frac{1}{r_0}(\Delta\dot{r} + 2T^{-1}\Delta r) \ ; \ q_\nu(t) = \nu\,\mathrm{e}^{-t/T}\int^{t}\nu\,\mathrm{e}^{t'/T}dt' \quad .$$

Thus, if, for instance, the amplitude is modulated by a periodic oscillation $\Delta r = \Delta r_m \cos \Omega_m t$, a resonance at $\Omega_m \approx r_0$ should be observed and the half-width of the resonance curve is $\approx \delta_+$. The steady state regime of population oscillations is described by

$$X \approx X_{\text{sat}} + \frac{\Delta r_m}{r_0}(2\tau\delta_+)^{-1}\frac{\sin(\Omega_m t + \varphi)}{[1 + (\Omega_m - r_0)^2\delta_+^{-2}]^{1/2}} \quad ; \tag{3.4.11}$$

$$\tan\varphi = \frac{\delta_+}{\Omega_m - r_0} \quad .$$

It is interesting to note that the population oscillation amplitude at the resonance $(\Delta r_m/r_0)(2\tau\delta_+)^{-1}$ can be much greater than the value of the constant component $X_{\text{sat}} = (1 + r_0^2 T\tau)^{-1}$ due to saturation (i.e., population can assume negative values), while the depth of the initial modulation is small. The condition required for this is $\Delta r_m > 2\delta_+(r_0 T)^{-1}$. The modulation frequency necessary for obtaining a resonance by means of modulation of frequency detuning $\nu = \nu_m \cos \Omega_m t$ is less by a factor of two ($\Omega_m \sim r_0/2$) because, as follows from (3.4.10), the effect is quadratic with respect to frequency. And if the frequency detuning has a constant nonzero component $\nu = \nu_0 + \nu_m$, cos $\omega_m t$, there is also a resonance at the frequency $\Omega_m \approx r_0$ whose amplitude is proportional to $\nu_0\nu_m$.

Such an approximation can be applied, for instance, to the analysis of population dynamics in the field of a single-mode laser that operates in the stationary regime (its noise being described by means of relatively weak fluctuations of amplitude and frequency), or in the field of a strong monochromatic signal with an arbitrary weak field superposed on it (the frequency drift of the resultant field is small, even if the noise spectrum width is large). In the simplest case this is a two-frequency signal with strong ($\nu_1 = \omega_1 - \nu_{21}$; $r_1^2 \gg \nu_1^2 + \delta_-^2$) and weak ($r_2 \ll r_1$; $\nu_2 = \omega_2 - \omega_{21}$) components, which can be presented as

$$r_1 \sin \omega_1 t + r_2 \sin \omega_2 t = [r_1 + \Delta r(t)] \sin\left(\omega_{21} t + \int \nu \, dt\right) \quad ,$$

where

$$\Delta r(t) \approx r_1 \cos(t \Delta\nu) \ ; \quad \nu(t) \approx \nu_1 - \frac{r_2}{r_1} \Delta\nu \cos(t\Delta\nu) \ ;$$
$$\Delta\nu = \omega_2 - \omega_1 = \nu_2 - \nu_1 \quad .$$

Substitution of these values of $\Delta r(t)$ and $\nu(t)$ into (3.4.10) results in an equation for the stationary regime of population oscillations with the frequencies $\Delta\nu$ and $2\Delta\nu$. This problem was considered in [3.12, 31] where the polarization spectrum was computed for the component $\Delta\nu$ only.

4) Sometimes the nature of the approximation is determined by the magnitude of the characteristic field variation T_{pulse}. In a quasi-stationary field ($T_{\text{pulse}} \gg \tau$, T), the population is given by (3.3.17) which follows from (3.1.13) when $\hat{D}_T \approx T^{-1}$, $\hat{D}_\tau \approx \tau^{-1}$. If the field pulse duration is much less than the lifetime but greater than the inverse line width (i.e., $\tau \gg T_{\text{pulse}} \gg T$ as is frequently the case for lasers [3.44]) then using $\hat{D}_\tau \approx d/dt$, $\hat{D}_T \approx T^{-1}$ in (3.1.13) we obtain

$$\dot{X} + T^{-1} r^2 (\nu^2 + T^{-2})^{-1} X = 0 \tag{3.4.12}$$

whence for $t - t_0 \lesssim T_{\text{pulse}}$

$$X \approx \exp\left[-T \int r^2 (1 + \nu^2 T^2)^{-1} \, dt\right] \quad .$$

Finally, if the pulse is very short ($T_{\text{pulse}} \ll T$, τ), then $\hat{D}_T \approx \hat{D}_\tau \approx d/dt$ can be assumed in (3.1.13) leading to (3.3.1), where the quantity v should be replaced by the population X and the rhs should be set equal to zero.

In this case, if the frequency detuning varies adiabatically slowly starting from the value $|\nu^2(t = -\infty)| \gg r^2$ and passes through zero, then for

$$r \frac{d(\nu/r)}{dt} \ll r^2 + \nu^2 \quad , \tag{3.4.13}$$

the solution of (3.3.3) in the WKB-approximation is

$$z = \exp\left(-\mathrm{i} \int \nu \, dt/2\right) \left(1 \pm \frac{f}{\sqrt{f^2+1}}\right)^{1/2} \exp\left(\pm \frac{\mathrm{i}}{2} \int \sqrt{r^2 + v^2} \, dt\right) \quad .$$

Substituting this expression into (3.3.2) and imposing the condition $|X| \to 1$ for $\nu^2 \gg r^2$, we obtain

$$X \approx \left[1 + \frac{r^2(t)}{\nu^2(t)}\right]^{-1/2} \quad , \tag{3.4.14}$$

where $X = 1$ for $t = -\infty$, and the population $X(t)$ changes its sign each time the frequency detuning passes through zero. In fact, (3.4.14) coincides with the Landau-Zener solution [3.52] in the adiabatic limit.

3.5 Relaxation in a Stationary Field

In a stationary field it is in principle also possible to observe nonstationary (in particular, oscillatory) regimes of population relaxation (when this field is turned on discontinuously or the equilibrium value of population in the field is disturbed in some way, etc.). In this connection, we shall give a brief outline of the characteristic frequencies and relaxation times describing these motions of population.

The complex frequencies Ω_{i} of the transition process (the imaginary part of which corresponds to partial logarithmic decrements) can be determined from an equation that follows from (3.1.13) when d/dt is replaced by $\mathrm{i}\Omega$:

$$D_\tau(\Omega)[D_T^2(\Omega) + \nu^2] + r^2 \hat{D}_\tau(\Omega) = 0 \quad ; \quad \text{where}$$
$$D_\tau(\Omega) = \mathrm{i}\Omega + \tau^{-1} \quad ; \quad D_T(\Omega) = \mathrm{i}\Omega + T^{-1} .$$

Let us introduce the notation $Z = \mathrm{i}\Omega + \delta_+$, where $\delta_+ = (T^{-1} + \tau^{-1})/2$, and rewrite this equation as

$$Z^2 + r^2 + \nu^2 \frac{Z - \delta_-}{Z - \delta_-} = \delta_-^2 \quad . \tag{3.5.1}$$

In the general case the solution of the cubic equation (3.5.1) can be obtained by applying the Cardan formula. Let us give a brief summary of typical cases.

1) When $T = \tau$, we have $\delta_- = 0$, or, $Z_1 = 0$ and $Z_{2,3} + r + \nu^2 = 0$, i.e.,

$$\Omega_1 = \mathrm{i}T^{-1} \; ;$$
$$\Omega_{2,3} = \mathrm{i}T^{-1} \pm (r^2 + \nu^2)^{1/2} \quad .$$

For $T^{-1} = 0$, these equations coincide with the well-known result in [3.3].

2) For $\nu \equiv 0$ (the exact frequency resonance), $Z_1 = -\delta_-$ and $Z_{2,3}^2 = \delta_-^2 - r^2$, or,

$$\Omega_1 = \mathrm{i}T^{-1} \quad ; \quad \Omega_{2,3} = \mathrm{i}\delta_+ \pm \sqrt{r^2 - \delta_-^2} \quad .$$

It should be noted that here the relaxation of population with the logarithmic decrement $-\mathrm{i}\Omega_1 = T^{-1}$ is caused by the relaxation of polarization that had existed in the system before the time t_0 at which the field has been switched on and is absent under the ordinary initial conditions (3.1.14). For $r^2 > \delta_-^2$, the logarithmic decrement of oscillations is the average value of the characteristic initial decrements T^{-1} and τ^{-1}, and for the oscillation relaxation time T_{eff} we have $2T_{\mathrm{eff}}^{-1} = T^{-1} + \tau^{-1}$. The amplitude $|r| = |\delta_|$ is the critical one for the appearance of oscillations. When $r^2 \ll \delta_-^2$,

$$\Omega \approx \mathrm{i}\left[\delta_+ \pm \left(\delta_- - \frac{r^2}{2\delta_-}\right)\right] \quad ; \quad \text{i.e.} \quad \Omega_2 \approx \mathrm{i}\left(\frac{1}{T} - \frac{r^2}{2\delta_-}\right) \quad ,$$

$$\Omega_3 = \mathrm{i}\left(\frac{1}{\tau} + \frac{r^2}{2\delta_-}\right) \quad ,$$

and it is possible here to speak of the change of relaxation time produced by the field [3.12]:

$$T_{\mathrm{eff}}^{-1} \approx T^{-1} - \frac{r^2}{2\delta_-} \quad ; \quad \tau_{\mathrm{eff}}^{-1} \approx \tau^{-1} + \frac{r^2}{2\delta_-} \quad . \tag{3.5.2}$$

For any T and τ, the quantities T_{eff} and τ_{eff} move "towards" each other as the field increases.

3) For a weak field ($r^2 \ll \nu^2 + \delta_-^2$),

$$\Omega_1 \approx \mathrm{i}\tau^{-1} \quad ; \quad \Omega_{2,3} \approx \mathrm{i}T^{-1} \pm \nu \quad ,$$

and under the initial conditions (3.1.14) there are no oscillations of the population difference with a frequency Ω_{23}.

4) When the oscillation frequency is greater than $|\delta_-|$, i.e., when $r^2 + \nu^2 \gg \delta_-^2$, in the first approximation $Z \approx \mathrm{i}(r^2 + \nu^2)^{1/2}$; whence in the next approximation for the two larger roots of (3.5.1) we obtain

$$Z_{1,2} = \mathrm{i}\Omega_{1,2} + \delta_+ \approx -\delta_- \frac{\nu^2}{r^2 + \nu^2}$$
$$\pm \mathrm{i}\left[r^2 + \nu^2 \delta_-^2 \frac{r^2(r^2 + 4\nu^2)}{(r^2 + \nu^2)^2}\right]^{1/2} \quad . \tag{3.5.3}$$

Therefore,

$$T_{\text{eff}}^{-1} = T^{-1} - r^2\delta_-(r^2+\nu^2)^{-1} \quad .$$

In a weak field ($r^2 \ll \nu^2$)$T_{\text{eff}} \approx T$, in a strong one (i.e. $r^2 \gg \nu^2$) $2T_{\text{eff}}^{-1} \approx T^{-1} + \tau^{-1}$.

The correction to the oscillation frequency is given by

$$-\delta_-^2 r^2(r^2+4\nu^2)(r^2+\nu^2)^{-2} \quad ,$$

thus, in the strong field the frequency is approximately equal to $(r^2+\nu^2-\delta_-^2)^{1/2}$ and in the weak one it is $\approx (r^2+\nu^2)^{1/2}$. The lower frequency logarithmic decrement is determined by

$$Z_3 = \mathrm{i}\Omega_3 + \delta_+ \approx \delta_-(\nu^2+r^2)^{-1}(\nu^2-r^2) \quad , \qquad \text{or}$$

$$\tau_{\text{eff}}^{-1} = -\mathrm{i}\Omega_3 \simeq (r^2T^{-1} + \nu^2\tau^{-1})(r^2+\nu^2)^{-1} \quad . \tag{3.5.4}$$

In the weak field $\tau_{\text{eff}} \approx \tau$, in the strong one $\tau_{\text{eff}} \approx T$.

5) The amplitude $r_{\text{cr}} = (\nu^2+\delta_-^2)^{1/2}$ is, so to speak, a border value between two regions. One of those is the region of the "weak" field ($r^2 < r_{\text{cr}}^2$) where, if there are any oscillations, they are mainly caused by the low intensity beats between the relaxing intrinsic polarization of the system and the field, while another is the region of the "strong" field ($r^2 > r_{\text{cr}}^2$) with the predominant Rabi oscillations. When $r^2 = r_{\text{cr}}^2$, the exact solution of (3.5.1) is

$$Z_1 = 0 \ ; \quad Z_{2,3} = -\tfrac{1}{2}\delta_- \pm \mathrm{i}(r^2+\nu^2-\tfrac{5}{4}\delta_-^2)^{1/2} \quad ,$$

which for $\nu^2 \gg \delta_-^2$ coincide with the approximate formulas of the previous item (4).

3.6 Polarization Dynamics in a Nonstationary Field

In the general case, the polarization $\boldsymbol{P}$ of a system in the dipole approximation is $\boldsymbol{P} = \mathrm{Tr}\{\hat{\varrho}\hat{d}\}$. This makes it possible to obtain a formula [in notations of (3.1.1) and (3.1.6)] for the resonance part of the two-level system polarization (per one atom):

$$\boldsymbol{P} = \boldsymbol{d}(\varrho_{12}+\varrho_{21}) = \tfrac{\mathrm{i}}{2}\eta_0\boldsymbol{d}(Y^*\,\mathrm{e}^{\mathrm{i}\omega_{21}t} - Y\,\mathrm{e}^{-\mathrm{i}\omega_{21}t}) \quad . \tag{3.6.1}$$

Thus, the quantity Y^* determines the complex amplitude of polarization (to an accuracy of $\mathrm{i}\eta_0 \boldsymbol{d}/2$).

Eliminating Y from (3.1.9, 11) and taking into account the definition (3.1.8), we obtain for Y^*

$$Y^* = \mathrm{e}^{\mathrm{i}\varphi}\left\{\frac{1}{r}\hat{D}_\tau(X-1) - \frac{\mathrm{i}r}{\nu}\left[X + \frac{1}{r}\hat{D}_\tau\left(\frac{1}{r}\hat{D}_\tau(X-1)\right)\right]\right\} \quad , \tag{3.6.2}$$

or, on the basis of (3.1.13)

$$Y^* = \mathrm{e}^{\mathrm{i}\varphi}\left\{\frac{1}{r}\hat{D}_\tau(X-1) + \mathrm{i}\hat{D}_T^{-1}\left[\frac{\nu}{r}\hat{D}_\tau(X-1)\right]\right\} \quad . \tag{3.6.3}$$

The last equation makes it possible to completely determine the dynamics of polarization if, for a given $r(t)$ and $\nu(t)$, the dynamics of the population $X(t)$ is found. One can easily see that the polarization amplitude $p = P_{(\mathrm{amp})}/\eta_0 d$ is

$$p \equiv |Y| = \left\{\left[\frac{\hat{D}_\tau(X-1)}{r}\right]^2 + \left[\hat{D}_T^{-1}\left(\frac{\nu}{r}\hat{D}_\tau(X-1)\right)\right]^2\right\}^{1/2} \tag{3.6.4}$$

and the polarization phase-shift φ_p with respect to the phase of the exciting field φ is given by

$$\tan(\varphi_\mathrm{p} - \varphi) = \hat{D}_\tau(X-1)/r\hat{D}_T^{-1}\left[\frac{\nu}{r}\hat{D}_\tau(X-1)\right] \quad . \tag{3.6.5}$$

For the exact resonance ($\nu \equiv 0$) and under the initial conditions (3.1.14)

$$p \equiv |Y| = \frac{1}{r}\hat{D}_\tau(X-1) \ ; \quad \varphi_\mathrm{p} - \varphi = \pi/2 \quad . \tag{3.6.6}$$

If the field parameters vary slowly in comparison with the transverse relaxation time T, we have $\hat{D}_T^{-1} \approx T$, whence

$$Y^* \approx \frac{1}{r}\mathrm{e}^{\mathrm{i}\varphi}(1 + \mathrm{i}\nu T)\hat{D}_\tau(X-1) \quad . \tag{3.6.7}$$

This makes it possible to speak of the "polarization follows the field" approximation having in mind that the polarization phase and frequency follow the phase and frequency of the field.

In the general case there is a simple relationship between the real amplitude of polarization $p \equiv |Y|$ and the population difference X which contains no field parameters and can be obtained directly from the equations (3.1.9, 10) for the density matrix. Multiplying (3.1.10) by Y^* and adding the resulting equation to its complex conjugate we arrive at

$$\frac{d}{dt}(YY^*) + 2T^{-1}(YY^*) = -X(aY + a^*Y^*) \quad ,$$

whence, if one takes $YY^* = p^2$ into account and makes use of (3.1.9),

$$p\hat{D}_T(p) + X\hat{D}_\tau(X-1) = 0 \quad , \tag{3.6.8}$$

follows.

If there is no relaxation or the field pulse is very short ($T_{\text{pulse}} \ll T$, τ), (3.6.8) leads to a simple invariant of motion:

$$p^2 + X^2 = \text{const} \quad . \tag{3.6.9}$$

Under the initial conditions (3.1.17) const $= 1$, therefore $p_{\max} = 1$ (when $X = 0$) or $(P_{\text{amp}})_{\max} = |\eta_0 d|$, which is natural.

For the quasistationary regime ($T_{\text{pulse}} \gg T$, τ)

$$p^2 = \frac{T}{\tau}X(1-X) \quad , \tag{3.6.10}$$

or,

$$p_{\max}^2 = \frac{1}{4} \cdot \frac{T}{\tau} \qquad (\text{for } X = 1/2) \quad ,$$

and in particular,

$$(p_{\max})_{\max} = \frac{1}{\sqrt{2}} \qquad (\text{for } T = 2\tau) \quad .$$

If $T_{\text{pulse}} \gg T$ (the polarization phase follows that of the field),

$$p^2 = -TX\hat{D}_\tau(X-1) \quad . \tag{3.6.11}$$

******Acknowledgements (A.E.K.):* **I discussed virtually all my new results with L.I. Gudzenko, a close friend of mine for about twenty years who deceased in 1978; a physicist with incredibly broad research interests, he contributed to a vast number of fields (for example, he was one of the first to predict the x-ray lasing scheme in 1965). His memory will always be dear to me. There are many other scientists and just friends whose moral support, professional expertise, and friendship were so indispensable and valuable to me in the country which I left in 1979 for the US. Although their names do not appear here, they know I remember them and miss them very much.**

I am greatly indebted to P.L. Kelley and H. Schlossberg for their encouragement and substantial moral support of my recent research done in the US, some results of which are included in this edition. This research was sponsored by the U.S. Air Force Office of Scientific Research, whose support I gratefully acknowledge. I wish also to thank P.W. Smith, W.J. Tomlinson, J.E. Bjorkholm, J.P. Gordon, P. Meystre, and my colleagues of MIT, Purdue University, and Johns Hopkins University for their help, collaboration, and many exciting discussions.

4. Polarization of Resonant Media

We begin this chapter by examining specific features of the nonlinear polarization that manifests itself in the presence of resonant parametric interactions in a medium with freely oriented molecules. For such medium it turns out to be necessary to introduce the parameter of local coherence that defines the efficiency of parametric interaction at a point. The transitions corresponding to the maximal value of this parameter are described.

Further, we analyze here three properties of the response of resonant media: dispersion (Sect. 4.2), polarization (Sect. 4.3), and spatial extension (Sect. 4.4). As a result, we obtain a generalization of the Kramers-Kronig relations for nonlinear resonant media (Sect. 4.2).

It is also demonstrated that the nonlinear susceptibility of resonant media cannot be arbitrary; it should obey a certain differential equation (Sect. 4.3). This equation makes it possible to estimate the response of matter to a field with elliptic polarization. This may be done without conducting any additional experiments, just on the basis of the response of a circularly polarized field. The polarization properties of the nonlinear susceptibility of resonant media discussed in this section have no linear analogs.

In Sect. 4.4 the connection between the capability of a nonlinear resonant medium to amplify a travelling wave and its ability to nonlinearly transformation (reflection) the latter into a counter propagating wave is established and treated in detail. This leads to a new conservation law for counter propagating waves in nonlinear resonant media (Chap. 5), which is not reducible to the well-known laws of conservation of energy and momentum. In particular, it permits the use of closed-form expressions in investigations into the three-dimensional structure of fields in media characterized by arbitrary nonlinearity.

The meaning of the discovered conservation law is illustrated by the example of a field propagating in a plane-parallel layer (Sects. 4.2, 3).

In Sect. 4.4 we present an analysis of the rotation of the polarization plane for counterpropagating waves, which, in fact, forms the foundation of polarization spectroscopy.

4.1 Nonlinear Polarization of Gaseous Media

4.1.1 Probability of Stimulated Multiphoton Transitions and Polarization of Freely Self-Orienting Systems

The energy levels of freely orienting molecules are degenerate with respect to projections m of the total angular momentum $\mathcal{J}$ onto the direction of the external field. Therefore, the expressions for nonlinear polarization obtained in this section, as well as the probabilities of stimulated transitions of a molecule into its excited state (see Sect. 2.1) corresponding to different quantum numbers m_1 and m_2 of the ground and the excited states, should be summed up with the weight $(2\mathcal{J}+1)$. In the classical limit (for large $\mathcal{J}$), this summation represents an averaging with respect to molecular orientations in an isotropic medium. Let us now consider the effect of the free orientation of molecules on resonant parametric interactions (RPI) which have been shown above to arise when more than one resonance condition of the type (1.2) is fulfilled. This is accompanied by the interference of absorption (emission) probability amplitudes that describe each individual incoherent multiphoton process. Because of the averaging, the interference terms in transition probabilities, i.e., the degree of coherence of parametric processes, may change. This will change the parametric susceptibility of matter as calculated in Sect. 2.4 and, therefore, also RPI dynamics.

As an example, the averaging of transition probability and polarization that characterize three- and four-wave RPI's will be performed. At first we shall neglect the saturation of populations of molecular lasing levels.

On the basis of (2.1.4), (1.3.15), (1.3.12), (1.2.16, 21), it is possible to find $\boldsymbol{P}(\omega_j)$ which describes the interaction of four waves with frequencies satisfying pairwise the second-order resonance conditions:

$$\omega_\alpha + \omega_\beta = \omega_\gamma + \omega_\delta = \omega_{21} \quad . \tag{4.1.1}$$

For instance, the single molecule polarization amplitude at the frequency ω_α has the form

$$\begin{aligned} \mathcal{P}_a^{\mathrm{res}}(\omega_\alpha) = & -\mathrm{i}\hbar^{-1} T \langle \eta \chi_{ab}^{21}(\omega_\alpha)\mathcal{E}_b^*(\omega_\beta)[\chi_{cd}^{12}(\omega_\alpha)\mathcal{E}_c(\omega_\alpha)\mathcal{E}_d(\omega_\beta) \\ & + \chi_{cd}^{12}(\omega_\gamma)\mathcal{E}_c(\omega_\gamma)\mathcal{E}_d(\omega_\delta)]\rangle \quad . \end{aligned} \tag{4.1.2}$$

The transition probability W_{12}^{in} is determined by (2.1.19) and (1.2.16, 21):

$$W_{12}^{\mathrm{in}} = 2T\hbar^{-2}|\chi_{ab}^{12}(\omega_\alpha)\mathcal{E}_a(\omega_\alpha)\mathcal{E}_b(\omega_\beta) + \chi_{cd}^{12}(\omega_\gamma)\mathcal{E}_c(\omega_\gamma)\mathcal{E}_d(\omega_\delta)|^2 \quad . \tag{4.1.3}$$

Here the Stark shift has been neglected.

Formulas (1.2.20) and (4.1.1) make it possible to express the second-order polarizability tensors occurring in (4.1.2, 3) as functions of one of the frequencies participating in the corresponding resonance condition (4.1.1), for example,

$$\chi^{12}_{ab}(\omega_\alpha) = \hbar^{-1}\sum_q \frac{(d_{1q})_a(d_{q2})_b}{\omega_{q1}-\omega_\alpha} + \frac{(d_{1q})_b(d_{q2})_a}{\omega_{q2}+\omega_\alpha} \quad . \tag{4.1.4}$$

As follows from (4.1.2, 3), the averaging of the terms in $\mathcal{P}^{\rm res}(\omega_j)$ and $W^{\rm in}_{12}$ that describe the parametric interaction of four waves can be reduced to the averaging of expressions of the type

$$\chi^{21}_{ab}(\omega_\alpha)\chi^{12}_{cd}(\omega_\gamma) = [\chi^*_{ab}(\omega_\alpha)]^{2\mathcal{J}_2m_2}_{1\mathcal{J}_1m_1}[\chi_{cd}(\omega_\gamma)]^{1\mathcal{J}_1m_1}_{2\mathcal{J}_2m_2} \quad , \tag{4.1.5}$$

where the subscripts and superscripts n, $\mathcal{J}$ and m denote the quantum numbers of the initial and final states of a molecule, respectively.

Polarization and probability of induced transitions corresponding to three-wave RPI's and the induced absorption due to a constant field $\mathcal{E}(\omega_0)$ superposed on the medium are also proportional to tensors of the type (3.1.5). Indeed, assuming, for instance, that in (4.1.1) $\omega_\beta = 0$, and calculating the resonant polarization at the frequency ω_α which is equal to the transition frequency ω_{21}, one can demonstrate that it coincides with the polarization (4.1.2)[1] if in the expression for the latter, the quantity $\mathcal{E}_{b,c}(\omega_\beta)$ is replaced by $\mathcal{E}_{b,c}(0)$ [note that $\chi^{12}_{ab}(\omega_\alpha) = \chi^{12}_{ba}(0)$].

Parametric polarization at the resonant field frequency and the respective part of W_{12} in the absence of a static field for dipole transitions are determined by the third rank tensor $d_a\chi_{bc}$ which vanishes when averaged in freely orienting media. In this case, nonlinear polarization with the frequency $\omega_\alpha = \omega_{21}$ in three-wave RPI's may only be due to the quadrupole moment of transition 2–1. Such a process is of no practical interest, and will not be considered.

Thus, the parametric part of polarization responsible for three and four-wave interactions in freely orienting systems is completely determined by the tensor (4.1.5) which has to be summed over m_1 and m_2. For Raman scattering and two-photon absorption this procedure is discussed in detail in [4.1, 2]. By applying it to the tensor (4.1.5) we obtain

$$\begin{aligned}\langle\chi^{21}_{ab}(\omega_\alpha)\chi^{12}_{cd}(\omega_\gamma)\rangle &= G^0(\omega_\alpha,\omega_\gamma)\delta_{a\bar{b}}\delta_{c\bar{d}} \\ &+ \tfrac{1}{10}(\delta_{a\bar{c}}\delta_{b\bar{d}} + \delta_{a\bar{d}}\delta_{b\bar{c}} - \tfrac{2}{3}\delta_{a\bar{b}}\delta_{c\bar{d}})G^s(\omega_\alpha,\omega_\gamma) \\ &+ \tfrac{1}{6}(\delta_{a\bar{c}}\delta_{b\bar{d}} - \delta_{a\bar{d}}\delta_{b\bar{c}})G^a(\omega_\alpha,\omega_\gamma) \quad ,\end{aligned} \tag{4.1.6}$$

[1] This result is valid under the assumption that the constant field strength is not sufficient for removal of the spatial degeneracy of levels.

where the invariants $G^{0,s,a}(\omega_\alpha, \omega_\gamma)$ are given by

$$G^0(\omega_\alpha, \omega_\gamma) = [\chi^{0*}(\omega_\alpha)]^{2\mathcal{J}_1 m'}_{1\mathcal{J}_1 m'} [\chi^0(\omega_\gamma)]^{1\mathcal{J}_1 m'}_{2\mathcal{J}_1 m'} \quad , \tag{4.1.7}$$

$$G^{s,a}(\omega_\alpha, \omega_\gamma) = \sum_{(a,b)=0,\pm 1} [\chi^{s,a*}_{ab}(\omega_\alpha)]^{2\mathcal{J}_2 m'+a+b}_{1\mathcal{J}_1 m'} [\chi^{s,a}_{ab}(\omega_\gamma)]^{1\mathcal{J}_1 m'}_{2\mathcal{J}_2 m'+a+b} \quad . \tag{4.1.8}$$

The quantities $\chi^{0,s,a}_{ab}(\omega_j)$ occurring in (4.1.7,8) are, respectively, the scalar, symmetric and antisymmetric parts of the polarizability tensor written in circular coordinates $a_j = 0, \pm 1$ and m' denotes any possible value of the ground state 1 angular momentum projection (for instance, 0 in a molecule with an integer $\mathcal{J}_1$ and 1/2 in a molecule with a half-integer $\mathcal{J}_1$). Formulas (60.11–13) in [4.2] express the irreducible parts of $\chi^{0,s,a}_{ab}$ through matrix elements of transition dipole moments.

Equations (4.1.7,8) have been derived by taking into account the following selection rules for the parts of the polarizability tensor:

for $\chi^0(\omega_j)$: $\quad \Delta\mathcal{J} = \mathcal{J}_2 - \mathcal{J}_1 = 0, \qquad \Delta m = m_2 - m_1 = 0;$

for χ^a_{ab}: $\quad \Delta\mathcal{J} = 0, \pm 1, \qquad \Delta m = 0, \pm 1$ and transitions

$\mathcal{J}_1 = 0 \rightarrow \mathcal{J}_2 = 0$ are forbidden (since $\mathcal{J}_1 + \mathcal{J}_2 \geq 1$); for χ^s_{ab}: $\Delta\mathcal{J} = 0, \pm 1, \pm 2$, $\Delta m = 0, \pm 1, \pm 2$ and transitions $\mathcal{J}_1 = 0 \rightarrow \mathcal{J}_2 = 0, 1$; $\mathcal{J}_1 = 1/2 \rightarrow \mathcal{J}_2 = 1/2$ are forbidden (since $\mathcal{J}_1 + \mathcal{J}_2 \geq 2$).

Note that when one of the resonances (4.1.1) is degenerate, the antisymmetric part of polarizability corresponding to it becomes equal to zero. This can be easily established with the aid of (4.1.4).

The Wigner-Eckart theorem (see Sect. 107 in [4.3]) makes it possible to express the invariants $G^{0,s,a}(\omega_\alpha, \omega_\gamma)$ through reduced matrix elements of dipole moments of transitions $1 - k$ and $2 - k$ which can be computed if the oscillator forces of these transitions are known. Otherwise, it is necessary to use the invariants (4.1.7,8) directly and to determine these one must study the characteristics of the parametric process for different states of polarization of interacting waves.

4.1.2 The Local Coherence of Parametric Interaction

As follows from (4.1.3), the measure of coherence of a parametric process is the quantity

$$\mathcal{F} = \frac{\langle \chi^{21}_{ab}(\omega_\alpha)\chi^{12}_{cd}(\omega_\gamma)\rangle}{\langle \chi^{21}_{ab}(\omega_\alpha)\chi^{12}_{cd}(\omega_\alpha)\rangle^{1/2} \langle \chi^{21}_{ab}(\omega_\gamma)\chi^{12}_{cd}(\omega_\gamma)\rangle^{1/2}} \quad . \tag{4.1.9}$$

Since $\mathcal{F}$ reflects the relative role of the parametric interaction of waves as compared to that of incoherent processes (single- or two-photon absorption or Raman scattering), one would expect that the effectiveness of RPI depends on $\mathcal{F}$. The quantity $\mathcal{F}$ represents, in a sense, a characteristic similar to the wave detuning of interacting fields. This parameter describes the local coherence of two resonant processes (4.1.1) and the detuning of the wave vectors $k_\alpha + k_\beta - k_\gamma - k_\gamma$, i.e., their spatial coherence.

The value of $\mathcal{F}$ depends both on the type of the resonant transition and on the state of polarization of the interacting fields. Using (4.1.6, 9), the expression for $\mathcal{F}$ can be presented as

$$\mathcal{F} = \frac{\sum\limits_{f=0,s,a} l_{ij}^{(f)} G^{(f)}(\omega_i, \omega_j)}{\left[\sum\limits_{f=0,s,a} l_{ii}^{(f)} G^{(f)}(\omega_i)\right]^{1/2} \left[\sum\limits_{f=0,s,a} l_{jj}^{(f)} G^{(f)}(\omega_j)\right]^{1/2}} . \tag{4.1.10}$$

The dependence on the field polarization is contained in the coefficients $l_{ij}^{(f)}$ which are determined by the corresponding factors near the invariants $G^{0,s,a}$. The invariants $G^{(f)}(\omega_i)$ describing incoherent two-photon processes are for $\omega_\alpha = \omega_\gamma$ obtained from (4.1.7, 8).

For an oriented molecule (4.1.9) yields $|\mathcal{F}| = 1$. It is obvious that in the general case in a freely orienting medium $|\mathcal{F}| < 1$ but there are situations when $|\mathcal{F}| = 1$. Since this equation represents one of the conditions for the maximal effectiveness of parametric interactions, we shall discuss this question in more detail.

Consider the case when $l_{ij}^{(f)} = l_{ii}^{(f)} = l_{jj}^{(f)}$. For $f = 0$ and $f = s$, this occurs, for instance, when the fields have identical states of linear polarization and for $f = a$, when the states of the linear polarization of fields participating in each resonance are orthogonal. Applying selection rules one can easily construct examples of transitions with $|\mathcal{F}| = 1$.

1) If the tensor χ_{ab} only contains a contribution from its scalar part, it follows from (4.1.10) and (4.1.6) that $|\mathcal{F}| = 1$. This is the case, for instance, with transitions $n_1^2S_{1/2} \to n_2^2S_{1/2}$ in alkali metals and with transitions $n_1^1S_0 \to n_2^1S_0$ in the elements of the second group of the Mendeleev table.

2) When $\chi_{ab} = \chi_{ab}^s$ or $\chi_{ab} = \chi_{ab}^a$, two cases are possible: the contribution to $\chi_{ab}^{s,a}$ comes from intermediate levels of q with identical $\mathcal{J}_q$ or from levels with different $\mathcal{J}_q$. In the former case, one can express the $\chi_{ab}^{s,a}$ by the aid of the Wigner-Eckart theorem through reduced matrix elements and factors depending on $\mathcal{J}_1$, $\mathcal{J}_2$ and $\mathcal{J}_q$ to find out that these factors are the same for $\chi_{ab}^a(\omega_\alpha)$ and $\chi_{ab}^a(\omega_\gamma)$, and also for $\chi_{ab}^s(\omega_\alpha)$ and $\chi_{ab}^s(\omega_\gamma)$. This is the situation for quadrupole transitions ($\mathcal{J}_2 = \mathcal{J}_1 + 2$), for example, $n_1^2S_{1/2} \to n_2^2D_{5/2}$ in the elements of the first group of the Mendeleev table and $n_1^1S_0 \to n_2^1D_2$ in the

elements of the second one, for Raman transitions between rotational levels of diatomic molecules. In all cases enumerated above $|\mathcal{F}| = 1$.

When the tensor $\chi_{ab} = \chi_{ab}^{s,a}$ contains a contribution from levels with different $\mathcal{J}_q$, it can easily be shown on the basis of (4.1.7,8) that $|\mathcal{F}|<1$.

3) One can assume with a sufficient degree of accuracy that $|\mathcal{F}| = 1$ for transitions with $\mathcal{J}_2 = \mathcal{J}_1$, $\mathcal{J}_2 = \mathcal{J}_1 + 1$ if the contribution to χ_{ab} comes from only one intermediate level. This happens when there is a quasi-resonance with this level and the contribution of other levels to polarizability can be neglected.

4) If the polarizability tensor contains not less than two irreducible parts then the general case corresponds to $|\mathcal{F}|<1$. An example of this is the vibrational transition $v = 0$, $\mathcal{J} = 1 \rightarrow v = 1$, $\mathcal{J} = 1$ in molecular hydrogen the polarizability of which contains scalar and symmetric parts.

5) If $\chi_{ab} = \chi_{ab}^{s}$, there are such combinations of the states of field polarization that $|\mathcal{F}|<1$ holds even when the condition

$$G^s(\omega_i,\omega_j) = [G^s(\omega_i)]^{1/2}[G^s(\omega_j)]^{1/2}$$

is satisfied. For instance, if the fields participating in one of the resonances are polarized along the x axis, and those participating in the other along the y axis, we have $l_{ij}^{(s)} = 0.5 \cdot [l_{ii}^{(s)} l_{jj}^{(s)}]^{1/2}$ and, therefore, $\mathcal{F} = 0.5$.

Finally, there are such combinations of the field polarization states for which $\mathcal{F} = 0$. For example, if all four fields participating in two Raman resonances ($\omega_1 - \omega_2 = \omega_3 - \omega_4 = \omega_{21}$) are circular polarized, with $E(\omega_1)$, $E(\omega_2)$ and $E(\omega_4)$ being clockwise polarized and $E(\omega_3)$ counterclockwise then $l_{13} = 0$[2] and thus $\mathcal{F} = 0$, see (4.1.10).

The above argument is based on the assumption that no saturation of populations of resonant levels occurs. Note that if populations of sublevels m of states 1 and 2 are the same (i.e., the removal of spatial degeneracy by a constant field or by a high frequency Stark effect is negligible), the population difference is independent of m and is determined by the averaged probability of induced transitions W_{12}^{st}, see (4.1.3) and (2.1.20). Under these conditions all formulas of Sect. 2.1 for η remain valid and it is only necessary to replace the corresponding products of polarizability tensors by their average values. The expression for the nonlinear polarization averaged with respect to orientations can be obtained by means of (4.1.2), where the quantity η has to be taken out of the averaging sign.

[2] When l_{13} is computed, one must take into account that for Raman resonance in (4.1.6), the quantity b should be replaced by $\bar{b}$, and d, by $\bar{d}$.

4.1.3 Influence of the Doppler Effect on the Shape of the Absorption Line for Multiphoton Interactions

In the above, a uniform shape of the absorption line of a resonant transition was assumed. In the case of rarefied gases, however, it is necessary to take into account the Doppler line broadening. Let us discuss briefly the way in which this effect influences the response of the nonlinear medium in the processes considered earlier.

In the equations for the density matrix (1.4.1) the translational motion is catered for the term $\boldsymbol{v}\nabla\sigma$ on the lhs of these equations ($\boldsymbol{v}$ being the particle velocity). Obviously, for plane waves participating in second-order resonances (4.1.1) this is equivalent to the replacement of the inverse half-line-width T in the expressions for stationary polarization of a uniformly moving particle by

$$\Gamma_{\alpha\beta} = \{T^{-1} + \mathrm{i}[\nu + (\boldsymbol{k}_\alpha + \boldsymbol{k}_\beta)\cdot\boldsymbol{v}]\}^{-1} \quad ,$$

where ν is the frequency detuning. When Γ is averaged by the aid of the one-dimensional Maxwell velocity distribution, we obtain

$$\langle\Gamma_{\alpha\beta}\rangle_v = \frac{\sqrt{\pi}}{(k_\alpha + k_\beta)\overline{v}}\Phi\left(-\frac{\nu}{(k_\alpha + k_\beta)\overline{v}} \quad , \quad \frac{T^{-1}}{(k_\alpha + k_\beta)\overline{v}}\right) \quad . \tag{4.1.11}$$

In this equation $\overline{v}$ is the average velocity, and $\Phi(x,y) = U(x,y) + \mathrm{i}V(x,y)$ is the probability integral depending on the complex argument $x + \mathrm{i}y$. If the homogeneous line width is much less than the Doppler one, i.e., if $T^{-1} \ll (k_\alpha + k_\beta)\overline{v}$, we have

$$\Phi(x,0) = \mathrm{e}^{-x^2}\left(1 - 2\sqrt{\pi}\mathrm{i}\int\limits_0^x \mathrm{e}^{t^2}\,dt\right)$$

and the transition line has a Gaussian shape.

The above parametric interactions of the light waves are effective if the detuning of the wave vectors is small:

$$\delta k = k_\alpha + k_\beta - k_\gamma - k_\delta \ll k_{\alpha,\beta,\gamma,\delta} \quad .$$

Therefore, under these conditions for the one-dimensional interaction of waves, it is possible to assume that $\langle\Gamma_{\alpha\beta}\rangle \approx \langle\Gamma_{\gamma\delta}\rangle = \Gamma$.

It is well known that for countertravelling waves the line of degenerate TPA contains a peak of homogeneous width [4.4]. We shall use this circumstance in the analysis of the soft mode of excitation of stimulated two-photon emission based on four-photon RPI's (see Sect. 7.4). Note in this connection that the line shape for countertravelling waves can be obtained in a similar way to that of (4.1.11):

$$\langle \Gamma_{\alpha\beta}\rangle = \frac{\sqrt{\pi}}{(k_\alpha + k_\beta)\overline{v}} U\left(\frac{\nu}{(k_\alpha + k_\beta)\overline{v}} \ , \ \frac{T^{-1}}{(k_\alpha + k_\beta)\overline{v}}\right)$$
$$+ \frac{\sqrt{\pi}}{|k_\alpha - k_\beta|\overline{v}} V\left(\frac{\nu}{|k_\alpha - k_\beta|\overline{v}} \ , \ \frac{T^{-1}}{|k_\alpha - k_\beta|\overline{v}}\right) \ .$$

If the condition $T^{-1} \ll |k_\alpha \pm k_\beta|\overline{v}$ is fulfilled, the line shape includes a stellar "pedestal" with a width $(k_\alpha + k_\beta)\overline{v}$ and a "peak" with a width $|k_\alpha - k_\beta|\overline{v}$. For degenerate TPA, the quantity $\langle \Gamma_{\alpha\beta}\rangle$ is given by

$$\langle \Gamma_{\alpha\alpha}\rangle = \frac{\sqrt{\pi}}{2k\overline{v}} e^{-(\nu/2k\overline{v})^2} + \frac{2T^{-1}}{\nu^2 + T^{-2}} \ . \tag{4.1.12}$$

Formula (4.1.12) coincides with the corresponding equation of [4.4] for the exact resonance $\nu = \omega_\alpha + \omega_\beta - \omega_{21} = 0$.

4.2 Dispersion Properties of the Resonant Susceptibility of Media with Identically Oriented Particles

It is usually quite difficult to conduct theoretical and experimental studies into the dispersion properties of the macroscopic susceptibility of a nonlinear medium. Dispersion relations of the Kramers-Krönig type make it possible to drastically decrease the amount of the accompanying research work, hence they acquire, in addition to their general theoretical value, practical significance.

It will be shown below that the integral over frequencies of the macroscopic susceptibility can in a very simple way be expressed through the microscopic parameters of a single molecule. We shall also establish the connection between the real and imaginary parts of susceptibility, which is similar to the Kramers-Krönig relations. Apparently, the presence of degeneracy makes the study much more complex. Therefore, in this section only the model of a medium with identically oriented particles will be considered.

The macroscopic susceptibility is obtained as the average of susceptibility of a single molecule (2.2.5) over the distribution of its microscopic parameters. The response determined in this manner may essentially differ from χ^{res} and may become a rather complex function of the frequency and the field strength. In a number of cases, however, the use of the causality principle and the resonant nature of the interaction with the field make it possible to obtain general equations for it [4.5].

The averaging of $\chi_{ik}^{\mathrm{res}}(\omega)$ over the distribution of eigenfrequencies can be written as

$$\chi_{ik}(\omega, \vec{\mathcal{E}}) = \frac{i d_i^+ d_k \eta_0}{\hbar} \int_{-\infty}^{\infty} f(\omega_0) \frac{[i(\omega_0 - \omega) + 1/T] d\omega_0}{[(\omega_0 - \omega)^2 + (1/T^2) + (4\tau |d\vec{\mathcal{E}}|^2/\hbar^2 T)]} \ . \tag{4.2.1}$$

Here, $f(\omega_0)$ is the normalized transition frequency distribution function of molecules. Let us calculate the integral of χ_{ik} over all frequencies:

$$\int_{-\infty}^{\infty} \chi_{ik}(\omega, \vec{\mathcal{E}})d\omega = \frac{\mathrm{i}d_i^+ d_k \eta_0}{\hbar} \int_{-\infty}^{\infty} f(\omega_0)d\omega_0$$

$$\times \int_{-\infty}^{\infty} \frac{[\mathrm{i}(\omega_0 - \omega) + 1/T]d\omega}{[(\omega_0 - \omega)^2 + (1/T^2) + 4(\tau|\boldsymbol{d}\vec{\mathcal{E}}|^2/\hbar^2 T)]} \ . \qquad (4.2.2)$$

In the rhs of this equation the order of integration over ω and ω_0 has been interchanged. The internal integral over ω is easily computed and turns out to be independent of ω_0:

$$\int_{-\infty}^{\infty} \frac{\mathrm{i}(\omega_0 - \omega) + 1/T}{(\omega - \omega_0)^2 + (1/T^2) + 4(\tau|\boldsymbol{d}\vec{\mathcal{E}}|^2/\hbar^2 T)} d\omega$$

$$= \pi \frac{1/T}{\sqrt{(1/T^2) + 4(\tau|\boldsymbol{d}\vec{\mathcal{E}}|^2/\hbar^2 T)}} \ ,$$

and the integral over ω_0 of the distribution function is equal to unity. Hence, regardless of the form of $f(\omega_0)$, we obtain

$$\int_{-\infty}^{\infty} \chi_{ik}(\omega)d\omega = \frac{\mathrm{i}d_i^+ d_k \pi \eta_0}{\hbar\sqrt{1 + 4\tau T|\boldsymbol{d}\vec{\mathcal{E}}|^2/\hbar^2}} \ . \qquad (4.2.3)$$

An alternative approach would be to divide χ_{ik} into its Hermitian and anti-Hermitian parts:

$$\int_{-\infty}^{\infty} \chi_{ik}^h(\omega)d\omega = 0 \ , \qquad (4.2.4a)$$

$$\int_{-\infty}^{\infty} \chi_{ik}^a(\omega)d\omega = \frac{\mathrm{i}\pi d_i^* d_k \eta_0}{\hbar\sqrt{1 + 4\tau T|\boldsymbol{d}\vec{\mathcal{E}}|^2/\hbar^2}} \ . \qquad (4.2.4b)$$

On the basis of (4.2.4b), it is possible to express a certain combination of the microscopic parameters of molecules through the integral over frequencies of the anti-Hermitian part of macroscopic polarizability.

It is obvious that for real d the Hermitian part of polarizability χ_{ik}^{h} and its Hermitian part χ_{ik}^{a} coincides with the real and imaginary parts of χ_{ik}, respectively. It should be noted that in the general case $\chi_{ik}^{\mathrm{h}}(\omega)$ is not an odd function of ω. Therefore, the validity of (4.2.4a) is not a priori evident.

Equation (4.2.3) can be written in another form:

$$\int_{-\infty}^{\infty} \chi_{ik}(\omega,\vec{\mathcal{E}})d\omega = \frac{1}{\sqrt{1+4\tau T|d\vec{\mathcal{E}}|^2/\hbar^2}} \int_{-\infty}^{\infty} \chi_{ik}(\omega,0)d\omega \quad . \tag{4.2.5}$$

This establishes the integral connection between the linear and nonlinear polarizabilities of a resonant medium.

The Kramers-Krönig relations do not hold for nonlinear polarizability of a single molecule. This also applies to macroscopic nonlinear polarizability χ_{ik}. Nevertheless, it is possible to find an equation connecting the real and imaginary parts of χ_{ik}, at least for the case under consideration. In order to do this, let us introduce a certain auxiliary function $\tilde{\chi}_{ik}(\omega)$:

$$\tilde{\chi}_{ik} = \frac{\mathrm{i}d_i^* d_k \eta_0}{\hbar} \int_{-\infty}^{\infty} f(\omega_0) \times \frac{d\omega_0}{\left[\mathrm{i}(\omega-\omega_0)+\sqrt{(1/T^2)+4(\tau|d\vec{\mathcal{E}}|^2/\hbar^2 T)}\right]} \quad . \tag{4.2.6}$$

It can be obtained from the expression for linear polarizability in which the parameter $1/T$ is replaced by

$$r = \sqrt{(1/T^2)+4\tau|d\vec{\mathcal{E}}|^2/\hbar^2 T} \quad .$$

One can easily find the relationship between the true function $\chi_{ik}(\omega)$ and the auxiliary one $\overline{\chi}_{ik}(\omega)$:

$$\tilde{\chi}_{ik}^{\mathrm{h}} = \chi_{ik}^{\mathrm{h}} \quad , \tag{4.2.7}$$

$$\tilde{\chi}_{ik}^{\mathrm{a}} = \mathrm{Tr}\,\chi_{ik}^{\mathrm{a}} \quad . \tag{4.2.8}$$

According to its definition, however, auxiliary polarizability $\overline{\chi}_{ik}(\omega)$ has no singularities in the lower[3] half-plane of complex ω, and, therefore, it satisfies the Kramers-Krönig equations

$$\tilde{\chi}_{ik}^{\mathrm{a}}(\omega) = \frac{1}{\pi} \int_{-\infty}^{+\infty} \frac{\tilde{\chi}_{ik}^{\mathrm{h}}(x)}{x-\omega}dx \quad , \tag{4.2.9}$$

[3] This is associated with our choice of the temporal dependence of the type $e^{i\omega t}$, instead of $e^{-i\omega t}$.

$$\tilde{\chi}^{\rm h}_{ik}(\omega) = -\frac{1}{\pi} \int\!\!\!\!\!\!- _{-\infty}^{+\infty} \frac{\tilde{\chi}^{\rm a}_{ik}(x)}{x-\omega} dx \quad . \tag{4.2.10}$$

Substituting (4.2.7, 8) into these formulas, we obtain:

$$\chi^{\rm h}_{ik}(\omega, \vec{\mathcal{E}}) = -\frac{\rm Tr}{\pi} \int\!\!\!\!\!\!- _{-\infty}^{+\infty} \frac{\chi^{\rm a}_{ik}(x, \vec{\mathcal{E}})}{x-\omega} dx \quad , \tag{4.2.11}$$

$$\chi^{\rm a}_{ik}(\omega, \vec{\mathcal{E}}) = \frac{1}{\pi {\rm Tr}} \int\!\!\!\!\!\!- _{-\infty}^{+\infty} \frac{\chi^{\rm h}_{ik}(x, \vec{\mathcal{E}})}{x-\omega} dx \quad . \tag{4.2.12}$$

It should be recalled that the coefficient Tr can be found from (4.2.4b). Thus, the knowledge of $\chi^{\rm a}_{ik}$ as a function of the frequency and the field strength is sufficient for the determination of both the parameters of the medium molecules (4.2.4b) and the Hermite part of the macroscopic polarizability $\chi^{\rm h}_{ik}$ (4.2.11).

We note one consequence of the above formulas, which is important for applications and is associated with the nature of the field dependence of nonlinear polarizability's Hermitian and antihermitian parts.

As follows from (4.2.12), the saturated field for $\chi^{\rm a}_{ik}$ coincides with the minimum of the saturated fields for $\chi^{\rm h}_{ik}$ and for the factor $(1/T^2 + 4\tau|\boldsymbol{d}\cdot\vec{\mathcal{E}}|^2/T\hbar^2)^{-1/2}$. The latter is obviously equal to the saturated field for the polarizability of a single molecule. At the same time, the saturated field for $\chi^{\rm h}_{ik}$ is always greater than or equal to the saturated field for the polarizability of a single molecule. Thus, the characteristic field that essentially changes $\chi^{\rm a}_{ik}$ and coincides with the saturated field for the polarizability of a single molecule,

$$\vec{\mathcal{E}}^{\rm a}_{\rm sat} \sim \hbar/d\sqrt{\tau T} \quad , \tag{4.2.13}$$

is less or even much less than the saturated field for $\chi^{\rm h}_{ik}$.

In [4.6] the above was demonstrated for the special case of a Lorentz shape of the nonuniform broadening of the lasing substance. In this rather frequent situation which corresponds for the fields that are not too strong, i.e., for $\mathcal{E} \ll \mathcal{E}^{\rm h}_{\rm sat}$, to

$$\vec{\mathcal{E}}^{\rm h}_{\rm sat} \gg \hbar/d\sqrt{\tau T} \quad , \tag{4.2.14}$$

the polarizability $\chi^{\rm h}_{ik}$ in (4.2.11, 12) can be replaced by its linear limit

$$\chi^{\rm a}_{ik}(\omega, \vec{\mathcal{E}}) = \frac{1}{\pi} \frac{1}{\sqrt{1 + 4\tau T|d\vec{\mathcal{E}}|^2/\hbar^2}} \int_{-\infty}^{\infty} \frac{\chi^{\rm h}_{ik}(x, 0)}{x-\omega} dx \quad . \tag{4.2.15}$$

Application of the Kramers-Krönig relations for $\chi^{\mathrm{h}}_{ik}(\omega,0)$ and $\chi^{\mathrm{a}}_{ik}(\omega,0)$ makes it possible to obtain an equation that connects the anti-Hermitian part of nonlinear polarizability χ^{a}_{ik} with its linear limit

$$\chi^{\mathrm{a}}_{ik}(\omega,\vec{\mathcal{E}}) = \frac{1}{\sqrt{1+4\tau T|\boldsymbol{d}\vec{\mathcal{E}}|^2/\hbar^2}}\chi^{\mathrm{a}}_{ik}(\omega,0) \quad . \tag{4.2.16}$$

The case when the distribution function of the molecular eigenfrequencies is symmetric with respect to a certain frequency ω_p is of some practical interest. For such $f(\omega_0)$, it follows directly from the definition of $\chi_{ik}(\omega,\vec{\mathcal{E}})$ that

$$\chi_{ik}(\omega-\omega_p,\mathcal{E}) = -\chi^*_{ik}(\omega_p-\omega,\mathcal{E}) \quad . \tag{4.2.17}$$

By separating the Hermitian and anti-Hermitian parts of this equation, we find that χ^{a}_{ik} and χ^{h}_{ik}, respectively, are the symmetric and antisymmetric functions of the frequency detuning $\Omega = \omega - \omega_p$.

Now one can derive for $\chi_{ik}(\omega,\mathcal{E})$ a number of new formulas that are similar to the well-known equations for $\mathcal{E}(\Omega)$ of linear media (see, for instance, [4.7], Sect. 62). As follows from the symmetry relations (4.2.17) and from (4.2.11, 12),

$$\chi^{\mathrm{h}}_{ik}(\Omega,\mathcal{E}) = \frac{2\mathrm{i}\Omega rT}{\pi}\int\limits_0^\infty \frac{\chi^{\mathrm{a}}_{ik}(x,\mathcal{E})}{x^2-\Omega^2}dx \quad , \tag{4.2.18}$$

$$\chi^{\mathrm{a}}_{ik}(\Omega,\mathcal{E}) = \frac{2\mathrm{i}}{\pi rT}\int\limits_0^\infty \frac{x\chi^{\mathrm{h}}_{ik}(x,\mathcal{E})}{x^2-\Omega^2}dx \quad . \tag{4.2.19}$$

It can be easily seen that in the typical case of the bell-shaped dependence of the imaginary part of susceptibility on frequency, as well as in the linear case, the behavior of the curve $\mathrm{Re}\{\chi\}$ does not depend on ω. Indeed, the quantity $\mathrm{Im}\{\chi(x)\}$ makes its main contribution to the integral (4.2.11) in the range $\omega_1<x<\omega_2$. Then it follows from (4.2.11) that for $\omega<\omega_1$, $\mathrm{Re}\{\chi\}>0$. At the same time, the quantity $\mathrm{Re}\{\chi\}$ has the order of magnitude $\mathrm{Re}\{\chi\}\sim r\,\mathrm{Im}\{\chi\}(\omega_2-\omega_1)/\pi\omega$ and increases. For $\omega>\omega_2$ it is negative and decreases. Thus, between ω_1 and ω_2 the curve $\mathrm{Re}\{\chi\}$ necessarily passes through zero and, as in the linear case, has an interval of anomalous dispersion.

4.3 The Equation for the Nonlinear Susceptibility for the Single-Photon Resonance

In the preceding section we have considered certain properties of χ that result from the averaging over the eigenfrequencies of molecules. What can be learned about the properties of susceptibility in the more general case of spread of both molecular eigenfrequencies and their orientations? It will be shown below that under these conditions the behavior of nonlinear susceptibility is governed by a certain differential equation (4.3.8).

Suppose that the averaging over the eigenfrequencies of molecules has been performed.

As was already noted, neglects of the dependence of the molecular eigenfrequencies on the field allows to describe the resonant interaction between field and molecule by the scalar product $\boldsymbol{d}\cdot\vec{\mathcal{E}}$, where the vector $\boldsymbol{d}$ is determined by the molecule and does not depend on the field. Thus, before the averaging over molecular orientations, the susceptibility tensor is a function of $|\boldsymbol{d}\cdot\vec{\mathcal{E}}|^2$

$$\hat{\chi}^0 = \hat{\chi}^0(|\boldsymbol{d}\vec{\mathcal{E}}|^2) \quad . \tag{4.3.1}$$

Let us choose the polar reference frame in such a way that the angle θ is measured from the normal to the plane of the field $\vec{\mathcal{E}}$ polarization ellipse, and the angle φ is measured from one of the main semiaxes of the ellipse in its plane. Then

$$|\boldsymbol{d}\cdot\vec{\mathcal{E}}|^2 = d^2 \sin^2\theta(\cos^2\varphi A^2 + \sin^2\varphi B^2) \quad ,$$

where $\vec{\mathcal{E}} = \boldsymbol{A} + \mathrm{i}\boldsymbol{B}$ and A, d, B are real quantities. The following relations between A, B and $\vec{\mathcal{E}}$

$$|\vec{\mathcal{E}}|^2 = A^2 + B^2 = \xi \quad , \quad \sqrt{\vec{\mathcal{E}}^2\cdot\vec{\mathcal{E}}^{*2}} = A^2 - B^2 = \mu \tag{4.3.2}$$

are valid. In this notation

$$|\boldsymbol{d}\cdot\vec{\mathcal{E}}|^2 = \frac{d^2\sin^2\theta}{2}(\xi + \mu\cos 2\varphi) \quad .$$

Consider some special cases. The field of circular polarization:

$$\mu = 0 \quad , \quad |\boldsymbol{d}\cdot\vec{\mathcal{E}}|^2 = \frac{d^2\sin^2\theta}{2}\xi \quad ,$$

and the field of linear polarization:

$$\xi = \pm\mu \quad , \quad |\boldsymbol{d}\cdot\vec{\mathcal{E}}|^2 = \frac{d^2 \sin^2\theta}{2}\xi(1\pm\cos 2\varphi) \quad .$$

Thus, the macroscopic polarizability $\hat{\chi}$ is equal to

$$\hat{\chi} = \iint f(\theta,\varphi)\hat{\chi}^0\left(\frac{d^2 \sin^2\theta}{2}(\xi + \mu \cos 2\varphi)\right)d\theta\, d\varphi \quad , \tag{4.3.3}$$

where $f(\theta,\varphi)$ is the distribution function for the directions of vector $\boldsymbol{d}$.

If the field polarization ellipse is now rotated around the z axis by an angle ψ, the result of averaging will certainly change. For the quantity χ we obtain:

$$\hat{\chi}(\xi,\mu,\psi) = \iint f(\theta,\varphi-\psi)\hat{\chi}^0\left(\frac{d^2 \sin^2\theta}{2}(\xi + \mu \cos 2\varphi)\right)d\theta\, d\varphi \quad . \tag{4.3.4}$$

Consider the Fourier transform of $\hat{\chi}$ with respect to the variables ξ and ψ. The Fourier transform of $\hat{\chi}(s,\mu,\lambda)$ will satisfy

$$\bar{\hat{\chi}}(s,\mu,\lambda) = \iint \bar{f}(\theta,\lambda)\mathrm{e}^{\mathrm{i}\lambda\varphi}\,\mathrm{e}^{\mathrm{i}s\mu\cos 2\varphi}\bar{\hat{\chi}}^0\left(\frac{2s}{d^2\sin^2\theta}\right)d\varphi\, d\theta \quad . \tag{4.3.5}$$

This expression can be easily averaged over φ:

$$\bar{\hat{\chi}}(s,\mu,\lambda) = \int \bar{f}(\theta,\lambda)J_{\lambda/2}(2\theta\mu)\bar{\hat{\chi}}^0\left(\frac{2s}{d^2\sin^2\theta}\right)d\theta \quad . \tag{4.3.6}$$

Here $J_{\lambda/2}$ is the Bessel function of the order $\lambda/2$ and $\bar{f}$ is the Fourier transform of f. It has become obvious that $\bar{\hat{\chi}}$ satisfies the Bessel equation with respect to the variable μ:

$$-s^2\bar{\hat{\chi}} - \frac{1}{\mu}\frac{d}{d\mu}\left(\mu\frac{d\bar{\hat{\chi}}}{d\mu}\right) + \frac{\lambda^2}{4\mu^2}\bar{\hat{\chi}} = 0 \quad . \tag{4.3.7}$$

Returning to original functions, we obtain the differential equation for $\hat{\chi}$:

$$\frac{\partial^2\hat{\chi}}{\partial\xi^2} - \frac{1}{\mu}\frac{\partial}{\partial\mu}\left(\mu\frac{\partial\hat{\chi}}{\partial\mu}\right) - \frac{1}{4\mu^2}\frac{\partial^2\hat{\chi}}{\partial\psi^2} = 0 \quad . \tag{4.3.8}$$

This equation establishes a connection between the dependence of polarizability on the energy of the field and on the type of field polarization. It also makes it possible to relate the responses for various types of field polarization to each other.

As an example, let us establish a connection between the responses to the circular and elliptic states of polarization of the field $\vec{\mathcal{E}}$ in an isotropic medium. The last term in (4.3.8) is equal to zero because of the isotropy, and

$$\frac{\partial^2\hat{\chi}}{\partial\xi^2} - \frac{1}{\mu}\frac{\partial}{\partial\mu}\left(\mu\frac{\partial\hat{\chi}}{\partial\mu}\right) = 0 \quad . \tag{4.3.9}$$

The solution of the Cauchy problem for this equation with the boundary condition

$$\hat{\chi}|_{\mu=0} = \hat{\chi}_{\text{cir}}(\xi) \quad ,$$

where $\hat{\chi}_{\text{cir}}$ is the response to the circular polarization, results in the wanted polarizability of an arbitrary field expressed through the polarizability of a field with circular polarization:

$$\hat{\chi}(\xi,\mu) = \frac{1}{\pi}\int_{-\mu}^{\mu}\frac{\hat{\chi}_{\text{cir}}(\xi - x)}{\sqrt{\mu^2 - x^2}}dx \quad .$$

In particular, for linear polarization ($\xi = \mu$)

$$\hat{\chi}(\xi,\xi) = \frac{1}{\pi}\int_{-\xi}^{\xi}\frac{\hat{\chi}_{\text{cir}}(\xi - x)}{\sqrt{\xi^2 - x^2}}dx \quad .$$

In conclusion, let us give a brief outline of some other situations when (4.3.8) is applicable. It can be applied:

1) If the dependence on the field can be presented as

$$\hat{\chi}^0 = \sum_\nu \hat{\chi}_\nu^0(|\boldsymbol{d}_\nu \cdot \vec{\mathcal{E}}|^2) \quad ,$$

i.e., if the interaction between the molecule and the field is equivalent to the interaction between several different two-level subsystems,

2) to one-dimensional molecules, when the vectors of the elements of matrix $\boldsymbol{d}$ are parallel and the interaction with the field is, as in the case of a two-level molecule, determined by the quantity $(\boldsymbol{d}_0 \cdot \vec{\mathcal{E}})$ where $\boldsymbol{d}_0$ is the unit vector of the direction of molecular orientation,

3) to the weak field, which is equivalent to (1) since

$$\hat{\chi}^0 = \sum_\nu \chi_\nu^0 |\boldsymbol{d}_\nu \cdot \vec{\mathcal{E}}|^2 \quad .$$

As we have seen, the isotropic quantity $\hat{\chi}$ depends in general both on $\xi = |\vec{\mathcal{E}}|^2$ and on $\mu^2 = |\vec{\mathcal{E}}^2|^2$. Yet, in the literature only the dependence of χ on $|\vec{\mathcal{E}}|^2$ is mentioned. When does this occur? Let us list certain rather obvious possibilities:

the field $\vec{\mathcal{E}}$ is linearly polarized ($\xi^2 = \mu^2$);
the field $\vec{\mathcal{E}}$ is circularly polarized ($\mu = 0$);
all molecules have the same orientation ($\chi = \chi(|\vec{\mathcal{E}}_{||}|^2)$;
in a weak field ($\chi \sim |\vec{\mathcal{E}}|^2$);
some nonresonant media (plasma, $\chi \sim \exp\{-|\vec{\mathcal{E}}|^2\}$).

4.4 The Properties of Spatial Harmonics of Susceptibility

Because of the inhomogeneity of the field, the susceptibility is not just a function of the field but also of the coordinates. The nonzero spatial harmonic of χ appearing under these conditions is related in an obvious manner to the ability of the nonlinear medium to amplify the travelling wave. The second spatial harmonic relates to its ability for nonlinear re-reflection of the travelling wave by the counterpropagating one. It will be established in this section that the properties of the nonlinear medium to amplify and re-reflect waves are not independent since there is a differential relationship between the zeroth and second spatial harmonics of χ [4.8].

We shall limit ourselves to the most widely used case of the quasi-harmonic dependence of the field on at least one coordinate. This means that there exists a coordinate z (not necessarily a Cartesian one) on which field $\boldsymbol{E}$ and polarization $\boldsymbol{P}$ depend via

$$\begin{aligned} \boldsymbol{E} &= (\boldsymbol{C}_1 \,\mathrm{e}^{ikz} + \boldsymbol{C}_2 \,\mathrm{e}^{-ikz})\mathrm{e}^{\mathrm{i}\omega t} + \mathrm{c.c.} \\ \boldsymbol{P} &= (\vec{\mathcal{P}}_1 \,\mathrm{e}^{ikz} + \vec{\mathcal{P}}_2 \,\mathrm{e}^{-ikz})\mathrm{e}^{\mathrm{i}\omega t} + \mathrm{c.c.} \quad , \end{aligned} \tag{4.4.1}$$

where $\vec{\mathcal{C}}_{1,2}$ and $\vec{\mathcal{P}}_{1,2}$ are functions of z that vary slowly in comparison with $\exp(\pm \mathrm{i}kz)$. Of course, this does not mean that there are no rapid variations of these quantities as functions of other spatial coordinates. The complex vector functions $\vec{\mathcal{C}}_{1,2}$ and $\vec{\mathcal{P}}_{1,2}$ can be interpreted as the amplitudes of waves of field and polarization that travel to the right and left. In this approximation, the material equation $\boldsymbol{P} = \hat{\chi}\boldsymbol{E}$ should be replaced by the system

$$\begin{aligned} \vec{\mathcal{P}}_1 &= \langle\hat{\chi}\rangle \boldsymbol{C}_1 + \langle\hat{\chi}\,\mathrm{e}^{-2ikz}\rangle \boldsymbol{C}_2 \ , \\ \vec{\mathcal{P}}_2 &= \langle\hat{\chi}\rangle \boldsymbol{C}_2 + \langle\hat{\chi}\,\mathrm{e}^{2ikz}\rangle \boldsymbol{C}_1 \quad . \end{aligned} \tag{4.4.2}$$

Here the angular brackets mean averaging over z during the interval $2\pi/k$. Although the field equations become much simpler, instead of polarizability $\hat{\chi}$ it is now necessary to know three functions that establish the connection between the amplitudes of the counterpropagating waves of field and polarization. Their dependence on $\boldsymbol{C}_1$ and $\boldsymbol{C}_2$ may be even more complex than the dependence of $\hat{\chi}$ on $\boldsymbol{E}$.

It can be shown, however, that there is a definite relationship between the averaged susceptibilities. We shall demonstrate it for an isotropic medium with its polarizability depending on $|\vec{\mathcal{E}}|^2$. The results to be obtained can easily be generalized to the case of arbitrary resonant media.

4.4.1 Relationships Between Direct and Mixed Susceptibilities

Let us introduce $\mathcal{J}_{1,2} = |C_{1,2}|^2$, $\mathcal{W} = \mathcal{J}_1 + \mathcal{J}_2 = |C_1|^2 + |C_2|^2$ as the average with respect to z over intervals that are much greater than k^{-1}. $u = (C_1 C_2^*)$ is the relative energy of countertravelling waves and $v = |u|^2$.

In terms of this notation the square of the modulus of the field is equal to

$$|\vec{\mathcal{E}}|^2 = \mathcal{W} + u\,\mathrm{e}^{2ikz} + u^*\,\mathrm{e}^{-2ikz} \quad .$$

One can see directly from the definition of the averaging operation that the result of averaging should not depend on the reference point of the averaging interval. Hence it follows that the "direct" susceptibility $\langle\chi\rangle$ and "mixed" susceptibilities $\langle\chi\exp(-2ikz)\rangle/u$, $\langle\chi\exp(2ikz)\rangle/u^*$ are only functions of $\mathcal{W}$ and v since there is such reference point of the averaging variable for which the result does not depend on the phase of u.

For the above functions the result of operations of averaging will not change if the averaging variable z is replaced by $-z$, and at the same time u is replaced by u^* and vice versa. On the other hand, if such an operation is applied to the quantity $\langle\chi\exp(-2ikz)\rangle/u$, it is transformed into $\langle\chi\exp(2ikz)\rangle/u^*$ which is an indication of their equivalence.

Thus, the three coefficients of the vectors C_1, C_2 in the (4.4.2) have been reduced to two functions of $\mathcal{W}$ and v: the direct susceptibility

$$a(\mathcal{W}, v) = \langle\chi\rangle \tag{4.4.3}$$

and the normalized mixed susceptibility

$$b(\mathcal{W}, v) = \langle\chi\,\mathrm{e}^{-2ikz}\rangle/u \quad . \tag{4.4.4}$$

To demonstrate that these two functions are also not independent we shall differentiate the direct susceptibility with respect to the mutual energy. It follows from the definition of v that

$$\frac{\partial a}{\partial u} = u^* \frac{\partial a}{\partial v} \quad .$$

On the other hand,

$$\frac{\partial a}{\partial u} = \frac{\partial\langle\chi\rangle}{\partial u} \quad .$$

The derivative with respect to u can obviously be performed before the averaging

$$\left\langle\frac{\partial\chi}{\partial u}\right\rangle=\left\langle\frac{\partial\chi}{\partial|\vec{\mathcal{E}}|^2}\frac{\partial|\vec{\mathcal{E}}|^2}{\partial u}\right\rangle=\left\langle\frac{\partial\chi}{\partial|\vec{\mathcal{E}}|^2}\mathrm{e}^{2ikz}\right\rangle \quad .$$

Further, replacing the differentiation with respect to $|\vec{\mathcal{E}}|^2$ by that with respect to $\mathcal{W}$ and taking the latter out of the angular brackets, we obtain

$$\frac{\partial\langle\chi\rangle}{\partial u}=\frac{\partial}{\partial\mathcal{W}}\langle\chi\,\mathrm{e}^{2ikz}\rangle \quad .$$

From this formula and the definition of the mixed susceptibility (4.4.4), we find

$$\frac{\partial a}{\partial u}=u^*\frac{\partial b}{\partial\mathcal{W}} \quad .$$

The comparison of this expression with the identity for $\partial a/\partial v$ yields

$$\frac{\partial a}{\partial v}=\frac{\partial b}{\partial\mathcal{W}} \quad . \tag{4.4.5}$$

Differentiating a with respect to $\mathcal{W}$ we obtain in a similar way the second relation

$$\frac{\partial a}{\partial\mathcal{W}}=\frac{\partial(vb)}{\partial v} \quad . \tag{4.4.6}$$

Formulas (4.4.5, 6) represent the desired equations connecting the direct and the mixed susceptibilities.

It should be noted that the derivation of these formulas is essentially based only on the interdependence between χ and $|\vec{\mathcal{E}}|^2$.

4.4.2 The Connection Between Susceptibilities χ, a and b

The relationships (4.4.5, 6) are sufficient for establishing the connection between the susceptibility of the travelling wave (or the orthogonal counterpropagating waves) $\chi(\mathcal{W})$ and the averaged susceptibilities $a(\mathcal{W},v)$ and $b(\mathcal{W},v)$ for arbitrary countertravelling waves. Note first of all that if (4.4.5, 6) are considered as equations with respect to functions a and b, they can easily be supplemented by boundary conditions. In the limit $v\rightarrow 0$ (purely travelling wave or the normal state of polarization of counterpropagating), the susceptibility $\chi(|\vec{\mathcal{E}}|^2)$ is a slowly varying function of z and coincides obviously with the direct susceptibility, i.e.,

$$a(\mathcal{W},0)=\chi(\mathcal{W}) \quad . \tag{4.4.7}$$

In the case of mixed susceptibility, the corresponding condition follows from (4.4.5, 6):

$$b(\mathcal{W},0) = \frac{d\chi(\mathcal{W})}{d\mathcal{W}} \quad . \tag{4.4.8}$$

Let us calculate the Fourier transform of (4.4.5–8) with respect to energy $\mathcal{W}$. Then (4.4.5, 6) turn into ordinary Bessel differential equations with respect to the variable v for $\overline{a}(s,v)$ and $\overline{b}(s,v)$ ($\overline{a}$, $\overline{b}$ being Fourier transforms of a and b with respect to $\mathcal{W}$) with the initial conditions

$$\overline{a}(s,0) = \overline{\chi}(s) \quad , \quad \overline{b}(s,0) = \mathrm{i}s\overline{\chi}(s) \quad ,$$

where $\overline{\chi}(s)$ is the Fourier transform of $\chi(\mathcal{W})$. The solution of the corresponding Bessel equation yields the following formulas for Fourier transforms of the direct and inverse susceptibilities and $\overline{\chi}(s)$[4]:

$$\overline{a}(s,v) = \overline{\chi}(s)J_0(2s\sqrt{v}) \quad , \tag{4.4.9}$$

$$\overline{b}(s,v) = \frac{\mathrm{i}}{\sqrt{v}}\overline{\chi}(s)J_1(2s\sqrt{v}) \quad . \tag{4.4.10}$$

The above relationships can often simplify the methods of obtaining the direct and mixed susceptibilities since they make it possible to replace the averaging by the Fourier transform.

By performing the inverse Fourier transformation in (4.4.9, 10), we obtain a new integral relationship between a, b and χ:

$$a = \frac{1}{\pi}\int\limits_{-2\sqrt{v}}^{2\sqrt{v}} \frac{\chi(\mathcal{W}-\xi)}{\sqrt{4v-\xi^2}}d\xi \quad , \quad b = \frac{1}{\pi v}\int\limits_{-2\sqrt{v}}^{2\sqrt{v}} \frac{\xi\chi(\mathcal{W}-\xi)}{\sqrt{4v-\xi^2}}d\xi \quad ,$$

$$\int\limits_{-2\sqrt{v}}^{2\sqrt{v}} \frac{\xi a(\mathcal{W}-\xi,v)}{\sqrt{4v-\xi^2}}d\xi = v\int\limits_{-2\sqrt{v}}^{2\sqrt{v}} \frac{b(\mathcal{W}-\xi,v)}{\sqrt{4v-\xi^2}}d\xi \quad ,$$

where $\mathcal{W} \geq 2\sqrt{v} \geq \xi$.

[4] It is of interest to note that the function $\overline{\chi}(s)$ satisfies the Kramers-Krönig relations written in terms of the variable s.

4.4.3 Potential Function for Susceptibilities

An important consequence of (4.4.5, 6) is the possibility to introduce potential functions $Y(\mathcal{W},v)$ and $K(\mathcal{W},v)$ from which one can obtain the direct and mixed susceptibilities through simple differentiation:

$$a=\frac{\partial Y}{\partial v} , \quad vb=\frac{\partial Y}{\partial \mathcal{W}} , \tag{4.4.11}$$

$$a=\frac{\partial K}{\partial \mathcal{W}} , \quad b=\frac{\partial K}{\partial v} . \tag{4.4.12}$$

The expressions for K and Y containing $\chi(|\vec{\mathcal{E}}|^2)$ follow directly from these formulas and from the definition of a and b:

$$K=\left\langle \int_0^{|\mathcal{E}|^2} \chi(\xi)d\xi \right\rangle , \quad Y=\frac{v}{u}\left\langle \mathrm{e}^{-2ikz} \int_0^{|\mathcal{E}|^2} \chi(\xi)d\xi \right\rangle . \tag{4.4.13}$$

Thus, in obtaining a and b, two averaging operations can be replaced by one for the derivation of K or Y with the subsequent differentiation with respect to energy and relative energy[5].

The role of potential functions is not exhausted by this result. It will be demonstrated in the next chapter that the quantity $\mathrm{Im}\{Y(\mathcal{W},v)\}$[6] is conserved in the stationary interaction of one-dimensional countertravelling waves.

For orthogonal counterpropagating waves, the relative energy is zero. In this case the form of the potential functions is obvious

$$Y=0 , \quad K=\int_0^{\mathcal{W}} \chi(\xi)d\xi .$$

These formulas, of course, also describe the purely travelling wave.

For nearly orthogonal counterpropagating waves, and for $\mathcal{J}_1 \gg \mathcal{J}_2$, when $\mathcal{W} \gg 4v$, (4.4.9, 10) enable one to easily obtain the following expansions

[5] For the sake of reference, let us give another relationship for Y. The averaging of the second formula (4.4.13) can be performed by integration by parts; then

$$Y=av-\frac{v}{u^2}\langle \chi^{-4ikz}\rangle .$$

In other words, there is an algebraic relationship between Y, the zero spatial harmonic of χ and its normalized fourth spatial harmonic.

[6] It is demonstrated in [4.9] that the potential function K is essential for the determination of the transverse structure of countertravelling waves.

$$
\begin{aligned}
a &\approx \chi(\mathcal{W}) + v\frac{d^2\chi(\mathcal{W})}{d\mathcal{W}^2} + \dots , \\
b &\approx \frac{d\chi(\mathcal{W})}{d\mathcal{W}} + \frac{1}{2}v\frac{d^3\chi(\mathcal{W})}{d\mathcal{W}^3} + \dots \quad .
\end{aligned}
\tag{4.4.14}
$$

The potential susceptibilities are in this case given by

$$
\begin{aligned}
Y &\approx v\chi(\mathcal{W}) + \frac{1}{2}v^2\frac{d^2\chi(\mathcal{W})}{d\mathcal{W}^2} + \dots , \\
K &\approx \int_0^{\mathcal{W}} \chi(\xi)d\xi + v\frac{d\chi(\mathcal{W})}{d\mathcal{W}} + \frac{1}{4}v^2\frac{d^3\chi(\mathcal{W})}{d\mathcal{W}^3} + \dots \quad .
\end{aligned}
\tag{4.4.15}
$$

There is a far-reaching analogy between the properties of analytic functions on one hand, and those of the functions a and b, on the other. For instance, the Euler-d'Alembert analyticity conditions correspond to relations (4.4.5, 6) and the Laplace equation for analytic functions corresponds to an equation of the Tricomy type for functions a and b, and so on. This analogy is based on quite an obvious source: all the properties of both groups of functions follow from a certain form of their arguments ($x + iy$ in the case of analytic functions, $|\vec{\mathcal{E}}|^2$ in our case).

5. Structure of One-Dimensional Waves for the Single-Photon Resonance

The theory developed in Chap. 5 leads to new conservation laws for counter travelling in nonlinear resonant media (Sect. 5.1) which are not reducible to the well-known laws of conservation of energy and momentum. They permit, in particular, the use of closed-form expressions for investigations into the structure of fields in media with an arbitrary degree of nonlinearity.

The meaning of the discovered conservation laws is illustrated by the aid of the example of a field propagating in a plane parallel layer (Sects. 5.2, 3).

Section 5.4 of this chapter presents an analysis of the rotation of the polarization plane for counter travelling. As a matter of fact, it forms the foundation of polarization spectroscopy.

5.1 Conservation Laws for One-Dimensional Waves in Resonant Media

Differential equations for the amplitudes of counter propagating waves are quite complicated. Therefore, it is rather difficult to find their solution by a change of variables. For instance, taking into account the linear attenuation of the field [5.1], i. e., polarizability of the type

$$\chi_{ik} = \frac{d_i d_k}{1 + |\vec{\mathcal{E}}|^2} - g \quad ,$$

requires a change of variables which is quite different from that appropriate for $g = 0$ [5.2, 3]. As will be demonstrated in this section, such an approach to solving the equations for the field amplitudes is not necessary, since the interaction of one-dimensional monochromatic waves implies specific conservation laws for the field which are essentially associated with the resonant nature of the interaction. Thus it is for any type of nonlinearity possible to reduce the determination of the structure of the one-dimensional stationary field to quadratures.

In the one-dimensional case, the system of equations for complex amplitudes of the field $C_{1,2}$ is

$$\frac{dC_1}{dz} = \frac{2\pi i k}{\varepsilon}(aC_1 + ubC_2) \ ,$$
$$\frac{dC_2}{dz} = -\frac{2\pi i k}{\varepsilon}(aC_2 + u^* bC_1) \ . \tag{5.1.1}$$

The lhs of these equations are all that is left from the wave equations after taking into account the one-dimensionality and the stationary nature of the field. The rhs are computed exploiting the known dependence of the polarization on the field amplitudes and the equality of normalized mixed susceptibilities for counter travelling (4.4.3, 4).

This system easily enables one to find the first conservation law for one-dimensional waves. Adding the vector product of the first equation with $\boldsymbol{C}_2$ to the vector product of $\boldsymbol{C}_1$ with the second equation we obtain

$$\frac{d}{dz}[C_1 C_2] = 0 \quad ,$$

whence the first invariant for the field is

$$[C_1 C_2] = \boldsymbol{A} \quad . \tag{5.1.2}$$

Here $\boldsymbol{A}$ is a constant vector. It is directly along the z-axis and its magnitude is, generally speaking, complex.

The conservation law (5.1.2) leads to the following relationship for $\mathcal{J}_1$, $\mathcal{J}_2$ and v :

$$\mathcal{J}_1 \mathcal{J}_2 - v = |\boldsymbol{A}|^2 = \text{const} \quad . \tag{5.1.3}$$

In particular, for orthogonal waves the product of their energies is conserved $\mathcal{J}_1\mathcal{J}_2 = \text{const}$ and for parallel $\boldsymbol{C}_1$ and $\boldsymbol{C}_2$ the quantity $\boldsymbol{A} = 0$ and, therefore, $\mathcal{J}_1\mathcal{J}_2 = v$. For linearly polarized waves (5.1.3) can be written as

$$\mathcal{J}_1 \mathcal{J}_2 \sin^2 \varphi = \text{const} \quad , \tag{5.1.4}$$

where φ is the angle between vectors $\boldsymbol{C}_1$ and $\boldsymbol{C}_2$.

We shall demonstrate that the second conservation law for the field is

$$\text{Im}\{Y(\mathcal{W}, v)\} = B = \text{const} \quad . \tag{5.1.5}$$

In order to do so, let us first derive the equations for the energy density $\mathcal{W}$, relative energy u and the quantity v.

By adding the scalar product of the first equation in (5.1.1) and $\boldsymbol{C}_1^*$ to its complex conjugate, we obtain for $\mathcal{J}_1$

$$\frac{d\mathcal{J}_1}{dz} = \frac{4\pi k}{\varepsilon}(\mathcal{J}_1 \,\text{Im}\{a\} + v\,\text{Im}\{b\}) \quad . \tag{5.1.6}$$

A similar procedure leads to an equation for $\mathcal{J}_2$

$$\frac{d\mathcal{J}_2}{dz} = -\frac{4\pi k}{\varepsilon}(\mathcal{J}_2 \operatorname{Im}\{a\} + v \operatorname{Im}\{b\}) \quad . \tag{5.1.7}$$

Making use of the definition of u and the identity

$$\frac{du}{dz} = C_2^* \frac{dC_1}{dz} + C_1 \frac{dC_2^*}{dz} \quad ,$$

one can easily obtain an equation for u

$$\frac{du}{dz} = \frac{2\pi i k}{\varepsilon} u(a + a^* + \mathcal{J}_2 b + \mathcal{J}_1 b^*) \quad , \tag{5.1.8}$$

from which the relationship for v

$$\frac{dv}{dz} = \frac{4\pi k}{\varepsilon} v(\mathcal{J}_2 - \mathcal{J}_1)\operatorname{Im}\{b\} \quad , \tag{5.1.9}$$

follows. On the other hand, (5.1.6, 7) leads to

$$\frac{d\mathcal{W}}{dz} = \frac{4\pi k}{\varepsilon}(\mathcal{J}_1 - \mathcal{J}_2)\operatorname{Im}\{a\} \quad . \tag{5.1.10}$$

From the above equations for v and $\mathcal{W}$ one obtains

$$v \operatorname{Im}\{b\} d\mathcal{W} + \operatorname{Im}\{a\} dv = 0 \quad .$$

Finally, making use of the properties of the potential function (4.4.11) we arrive at the wanted conservation law (5.1.5).

Thus, in an isotropic weakly nonlinear medium the energy density, mutual orientation and phases of two monochromatic counter propagating waves are changed in such a way that the quantities

$$\boldsymbol{A} = [\boldsymbol{C}_1 \boldsymbol{C}_2] \ , \quad B = v \operatorname{Im}\left\{\frac{1}{u}\left\langle e^{-2ikz} \int_0^{|\vec{\mathcal{E}}|^2} \chi(\xi) d\xi \right\rangle\right\} \quad ,$$

are conserved. Since the invariants $|\boldsymbol{A}|^2$ and B represent two first integrals of the system of equations for $\mathcal{J}_1$, $\mathcal{J}_2$ and v, they make it in principle possible to reduce for any $\chi(|\vec{\mathcal{E}}|^2)$ the problem of determining $\mathcal{J}_{1,2}(z)$ and $v(z)$ to quadratures. The knowledge of the quantities $\mathcal{J}_{1,2}$ and v, in its turn, enables one to determine all the characteristics of the vectors $\boldsymbol{C}_1$ and $\boldsymbol{C}_2$.

Note that we do not know about another method of obtaining first integrals of the system (5.1.1) with arbitrary $\chi(|\vec{\mathcal{E}}|^2)$. Even in the special case when

$\chi = [\alpha/(1+|\vec{\mathcal{E}}|^2)] - g$ and $C_1 \uparrow\uparrow C_2$, the first integral of the corresponding system has been calculated in a primitive way after a rather artificial change of variables [5.1] (see Sect. 5.3).

The following example seems to be even more convincing. Assume that

$$\chi(|\vec{\mathcal{E}}|^2) = \alpha + \beta \exp\{-|\vec{\mathcal{E}}|^2\} \quad ,$$

where α and β are complex numbers. Such nonlinear susceptibility can also be found in nonresonant media, for instance, in a plasma in a high frequency electromagnetic field (see [5.4]). In this case the equations for the intense counter travelling waves have the form

$$\begin{aligned}\frac{d\mathcal{J}_1}{dz} &= 2\,\mathrm{Im}\{\alpha\}\mathcal{J}_1 + 2\,\mathrm{Im}\{\beta\}\sqrt{\mathcal{J}_1}[\sqrt{\mathcal{J}_1}I_0(2\sqrt{\mathcal{J}_1\mathcal{J}_2}) \\ &\quad - \sqrt{\mathcal{J}_2}I_1(2\sqrt{\mathcal{J}_1\mathcal{J}_2})]\exp\{-(\mathcal{J}_1+\mathcal{J}_2)\} \quad , \\ \frac{d\mathcal{J}_2}{dz} &= -2\,\mathrm{Im}\{\alpha\}\mathcal{J}_2 - 2\,\mathrm{Im}\{\beta\}\sqrt{\mathcal{J}_2}[\sqrt{\mathcal{J}_2}I_0(2\sqrt{\mathcal{J}_1\mathcal{J}_2}) \\ &\quad - \sqrt{\mathcal{J}_1}I_1(2\sqrt{\mathcal{J}_1\mathcal{J}_2})]\exp\{-(\mathcal{J}_1+\mathcal{J}_2)\} \quad ,\end{aligned}$$

where $I_{0,1}$ are modified Bessel functions. In our opinion it is for this system even more difficult to find the first integral in a direct manner than in the previous case. But the conservation laws (5.1.3, 5) immediately lead to

$$\sqrt{\mathcal{J}_1\mathcal{J}_2}[\mathrm{Im}\{\alpha\}\sqrt{\mathcal{J}_1\mathcal{J}_2} + \mathrm{Im}\{\beta\}\,\mathrm{e}^{-(\mathcal{J}_1+\mathcal{J}_2)}I_1(2\sqrt{\mathcal{J}_1\mathcal{J}_2})] = \mathrm{const} \quad .$$

One remarkable property of invariant B that has to be mentioned, is its additivity with respect to polarizability. When the effective field can be assumed to be equal to a macroscopic one, this property of B can be formulated in the following way. For a mixture of nonlinear media the first integral B equals the correspondingly weighed "mixture" of the first integrals B of these media. Because of the additivity of B, it is possible to conduct both analysis and "synthesis" of the conservation law by presenting χ in the form of "elementary nonlinearities" with their well-studied invariants. For the relation of $\chi(|\vec{\mathcal{E}}|^2)$ with B see Table 5.1.

For media with polarizability equal to the linear combination of the quantities χ from the left column of the table, the first integral equals to the same linear combination of the corresponding quantities B from its right column. For instance, the first integral obtained for $\chi = (\alpha/|1+\vec{\mathcal{E}}|^2) - g$ is equal to the linear combination of the first and third invariants taken from the table

$$B = \mathrm{Im}\{\alpha\}\left[(1+\mathcal{W}) - \sqrt{(1+\mathcal{W})^2 - 4\mathcal{J}_1\mathcal{J}_2} - g\mathcal{J}_1\mathcal{J}_2\right] = \mathrm{const} \quad .$$

Table 5.1. The most wide-spread expressions for susceptibility χ and the corresponding invariants are given. The constant factor α near χ and B is omitted

χ	$B = \mathrm{Im}\{Y\}$
1	ν
$\lvert\vec{\mathcal{E}}\rvert^{2n}$	$\nu \sum_{k=0}^{[n/2]} \mathcal{W}^{n-2k} \nu^k \frac{n!}{k!(k+1)!(n-2k)!}$
$n = 0, 1, 2, \ldots$	$k = 0, 1, 2, \ldots, [n/2]$
$\frac{1}{1+\lvert\vec{\mathcal{E}}\rvert^2}$	$\mathcal{W} + 1 - \sqrt{(\mathcal{W}+1)^2 - 4\nu}$
$1/\lvert\vec{\mathcal{E}}\rvert^2$	$\mathcal{W} - \sqrt{\mathcal{W}^2 - 4\nu}$
$\exp\{-\lvert\vec{\mathcal{E}}\rvert^2\}$	$\mathrm{e}^{-\mathcal{W}} I_1(2\sqrt{\nu})\sqrt{\nu}$

In this formula we have also taken into account the fact that for identical molecules $v = \mathcal{J}_1 \mathcal{J}_2$.

5.2 Stationary Oscillations in a Layer of Identical Molecules Without Distributed Losses

Consider stationary states of a one-dimensional field in a plane-parallel layer of an active substance (the one-dimensional model of a laser).

According to (2.2.5), the polarizability of a medium consisting of identical molecules is given by

$$\chi_{ik} = -\mathrm{i} d_i d_k \eta_0 \frac{-\mathrm{i}\nu + 1/T}{\hbar(\nu^2 + (1/T^2) + 4\tau(\lvert d\vec{\mathcal{E}}\rvert^2/\hbar^2 T))} \quad . \tag{5.2.1}$$

To make the case simpler, $\boldsymbol{d}$ will be supposed to be real. Assume that the field is parallel to the layer boundary plane. The reference plane will be chosen in such a way that the y axis coincides with the direction of $\boldsymbol{d}$. Assume, further, that $\vec{\mathcal{E}} = \mathcal{E}\boldsymbol{y}^0$ depends only on a coordinate z. Then the equation for the field will have the form

$$\frac{d^2\tilde{\mathcal{E}}}{dz^2} + k^2\tilde{\mathcal{E}} = -\mathrm{i}\frac{\alpha k^2 \tilde{\mathcal{E}}}{1 + \lvert\tilde{\mathcal{E}}\rvert^2} \quad , \tag{5.2.2}$$

where the symbol $\tilde{\mathcal{E}}$ denotes the dimensionless function $\mathcal{E}/\mathcal{E}_{\mathrm{sat}}(\nu)$;

$$\alpha = -\frac{4\pi T\eta_0 \lvert\boldsymbol{d}\rvert^2(1 - \mathrm{i}\Delta)}{\hbar(1 + \Delta^2)} \quad .$$

Since the rhs of this equation is small,

$$\tilde{\mathcal{E}} = \tilde{C}_1(z)\,\mathrm{e}^{ikz} + \tilde{C}_2(z)\,\mathrm{e}^{-ikz} \quad ,$$

where $\tilde{C}_1$ and $\tilde{C}_2$ are slowly varying dimensionless complex functions that determine amplitude and phase of waves travelling, respectively, in positive and negative directions. In this approximation the quantities $\tilde{C}_1$ and $\tilde{C}_2$ are given by (5.1.1). For the particular polarizability (5.2.1) they can be written as

$$\frac{d\tilde{C}_{1,2}}{dz} = \pm\frac{k\alpha}{4}\tilde{C}_{1,2}\frac{[1+2(\mathcal{J}_1+\mathcal{J}_2)+(\mathcal{J}_1-\mathcal{J}_2)^2]^{1/2}\pm\mathcal{J}_2\mp\mathcal{J}_1-1}{\mathcal{J}_{1,2}[1+2(\mathcal{J}_1+\mathcal{J}_2)+(\mathcal{J}_1-\mathcal{J}_2)^2]^{1/2}} \quad . \tag{5.2.3}$$

Hence one can easily obtain equations for $\mathcal{J}_{1,2} = |\tilde{C}_{1,2}|^2$:

$$\frac{d\mathcal{J}_{1,2}}{dz} = \pm\frac{k\,\mathrm{Re}\{\alpha\}}{2}\frac{[1+2(\mathcal{J}_1+\mathcal{J}_2)+(\mathcal{J}_1-\mathcal{J}_2)^2]^{1/2}\pm\mathcal{J}_2\mp\mathcal{J}_1-1}{[1+2(\mathcal{J}_1+\mathcal{J}_2)+(\mathcal{J}_1-\mathcal{J}_2)^2]^{1/2}} \quad . \tag{5.2.4}$$

It follows from (5.2.3, 4) that the phases $\varphi_{1,2}$ are related to $\mathcal{J}_{1,2}$ through simple equations:

$$\frac{d\varphi_{1,2}}{dz} = \frac{\mathrm{Im}\{\alpha\}}{2\,\mathrm{Re}\{\alpha\}}\frac{d}{dz}\ln\mathcal{J}_{1,2} \quad . \tag{5.2.5}$$

By substituting $\mathrm{Im}\{\alpha\}/\mathrm{Re}\{\alpha\} = -\Delta$ and making use of (5.2.5) we obtain

$$\varphi_{1,2} = -\frac{\Delta}{2}\ln\mathcal{J}_{1,2} + D_{1,2} \quad , \tag{5.2.6}$$

where $D_{1,2}$ are arbitrary constants.

Consider first the variation of the amplitudes. According to the general theory, the first integral of the system (5.2.4) is equal to [see Table 5.1]

$$1+\mathcal{J}_1+\mathcal{J}_2-\sqrt{(1+\mathcal{J}_1+\mathcal{J}_2)^2-4\mathcal{J}_1\mathcal{J}_2} = 2B \quad . \tag{5.2.7}$$

On the basis of this invariant we can obtain the solution of the system (5.2.4):

$$\begin{aligned}&\mathcal{J}_{1,2}-B-\frac{B}{\mathcal{J}_{1,2}-B}+\ln(\mathcal{J}_{1,2}-B)+\frac{\ln B}{2}\\ &\quad = \pm k\,\mathrm{Re}\{\alpha\}z \quad ,\end{aligned} \tag{5.2.8}$$

where $\mathcal{J}_{1,2}>B$. This formula implies that the reference point is placed at a location with respect to which the whole picture is symmetric. At this point

$\mathcal{J}_1 = \mathcal{J}_2 = B + \sqrt{B}$, and the sum $\mathcal{J}_1 + \mathcal{J}_2$, which is proportional to the energy density of the field averaged over the period $2\pi/k$, is minimal.

The constant B is determined from the boundary conditions. Assume that the inequalities $-L_1<z<L_2$ define an active layer in the generation regime, i.e., outside of the layer there are only diverging waves. Then, if one takes into account that the amplitudes vary slowly, one can easily show that $\mathcal{J}_1/\mathcal{J}_2 = r_1$ holds for $z = -L_1$ and $\mathcal{J}_2/\mathcal{J}_1 = r_2$ for $z = L_2$. The $r_{1,2}$ are the power reflection coefficients of the layer boundaries under linear conditions. When these formulas are substituted into (5.2.8), the quantity B can be expressed through r_1, r_2 and the layer thickness $L = L_1 + L_2$.

Equation (5.2.8) determines $\mathcal{J}_{1,2}$ only when $B>0$. On the basis of this it can be demonstrated that stationary oscillations are possible if

$$L>L_{\mathrm{cr}} = \frac{\ln(r_1 r_2)^{-1/2}}{k\,\mathrm{Re}\{\alpha\}} \quad . \tag{5.2.9}$$

Since for $L \to L_{\mathrm{cr}}$ we have $\mathcal{J}_{1,2} \to 0$, this expression naturally coincides with the self-excitation condition obtained from the linear theory [5.5].

According to (5.2.6), the spatial variation of the wave phase depends on $\mathcal{J}_{1,2}$. If there are no losses in the medium outside the layer, the boundary conditions for $\varphi_{1,2}$ are

$$(\varphi_1 - \varphi_2 - 2kz)|_{z=-L_1,L_2} = \pi p_{1,2} \quad , \tag{5.2.10}$$

where $p_{1,2} = 0, \pm 2, \pm 4, \dots$, if the dielectric constant of the external medium $\varepsilon_0>\varepsilon$, and $p_{1,2} = \pm 1, \pm 3, \dots$, if $\varepsilon_0<\varepsilon$. If (5.2.10) is taken into account, the detuning $\Delta\omega$ in (5.2.6) corresponding to a given p, can easily be found. The generation frequencies are then given by

$$\omega_p = \omega_{21} \frac{\pi p + \omega_{21} \ln(r_1 r_2)^{-1/2}}{2k_0 L + \omega_{21} T \ln(r_1 r_2)^{-1/2}} \;, \quad k_0 = \frac{\omega_{21}\sqrt{\varepsilon}}{c} \;,$$
$$p = p_1 - p_2 \quad , \tag{5.2.11}$$

and the solution is completely determined by (5.2.8) and (5.2.6).

Formulas (5.2.6, 8, 11) make it possible to find the spectrum of steady-state nonlinear oscillations (modes) of a laser (a linear superposition of such modes is certainly not a solution). As follows from (5.2.11), because of the presence of an active substance the mode spectrum becomes denser while it still remains equidistant. The values of ω_p depend neither on η_0 nor on τ[1]. If the quantity $2k_0L$ is a multiple of π, one of the frequencies ω_p (corresponding to $2k_0L = \pi p$)

[1] Note that the frequencies (5.2.11) do not coincide with the self-excitation frequencies of the corresponding laser modes (except for the threshold modes for which $L = L_{\mathrm{cr}}$ and $\mathcal{J}_{1,2} = 0$).

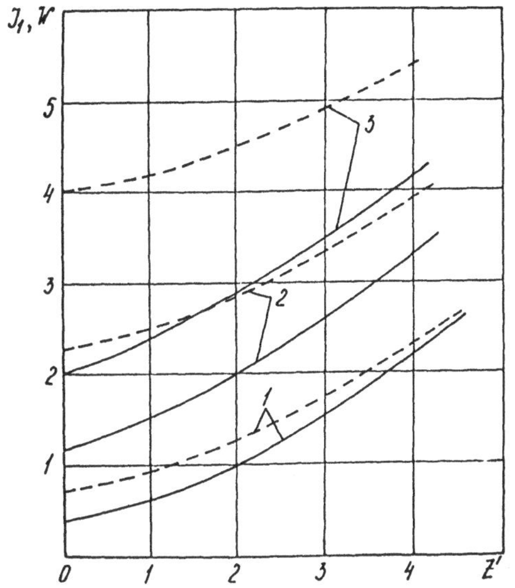

Fig. 5.1. The distribution inside the layer of the dimensionless energy density $\mathcal{W}$ (– – –) and the dimensionless energy density of a wave propagating to the right $\mathcal{J}_1$ (—); $z' = k\,\mathrm{Re}\{\alpha z\}$ is the dimensionless coordinate. Curves *1, 2, 3* correspond to parameter values $B = 0.1$, 0.5, 1, respectively. For $L' = 1$ the curves *1, 2, 3* correspond to the reflection coefficient $r = 0.4$, 0.62 and 0.77, respectively

is equal to ω_{21}. One must bear in mind that the number of stationary modes in a layer is finite. Indeed, as $\Delta\omega$ increases, the quantity $\mathrm{Re}\{\alpha\}$ diminishes until finally, condition (5.2.9) is no longer valid.

Not all modes for which conditions (5.2.9) are satisfied are characterized by large oscillation amplitudes. The number of predominant modes depends on the length of the interval $\omega_{p+1} - \omega_p$ relative to the spectral line width T^{-1}. To obtain an estimate, let us choose the values of parameters that are typical for solids: $r = 1$, $k_0 L = 10^5$ and $\omega_{21} T = 3 \times 10^3$; then $(\omega_{p+1} - \omega_p)T \sim 0.1$. In this case there is only one nonlinear mode.

Now consider the question of the spatial structure of the field inside the generating layer. For each mode the quantity $\mathcal{J}_1$ increases, and $\mathcal{J}_2$ diminishes with the increase of z. Note that the minimum of the average energy density which is proportional to $\mathcal{W} = \mathcal{J}_1 + \mathcal{J}_2$ is closer to the less transparent boundary of the layer, while the maximal value of $\mathcal{W}$ is located at the more transparent boundary. Figure 5.1 depicts the dimensionless variables $\mathcal{J}_1$ and $\mathcal{W}$ as functions of the dimensionless coordinate $z' = k\,\mathrm{Re}\{\alpha\}z$ and qualitatively illustrates the nature of the energy distribution inside the layer. It is assumed that the left boundary of the layer is ideally reflecting (it contains the minimum of $\mathcal{W}$), i.e., that $r_1 = 1$. The constant B determined by the quantity $r_2 = r$ and the dimensionless layer thickness $L' = k\,\mathrm{Re}\{\alpha\}L$ plays the role of a parameter. For a fixed B the plotted curves are the same for all modes. In a given layer, however, the mode amplitudes have different distributions since they correspond to different values of L' that specify B.

The laser efficiency is characterized by the power emitted. For a layer with $r_1 = 1$ the density of the energy flux from the right boundary is proportional to $S = \mathcal{J}_1(1 - r)$. For this case the elimination of B from (5.2.8) yields for

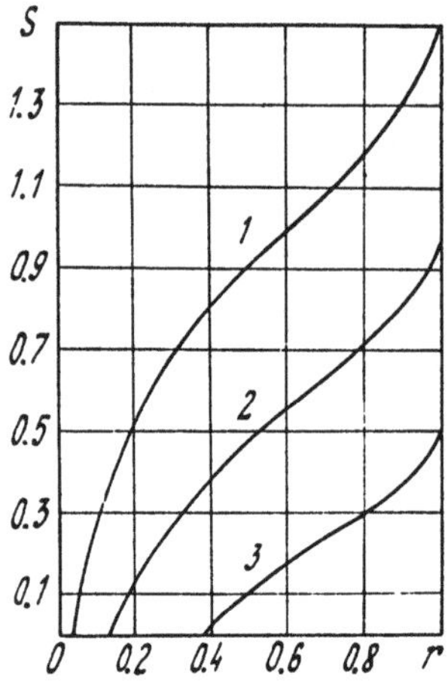

Fig. 5.2. The emitted power S as a function of r, the layer boundary reflection coefficient. Curves *1, 2, 3* correspond to values $L' = 1.5$, 1 and 0.5

$z = L_2$ a relationship between r and S

$$r = \frac{S + 1 - \exp[-2(L' - S)]}{S - 1 + \exp[2(L' - S)]} \quad . \tag{5.2.12}$$

It follows from this formula that the function $S(r)$ increases monotonically from zero at $r = r_{\rm cr}$ corresponding to (5.2.9), to the value L' at $r = 1$. The behavior of (5.2.12) for different L' is shown on Fig. 5.2. The monotonic increase of the output power when $r \to 1$ can be explained by the neglect of all losses except the one due to radiation through the boundary.

A number of authors who dealt with the question of stationary states of a plane layer of an active substance (see, for instance, [5.6, 7]) used instead of Maxwell equations for the field phenomenological ones for the values of the radiation flux and energy density averaged over space and time. The corresponding expressions can be obtained in a formal way from (5.2.2) if in the averaging process the oscillating terms [proportional to $\exp(\pm 2ikz)$] are discarded. One can easily see that in the nonlinear case of identical molecules such solutions are only correct for a travelling wave in an infinite medium, provided one of the quantities $C_{1,2}$ is equal to zero. (Here spatial dispersion is not taken into account. When there is strong spatial dispersion, the above solutions may be valid [5.8]).

5.3 Stationary Oscillations in a Layer of Identical Molecules in the Presence of Distributed Losses

In most cases, it is unrealistic to assume that there are no volume losses inside a layer. The losses that are caused, for instance, by the conductivity of the medium or light scattering by random inhomogeneities may turn out to be essential or even predominant as compared to the losses due to radiation through the layer boundaries. This occurs, because the dimensions of the system are much larger than the wavelength of light. It will be shown below that the inclusion of such dissipation leads to qualitatively new types of solutions.

One can allow for volume losses by introducing an additional term into (5.2.2), which now assumes the form

$$\frac{d^2\mathcal{E}}{dz^2} + k^2\mathcal{E} = \mathrm{i}k^2\alpha\frac{\mathcal{E}}{1+|\mathcal{E}|^2} + \mathrm{i}k^2 g\mathcal{E} \quad . \tag{5.3.1}$$

Generally speaking, the small attenuation coefficient g may depend on ω.

Applying the same procedure as used with (5.2.2) we arrive at the equations for dimensionless intensities of travelling waves

$$\begin{aligned}\frac{d\mathcal{J}_{1,2}}{dz} &= \pm\frac{k\,\mathrm{Re}\{\alpha\}}{2}\\ &\times\frac{\sqrt{1+2(\mathcal{J}_1+\mathcal{J}_2)+(\mathcal{J}_1-\mathcal{J}_2)^2}\pm\mathcal{J}_2\mp\mathcal{J}_1-1}{\sqrt{1+2(\mathcal{J}_1+\mathcal{J}_2)+(\mathcal{J}_1-\mathcal{J}_2)^2}} - kg\mathcal{J}_{1,2} \quad .\end{aligned} \tag{5.3.2}$$

According to the general theory (Sect. 5.1), the first integral of this system is

$$2B = \mathrm{Re}\{\alpha\}[(1+\mathcal{W}) - \sqrt{(1+\mathcal{W})^2 - 4\mathcal{J}_1\mathcal{J}_2}] - g\mathcal{J}_1\mathcal{J}_2 = \mathrm{const} \quad . \tag{5.3.3}$$

It would be of methodological interest to obtain it in a simple way. To this end, let us perform a change of variables

$$\begin{aligned} u &= \mathcal{J}_1 + \mathcal{J}_2 + 1 - \frac{\mathrm{Re}\{\alpha\}}{g} \quad ,\\ v &= \sqrt{1+2(\mathcal{J}_1+\mathcal{J}_2)-(\mathcal{J}_1-\mathcal{J}_2)^2} - \frac{\mathrm{Re}\{\alpha\}}{g} \quad .\end{aligned} \tag{5.3.4}$$

The first integral of (5.3.2) can be written as

$$v^2 - u^2 = 4\frac{\mathrm{Re}\{\alpha\}}{g}B \quad , \tag{5.3.5}$$

or, in the former notation,

$$\begin{aligned}&(\mathcal{J}_1 - B)(\mathcal{J}_2 - B)\\ &\quad B + \frac{g}{\mathrm{Re}\{\alpha\}}\mathcal{J}_1\mathcal{J}_2\left(1+\mathcal{J}_1+\mathcal{J}_2-2B-\frac{g}{\mathrm{Re}\{\alpha\}}\mathcal{J}_1\mathcal{J}_2\right) \quad ,\end{aligned} \tag{5.3.6}$$

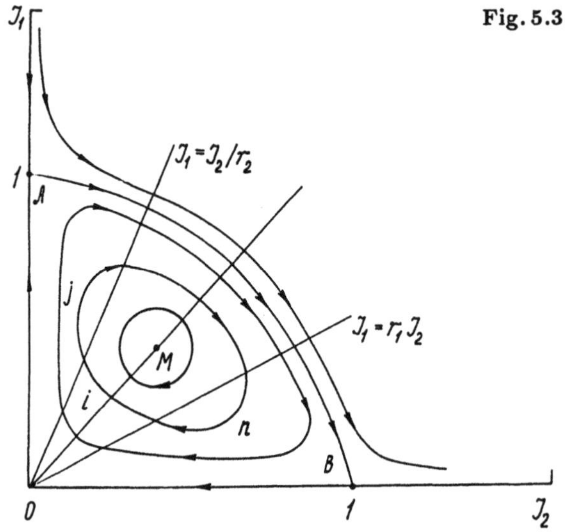

Fig. 5.3. The phase plane of system (5.3.2)

where B is an arbitrary constant. This first integral is identical to integral (5.3.3). The second integral of (5.3.2) can also be easily obtained

$$z - z_0 = -g \int \frac{\left(v + \frac{\mathrm{Re}\{\alpha\}}{g}\right)\left(v^2 - 4\frac{\mathrm{Re}\{\alpha\}}{g}B\right)^{-1/2} dv}{\left[\left(v + \frac{\mathrm{Re}\{\alpha\}}{g}\right)^2 - 2\sqrt{v^2 - 4\frac{\mathrm{Re}\{\alpha\}}{g}B} - 2\frac{\mathrm{Re}\{\alpha\}}{g} + 1\right]^{1/2}}, \tag{5.3.7}$$

where z_0 and B are constants. The dependence of $\mathcal{J}_{1,2}$ on z can be determined from (5.3.7) by calculating elliptic integrals.

Omitting relatively cumbersome formulas, let us study qualitatively the solution in the phase plane, i. e., in the $\mathcal{J}_1\mathcal{J}_2$ plane. First of all, setting the rhs of (5.3.2) equal to zero, we shall determine the locations of equilibrium points (singularities). If $\mathrm{Re}\{\alpha\}/g<1$, there is only one such singularity ($\mathcal{J}_1 = 0$, $\mathcal{J}_2 = 0$) of the saddle point type. In this case the solution is qualitatively similar to that at $\alpha = 0$, i. e., to damped waves in a linear medium.

The most interesting case is when $\mathrm{Re}\{\alpha\}/g>1$. Figure 5.3 depicts the structure of trajectories in the phase plane under these conditions, corresponding to (5.3.2). The arrows point into the directions of increasing z. System (5.3.2) has four singularities:

$$O(\mathcal{J}_1 = 0\ ,\quad \mathcal{J}_2 = 0\ ,\quad \text{saddle point})\ ;$$

$$A\left(\mathcal{J}_1 = \frac{\mathrm{Re}\{\alpha\}}{g} - 1\ ,\quad \mathcal{J}_2 = 0\ ,\quad \text{saddle point}\right)\ ;$$

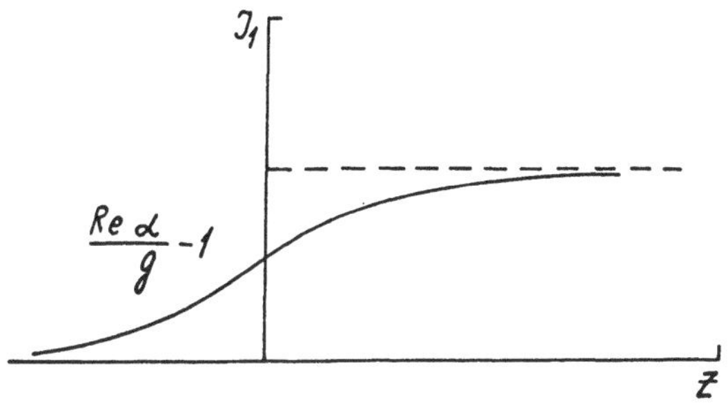

Fig. 5.4. The qualitative dependence of the field intensity on the coordinate for half-space

$$B\left(\mathcal{J}_1=0\ ,\quad \mathcal{J}_2=\frac{\mathrm{Re}\{\alpha\}}{g}-1\ ;\quad \text{saddle point}\right)\ ;$$

$$M\left(\mathcal{J}_1=\mathcal{J}_2=\frac{1}{2}\left[\frac{\mathrm{Re}\{\alpha\}}{g}-\frac{1}{4}-\sqrt{\frac{\mathrm{Re}\{\alpha\}}{g}+\frac{1}{16}}\right]\ ;\quad \text{center}\right)\ .$$

The singular points correspond to solutions with no spatial variation of the amplitude A and B to travelling waves and M to a standing wave. Note that when $g=0$ there are no solutions for an infinite medium that are finite over the whole interval.

The waves of finite amplitude that are possible in an infinite layer are described, in addition to the singular points, by separatrices OA and OB (see Fig. 5.3). These separatrices correspond to travelling waves whose amplitudes depend on z. For the separatrix OA Fig. 5.4 depicts the amplitude as a function of z. Such waves are generated, when there is a wave incident on the boundary of a semiinfinite active layer. The curve AB (which corresponds to the interaction between two waves converging from infinity) divides the phase plane into two regions with qualitatively different types of solutions. The region outside OAB describes strong fields the nonlinear effects being small due to emission. The behavior of the solution is mainly determined by the influence of dissipation. In this case the solutions are qualitatively similar to those existing over the whole plane when $\mathrm{Re}\{\alpha\}<g$.

Let us consider now trajectories that are inside the region OAB. The main specific feature due to volume losses is the appearance of solutions that are periodic in space. Both the amplitudes of the travelling waves $\mathcal{J}_1$ and $\mathcal{J}_2$ and the average energy density of the field $\mathcal{W}$ proportional to $\mathcal{J}_1+\mathcal{J}_2$, exhibit periodic variations[2]. The average energy flux periodically changes its direction depending on which of the quantities $\mathcal{J}_{1,2}$ is predominant.

Generally speaking, in order to determine the spatial period of modulation one has to calculate integral (5.3.7). We shall limit ourselves here to find the period that corresponds to small deviations of amplitudes from their equilib-

[2] The nature of variation of $\mathcal{W}$ can be easily determined from Fig. 5.3 if one allows for the fact that this quantity is constant on the straight lines $\mathcal{J}_1+\mathcal{J}_2=\text{const}$.

rium value $\mathcal{J}_M$. Linearizing (5.3.2) near $\mathcal{J}_1 = \mathcal{J}_2 = \mathcal{J}_M$ and presenting the solution in the form $\mathcal{J}_{1,2} = \mathcal{J}_M + \mathcal{J}_{1,2}\exp(\mathrm{i}\kappa kz)$ we obtain

$$\kappa^2 = g^2\left(1 - \frac{2\mathcal{J}_M}{\sqrt{1+4\mathcal{J}_M}-1}\right)\left[\frac{2\mathcal{J}_M}{(1+4\mathcal{J}_M)(\sqrt{1+4\mathcal{J}_M}-1)} - 1\right] \quad . \tag{5.3.8}$$

This value of κ is maximal. As the amplitudes deviate more and more from the constant value, the period Λ of the process increases (for trajectories that are close to the boundaries of the region $\Lambda \to \infty$).

Note that when $g \to \alpha$ all singular points approach the origin of coordinates; in particular,

$$\mathcal{J}_M \approx \frac{1}{3}\left(\frac{\mathrm{Re}\{\alpha\}}{g} - 1\right) \quad \text{and} \quad \kappa = \left(\frac{\mathrm{Re}\{\alpha\}}{g} - 1\right)\frac{g}{\sqrt{3}} \quad .$$

And if $g \to 0$, the singular points A, B, M move away to infinity. For $g \ll \mathrm{Re}\{\alpha\}$

$$\mathcal{J}_M \approx \frac{1}{2}\left(\frac{\mathrm{Re}\{\alpha\}}{g} - \sqrt{\frac{\mathrm{Re}\{\alpha\}}{2g}}\right) \ , \qquad \kappa \approx g\mathcal{J}_M^{1/4} \approx 2^{1/4}g^{3/4}\alpha^{1/4} \quad .$$

Therefore, both for $g \to \mathrm{Re}\{\alpha\}$ and $g \to 0$ the process is no longer periodic; the function $\kappa(g)$ has a maximum in the interval $0 < g < \mathrm{Re}\{\alpha\}$.

As follows from (5.3.6), when $g = 0$ the phase trajectories represent a family of hyperbolas $(\mathcal{J}_1 - B)(\mathcal{J}_2 - B) = B$ that describe the lossless medium considered above.

In light of the above results we study now the generation of an active layer of finite thickness which is limited by the planes $z = -L_1$ and $z = L_2$. Suppose that outside of the layer there are only outgoing waves. Then the boundary conditions for the amplitude moduli, as before (see Sect. 5.2), can be written as

$$\mathcal{J}_1/\mathcal{J}_2 = r_1 \quad \text{for} \quad z = -L_1 \ ,$$
$$\mathcal{J}_2/\mathcal{J}_1 = r_2 \quad \text{for} \quad z = L_2 \ ,$$

where $r_{1,2}$ are the power reflection coefficients of the layer boundaries.

Let us draw the boundary lines $\mathcal{J}_1 = r_1\mathcal{J}_2$ and $\mathcal{J}_2 = r_2\mathcal{J}_1$ in the phase plane (Fig. 5.3). The phase trajectories corresponding to a given layer should have their points of origin on the first of these straight lines and their end points on the second. Thereby, all trajectories that are outside of the so-called generating region OAB are excluded. If, in addition to r_1 and r_2 the layer thickness $L = L_1 + L_2$ is specified, the solution is determined by those curves for which the interval of variation of z between the boundary lines is equal to

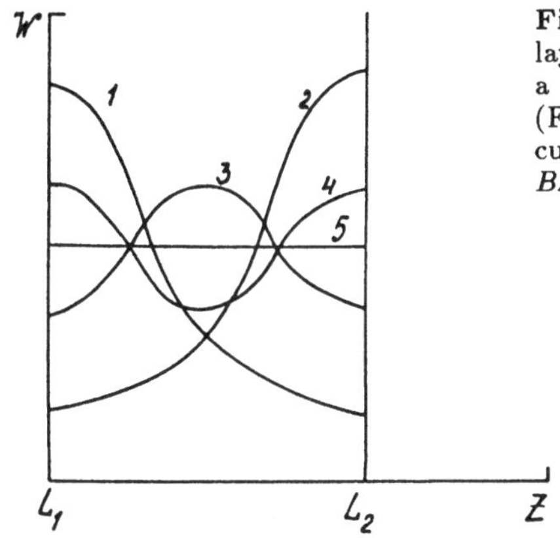

Fig. 5.5. Distribution of the field energy inside the layer ($g \neq 0$). Curves *1* and *2* correspond to half a turn along the trajectory in the phase plane (Fig. 5.3), curves *3* and *4* to a complete turn and curve *5* to a singular point M. $B_1 = B_2 < B_3 = B_4 < B_5$

L (the analytical aspect of the choice of these trajectories is the determination of constants z_0 and B in solution (5.3.7) on the basis of known r_1, r_2 and L).

When $g = 0$ holds for a given layer then there exists only the solution studied in Sect. 5.2. Let us first analyze the peculiarities caused by the presence of volume losses. For that purpose we use the example of a layer with opaque walls and assume $r_1 = r_2 = 1$ (radiation losses are here of no fundamental importance). In this case the allowed trajectories begin and end on the same straight line, the bisectrix that passes through the singular point M. It is obvious that this point itself is always a solution. If the layer thickness L, however, is greater than π/κ [κ being defined by (5.3.8)], two possible additional asymmetric solutions appear (see Fig. 5.3). For $L > 2\pi/\kappa$, there are two more solutions corresponding to full turns on the phase trajectory. As L increases further, a greater number of solutions emerge, e. g., those resulting from one and a half turn on the given trajectory and a half turn of one of the trajectories of the large radius, etc. Figure 5.5 depicts several possible spatial distributions of $\mathcal{W} = \mathcal{J}_1 + \mathcal{J}_2$.

Thus, for a layer with $L > \pi/\kappa$, there is a finite number of different stationary nonlinear oscillations possible that are characterized by the same frequency, which grows with increasing layer thickness L.

All the above also applies to a layer with radiation at the boundaries. In this case, however, stationary oscillations only arise when the layer thickness is sufficiently large, namely, when

$$L > L_{\mathrm{cr}} = \ln(r_1 r_2)^{-1/2} / k(\mathrm{Re}\{\alpha\} - g) \qquad (5.3.9)$$

[compare with (5.2.9) for a layer without distributed losses]. When $L \to L_{\mathrm{cr}}$, we have $\mathcal{J}_{1,2} \to O$. Therefore, (5.3.9) can be obtained from the linear theory. For a sufficiently large thickness of the layer the solution is also many-valued.

Let us consider the output power of the laser. It will be assumed, for the sake of simplicity, that one of the layer boundaries is opaque ($r_1 = 1$); then the output power is proportional to

$$S = \mathcal{J}_1(L_2) - \mathcal{J}_2(L_2) = (1 - r)\mathcal{J}_2(L_2) \quad .$$

It has been demonstrated in Sect. 5.2 that when $g = 0$ this function reaches its maximum $S_{\max} = kl\,\mathrm{Re}\{\alpha\}$ at $r = 1$. If $g \neq 0$, the field in the layer is always finite and $S = 0$ both for $r = r_{\mathrm{cr}}$ and for $r = 1$. For a certain $r = r_{\mathrm{opt}}$, the output power has a maximal value. As we have mentioned before, in this case it turns out to be impossible to determine the dependence of S on r in a closed form. The qualitative aspects of this dependence, however, can be traced on the phase plane (Fig. 5.3). The numerical value of S is equal to the distance along the vertical between the point on Fig. 5.3 that corresponds to the right boundary of the layer and the bisectrix. In the case of several possible solutions each of them is characterized by its own value of the output power. It can be shown that for all types of solutions the magnitude of S is always less than the quantity $S_{\max}$ which is reached when $g = 0$, $r = 1$.

Let us now find the frequency spectrum of the layer oscillations. Returning to (5.3.1, 2), we obtain [by analogy with (5.2.5)] the following equations connecting the variations of phases $\varphi_{1,2}$ of waves and their amplitudes $\mathcal{J}_{1,2}$ with each other:

$$\frac{d\varphi_{1,2}}{dz} = -\frac{T\Delta\omega}{2}\left(\frac{d\ln \mathcal{J}_{1,2}}{dz} \pm gk\right) \quad . \tag{5.3.10}$$

For given boundary conditions these equations determine the spectrum of possible frequencies of monochromatic oscillations:

$$\omega_p = \omega_{21}\frac{\pi p + \omega_{21}T[\ln(r_1 r_2)^{-1/2} + gk_0 L]}{2k_0 L + \omega_{21}T[\ln(r_1 r_2)^{-1/2} + gk_0 L]} \quad . \tag{5.3.11}$$

If g depends on ω, (5.3.11) should be regarded as an equation specifying ω_p. In this case the spectrum of laser frequencies is not equidistant. As follows from comparison of (5.3.11) and (5.2.11), the effect of volume losses on the spectrum of laser oscillations coincides qualitatively with their effect on radiation.

If the reflection coefficients $r_{1,2}$ of the layer boundary are complex, the quantities p in (5.3.11) will have integer values which crucially affect the oscillation spectrum. The field amplitude distribution, as well as the power emitted by the layer, depend only on moduli of $r_{1,2}$.

In conclusion, let us mention some interesting peculiarities of the field behavior that are associated with the fact that the medium represents a distributed and nonlinear system, and contradicts in a certain sense our intuition based on linear problems.

In a system with linear attenuation ($g \neq 0$) under symmetric boundary conditions, solutions with an asymmetric distribution of the field amplitude are possible (as was shown in Sect. 5.2, in the absence of attenuation there are no such solutions).

In the case of linear attenuation ($g \neq 0$) degeneracy can occur, i.e., a single frequency can correspond to several stationary states (there is no such nonlinear degeneracy for $g = 0$). In linear systems we are accustomed to situations that are directly opposed to this: the introduction of attenuation removes the degeneracy.

5.4 Rotation of Polarization Planes of Countertravelling Waves in an Isotropic Nonlinear Medium

As is well known, the direction of polarization of countertravelling waves in an isotropic homogeneous linear medium does not vary with the distance along the propagation axis. In order to make sure that the situation for the nonlinear medium is different, it is sufficient to consider linearly polarized countertravelling waves (with the real vectors C_1 and C_2) in a purely active medium (imaginary χ) [5.9].

Under different conditions (real or complex χ, elliptic polarization of countertravelling waves, etc.) the interpretation of this phenomenon is more complicated, hence we are not going to discuss these cases here.

Taking the vector product of the first equation of (5.1.1) with C_1 we obtain an equation for the angular velocity of the polarization of the vector C_1

$$\mathcal{J}_1 \frac{d\varphi_1}{dz} = -\frac{2\pi k}{\varepsilon} A\sqrt{v}\,\mathrm{Im}\{b\} \quad . \tag{5.4.1}$$

In this expression φ_1 is the angle of rotation of vector C_1 and is measured clockwise for a given direction of the axis. Similarly, the rotation of vector C_2 is determined by

$$\mathcal{J}_2 \frac{d\varphi_2}{dz} = -\frac{2\pi k}{\varepsilon} A\sqrt{v}\,\mathrm{Im}\{b\} \quad . \tag{5.4.2}$$

It is of interest to note that the polarization planes of the countertravelling waves rotate in the same direction. This becomes especially obvious when one considers the rotation of the bisectrix of the angle between C_1 and C_2 [the variation of the angle itself is described by (5.1.4)]. The behavior of such an average direction of polarization states is governed by

$$\frac{d\theta}{dz} = -\frac{2\pi k}{\varepsilon} \frac{AW\sqrt{v}\,\mathrm{Im}\{b\}}{A^2 + v} \quad , \quad \text{where}$$
$$2\theta = \varphi_1 + \varphi_2 \quad . \tag{5.4.3}$$

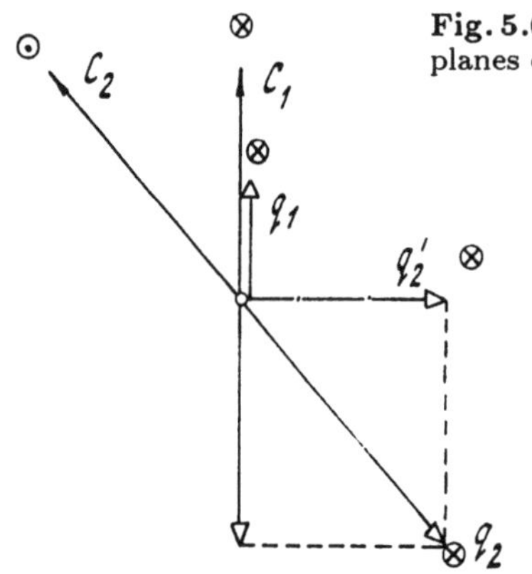

Fig. 5.6. A diagram explaining rotation of the polarization planes of counterwaves in an isotropic nonlinear medium

One can see from this formula that even in a symmetric system of length $2L$ with the symmetric distribution of energy $\mathcal{W}(z)$ and with relative energy $\sqrt{v(z)}$ there will be a twisting of the average orientation of polarization states to one side by an angle

$$\Delta\theta = -\frac{4\pi k}{\varepsilon}\int_0^L \frac{A\mathcal{W}\sqrt{v}\,\mathrm{Im}\{b\}}{A^2+v}dz \quad . \tag{5.4.4}$$

The rotation of the polarization plane can be explained qualitatively in the following way. The increment of the vector C_1 per unit of length consists of two vectors (see Fig. 5.6): the amplification (attenuation) of the wave C_1 associated with its self-action (vector $\boldsymbol{q}_1$ parallel to C_1) and the reflection of the wave C_2 from modulation of χ caused by the dependence of χ on the standing field (vector $\boldsymbol{q}_2$ parallel to C_2). It is precisely the component of $\boldsymbol{q}_2$ which is normal to C_1 (vector $\boldsymbol{q}_2'$ on Fig. 5.6) that is responsible for the rotation of the latter. The absence of the effect for parallel C_1 and C_2 and for perpendicular C_1 and C_2 can be easily explained. In the first case the additional vector $\boldsymbol{q}_2$ is parallel to C_1 and, therefore, $\boldsymbol{q}_2' = 0$. As far as (5.4.1,2) are concerned, this corresponds to the quantity A being equal to zero. In the second case there is no modulation of χ, and therefore $\boldsymbol{q}_2 = 0$. In (5.4.1,2) it is the quantity $\sqrt{v}\,\mathrm{Im}\{b\}$ which becomes equal to zero.

The angle of rotation of the field polarization state can in two special cases be determined in a simpler way: when the field is spatially homogeneous and when the energy of one wave is much greater than that of the other.

In the first case the increment of angles over the length L is equal to

$$\Delta\varphi_{1,2} = -\frac{\pi k}{\varepsilon}\sin\varphi^{\mathrm{st}}\cos\varphi^{\mathrm{st}}\mathcal{W}^{\mathrm{st}}\,\mathrm{Im}\{b^{\mathrm{st}}\}L \quad . \tag{5.4.5}$$

The quantities $\mathcal{W}^{\mathrm{st}}$, φ^{st}, $\mathrm{Im}\{b^{\mathrm{st}}\}$ satisfy an algebraic equation (the condition of stationarity with respect to z)

$$\mathrm{Im}\{a^{\mathrm{st}}\} + \frac{\mathcal{W}^{\mathrm{st}}}{2}\cos^2\varphi^{\mathrm{st}}\,\mathrm{Im}\{b^{\mathrm{st}}\} = 0 \quad . \tag{5.4.6}$$

Therefore, $\Delta\varphi_{1,2}$ is also equal to

$$\Delta\varphi_{1,2} = \frac{2\pi k}{\varepsilon}\tan\varphi^{\mathrm{st}}\,\mathrm{Im}\{a^{\mathrm{st}}\}L \quad . \tag{5.4.7}$$

If, in order to estimate $\Delta\varphi_{1,2}$ we assume that $\mathrm{Im}\{a^{\mathrm{st}}\} \approx 1/2\,\mathrm{Im}\{\chi\}|_{\varepsilon^2=0}$, $\varphi^{\mathrm{st}} = \pi/4$, then

$$\Delta\varphi_{1,2} \approx \tfrac{1}{4}\ln\Gamma \quad , \tag{5.4.8}$$

where Γ is the linear amplification coefficient of the sample. Under these conditions the magnitude of the effect is proportional to its length. For glass with neodymium impurities over a length of 1 cm, the quantity $\Delta\varphi_{1,2} \approx 1.6°$.

Consider the second case, when the energy of one wave is much greater than that of the other, let us take

$$\mathcal{J}_1 \ll \mathcal{J}_2 \quad , \tag{5.4.9}$$

then, according to (4.4.14),

$$\mathrm{Im}\{b\} \approx \frac{d}{d\mathcal{J}_2}\mathrm{Im}\{\chi\}\,|_{\mathcal{J}_1=0} \quad .$$

Substituting this formula into (5.4.1), we obtain

$$\frac{d\varphi_1}{dz} = -\frac{\pi k}{\varepsilon}\mathcal{J}_2 \sin 2\varphi \frac{d}{d\mathcal{J}_2}\mathrm{Im}\{\chi\}|_{\mathcal{J}_1=0} \quad . \tag{5.4.10}$$

If one takes (5.4.9) into account, (5.1.7) for $\mathcal{J}_2$ can be reduced to

$$\frac{d\mathcal{J}_2}{dz} = -\frac{4\pi k}{\varepsilon}\mathcal{J}_2\,\mathrm{Im}\{\chi\}|_{\mathcal{J}_1=0} \quad . \tag{5.4.11}$$

Comparison of this equation with (5.4.10) yields

$$\frac{d\varphi_1}{d\mathcal{J}_2} = \frac{\sin 2\varphi}{4}\frac{d}{d\mathcal{J}_2}\ln\mathrm{Im}\{\chi\}|_{\mathcal{J}_1=0} \quad . \tag{5.4.12}$$

In the accepted approximation it can be assumed that the direction of polarization of the strong field $\boldsymbol{C}_2$ does not change, and all angles can be measured from this direction. Then φ will be equal to φ_1, and (5.4.12) can be easily solved. This leads to

$$\tan\varphi_1(L) = \sqrt{\frac{\mathrm{Im}\{\chi(L)\}}{\mathrm{Im}\{\chi(0)\}}}\tan\varphi_1(0) \quad . \tag{5.4.13}$$

The tangent of the increment of the angle φ_1 is obtained from (5.4.13) after elementary transformations[3]:

$$\tan\Delta\varphi_1 = \tan\varphi_1(0)\frac{1-y}{1+y\tan^2\varphi_1(0)} \ , \quad \text{where}$$

$$y = \sqrt{\frac{\mathrm{Im}\{\chi(0)\}}{\mathrm{Im}\{\chi(L)\}}} \quad . \tag{5.4.14}$$

The value of the angle φ_{opt} for which the quantity $\Delta\varphi_1$ is maximal can be easily determined

$$\tan\varphi_{1\mathrm{opt}}(0) = y^{-1/2} \quad . \tag{5.4.15}$$

Then we have

$$\tan\Delta\varphi_{1\mathrm{max}} = (1-y)/2\sqrt{y} \quad . \tag{5.4.16}$$

For instance, when χ changes over the length of the sample by a factor of 2 then $\Delta\varphi_{1\mathrm{max}} \approx 10°$.

Let us demonstrate that there is a simple relationship between the ratio y^2 of polarizabilities at different ends of the sample and the values of the field at entrance and exit boundaries of the sample.

Since in the case of counterpropagating waves the intensity of one of them is much less than the intensity of the other, the wanted relationship can be derived by considering the nonlinear amplification of the travelling wave. Assume that a light flux with the energy density $\mathcal{W}_0$ is incident on the entrance of a plane parallel sample. It will also be supposed that the transverse structure of the field is homogeneous and there are no reflections from the sample boundaries. Then the energy variation along the propagation axis z is described by [see, for instance (5.1.19)]

$$\frac{d\mathcal{W}}{dz} = \frac{4\pi k}{\varepsilon}\mathcal{W}\,\mathrm{Im}\{\chi(\mathcal{W})\} \quad . \tag{5.4.17}$$

[3] It can be shown that for a medium in which the modulation of polarization (the second harmonic of χ) is attenuated by a factor of γ^{-1}:

$$\tan\Delta\varphi_1 = \frac{\tan\varphi_1(0)(1-y^{\nu})}{y^{\nu}+\tan^2\varphi_1(0)} \quad .$$

At the output of the nonlinear amplifier the energy will, obviously, be determined by

$$\frac{\varepsilon}{4\pi k}L = \int\limits_{\mathcal{W}_0}^{\mathcal{W}} \frac{d\xi}{\xi \operatorname{Im}\{\chi(\xi)\}} \quad . \tag{5.4.18}$$

Differentiating this equation with respect to the input energy $\mathcal{W}_0$ we obtain

$$[\mathcal{W} \operatorname{Im}\{\chi(\mathcal{W})\}]^{-1}\frac{d\mathcal{W}}{d\mathcal{W}_0} - [\mathcal{W}_0 \operatorname{Im}\{\chi(\mathcal{W}_0)\}]^{-1} = 0 \ , \quad \text{or}$$

$$\operatorname{Im}\{\chi(\mathcal{W})\} = \operatorname{Im}\{\chi(\mathcal{W}_0)\}\frac{\mathcal{W}_0}{\mathcal{W}}\frac{d\mathcal{W}}{d\mathcal{W}_0} \quad . \tag{5.4.19}$$

If the input signal is sufficiently weak, the quantity $\mathrm{O}\operatorname{Im}\{\chi(\mathcal{W}_0)\}$ can be replaced by its linear limit:

$$\operatorname{Im}\{\chi(\mathcal{W})\} = \operatorname{Im}\{\chi(0)\}\frac{\mathcal{W}_0}{\mathcal{W}}\frac{d\mathcal{W}}{d\mathcal{W}_0} \quad . \tag{5.4.20}$$

The magnitude of $\operatorname{Im}\{\chi(\mathcal{W})\}$ is determined by the dependence between the output energy of the amplifier and its input energy. The formula obtained can have an independent significance for the experimental measurement of nonlinear $\operatorname{Im}\chi$. As a result,

$$y^2 = \frac{\mathcal{W}(L)}{\mathcal{W}(0)}\frac{d\mathcal{W}(L)}{d\mathcal{W}(0)} \quad .$$

Finally, (5.4.14) for the rotation angle assumes the form

$$\tan\Delta\varphi_1 = \tan\varphi_1(0)\left[\sqrt{\mathcal{W}(0)\frac{d\mathcal{W}(L)}{d\mathcal{W}(0)}} - \sqrt{\mathcal{W}(L)}\Big/\sqrt{\mathcal{W}(0)\frac{d\mathcal{W}(L)}{d\mathcal{W}(0)}}\right.$$
$$\left. + \sqrt{\mathcal{W}(L)}\tan^2\varphi_1(0)\right] \quad . \tag{5.4.21}$$

It is important to note that in the above approximation ($\mathcal{J} \ll \mathcal{J}_2$) the rotation angle does not depend explicitly either on the type of nonlinearity or on the value of polarizability inside the sample, or even on concentration of the nonlinear medium. This circumstance is of special significance for estimating the possibility to observe the effect experimentally.

The rotation of the polarization plane of countertravelling waves in a resonant medium was discovered experimentally in [5.10]. At present this phenomenon is widely applied in the so-called polarization spectroscopy [5.11].

6. Three-Photon Resonant Parametric Processes

This chapter deals with three-photon parametric processes in which the pumping frequency or one of the generated frequencies are in resonance with the transition frequency of the electron or vibrational levels. It is discussed to what extent it is feasible to use such processes in gases to produce coherent radiation at VUV and molecular vibration frequencies. Spatial modulation of the electrostatic field applied to the medium is shown to lead to a considerable improvement in the parametric conversion efficiency.

The simplest parametric (coherent) processes involve three-photons leading to the doubling or division of radiation frequency and to the generation of sums or differences of frequencies. When the frequency of one interacting field coincides with the transition frequency in matter, we are dealing with a resonant parametric process. It is obvious that the two-photon resonance condition is also satisfied in this case. Therefore, in addition to the parametric interaction of fields, one observes single-photon and two-photon absorption or stimulated Raman scattering[1]. Sometimes interaction processes may be affected by saturation of populations of the resonant transition. The authors of the first theoretical papers (see, for instance, [6.1–4]) devoted to three-photon resonant parametric phenomena did not take into account two-photon transitions and saturation. On the basis of expressions for polarization as obtained and analyzed in the previous chapters, we shall be able to describe resonant parametric processes allowing for all essential factors. We shall see how the effects of parametric self-induced transparency and phase locking of interacting waves discussed in Sect. 2.4 manifest themselves on the macroscopic scale.

Resonant parametric processes are not only interesting as a group of phenomena of nonlinear optics with their own specifics but also in view of their possible applications. It will be shown below that, for example, the resonant doubling of light frequency in vapors and gases subject to the action of an electrostatic field can be used for the construction of powerful coherent radiation sources in the vacuum ultraviolet (VUV) range, where the traditional methods of quantum electronics and nonlinear optics are very difficult to apply. The transition to resonance doubling and the addition of frequencies in image vi-

[1] If there is an inversion of populations in the medium, the emission of fields with the corresponding frequencies will occur instead of single- and two-photon absorption.

sualization systems may improve the resolving power of such systems because resonant parametric interactions are less critical with respect to wave detuning.

The generation of a difference frequency accompanying stimulated Raman light scattering by vibrational excitations of gas molecules subject to the action of an electrostatic field may also turn out to be useful. In this case it is possible to obtain in the infrared (IR) range intensities of radiation that are sufficient for the effective population of excited vibrational levels of molecules with a dipole-allowed transition [6.5]. This is necessary, particularly for resonant excitation of individual chemical bonds.

The behavior of fields will be described on the basis of (1.1.2) or (1.1.4, 5); the following simplifying assumptions concerning the quantity $\vec{\mathcal{P}}^{\mathrm{nl}}$ will be made. First of all, it will be supposed that for molecules at rest the resonance conditions are completely satisfied. Then the fact that in most experimental studies, interacting fields are much less than Stark ones must be taken into account. Therefore, the change of transition frequency produced by radiation can be neglected, and the polarization can be calculated under the assumption that $\Omega = 0$. We shall also discard the nonresonant term in nonlinear susceptibility. Usually it is by several orders of magnitude smaller than the resonant term, and for this reason its influence on the processes of field conversion is indeed small. Finally, it will be assumed that the change of the refractive index of the medium, associated with the redistribution of populations between resonance levels (see Sect. 2.2) does not affect the phase velocities of waves, i.e., that the latter only vary as a result of parametric interactions[2].

The above suppositions enable one to analyze the main regularities of resonant parametric interactions in their pure form, the abstract from numerous accompanying phenomena, and also to obtain the necessary estimates. Certainly, sometimes in order to produce a complete description and to compare the results with experimental data, it is necessary to conduct a study which is not based on the above assumptions.

Thus, in [6.6–10], equations were obtained that allow for frequency detunings and nonresonant susceptibility in the case of three-photon resonant parametric processes. These equations were used [6.9] to investigate the resonant doubling of frequency in molecular crystals. In [6.11], the generation of second harmonic has been studied under the condition that the sum of the pumping and harmonic frequencies coincides with the transition frequency in matter. Not only the nonlinear absorption of waves, but also the nonlinearity of the refractive index caused by the transition of matter into its excited state was taken into account. In spite of this nonlinearity, there are no spatial oscillations of the second harmonic intensity. The reason for this is that the above nonlinearity of the refractive index is compensated for by its parametric nonlinearity, which leads to phase locking of interacting waves.

---

[2] Section 6.2, where the nonlinear detuning of wave phase velocities produced by a stimulated Raman process is taken into account, represents an exception.

The group of papers [6.12–14] in which some resonant parametric interactions were considered in situations where the interacting waves had different states of polarization (linear and circular) are also of interest. Though the dynamics of such processes might exhibit curious peculiarities, the estimate of the efficiency of wave conversion under optimal conditions demonstrates no essential change.

6.1 Addition and Doubling of Frequencies for a Transition Frequency in Matter That Coincides with the Sum Frequency or the Frequency of the Harmonic

Consider the interaction between waves whose frequencies satisfy the condition

$$\omega_1 + \omega_3 = \omega_2 = \omega_{21} \quad . \tag{6.1.1}$$

By making use of the polarization of matter calculated on the basis of resonance conditions (6.1.1) one can obtain from (1.1.4, 5) a system of equations for real amplitudes and the difference between phases of interacting waves $\theta = \varphi_1 + \varphi_3 - \varphi_2 + \delta k z$ (it is assumed that all waves propagate along the z axis):

$$\begin{aligned} \frac{d\mathcal{A}_{1,3}}{dz} &= -g_{1,3}[\langle (\boldsymbol{d}\cdot\boldsymbol{e}_2)\kappa\eta\rangle \mathcal{A}_{3,1}\mathcal{A}_2 \cos\theta + \langle \kappa^2\eta\rangle \mathcal{A}_{3,1}^2\mathcal{A}_{1,3}] - \alpha_{1,3}\mathcal{A}_{1,3} \quad , \\ \frac{d\mathcal{A}_2}{dz} &= -g_2[\langle (\boldsymbol{d}\cdot\boldsymbol{e}_2)\kappa\eta\rangle \mathcal{A}_1\mathcal{A}_3 \cos\theta + \langle (\boldsymbol{d}\cdot\boldsymbol{e}_2)^2\eta\rangle \mathcal{A}_2] - \alpha_2\mathcal{A}_2 \quad , \\ \frac{d\theta}{dz} &= \delta k + \langle (\boldsymbol{d}\cdot\boldsymbol{e}_2)\chi\eta\rangle \left[g_1\frac{\mathcal{A}_2\mathcal{A}_3}{\mathcal{A}_1} + g_3\frac{\mathcal{A}_1\mathcal{A}_2}{\mathcal{A}_3} + g_2\frac{\mathcal{A}_1\mathcal{A}_3}{\mathcal{A}_2}\right] \sin\theta \quad . \end{aligned} \tag{6.1.2}$$

Here $g_j = (2\pi N\omega_j^2 T/\hbar c^2 k_j)$ is introduced and N is the particle number density of the nonlinear substance. T is the inverse line width (in the presence of Doppler line broadening, the quantity T should be replaced by

$$T' = \frac{\sqrt{\pi}}{\omega_{21}\sqrt{\varepsilon(\omega = \omega_{21})}}\frac{c}{\overline{v}} \quad ,$$

where $\overline{v}$ is the rms velocity of the thermal motion of gas molecules). The angular brackets imply that the distribution function of molecules has to be used to average over its orientations. The quantity $\delta k = k_1 - k_2 + k_3$ denotes the wave detuning or detuning of synchronism, and the α_j describe those linear losses of the j-th field in the medium which are not related to the transition $2 \to 1$.

In the stationary case under consideration the population difference is given by

$$\eta = \eta_0 \left\{ 1 + \frac{4\tau T}{\hbar^2} [(\boldsymbol{d}\cdot \boldsymbol{e}_2)^2 \mathcal{A}_2^2 \right.$$

$$\left. + 2(\boldsymbol{d}\cdot \boldsymbol{e}_2)\kappa \mathcal{A}_1 \mathcal{A}_2 \mathcal{A}_3 \cos\theta + \kappa^2 \mathcal{A}_1^2 \mathcal{A}_3^2] \right\}^{-1} \quad . \tag{6.1.3}$$

One can easily see that it depends on the molecule's orientation and, generally speaking, cannot be taken out of the angular brackets.

Here we are going to deal with two models of matter: (i) a solid with resonant impurity molecules of identical orientation and (ii) a gas subject to the action of an electrostatic field[3]. The advantage of the first model is that it makes it possible to study the resonant parametric interaction with the saturation of populations of matter resonant levels taken into account.

6.1.1 Addition and Doubling of Frequencies in a Medium with Identically Oriented Molecules

If the component of a medium that doubles the frequency contains identically oriented molecules, the averaging over orientations in (6.1.2) can be omitted (i.e., the angular brackets can be discarded). Let us first consider the situation when $\delta k = 0$. This example enables one to trace easily the qualitative difference between a resonant parametric interaction and a nonresonant one and to estimate the maximal efficiency of the resonant parametric conversion.

From the last equation in (6.1.2), it follows that for $\delta k = 0$ and $\eta > 0$, the plane $\theta = \pi$ in the phase space $\{\mathcal{A}_1, \mathcal{A}_2, \mathcal{A}_3, \theta\}$ is stable. We assume that $(\boldsymbol{d}\cdot\boldsymbol{e}_2)\kappa > 0$; when $(\boldsymbol{d}\cdot\boldsymbol{e}_2)\kappa < 0$, the plane $\theta = 0$ is stable and the other results of this section remain the same.

It would be of interest to consider field generation under the conditions when its strength at the boundary of the nonlinear medium is close to zero or, to put it more precisely, is determined by noise. Then, as one can see from (6.1.2), the velocity of motion of the phase difference to the plane $\theta = \pi$ is much greater than the rates of variation of the field amplitudes $\mathcal{A}_j$. Therefore, as far as the first three equations of system (6.1.2) are concerned, it can be assumed that $\theta = \pi$. Introducing new variables (normalized amplitudes)

$$a_{1,3} = \frac{\kappa}{(\boldsymbol{d}\cdot \boldsymbol{e}_2)} \sqrt{\frac{g_{3,1}}{g_2}} \mathcal{A}_{1,3} \quad ,$$

$$a_2 = \frac{\kappa}{(\boldsymbol{d}\cdot \boldsymbol{e}_2)} \frac{\sqrt{g_1 g_3}}{g_2} \mathcal{A}_2 \tag{6.1.4}$$

[3] A description of experiments on resonant frequency doubling in a solid can be found in [6.15] (on exciton levels) and [6.16] (on the electron transition in molecular crystals). Nothing is known about the observation of this process in gases.

and the normalized coordinate

$$\zeta = g_2\eta_0(\boldsymbol{d}\cdot\boldsymbol{e}_2)^2 z \quad , \tag{6.1.5}$$

and making use of (6.1.3), we can rewrite (6.1.2) as

$$\begin{aligned}\frac{da_{1,3}}{d\zeta} &= \frac{a_{3,1}(a_2 - a_1a_3)}{1+\mathcal{Y}(a_2-a_1a_3)^2} - \alpha^{\rm n}_{1,3}a_{1,3} \quad , \\ \frac{da_2}{d\zeta} &= \frac{a_1a_3 - a_2}{1+\mathcal{Y}(a_2-a_1a_3)^2} - \alpha^{\rm n}_2 a_2 \quad .\end{aligned} \tag{6.1.6}$$

$$\mathcal{Y} = \frac{4\tau T}{\hbar^2}\frac{g_2^2}{g_1g_3}\frac{(\boldsymbol{d}\cdot\boldsymbol{e}_2)^4}{\kappa^2} \tag{6.1.7}$$

is the saturation parameter written for normalized amplitudes a_j. The $\alpha^{\rm n}_j$ are normalized attenuation coefficients:

$$\alpha^{\rm n}_{1,3} = \alpha_{1,3}\frac{\sqrt{g_2}}{\sqrt{g_{1,3}}}\frac{(\boldsymbol{d}\cdot\boldsymbol{e}_2)}{\kappa} \quad ; \quad \alpha^{\rm n}_2 = \alpha_2\frac{g_2}{\sqrt{g_1g_3}}\frac{(\boldsymbol{d}\cdot\boldsymbol{e}_2)}{\kappa} \quad .$$

The frequency doubling represents a special case of the addition of frequencies when the resonance condition is degenerate:

$$\omega_1 = \omega_3 = \omega \quad , \quad \omega_2 = 2\omega = \omega_{21} \quad . \tag{6.1.8}$$

In this case $a_1 = a_3$.

Note that in the absence of degeneration the plane $a_1 = a_3$ is also stable when $\alpha^{\rm n}_1 = \alpha^{\rm n}_3$. This can be easily demonstrated by multiplying the equations for a_1 and a_3 from (6.1.6) by a_1 and a_3, respectively, and subtracting one equation from the other. Condition $\alpha^{\rm n}_1 = \alpha^{\rm n}_3$ is satisfied, for instance, when the difference between ω_1 and ω_3 is small. Therefore, all the relations for variables a_1, a_2, ζ obtained when conditions (6.1.8) are satisfied, are also valid for the interaction of fields with nearly equal frequencies ω_1 and ω_3. The amplitudes $\mathcal{A}_j$ must be determined by means of (6.1.4).

Thus, we assume that $a_3 = a_1$ and neglect attenuation which is not caused by the resonant transition $2 \to 1$. Then for a_1 and a_2,

$$\begin{aligned}\frac{da_1}{d\zeta} &= \frac{a_1(a_2 - a_1^2)}{1+\mathcal{Y}(a_2-a_1^2)^2} \quad , \\ \frac{da_2}{d\zeta} &= \frac{a_1^2 - a_2}{1+\mathcal{Y}(a_2-a_1^2)^2} \quad .\end{aligned} \tag{6.1.9}$$

In order to obtain equations for frequency doubling in dimensional quantities one has to use (6.1.4, 5, 7) with

$$g_1 = g_3 = \frac{4\pi N\omega^2 T}{\hbar c^2 k_1} \quad , \quad g_2 = \frac{8\pi N\omega^2 T}{\hbar c^2 k_2} \quad . \tag{6.1.10}$$

Dividing the first equation of (6.1.9) by the second we can find the first integral:

$$a_2 - a_2(0) = \ln\left[a_1(0)/a_1\right] \quad . \tag{6.1.11}$$

In this formula $a_{1,2}(0)$ are the values of normalized amplitudes at the boundary of the nonlinear layer (at $z = 0$).

It also follows from (6.1.9) that regardless of the boundary values $a_j(0)$, the amplitudes of the second and first harmonics satisfy a limiting relation[4]

$$a_2(\infty)/a_1^2(\infty) = 1 \quad , \tag{6.1.12}$$

for which $\eta = \eta_0$. Thus, when a field with the frequency ω is incident on matter, the population difference η near the boundary is less than its equilibrium value because of the two-photon absorption of the field $\mathcal{A}_\omega$. Under the conditions of synchronism, the field $\mathcal{A}_{2\omega}$ that appears in the system increases as one moves along the z axis; its parametric interaction with $\mathcal{A}_\omega$ leads to the increase of η and its return to the equilibrium value. When $z \to \infty$, the system approaches a state characterized by parametric self-induced transparency (see Sects. 2.1, 2.4).

Let us now apply (6.1.11, 12) to determine the limiting values of the field amplitudes resulting from interaction:

$$a_2(\infty) + \tfrac{1}{2}\ln a_2(\infty) = a_1^2(\infty) + \ln a_1(\infty) = a_2(0) + \ln a_1(0) \quad . \tag{6.1.13}$$

In frequency multipliers the field of the higher harmonic at the input is zero $a_2(0) = 0$. Taking this into account, we can reduce (6.1.13) to

$$a_1(0) = \sqrt{a_2(\infty)}\exp a_2(\infty) \quad ,$$

or, for dimensional amplitudes

$$\mathcal{A}_1(0) = \left[\frac{(\boldsymbol{d}\cdot\boldsymbol{e}_2)}{\kappa}\mathcal{A}_{2\,\mathrm{lim}}\right]^{1/2} \exp\left[\frac{\kappa}{(\boldsymbol{d}\cdot\boldsymbol{e}_2)}\mathcal{A}_{2\,\mathrm{lim}}\right] \quad , \tag{6.1.14}$$

where $\mathcal{A}_{2\,\mathrm{lim}}$ is the limiting value of the second harmonic amplitude (corresponding to $z \to \infty$).

If

$$\mathcal{A}_{2\,\mathrm{lim}} \ll (\boldsymbol{d}\cdot\boldsymbol{e}_2)/\kappa \quad , \tag{6.1.15}$$

[4] It will be shown below that when $a_2(0)/a_1^2(0) \neq 1$, we have $a_2 \to a_2(\infty)$, $a_1 \to a_1(\infty)$ for $z \to \infty$, see (6.1.18).

(6.1.14) can be replaced by the approximate formula

$$\mathcal{A}_{2\,\mathrm{lim}} \approx \frac{\kappa}{(\boldsymbol{d}\cdot\boldsymbol{e}_2)}\mathcal{A}_1^2(0) \quad . \tag{6.1.16}$$

This expression can be applied when $\mathcal{A}_1(0) \lesssim 0.3(\boldsymbol{d}\cdot\boldsymbol{e}_2)/\kappa$. If $(\boldsymbol{d}\cdot\boldsymbol{e}_2) \approx 10^{-20}$ cgse, $\kappa = 10^{-24}$ cgse, we have $(\boldsymbol{d}\cdot\boldsymbol{e}_2)/\kappa = 10^4$, and the pumping field strength at the input should be no greater than 10^6 W/cm. In this case the conversion coefficient

$$\beta = \frac{\mathcal{A}_{2\,\mathrm{lim}}}{\mathcal{A}_1(0)} = \frac{\kappa}{(\boldsymbol{d}\cdot\boldsymbol{e}_2)}\mathcal{A}_1(0) \tag{6.1.17}$$

increases linearly with the increase of the pumping field in the same way as during the nonresonant frequency doubling (for $z<1/\delta k$, l_{lim}) [6.17]. From the theoretical point of view it is well known that the nonresonant mechanism of frequency doubling facilitates the complete transfer of energy from the first harmonic into the second, if pumping field strength and thickness of the nonlinear medium are increased. On the basis of (6.1.14) it can easily be shown that in our situation the conversion coefficient reaches its maximum when $\mathcal{A}_1(0) \approx 1.16(\boldsymbol{d}\cdot\boldsymbol{e}_2)/\kappa$ and $(\beta_{\mathrm{lim}})_{\mathrm{max}} \approx 0.43$. Thereby $\mathcal{A}_{2\,\mathrm{lim}} = 0.5(\boldsymbol{d}\cdot\boldsymbol{e}_2)/\kappa$ holds and the ratio of the energy transferred into the second harmonic to that absorbed by matter is approximately equal to 0.23.

Let us analyze the second harmonic amplitude as a function of a coordinate. Combining the second equation of (6.1.9) and (6.1.11) we obtain

$$\zeta = \int\limits_{a_2(0)}^{a_2} \{[a_1^2(0)e^{2[a_2(0)-a]} - a]^{-1} + \mathcal{Y}[a_1^2(0)e^{2[a_2(0)-a]} - a]\}da \quad . \tag{6.1.18}$$

In the general case the integral (6.1.18) cannot be expressed through elementary functions. When $a_2(\infty)\ll 1$ [condition (6.1.15) is satisfied], however, an approximate formula can be derived for it. We shall limit ourselves here to the case when the first harmonic field amplitude can be regarded as constant. For $a_2(0) = 0$ we obtain

$$\zeta = \mathcal{Y}a_1^2(0)a_2\left(1 - \frac{a_2}{2a_1^2(0)}\right) - \ln\left(1 - \frac{a_2}{a_1^2(0)}\right) \quad ,$$

or, in dimensional variables,

$$z = [g_2\eta_0(\boldsymbol{d}\cdot\boldsymbol{e}_2)^2]^{-1}\left\{\frac{4\tau T\kappa^2}{\hbar^2}\mathcal{A}_1^2(0)\frac{(\boldsymbol{d}\cdot\boldsymbol{e}_2)}{\kappa}\mathcal{A}_2 \times\left(1 - \frac{(\boldsymbol{d}\cdot\boldsymbol{e}_2)}{2\kappa}\frac{\mathcal{A}_2}{\mathcal{A}_1^2(0)}\right) - \ln\left(1 - \frac{(\boldsymbol{d}\cdot\boldsymbol{e}_2)}{\kappa}\frac{\mathcal{A}_2}{\mathcal{A}_1^2(0)}\right)\right\} \quad . \tag{6.1.19}$$

The first term in curly brackets in (6.1.19) is due to saturation of the difference between populations of resonance levels. It can be neglected if [see also (6.1.3)]

$$\mathcal{A}_1^2(0) \ll \frac{\hbar}{2\kappa\sqrt{\tau T}} = \mathcal{J}_{\mathrm{sat}} \quad . \tag{6.1.20}$$

In the case of nonresonant frequency doubling in the approximation in which the field of the first harmonic is constant, the distance in which the second harmonic amplitude reaches a certain value $\overline{a}_2$ is inversely proportional to $a_1^2(0)$ [6.17]. As one can easily see from (6.1.19), when frequency doubling is resonant, there is an optimal value of the pumping field

$$a_1^2(0) = \tfrac{1}{2}(\overline{a}_2 + \sqrt{\overline{a}_2^2 + 4/\mathcal{Y}}) \quad , \tag{6.1.21}$$

for which $z(\overline{a}_2) = z_{\mathrm{min}}$. The quantity z_{min} can be found from (6.1.21) and (6.1.19).

Let us make some numerical estimates. We assume the nonlinear substance to consist of the impurities in the dielectric for which the synchronism conditions can be satisfied. Suppose that $\mathcal{A}_1(0) = 2 \times 10^3$ cgse, $(\boldsymbol{d} \cdot \boldsymbol{e}_2) = 10^{-20}$ cgse, $\kappa = 10^{-24}\,\mathrm{cm}^3$, $\omega = 10^{15}\,\mathrm{s}^{-1}$, $T = 10^{-11}$ s, $N\eta_0 = 10^{20}\,\mathrm{cm}^{-3}$, and the life time $\tau \ll 10^{-7}$ s. Then saturation can be neglected, and it follows from (6.1.19) that one obtains at the output of the nonlinear layer [whose thickness $z = 0.01$ cm] the values of $\mathcal{A}_2 \approx 2 \times 10^2$ cgse $\approx 0.5\,\mathcal{A}_{2\,\mathrm{lim}}$ and of the conversion coefficient $\beta \approx 10\%$. Such a magnitude of the conversion coefficient for the intensity of the same first harmonic can be achieved in a KDP crystal of the length 0.1 cm, see Eq. (3.36a) in [6.17].

Thus, a substance which is (because of the small value of the nonresonant susceptibility) not suitable for nonresonant parametric conversion, may turn out to facilitate resonant conversion, provided one inserts impurities with the following properties into the material: they should allow for two-photon transitions between the levels 2 and 1 (for which the resonance condition (6.1.1) is fulfilled) and have a sufficiently small line width of the transition $2 \rightarrow 1$.

For the density of matter $N \approx 10^{20}\,\mathrm{cm}^{-3}$, the value of the transformation coefficient $\beta \approx 10\%$ can be obtained with the lengths of the same order of magnitude as, or even shorter than, in KDP crystals. Note, however, that the above values of $(\boldsymbol{d} \cdot \boldsymbol{e}_2)$ and κ are optimal in the sense that when the quantity $|(\boldsymbol{d} \cdot \boldsymbol{e}_2)|$[5] becomes much smaller than 10^{-20} cgse, the conversion length increases. At the same time, as we have demonstrated earlier, β cannot be greater than $\beta_{\mathrm{lim}} = 0.43$. The increase of $|(\boldsymbol{d} \cdot \boldsymbol{e}_2)|$ leads to the diminishing of the conversion coefficient, see (6.1.17).

[5] Note that the decrease of $|(\boldsymbol{d} \cdot \boldsymbol{e}_2)|$ can be achieved if the field $\vec{\mathcal{E}}_2$ is oriented in a certain way with respect to the direction of $\boldsymbol{d}$.

It is also of interest to determine the distance over which the magnitude of the second harmonic field constitutes a considerable part of $\mathcal{A}_{2\,\mathrm{lim}}$, for instance, $\mathcal{A}_2 = 0.9\,\mathcal{A}_{2\,\mathrm{lim}}$. Assume that at the same time $\mathcal{A}_{2\,\mathrm{lim}} = \mu(\boldsymbol{d}\cdot\boldsymbol{e}_2)/\kappa$, where $\mu \ll 1$. Then, according to (6.1.16, 19),

$$z_{0,9} \approx [g_2\eta_0(\boldsymbol{d}\cdot\boldsymbol{e}_2)^2]^{-1}\left\{2,3 + 0,5\mu^2\frac{4\tau T(\boldsymbol{d}\cdot\boldsymbol{e}_2)^4}{\hbar^2\kappa^2}\right\} \quad ,$$

$$A_1(0) = \sqrt{\mu}\frac{(\boldsymbol{d}\cdot\boldsymbol{e}_2)}{\kappa} \quad . \tag{6.1.22}$$

As one can see from (6.1.22), the quantity $z_{0.9}$ strongly depends on the lifetime of the molecules of the nonlinear substance at level 2 if

$$\tau \gtrsim \tau_{\mathrm{cr}} = 1.14\hbar^2\kappa^2/\mu^2 T(\boldsymbol{d}\cdot\boldsymbol{e}_2)^4 \quad . \tag{6.1.23}$$

For the above values of the parameters of matter and $\mathcal{A}_1(0) = 2\times10^3$ cgse, we have $\beta_{\mathrm{lim}} = 0.2$, $\mu = 4\times10^{-2}$ and $\tau_{\mathrm{cr}} = 0.80\times10^{-8}$ s. When $\tau \approx \tau_{\mathrm{cr}}$, the quantity $z_{0.9} \approx 0.1$ cm.

Consider the qualitative aspects of the phenomena for incomplete synchronism with $\delta k \neq 0$. Introducing the new variable $\tilde{\theta} = \theta - \pi$ the last equation of (6.1.2) can be written as

$$\frac{d\tilde{\theta}}{dz} = -f(z)\sin\tilde{\theta} + \delta k \quad , \tag{6.1.24}$$

where $f(z)$ is the expression in front of $\sin\theta$ in (6.1.2).

We shall demonstrate that when $\delta k < \min f(z) = f_{\min}$ the function $\tilde{\theta}(z)$ is bounded. It will be assumed that $\delta k > 0$ (the situation when δk is negative can be treated in a similar manner if one substitutes $\tilde{\theta} \to -\tilde{\theta}$) and it should be borne in mind that $f(z)$ is also a positive function.

Let us introduce two functions $\Psi_1(z)$ and $\Psi_2(z)$ such that

$$\frac{d\Psi_1}{dz} \leq -f(z)\sin\Psi_1 + \delta k \quad , \quad \frac{d\Psi_2}{dz} \geq -f(z)\sin\Psi_2 + \delta k \quad . \tag{6.1.25}$$

Then, of $\Psi_1(0) \leq \tilde{\theta}(0) \leq \Psi_2(0)$, we have $\Psi_1(z) \leq \tilde{\theta}(z) \leq \Psi_2(z)$ [6.18]. Suppose that $\Psi_1 \equiv 0$ and Ψ_2 is equal to the solution of (6.1.24) in which $f(z)$ is replaced by $f_{\min}$. The following equations

$$\frac{d\Psi_2}{dz} = -f_{\min}\sin\Psi_2 + \delta k \geq -f(z)\sin\Psi_2 + \delta k \quad ,$$

$$\Psi_2(z) = 2\arctan\frac{f_{\min} + r - (f_{\min} - r)\mathcal{B}\exp(rz)}{\delta k[1 - \mathcal{B}\exp(rz)]} \quad , \quad \text{where} \tag{6.1.26}$$

$$r = \sqrt{f_{\min}^2 - (\delta k)^2} \quad , \mathcal{B} = \frac{\delta k \tan 0.5\Psi_2(0) - f_{\min} - r}{\delta k \tan 0.5\Psi_2(0) - f_{\min} + r} \quad ,$$

are valid.

It can be shown on the basis of (6.1.26) that Ψ_2 is approaching the limit

$$\Psi_2(\infty) = 2 \arctan \frac{f_{\min} - r}{\delta k} \quad .$$

If

$$(\delta k)^2 \ll f_{\min}^2 \quad , \tag{6.1.27}$$

this limit is $\Psi_2(\infty) \approx \delta k / f_{\min}$. Thus, if inequality (6.1.27) is satisfied, the quantity $\tilde{\theta}$ tends to be inside the range

$$0 \leq \tilde{\theta} \leq \delta k / f_{\min} \quad , \tag{6.1.28}$$

and in addition $\theta \approx \pi + \delta k f^{-1}$. At the same time, the rate of variation of $\tilde{\theta}$ with the increase of coordinate z, equal to $[f^2 - (\delta k)^2]^{1/2}$ [see (6.1.27)], is the minimal rate of this variation. As already noted, this rate is in the case of frequency doubling (or addition) proportional to $\mathcal{A}_1^2(z)/\mathcal{A}_2(z)$ (or $\mathcal{A}_1\mathcal{A}_3/\mathcal{A}_2$). In the present context the rate is at first (where $\mathcal{A}_2(0) \approx 0$) much greater than the minimal one.

In order to estimate $f_{\min}$, one can make use of the result of the analysis performed for $\delta k = 0$. When (6.1.15, 20) are valid, it follows from (6.1.2, 14) that

$$\min f \approx g_2(\boldsymbol{d} \cdot \boldsymbol{e}_2) \kappa \eta_0 \frac{\mathcal{A}_1^2(0)}{\mathcal{A}_{2\,\mathrm{lim}}} \approx g_2(\boldsymbol{d} \cdot \boldsymbol{e}_2)^2 \eta_0 \quad .$$

For the above values of the parameters of matter we find $f_{\min} \approx 10 - 100$. Therefore, when $|\delta k| < (3 - 10)\,\mathrm{cm}^{-1}$, in the range (6.1.28)

$$\cos\theta \approx -[1 - (\delta k / f)^2]^{1/2} \quad .$$

Thus, the following conclusion can be made: the fact that δk differs from zero does not lead to the breakdown of the phase but to its locking, however, it turns out to be imcomplete ($\cos\theta \neq -1$). Let us investigate the influence of this phenomenon on the amplitudes of the first and second radiation harmonics.

When one takes into account the difference of the phase value θ from π, instead of equations (6.1.9), one obtains in the case of negligible saturation, the system of equations ($\tilde{\theta} \ll 1$)

$$\frac{da_2}{d\zeta} = a_1^2\left(1 - \frac{\tilde{\theta}^2}{2}\right) - a_2 \quad , \quad \frac{da_1^2}{d\zeta} = 2a_1^2\left[a_2\left(1 - \frac{\tilde{\theta}^2}{2}\right) - a_1^2\right] \quad . \tag{6.1.29}$$

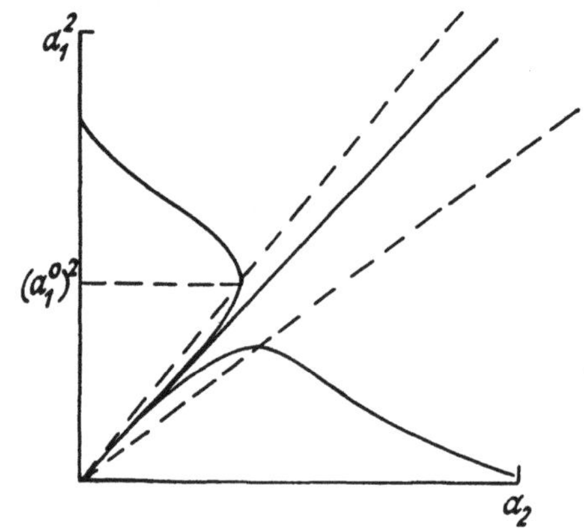

Fig. 6.1. The phase plane in the case of incomplete synchronism

If $a_1^2 \ll 1$ (i.e., if $\mathcal{A}_1^2 \ll (\boldsymbol{d}\cdot\boldsymbol{e}_2)^2/\kappa^2$), the quantity a_2 quickly attains the slow motion range, where

$$a_2 \approx a_1^2\left(1-\frac{\tilde{\theta}^2}{2}\right) \quad , \tag{6.1.30}$$

and remains there henceforth. Because of this,

$$\frac{da_1^2}{d\zeta} \approx -2a_1^4\tilde{\theta}^2 \quad , \quad a_1^2 \approx \frac{(a_1^0)^2}{1+2(a_1^0)^2\tilde{\theta}^2\zeta} \quad . \tag{6.1.31}$$

In these formulas a_1^0 is the value of a_1 for which the representing point falls into the range $a_1^2 - a_2 \approx a_1^2\tilde{\theta}^2/2$ (Fig. 6.1).

Expressions (6.1.30, 31) show that a_1 and a_2 approach a stable state of equilibrium $a_1 = a_2 = 0$ and not a locus of singular points $a_1^2 = a_2$, as in the case when $\delta k = 0$. According to (6.1.31), the velocity of this motion does not exceed $2a_1^4(0)\tilde{\theta}^2$.

In conclusion, we would like to make the following remark. If the only radiation in the medium is that of the frequency 2ω, its amplitude decreases by a factor of e as it passes the normalized path length $\zeta_1 = 1$. If at the boundary the fields of the first and second harmonics are such that $a_1^2/a_2 = 1$ holds, then the resonant field of the frequency 2ω propagates without absorption when $\delta k = 0$. In the presence of a wave detuning that satisfies condition (6.1.27), the parametric self-induced transparency is disturbed. However, it can still be observed, since, according to (6.1.30, 31), the field of the second harmonic will be attenuated by a factor less than two as radiation passes the normalized path $\zeta_2 = (2\tilde{\theta}^2 a_1^2(0))^{-1} \gg \zeta_1 = 1$. For $\mathcal{A}_1(0) = 2\times10^3$ cgse, $(\boldsymbol{d}\cdot\boldsymbol{e}_2)/\kappa \approx 10^4$ cgse and $\tilde{\theta} \approx 0.1$, the quantity ζ_2 corresponds to a distance of ≈ 20 cm.

6.1.2 On Resonant Frequency Doubling in Vapors and Gases

Gaseous media considered as active substances for frequency doubling attract the attention of researchers first of all because of the problem of constructing sources of coherent radiation in the ultraviolet (UV) and particularly vacuum ultraviolet (VUV) ranges. The use of resonant parametric interactions in vapors and gases in the development of new sources in the VUV range is caused by the fact that it is rather difficult to apply traditional methods of generation of intense coherent radiation in this range. The difficulties are associated with two factors. Firstly, the amplification coefficient of the active medium decreases with the third power of the wavelength and for these wavelengths practically there exist no mirrors with sufficiently high reflection coefficients. Secondly, in the VUV range lasers and nonlinear solid state elements give rise to strong absorption.

We shall dwell on the case of complete synchronism ($\delta k = 0$). According to the fourth equation of (6.1.2) the plane $\theta = \pi$ of the phase space is stable for both gaseous media and media containing identically oriented molecules. It is also obvious that at the initial interval of the nonlinear medium the velocity at which θ approaches π is much greater than the rate of variation of the field amplitudes $\mathcal{A}_j$. For this reason we follow the previous section in assuming $\theta = \pi$ to hold in the first three equations of (6.1.2).

Moreover, the population variation will be supposed to be negligible ($\eta = \eta_0$). In this case the population difference can be taken out of the angular brackets averaging only the molecular parameters over orientations. Note that the absence of the static field implies for gases an isotropic distribution of molecules over the possible orientations. Therefore, $\langle(\boldsymbol{d}\cdot\boldsymbol{e}_2)\kappa\rangle = 0$, and in (6.1.2) there are no terms giving rise to parametric conversion. Hence, the presence of an electrostatic field $\boldsymbol{e}_0 E_0$ is a necessary requirement for the parametric interaction of waves in a gas.

Taking the above into account and introducing normalized quantities

$$a_{1,3} = \sqrt{\frac{g_{3,1}}{g_2}\frac{\langle\kappa^2\rangle}{\langle(\boldsymbol{d}\cdot\boldsymbol{e}_2)^2\rangle}}\mathcal{A}_{1,3} \ , \quad a_2 = \sqrt{\frac{g_1 g_3}{g_2^2}\frac{\langle\kappa^2\rangle}{\langle(\boldsymbol{d}\cdot\boldsymbol{e}_2)^2\rangle}}\mathcal{A}_2 \ ,$$

$$\zeta = g_2\eta_0\langle(\boldsymbol{d}\cdot\boldsymbol{e}_2)^2\rangle z \ , \qquad \mathcal{F} = \frac{\langle(\boldsymbol{d}\cdot\boldsymbol{e}_2)\kappa\rangle}{\sqrt{\langle(\boldsymbol{d}\cdot\boldsymbol{e}_2)^2\rangle\langle\kappa^2\rangle}} \ , \tag{6.1.32}$$

we arrive at

$$\frac{da_{1,3}}{d\zeta} = -a_{3,1}(a_1 a_3 - \mathcal{F}a_2) \ , \quad \frac{da_2}{d\zeta} = -a_2 + \mathcal{F}a_1 a_3 \tag{6.1.33}$$

(linear absorption described by the coefficients α_j is neglected).

For a medium consisting of identically oriented molecules $\mathcal{F} = 1$ holds and, up to the accuracy of terms describing the saturation of populations and linear attenuation, (6.1.32, 33) reduce to (6.1.4, 5, 6). For gases, $\mathcal{F} \leq 1$ is valid.

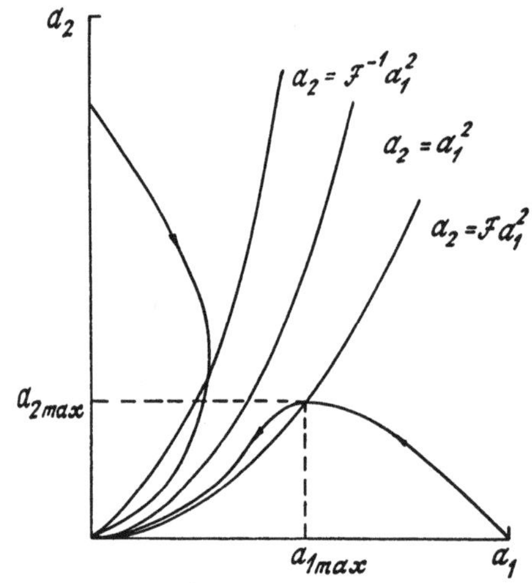

Fig. 6.2. The phase plane corresponding to (6.1.34)

Consider the doubling of frequency. If it is assumed that in (6.1.33) $a_3 \equiv a_1$, the following equations

$$a_1' = -a_1^3 + \mathcal{F} a_1 a_2 \ , \quad a_2' = -a_2 + \mathcal{F} a_1^2 \tag{6.1.34}$$

result. The phase plane corresponding to them is depicted in Fig. 6.2. Note first of all that the representing point, instead of approaching the locus of singular points $a_1^2 = a_2$, moves towards a stable stationary state $a_1 = a_2 = 0$. In other words, in gases there is even for completely synchronised interacting waves no complete self-induced transparency observed. Because of this, the second harmonic amplitude reaches its maximal value not at $z = \infty$, but at a certain point for which $a_{2\max} = \mathcal{F} a_{1\max}^2$. Introducing dimensional amplitudes, we obtain the following expression for the maximal amplitude of the second harmonic:

$$\mathcal{A}_{2\max} = \frac{\langle (\boldsymbol{d} \cdot \boldsymbol{e}_2)\kappa \rangle}{\langle (\boldsymbol{d} \cdot \boldsymbol{e}_2)^2 \rangle} \mathcal{A}_{1\max}^2 \quad . \tag{6.1.35}$$

This relationship enables one to formulate the requirements for the levels of matter, the transition between which would ensure the maximal conversion coefficient, and also to approximately evaluate this coefficient by replacing the quantity $\mathcal{A}_{1\max}$ by the input value of the pumping field amplitude $\mathcal{A}_1(0)$. To this end, let us compute the average quantities in (6.1.35) using the molecular orientation distribution function of the form

$$\frac{1}{4\pi} \exp\left(\frac{d_{11} E_0}{k_B T^0}\right) \approx \frac{1}{4\pi}\left(1 + \frac{d_{11} E_0}{k_B T^0}\right) \quad ,$$

where it is assumed that the matrix element d_{11} of the ground state dipole moment is the same for all sublevels. If we take $(\boldsymbol{d}\cdot\boldsymbol{e}_2) = d_{12} + \kappa_{20}E_0$ into account (where d_{12} and $\kappa_{20}E_0$ are, respectively, the projections on $\boldsymbol{e}_2$ of the intrinsic and the induced dipole moments of transition 1–2) we can obtain from (6.1.35) an approximate expression for the maximal conversion coefficient

$$\begin{aligned}\beta_{\max} &= \mathcal{A}_{2\max}/\mathcal{A}_1(0)\\ &\approx \frac{\langle\kappa_{11}d_{12}d_{11}/k_\mathrm{B}T^0\rangle_\mathrm{is} + \langle\kappa_{11}\kappa_{20}\rangle_\mathrm{is}}{\langle d_{12}^2\rangle_\mathrm{is} + \langle 2\kappa_{20}d_{12}d_{11}/k_\mathrm{B}T^0\rangle_\mathrm{is}E_0^2 + \langle\kappa_{20}^2\rangle_\mathrm{is}E_0^2}E_0\mathcal{A}_1(0) \quad .\end{aligned} \tag{6.1.36}$$

In this formula the symbol $\langle\,\rangle_\mathrm{is}$ means the averaging that includes the isotropic distribution function. For small values of the angular momentum it is equivalent to a summation over the magnetic quantum numbers of the ground and excited states (see Sect. 4.1).

In the case of polarized molecules, the term in the nominator of (6.1.36) caused by the orientation of particles in a constant field may exceed the contribution of the induced dipole moment in the doubling coefficient. Thus, when $d_{11} \sim d_{12} \sim 10^{-18}$ cgse, polarizability $\kappa_{20} \sim 10^{-24}\,\mathrm{cm}^3$ and $T^0 = 500\,\mathrm{K}$, the ratio $d_{12}d_{11}/\kappa_{20}k_\mathrm{B}T^0 \approx 10^2$. To estimate $\beta_{\max}$ in such molecules it can be assumed that the main contribution to the parametric part of nonlinear susceptibility only comes from resonant levels 1 and 2, i.e., that $\kappa_{11} \sim d_{12}|d_{22} - d_{11}|/\hbar\omega_1$. If the contribution of polarizability κ_{20} in (6.1.36) is neglected, the following estimate can be obtained:

$$\beta_{\max} \approx \frac{d_{11}|d_{22} - d_{11}|}{k_\mathrm{B}T^0\hbar\omega_1}E_0\mathcal{A}_1(0) \quad . \tag{6.1.37}$$

Consider a gas of nonpolar molecules. For this medium the quantity $\langle(\boldsymbol{d}\cdot\boldsymbol{e}_2)\kappa\rangle$ is proportional to the induced dipole moment of the resonant transition, and the maximal conversion coefficient is given by

$$\beta_{\max} \approx \frac{\langle\kappa_{20}\kappa_{11}\rangle_\mathrm{is}}{\langle\kappa_{20}^2\rangle_\mathrm{is}E_0^2 + \langle d_{12}^2\rangle_\mathrm{is}}E_0\mathcal{A}_1(0) \quad . \tag{6.1.38}$$

According to this formula, the maximal conversion in such a gas requires the existence of levels of matter between which the two-photon transition is allowed and $\langle\kappa_{20}^2E_0^2\rangle_\mathrm{is} \sim \langle d_{12}^2\rangle_\mathrm{is}$, i.e., the induced dipole moment of this transition being comparable with its intrinsic moment. It only makes sense to meet this requirement when the length $z_{\max}$, over which the optimal value $\beta_{\max}$ is reached, is not greater than the gaseous cell that produces the resonant doubling (or the length of the focal region formed during focussing of the pumping beam). Otherwise, as will be shown below (see Fig. 6.4), the induced dipole moment should be increased until the quantity $z_{\max}$ becomes equal to the cell length.

Expression (6.1.36) is based on the assumption that the generation of the second harmonic leaves the energy of the first harmonic practically unchanged. In a real situation it is necessary to take into account the two-photon absorption of the main frequency wave and the inverse parametric conversion of the field of double frequency in the pumping field. As one can easily see from (6.1.34), the latter process cannot be neglected if $\mathcal{F}^2>0.1$. In order to prove this, it is sufficient to substitute the maximal value $\mathcal{A}_{2\max} = \mathcal{F}a_{1\max}^2$ into the first of equations (6.1.34). In this case it is no longer possible to obtain a closed-form solution of the system (6.1.34), yet there exists a method for obtaining a simple graphical solution. Indeed, performing the change of variables $a_1 = a_{1n}$ and $a_2 = a_{2n}\sqrt{2}$, we arrive at the conclusion that the "velocity" of motion of the point with radius-vector ϱ (a_{1n}, a_{2n}) in "time" ζ is equal to $-\nabla\mathcal{U}$, where

$$\begin{aligned}\mathcal{U} &= 0.25(a_1^4 - 2\mathcal{F}a_1^2a_2 + a_2^2)\\ &= 0.25(a_{1n}^4 - 2\sqrt{2}\mathcal{F}a_{1n}^2a_{2n} + 2a_{2n}^2) \quad .\end{aligned} \tag{6.1.39}$$

Plotting the equipotential curves $\mathcal{U} =$ const and specifying the initial conditions it is possible to construct the phase trajectory of motion in the coordinates a_2, a_1^2. Whence the maximal value of the double frequency $a_{2\max}$ and the conversion length

$$\begin{aligned}\zeta_{\max} &= -\frac{1}{2}\int\limits_{a_1(0),a_2(0)}^{a_{1\max},a_{2\max}} \frac{d(\varrho^2)}{(\varrho\cdot\nabla\mathcal{U})}\\ &= -\int\limits_{a_1(0),a_2(0)}^{a_{1\max},a_{2\max}} \frac{2a_1da_1 + a_2da_2}{2(a_1^2 - \mathcal{F}a_2)a_1^2 + a_2(a_2 - \mathcal{F}a_1^2)} \quad ,\end{aligned} \tag{6.1.40}$$

are determined. Curve 1 in Fig. 6.3 depicts the trajectory of the motion arrived at this method for $\mathcal{F} = 0.5$. When $\mathcal{F}^2>0.1$, the integral in (6.1.40) can be computed with a sufficient accuracy if the trajectory is interpolated by a parabola $a_2 = c_1 - c_2a_1^2$, where the coefficients c_1 and c_2 can be found during the trajectory construction.

Figure 6.4 presents the qualitative dependence of the conversion coefficient β on the length of the gaseous cell.

As has been demonstrated above for $\mathcal{F}^2<0.1$ the parametric conversion of the field of doubled frequency in the pumping field [i.e., the second term in the rhs of the first of equations (6.1.34)] can be neglected. A similar assumption can also be made in the more general case of the addition of frequencies, see (6.1.33). In this approximation the solution of (6.1.33) has the form

$$a_2 = \mathcal{F}a_1(0)a_3(0)\zeta_0 \exp[-(\zeta+\zeta_0)][\mathrm{Ei}^*(\zeta+\zeta_0) - \mathrm{Ei}^*(\zeta_0)] \quad , \tag{6.1.41}$$

where

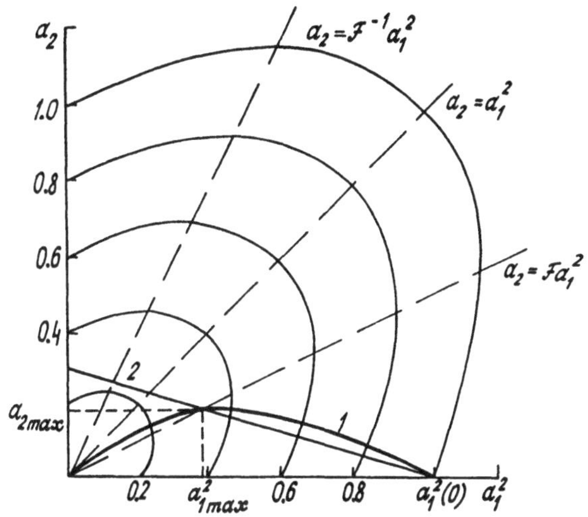

Fig. 6.3. The graphic solution of (6.1.34) by means of the potential function (6.1.39) for $\mathcal{F} = 0.5$. (*1*) is the phase trajectory for the boundary value of reduced pumping field, $a_1(0) = 1$ and (*2*) the interpolation straight line for calculating the conversion length (6.1.40)

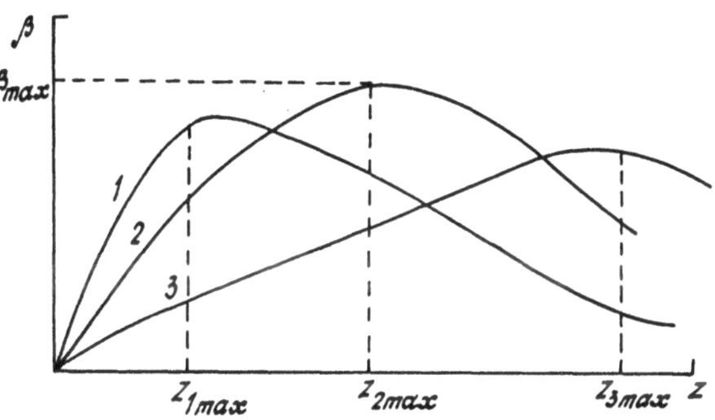

Fig. 6.4. The qualitative dependence of the coefficient of the pumping field conversion into the second harmonic on the conversion length in a gas consisting of nonpolar molecules. Curves *1, 2, 3* correspond to cases when the quantity $\langle \kappa_{20}^2 E_0^2 \rangle_{\text{is}}$ is greater than $d_{12}^2/3$, equal to $d_{12}^2/3$ and less than $d_{12}^2/3$

$$\zeta_0 = [a_1^2(0) - a_3^2(0)]^{-1} \quad , \tag{6.1.42}$$

and $\mathrm{Ei}^*(\zeta)$ is the integral exponential function. The solution obtained is valid for frequency doubling and for the addition of frequencies of fields with boundary values satisfying the inequality $|a_1^2(0) - a_3^2(0)| \ll 1$.

The amplitude of the second harmonic reaches its maximum after the wave passes the normalized path $\zeta_{\max}$ which is determined by the solution of

$$\mathrm{Ei}^*(\zeta_{\max} + \zeta_0) - \mathrm{Ei}^*(\zeta_0) = \frac{\exp(\zeta_{\max} + \zeta_0)}{\zeta_{\max} + \zeta_0} \quad . \tag{6.1.43}$$

According to (6.1.32), the cell length should be equal to

$$z_{\max} = \zeta_{\max}/g_2\langle(\boldsymbol{d}\cdot\boldsymbol{e}_2)^2\rangle\eta_0 \quad . \tag{6.1.44}$$

Let us give a few numerical estimates for the frequency doubling in a number of gases and vapors. As follows from (6.1.37), in gases consisting of polar molecules it is impossible to achieve a sufficiently high value of the coefficient of the pumping field conversion into the second harmonic. Indeed, for $d_{12} \approx (d_{22} - d_{11}) \approx 10^{-17}$ cgse, $d_{11}E_0/(k_B T^0) \lesssim 10^{-3}$, $\omega_1 = 10^{15}\,\mathrm{s}^{-1}$, we have $\beta_{\max} \lesssim 10^{-6}$.

Below we shall estimate the conversion coefficients in a gas consisting of nonpolar molecules. The magnitude of the dipole moment d_{12} of the transition between levels for which two-photon absorption is allowed ($\kappa \approx 10^{-24}\,\mathrm{cm}^3$) and which is necessary for calculating $\beta_{\max}$ is known for a few atoms and molecules. In our estimates it will be assumed that $d_{12} \approx 10^{-21}$ cgse, as is the case, for instance, with the transition $7D_{3/2} \rightarrow 6S_{1/2}$ in Cs [6.19] (the energy difference $\approx 25\,200\,\mathrm{cm}^{-1}$).

The Visible Range. The radiation frequency of a generator using the solution of rhodamine 6 G dye (corresponding to the wave length of 570–620 nanometers) can be doubled in vapors of K (transitions of $7d\,^2D$, $8d\,^2D$, $9s\,^2S$, $9d\,^2D \rightarrow 4s\,^2S$ [6.20]). Let us estimate the energy conversion coefficient $\beta^2_{\max}$ for $\mathcal{F} = 0.5$, $\kappa_{20}E_0 \approx 6 \times 10^{-22}$cgse, the pumping power 5 MW and the confocal parameter of the pumping radiation beam $b = 30$ cm [6.21]. Making use of the graphical integration method discussed above, the boundary value of the pumping field in the focal region $a_1^2(0) = 6$ gives rise to $\beta^2_{\max} = 0.16$. For the vapor pressure ≈ 2 mm Hg ($N_0 = 4 \times 10^{16}\,\mathrm{cm}^{-3}$), the conversion length is $L_{\max} \approx 8$ cm which is much less than the dimension b of the focal region.

The same magnitude of the conversion coefficient can be achieved in Cs when the frequency of the ruby generator radiation on the transition $9d\,^2D \rightarrow 6s\,^2S$ is doubled [6.20].

The Vacuum Ultraviolet Range. Consider the frequency doubling of the fourth harmonic via a neodymium laser in vapors of Cu (transition $6s\,^2D \rightarrow 4s\,^2S$ with the energy difference $75170.25\,\mathrm{cm}^{-1}$ [6.20]). In order for the resonance condition (6.1.8) to be exactly satisfied, it is necessary to tune the neodymium laser radiation frequency to the value $2\,\mathrm{cm}^{-1}$, which is easily done [6.22]. If the pumping field has the power 1 MW we have $\mathcal{F} = 0.7$, $\kappa_{20}E_0 \approx 10^{-21}$ cgse, the confocal beam parameter $b = 17$ cm and the quantity $\beta^2_{\max}$ is approximately equal to 0.1. For the vapor pressure of 1 mm Hg ($T^0 = 1600°$C [6.23]), we have $L_{\max} \approx 15$ cm.

Let us investigate the possibility of doubling the ninth harmonic of the neodymium laser as accomplished in [6.24]. In this case the suitable transition is $5p^{11}\,[I^{1/2}] \rightarrow 2p^6\,^1S$ in Ne provided the laser frequency is tuned to $32\,\mathrm{cm}^{-1}$ [6.22]. For the power of 0.1 MW, $b = 4$ cm and for the pressure of 6 mm Hg, the approximate values of $\beta^2_{\max}$ and $L_{\max}$ are 0.15 and 1.5 cm respectively.

6.2 Generation of the Second Harmonic of Resonant Pumping

Consider the frequency doubling for the case in which the frequency ω_{21} of one of the transitions of the matter molecules is nearly equal to the pumping field frequency

$$\omega_1 = \omega_{21} + \nu \ , \quad \nu \ll \omega_{21} \quad .$$

Under the assumption that the nonlinear component of the medium consists of identically oriented molecules, the equations that describe the behavior of the amplitudes of the first ($\mathcal{A}_1$) and the second ($\mathcal{A}_2$) harmonics and of the phase difference $\theta = 2\varphi_1 - \varphi_2$ are

$$\begin{aligned}
\frac{d\mathcal{A}_1}{dz} &= -g_1[(\boldsymbol{d}\cdot\boldsymbol{e}_2)^2\mathcal{A}_1 - \kappa^2\mathcal{A}_1\mathcal{A}_2^2 + 2\Delta(\boldsymbol{d}\cdot\boldsymbol{e}_2)\kappa\mathcal{A}_1\mathcal{A}_2 \sin\theta]\eta \ , \\
\frac{d\mathcal{A}_2}{dz} &= -g_2[\kappa^2\mathcal{A}_1^2\mathcal{A}_2 + (\boldsymbol{d}\cdot\boldsymbol{e}_2)\kappa(\cos\theta - \Delta\sin\theta)\mathcal{A}_1^2]\eta \ , \\
\frac{d\theta}{dz} &= [\delta k - 2g_1(\boldsymbol{d}\cdot\boldsymbol{e}_2)^2\Delta] + \Big\{\Delta\kappa^2(g_2\mathcal{A}_1^2 - 2g_1\mathcal{A}_2^2) \\
&\quad + \Delta(\boldsymbol{d}\cdot\boldsymbol{e}_2)\kappa\left(g_2\frac{\mathcal{A}_1^2}{\mathcal{A}_2} - 4g_1\mathcal{A}_2\right)\cos\theta + (\boldsymbol{d}\cdot\boldsymbol{e}_2)\kappa g_2\frac{\mathcal{A}_1^2}{\mathcal{A}_2}\sin\theta\Big\}\eta \quad .
\end{aligned} \tag{6.2.1}$$

The stationary population difference η which occurs on the rhs of (6.2.1) is equal to

$$\eta = \eta_0\left\{1 + \frac{4\tau T}{\hbar^2(1+\Delta^2)}[(\boldsymbol{d}\cdot\boldsymbol{e}_2)^2 + 2(\boldsymbol{d}\cdot\boldsymbol{e}_2)\kappa\mathcal{A}_2\cos\theta + \kappa^2\mathcal{A}_2^2]\mathcal{A}_1^2\right\}^{-1} \quad . \tag{6.2.2}$$

Expressions (6.2.1, 2) also contains the dimensionless detuning $\Delta = \nu T$ and the coefficients g_1 and g_2:

$$g_1 = \frac{4\pi\omega^2 NT}{\hbar k_1 c^2(1+\Delta^2)} \ , \quad g_2 = \frac{8\pi\omega^2 NT}{\hbar c^2 k_2(1+\Delta^2)} \quad . \tag{6.2.3}$$

Let us make a few remarks on the physical meaning of terms on the rhs of (6.2.1). The term which is proportional to $(\boldsymbol{d}\cdot\boldsymbol{e}_2)^2$ is associated with the single-photon absorption of the pumping radiation and the terms with κ^2 are responsible for Raman conversion of the second harmonic radiation into that of the first one. Parametric interaction is described by terms that are proportional to the product $(\boldsymbol{d}\cdot\boldsymbol{e}_2)\kappa$.

Note a specific feature of (6.2.1): the term which describes the variation of the first harmonic amplitude produced by parametric interaction, is proportional to the frequency detuning and vanishes under the conditions of exact resonance. In this regime the contribution of the resonant transition into the linear wave detuning $-2g_1(\boldsymbol{d}\cdot\boldsymbol{e}_2)^2\Delta$ vanishes too, as well as the nonlinear wave detuning $\Delta\kappa^2(g_2\mathcal{A}_1^2 - 2g_1\mathcal{A}_2^2)$ caused by the nonlinearity of the real susceptibility of matter during SRS, (see Chap. 2).

For large values of detuning ($\Delta \gg 1$) the parametric terms in the first two equations of (6.2.1) decrease as Δ^{-1}, i.e., slower than the terms associated with single-photon absorption and SRS (these decrease as Δ^{-2}). When $\Delta \sim \omega$, (6.2.1) become similar to the well-known truncated equations for the fields of the first and second harmonics that interact in a nonresonant quadratic medium without losses [6.17].

In general it is not possible to obtain a closed-form solution of (6.2.1). The maximal efficiency of the resonant field frequency doubling can be estimated however, under the conditions of complete synchronism, i.e., when $\delta k - 2g_1(\boldsymbol{d}\cdot\boldsymbol{e}_2)^2\Delta = 0$. The boundary conditions for the amplitudes at the input of the frequency doubler are

$$\mathcal{A}_1|_{z=0} = \mathcal{A}_1(0) \ , \quad \mathcal{A}_2|_{z=0} = 0 \ . \tag{6.2.4}$$

According to the last equation of (6.2.1), such boundary conditions ensure the phase locking in the second harmonic: the phase difference θ approaches its stable value

$$\theta_1 = \pi - \arctan\Delta \ , \tag{6.2.5}$$

with θ coming approaching θ_1 much faster than variations of the amplitude. Therefore, in the case of frequency doubling, θ can be replaced by θ_1 in the first two equations (6.2.1). This substitution results in equations, which in terms of normalized variables

$$\begin{aligned}&\zeta = g_1(\boldsymbol{d}\cdot\boldsymbol{e}_2)^2\eta_0 z \ , \quad a_1 = \sqrt{g_2/g_1}\,\kappa(\boldsymbol{d}\cdot\boldsymbol{e}_2)^{-1}\mathcal{A}_1 \ ,\\ &a_2 = \kappa(\boldsymbol{d}\cdot\boldsymbol{e}_2)^{-1}\mathcal{A}_2 \ , \mathcal{Y} = \frac{4\tau T(\boldsymbol{d}\cdot\boldsymbol{e}_2)^4 g_2}{\hbar^2\kappa^2(1+\Delta^2)g_1} \ ,\end{aligned} \tag{6.2.6}$$

have the form

$$\begin{aligned}\frac{da_1}{d\zeta} &= \frac{a_1[a_2^2 - 2a_2\Delta^2(1+\Delta^2)^{-1/2} - 1]}{1+\mathcal{Y}a_1^2[1-2a_2(1+\Delta^2)^{-1/2}+a_2^2]} \ ,\\ \frac{da_2}{d\zeta} &= \frac{a_1^2[(1+\Delta^2)^{1/2} - a_2]}{1+\mathcal{Y}a_1^2[1-2a_2(1+\Delta^2)^{-1/2}+a_2^2]} \ .\end{aligned} \tag{6.2.7}$$

The first integral of this system of equations

$$a_1^2 = a_1^2(0) + 2\Delta^2 \ln\left(1 - \frac{a_2}{\sqrt{1+\Delta^2}}\right) - 2\frac{1-\Delta^2}{\sqrt{1+\Delta^2}}a_2 - a_2^2 \quad , \tag{6.2.8}$$

makes it possible to reduce its integration to quadratures. However, it is only for $\Delta = 0$ that the exact solution of (6.2.7) can be expressed through elementary functions:

$$\begin{aligned}\zeta = (3 - a_1^2(0))^{-1}\Bigg\{&\left[\frac{1}{\sqrt{a_1^2(0)+1}} - \frac{1}{2}\right]\ln\left(\frac{a_2 + 1 + \sqrt{a_1^2(0)+1}}{1+\sqrt{a_1^2(0)+1}}\right)\\ &+ \ln(1-a_2) - \left[\frac{1}{\sqrt{a_1^2(0)+1}} + \frac{1}{2}\right]\ln\left|\frac{\sqrt{a_1^2(0)+1} - 1 - a_2}{\sqrt{a_1^2(0)+1}-1}\right|\Bigg\}\\ &+ \mathcal{Y}\left(a_2 - \frac{a_2^2}{2}\right) \quad .\end{aligned} \tag{6.2.9a}$$

In this formula, as well as below in (6.2.9b), the term proportional to $\mathcal{Y}$ describes the increase of length, which is necessary for the second harmonic amplitude to reach a given value due to saturation.

It follows from the solution of (6.2.9a) that if

$$a_1^2(0) < 3 \quad , \tag{6.2.10}$$

and $z \to \infty$, the sytem approaches a point in the phase space with coordinates

$$a_1(\infty) = 0 \ , \quad a_2(\infty) = \sqrt{a_1^2(0)+1} - 1 \quad , \tag{6.2.11}$$

and if at the doubler input the first harmonic amplitude satisfies an inequality inverse to (6.2.10), the integral curve terminates at the point

$$a_1(\infty) = \sqrt{a_1^2(0) - 3} \ , \quad a_2(\infty) = 1 \quad . \tag{6.2.12}$$

Thus, there are two straight lines on the phase plane, $a_1 = 0$ and $a_2 = 1$, which represent the geometric locus of equilibrium states. According to (6.2.7), the phase plane contains no other singularities. All integral curves lead to one of the above straight lines. The solution that corresponds to their intersection point results when $a_1(0) = \sqrt{3}$ and $a_2(0) = 0$. It is given by

$$\zeta = \frac{a_2}{4(1-a_2)} + \frac{1}{16}\ln\frac{3+a_2}{3(1-a_2)} + \mathcal{Y}\left(a_2 - \frac{a_2^2}{2}\right). \tag{6.2.9b}$$

It follows from the above that there is a limiting value of the second harmonic amplitude

$$\mathcal{A}_{2\mathrm{lim}} = (\boldsymbol{d}\cdot\boldsymbol{e}_2)\kappa^{-1} \quad , \tag{6.2.13}$$

that can be achieved with a pumping field $\mathcal{A}_1(0)$ exceeding the threshold amplitude

$$\mathcal{A}^{\mathrm{th}} = \sqrt{3g_1/g_2}(\boldsymbol{d}\cdot\boldsymbol{e}_2)\kappa^{-1}. \tag{6.2.14}$$

In this case, as in the case of the resonant second harmonic generation (see the previous section), the population difference η tends for $z \to \infty$ to its equilibrium value. Therefore, the quantity $\mathcal{A}^{\mathrm{th}}$ can be regarded as the threshold for parametric self-induced transparency for the doubling of the resonant pumping field frequency. But it is also possible to realize the parametric self-induced transparency for an arbitrary value of $a_1(0)$ if at the input $a_2 = 1$ holds.

Here are some numerical estimates. Assume that $N = 10^{20}\,\mathrm{cm}^{-3}$, $(\boldsymbol{d}\cdot\boldsymbol{e}_2) = 10^{-21}\,\mathrm{cgse}$, $\kappa = 10^{-24}\,\mathrm{cm}^3$. Then $\mathcal{A}^{\mathrm{th}} \to 1.1\times10^3\,\mathrm{cgse}$ and $\mathcal{A}_{2\mathrm{lim}} = 10^3\,\mathrm{cgse}$. Let us evaluate the length $z_{0.9}$ over which the field of the second harmonic becomes equal to $0.9\,\mathcal{A}_{2\infty}$. When $\tau = 10^{-8}\,\mathrm{s}$, $T = 10^{-11}\,\mathrm{s}$, it follows from (6.2.9b) that $z_{0.9} \to 6.4\,\mathrm{cm}$ if $\mathcal{A}_1(0) = \mathcal{A}^{\mathrm{th}}$. For $a_1^2(0) = 8$, the pumping field amplitude $\mathcal{A}_1(0) = 2\times10^3\,\mathrm{cgse}$ and, according to (6.2.9a), $z_{0.9} \to 1.2\,\mathrm{cm}$. Thus, as the quantity $\mathcal{A}_1(0)$ increases from the value $\mathcal{A}^{\mathrm{th}}$, the length $z_{0.9}$ decreases, and the amplitude $\mathcal{A}_{2\infty}$ does not depend on $\mathcal{A}_1(0)$. If, on the other hand, $\mathcal{A}_1(0) \ll \mathcal{A}^{\mathrm{th}}$, the second harmonic amplitude is proportional to the pumping field intensity

$$\mathcal{A}_{2\infty} \approx \frac{\kappa}{(\boldsymbol{d}\cdot\boldsymbol{e}_2)}\mathcal{A}_1^2(0) \quad . \tag{6.2.15}$$

One can see from (6.2.14, 15) that the magnitude of the conversion coefficient for low levels of the pumping field, as well as the rate at which the limiting value of the second harmonic amplitude (less than $\mathcal{A}^{\mathrm{th}}$) is reached, depend directly on the degree to which the two-photon transition between the resonance levels is allowed and the single-photon one is forbidden. As this degree increases, however, the limiting value of the amplitude (6.2.13) diminishes, while the length of conversion up to a given level increases.

Here are a few remarks on the effect of frequency detuning. It follows from the first equation of (6.2.7) that the parametric self-induced transparency is disturbed, i.e., the straight line $a_2 = \sqrt{1+\Delta^2}$ on which $da_2/d\zeta = 0$ (the analog of the straight line $a_2 = 1$ for $\Delta = 0$) does not represent the geometric locus of the equilibrium states. It is, however, an integral curve along which the amplitude a_1 attenuates. Because of this, all integral curves that begin at the axis $a_2 = 0$ should not intersect the line $a_2 = \sqrt{1+\Delta^2}$; they will instead terminate at the interval of the equilibrium state straight line $a_1 = 0$, with the end points $a_2 = 0$ and $a_2 = \sqrt{1+\Delta^2}$. Therefore, the limiting value of the second harmonic amplitude increases with the increase of frequency detuning:

$$\mathcal{A}_{2\lim} = (\boldsymbol{d} \cdot \boldsymbol{e}_2)\kappa^{-1}\sqrt{1+\Delta^2} \quad ,$$

compare with (6.2.13). In order to find the condition for obtaining the second harmonic with its amplitude close to its limiting value, one has to integrate equations (6.2.7). This can be done approximately, by decomposing the rhs of (6.2.8) into a power series with respect to a_2 and retaining two first terms of the expansion. We shall not present the resulting expression here because it is very cumbersome. It leads to the result that the condition for $\mathcal{A}_2 \sim \mathcal{A}_{2\lim}$ is that the pumping field be given by

$$a_1^2(0) \gtrsim 3 + 2\Delta^2 \quad .$$

6.3 Resonant Division of Frequency

The division of frequencies can be regarded as the simplest special case of the so-called decay processes, in which the pumping radiation energy with the frequency ω_2 goes over into the radiation energy at frequencies ω_1 and ω_3 satisfying condition (6.1.1). In the case of frequency division $\omega_3 = \omega_1$ holds. This process becomes resonant when one of the conditions

$$\omega_1 = \omega_{21} + \nu \quad , \tag{6.3.1}$$

$$\omega_2 = 2\omega_1 = \omega_{21} + \nu \tag{6.3.2}$$

is fulfilled. The phenomenon that occurs when the first of these equations is valid is opposed to the generation of the second harmonic by resonant pumping as described in the previous section. The process which is inverse to the frequency doubling under the conditions of resonance at the harmonic frequency (Sect. 6.1) is the division of frequency during the two-photon resonance (6.3.2) at the subharmonic frequency.

In accordance with the classification given in Sect. 2.4, both of the above types of frequency division belong to the second group of resonant parametric phenomena. It is characterized by a high pumping field intensity, as well as the fulfilment of certain additional conditions for the parameters of matter and frequency detuning which are necessary for phase locking. In every given case these conditions can be obtained from the equation for the phase difference. When the conditions for phase locking are not valid, it is necessary to analyze the equations for the field amplitudes together with those for the phase difference. In the general case this turns out to be quite difficult. However, for the frequency division during the exact resonance at the subharmonic frequency [when $\nu = 0$ in (6.3.1)] an exact particular solution of the problem is

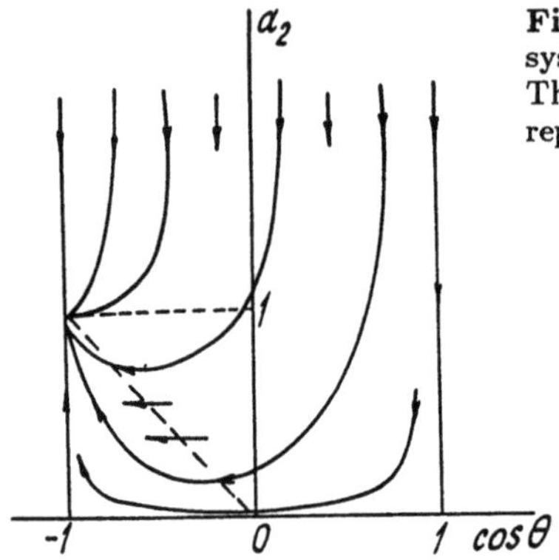

Fig. 6.5. The projection of the phase space for the system of equations (6.3.3) on the plane a_1 = const>0. The arrows indicate the direction of motion of the representing points projections

known [6.7]. Assuming a constant pumping field, the quasi-resonant case for this process ($\omega_{1,2} \gg \nu > T^{-1}$) is discussed in [6.25].

The authors of [6.6] investigated the frequency division of the resonant pumping field [see (6.3.2)] under the assumption that there is a resonator at the subharmonic frequency. This leads to the exclusion of the equation for the phase difference, because as the field in the resonator reaches its stationary distribution, those fluctuations for which θ corresponds to the stable value $\theta = \pi$, will increase.

Let us consider as an example, resonant subharmonic generation. When the frequency of matter coincides exactly with that of the subharmonic, (6.2.1) assume the form

$$\frac{da_1}{d\zeta} = \frac{a_1(a_2^2 - 1)}{1 + \mathcal{Y}a_1^2(1 + 2a_2 \cos\theta + a_2^2)} ,$$
$$\frac{da_2}{d\zeta} = -\frac{a_1^2(\cos\theta + a_2)}{1 + \mathcal{Y}a_1^2(1 + 2a_2 \cos\theta + a_2^2)} ,$$
$$\frac{d\theta}{d\zeta} = \frac{a_1^2/a_2}{1 + \mathcal{Y}a_1^2(1 + 2a_2 \cos\theta + a_2^2)} , \qquad (6.3.3)$$

where the notations are the same as in the preceding section, see (6.2.6).

The boundary conditions are

$$a_1|_{z=0} = a_{10} , \quad a_2|_{z=0} = a_{20} , \quad \theta|_{z=0} = \theta_0 . \qquad (6.3.4)$$

The phase space of system (6.3.3) in the space of variables $\{a_1, a_2, \cos\theta\}$ is bounded by the planes $\cos\theta = \pm 1$ (Fig. 6.5). The equilibrium states that can be physically realized occupy a plane $a_1 = 0$, and a straight line which represents the intersection of the planes $a_2 = 1$ and $\cos\theta = -1$ in this space. As follows from the first equation of (6.3.3), the quantity a_1 remains positive for any values of $a_2 \geq 0$ and $\cos\theta$. According to the last equation of (6.3.3) the plane $\cos\theta = 1$ is unstable for any positive a_1, a_2 and the phase difference θ approaches π ($\cos\theta \to -1$) as ζ increases. When $a_2 = +0$, the rate of θ coming closer to π becomes much greater than that of a_1 and a_2 change. For $a_2 < -\cos\theta$, the amplitude a_2 increases for any $a_1 > 0$ [see the second equation of (6.3.3)] since $\cos\theta$ assumes immediately a value close to -1 (and becomes $a_2 = +0 < -\cos\theta \approx 1$), the amplitude a_2 also remains always positive. For $a_2 > -\cos\theta$ and $a_1 > 0$, the quantity a_2 decreases regardless of the behavior of a_1.

To summarize, one can make the following conclusion: the representing point with the initial coordinates $\{\infty > a_1 > 0, \infty > a_2 > 0, \cos\theta_0 \neq 1\}$ should move towards the straight line ($\cos\theta = -1$, $a_2 = 1$) and when ζ is sufficiently large, the representing point should get into the infinitesimal vicinity of the line.

Let us solve (6.3.3). Multiplying the third equation of (6.3.3) by $\sin\theta$ and dividing the result by the second equation, we obtain a special case of a Riccati equation:

$$\frac{da_2}{d(\cos\theta)} = a_2 \frac{a_2 + \cos\theta}{\sin^2\theta} \quad . \tag{6.3.5}$$

Its solution can be found by standard methods:

$$a_2 = (\mathcal{T}_0 \sin\theta - \cos\theta)^{-1} \quad . \tag{6.3.6}$$

The constant $\mathcal{T}_0$ is determined by the initial conditions (6.3.4):

$$\mathcal{T}_0 = (1 + a_{20}\cos\theta_0)/a_{20}\sin\theta_0 \quad .$$

Depending on the relationship between θ_0 and a_{20}, the constant $\mathcal{T}_0$ can take on any value from $-\infty$ to $+\infty$. It follows from (6.3.6) that

$$\cos\theta = -\frac{1}{a_2(1+\mathcal{T}_0^2)} - \frac{|\mathcal{T}_0|}{a_2\sqrt{1+\mathcal{T}_0^2}} \times \sqrt{a_2^2 - \frac{1}{1+\mathcal{T}_0^2}} \quad . \tag{6.3.7}$$

We have introduced "$-|\mathcal{T}_0|$" because (6.3.7) should in phase space describe a cylindrical surface which is the geometric locus of phase trajectories that begin at the straight line ($a_2 = a_{20}$, $\cos\theta = \cos\theta_0$). For any a_{10}, a_{20}, $\mathcal{T}_0$, all these trajectories should pass through the straight line of equilibrium states ($a_2 = 1$, $\cos\theta = -1$). For this reason, the latter should lie on the above cylindrical surface regardless of sign and magnitude of $\mathcal{T}_0$. The symbol "$-|\mathcal{T}_0|$" in (6.3.7) ensures that this requirement is met.

Let us substitute the expression obtained for $\cos\theta$ into the first two equations of (6.3.3). By subsequently dividing one of them by the other and integrating the result, we arrive at another integral of motion:

$$a_1^2 + \left(\sqrt{a_2^2 - \frac{1}{1+\mathcal{T}_0^2}} + \frac{|\mathcal{T}_0|}{\sqrt{1+\mathcal{T}_0^2}}\right)^2$$
$$= a_{10}^2 + \left(\sqrt{a_{20}^2 - \frac{1}{1+\mathcal{T}_0^2}} + \frac{|\mathcal{T}_0|}{\sqrt{1+\mathcal{T}_0^2}}\right)^2 \quad . \tag{6.3.8}$$

Formulas (6.3.7, 8) and the first equation of (6.3.3) enable one to express the solution for a_1 via quadratures. The corresponding integrals can easily be computed, and the solution assumes the form:

$$2\zeta = \mathcal{Y}(a_1^2 - a_{10}^2) + 4\mathcal{Y}\Theta\left(\sqrt{B - a_1^2} - \sqrt{B - a_{10}^2}\right) + (B - 4\Theta^2)^{-1}$$
$$\times \ln\left\{\left|\frac{1-\sqrt{1-(a_1^2/B)}}{1-\sqrt{1-(a_{10}^2/B)}}\right|^{1+(2\Theta/\sqrt{B})}\right.$$
$$\times \left.\left|\frac{\sqrt{1-(a_{10}^2/B)} - (2\Theta/\sqrt{B})}{\sqrt{1-(a_1^2/B)} - (2\Theta/\sqrt{B})}\right|^2 \left|\frac{1+\sqrt{1-(a_1^2/B)}}{1+\sqrt{1-(a_{10}^2/B)}}\right|^{1-(2\Theta/\sqrt{B})}\right\} . \tag{6.3.9}$$

The boundary conditions for a_{20} and θ_0 are incorporated into this solution through constants

$$B = \left(\sqrt{a_{20}^2 - \frac{1}{1+\mathcal{T}_0^2}} + \frac{|\mathcal{T}_0|}{\sqrt{1+\mathcal{T}_0^2}}\right)^2 + a_{10}^2 \ ,$$
$$\Theta = \frac{|\mathcal{T}_0|}{\sqrt{1+\mathcal{T}_0^2}} \quad . \tag{6.3.10}$$

When $\mathcal{T}_0 = 0$, (6.3.9) reduces to the solution obtained in [6.7].

The first two terms in (6.3.9), proportional to $\mathcal{Y}$, reflect the influence of saturation on the coordinate dependence of the subharmonic intensity. These terms are finite for any a_1 and therefore they do not determine the asymptotic behavior of a_1^2 for large distances. The rhs of (6.3.9) may become infinite if one of the factors in the argument of logarithm vanishes or goes to infinity. There are two asymptotic values of the first harmonic amplitude.

1) When $\sqrt{B}<2\Theta$, the increase of ζ is accompanied by the quantity a_1^2 approaching zero (the representing point in the phase space reaches the plane of stationary states $a_1 = 0$).

2) If $\sqrt{B}>2\Theta$ and $\zeta \to \infty$ ($z \to \infty$), the quantity a_1^2 tends to $B-4\Theta^2$. As follows from (6.3.8) and (6.3.7), in this case $a_2 \to 1$ and $\cos\theta \to -1$ [the representing point arrives at the straight line of stationary states ($a_2 = 1$, $\cos\theta = -1$)]. Note that under such conditions at large distances from the input, the population difference comes close to its equilibrium value – in spite of the fact that the amplitudes of both fields differ from zero. This means that parametric self-induced transparency of macroscopic volumes of matter can also be observed in parametric processes of the second group (according to the classification of Sect. 2.4). The condition

$$\sqrt{B}>2\Theta \tag{6.3.11}$$

can be called the threshold of parametric self-induced transparency. It is of interest to note that this threshold depends on the difference of phases of interacting waves at the boundary (θ_0) as well.

If $\sqrt{B}>2\Theta$, the asymptotic value of a_1^2 is

$$\overline{a}_1^2 = a_{10}^2 + \left(\sqrt{a_{20}^2 - \frac{1}{1+\mathcal{T}_0^2}} + \frac{|\mathcal{T}_0|}{\sqrt{1+\mathcal{T}_0^2}}\right)^2 - \frac{4\mathcal{T}_0^2}{1+\mathcal{T}_0^2} \quad . \tag{6.3.12}$$

The subharmonic is amplified (i.e., $\overline{a}_1^2>a_{10}^2$) when

$$a_{20}^2>1 \quad . \tag{6.3.13}$$

The corresponding value of a_{20}^2 can be called the threshold of sub-harmonic amplification. This threshold does not depend on $\mathcal{T}_0$ and, therefore, on the input phase difference. The quantity $\mathcal{T}_0$ determines the value of $\overline{a}_1^2$, i.e., the efficiency of conversion of pumping energy (proportional a_2^2) into subharmonic energy (proportional a_1^2). It can easily be shown that the condition of the most intense conversion is

$$|\mathcal{T}_0| = \infty \quad . \tag{6.3.14}$$

In this case

$$\overline{a}_1^2 - a_{10}^2 = (a_{20}-1)(a_{20}+3) \quad . \tag{6.3.15}$$

The conversion efficiency is minimal for $|\mathcal{T}_0| = 0$ when a_1^2 is increased by a_{20}^2-1. If $a_{20}^2 \gg 1$, the increment of the subharmonic intensity can be approximately estimated (from above) regardless of $\mathcal{T}_0$:

$$\overline{a}_1^2 - a_{10}^2 \lesssim a_{20}^2 \quad . \tag{6.3.16}$$

In dimensional variables this estimate (for $a_{10}^2 \ll a_{20}^2$) and the condition of amplification threshold can be written as

$$\overline{\mathcal{A}}_1^2 \lesssim 0.5\,\mathcal{A}_{20}^2 \ , \quad \mathcal{A}_{20}^2 > (\boldsymbol{d}\cdot\boldsymbol{e}_2)^2\kappa^{-2} \quad .$$

Numerical estimates of the quantities appearing on the rhs of the last inequality have been given in the preceding sections. As follows from these results, the amplification of the subharmonic is only possible if the pumping field has a very large intensity; then the limiting value of the subharmonic energy is about half the energy of the pumping field.

6.4 Generation of the Difference Frequency During Stimulated Raman Scattering

In a substance in which stimulated Raman scattering occurs there are two waves: the pumping radiation wave with a frequency ω_p and the wave of the Stokes component of scattered radiation with a frequency ω_s. Obviously, in this case the difference frequency coincides with the frequency of molecular vibrations ω_{21} in the medium:

$$\omega_\mathrm{p} - \omega_\mathrm{s} = \omega_\mathrm{v} = \omega_{21} \quad . \tag{6.4.1}$$

Therefore, generation of the difference frequency during stimulated Raman scattering is a resonant parametric process.

The possibility of such a phenomenon was first mentioned in [6.26, 27]. The authors of [6.8, 12, 14, 25, 28, 29] conducted its theoretical analysis. In [6.25, 28], generation of the resonant field was considered in the approximation of the constant pumping field, which did not allow the authors to obtain certain important characteristics, e.g., the limiting value of the resonant field and the optimal conditions of the process. An analysis of equations that take into account variations of all three fields can be found in [6.8]. It is based on the example of a medium consisting of identically oriented particles. Reference [6.12] presents a theoretical study for the case when ω_{21} is the rotational transition frequency. This paper, as well as [6.14], is devoted to the question of the effect of the pumping wave and the Stokes component polarization on generation of the resonant field. The purpose of [6.29] is the analysis of the possibility and the conditions of effective generation of IR radiation during SRS on the vibrational levels of molecules in a gas subject to the action of an electrostatic field. (In [6.8, 28] it has been suggested to use a constant electric field to break the central symmetry of a gas or a liquid.) Reference [6.29] also describes experiments, the results of which agree with the conclusions obtained in the theoretical part of the paper.

Generation of the resonant IR field during SRS was first observed by the authors of [6.30] who reported that a wavelength of 2.41 microns, a power of about 40 W had been obtained in hydrogen. Reference [6.31] demonstrates the possibility of a substantial improvement of the efficiency of the resonant IR field generation by means of spatial modulation of the electrostatic field. As far as experimental studies are concerned, we should also mention [6.32] which describes the generation of a coherent IR field of the above type as applied to the analysis of the vibrational-rotational transition $Q_1(1)$ of the hydrogen molecule.

In this section two models of the medium will be considered: a solid with identically oriented impurity molecules and a gas placed in a constant electric field; the latter can be either uniform or spatially modulated.

6.4.1 Generation of Resonant Radiation During SRS in a Medium Consisting of Identically Oriented Molecules

The equations for complex amplitudes of the pumping field C_p, the Stokes component C_s and the light wave at the molecular vibration frequency C_v read

$$\begin{aligned}
&\frac{dC_\mathrm{v}}{dz} + g_\mathrm{v}[\langle(\boldsymbol{d}\cdot\boldsymbol{e}_\mathrm{v})^2\eta\rangle C_\mathrm{v} + \langle(\boldsymbol{d}\cdot\boldsymbol{e}_\mathrm{v})\kappa\eta\rangle C_\mathrm{s}^* C_\mathrm{p}\,\mathrm{e}^{\mathrm{i}\delta kz}] = 0\\
&\frac{dC_\mathrm{s}}{dz} - g_\mathrm{s}[\langle(\boldsymbol{d}\cdot\boldsymbol{e}_\mathrm{v})\kappa\eta\rangle C_\mathrm{v}^* C_\mathrm{p}\,\mathrm{e}^{\mathrm{i}\delta kz} + \langle\kappa^2\eta\rangle C_\mathrm{s}|C_\mathrm{p}|^2] = 0\\
&\frac{dC_\mathrm{p}}{dz} + g_\mathrm{p}[\langle(\boldsymbol{d}\cdot\boldsymbol{e}_\mathrm{v})\kappa\eta\rangle C_\mathrm{v} C_\mathrm{s}\,\mathrm{e}^{-\mathrm{i}\delta kz} + \langle\kappa^2\eta\rangle C_\mathrm{p}|C_\mathrm{s}|^2] = 0 \quad .
\end{aligned} \tag{6.4.2}$$

In the case under consideration, the angular brackets that denote the averaging over molecular orientations can be omitted. In the above formulas, $\delta k = k_\mathrm{v} + k_\mathrm{s} - k_\mathrm{p}$ is the wave detuning, κ is the component of the second order polarizability tensor for the resonant transition corresponding to the process under consideration, $\boldsymbol{e}_\mathrm{v}$ is the unit polarization vector of the field with the frequency ω_v, $g_j = 2\pi N\omega_j^2 T/c^2 k_j\hbar$, and the spatial distribution of the stationary population difference is given by

$$\eta = \eta_0\left\{1 + \frac{4\tau T}{\hbar^2}|(\boldsymbol{d}\cdot\boldsymbol{e}_\mathrm{v})C_\mathrm{v} + \kappa C_\mathrm{s}^* C_\mathrm{p}\mathrm{e}^{\mathrm{i}\delta kz}|^2\right\}^{-1} \quad . \tag{6.4.3}$$

In terms of normalized real amplitudes

$$\begin{aligned}
&a_\mathrm{v} = \frac{\sqrt{g_\mathrm{s}g_\mathrm{p}}}{g_\mathrm{v}}\frac{\kappa}{(\boldsymbol{d}\cdot\boldsymbol{e}_\mathrm{v})}|C_\mathrm{v}| \ , \quad a_\mathrm{s} = \sqrt{\frac{g_\mathrm{p}}{g_\mathrm{v}}}\frac{\kappa}{(\boldsymbol{d}\cdot\boldsymbol{e}_\mathrm{v})}|C_\mathrm{s}| \ ,\\
&a_\mathrm{p} = \sqrt{\frac{g_\mathrm{s}}{g_\mathrm{v}}}\frac{\kappa}{(\boldsymbol{d}\cdot\boldsymbol{e}_\mathrm{v})}|C_\mathrm{p}| \ ,
\end{aligned} \tag{6.4.4}$$

normalized distance

$$\zeta = g_{\mathrm{v}}\eta_0(\boldsymbol{d}\cdot\boldsymbol{e}_{\mathrm{v}})^2 z \quad , \tag{6.4.5}$$

and the phase difference

$$\theta = \delta k z - \varphi_{\mathrm{p}} + \varphi_{\mathrm{s}} + \varphi_{\mathrm{v}} \quad , \tag{6.4.6}$$

(6.4.2, 3) have the form

$$\begin{aligned}
\frac{da_{\mathrm{v}}}{d\zeta} &= -(a_{\mathrm{v}} + a_{\mathrm{s}}a_{\mathrm{p}} \cos\theta)\frac{\eta}{\eta_0} \ , \\
\frac{da_{\mathrm{s}}}{d\zeta} &= a_{\mathrm{p}}(a_{\mathrm{v}} \cos\theta + a_{\mathrm{p}}a_{\mathrm{s}})\frac{\eta}{\eta_0} \ , \\
\frac{da_{\mathrm{p}}}{d\zeta} &= -a_{\mathrm{s}}(a_{\mathrm{v}} \cos\theta + a_{\mathrm{s}}a_{\mathrm{p}})\frac{\eta}{\eta_0} \ , \\
\frac{d\theta}{d\zeta} &= \delta + \frac{\eta}{\eta_0}\left[\frac{a_{\mathrm{s}}^2 - a_{\mathrm{v}}^2}{a_{\mathrm{s}}a_{\mathrm{v}}}a_{\mathrm{p}} + \frac{a_{\mathrm{v}}a_{\mathrm{s}}}{a_{\mathrm{p}}}\right] \sin\theta \ , \\
\frac{\eta}{\eta_0} &= \{1 + \mathcal{Y}(a_{\mathrm{v}}^2 + 2a_{\mathrm{v}}a_{\mathrm{s}}a_{\mathrm{p}} \cos\theta + a_{\mathrm{p}}^2a_{\mathrm{s}}^2)\}^{-1} \quad .
\end{aligned} \tag{6.4.7}$$

Here

$$\delta = \delta k / g_{\mathrm{v}}\eta_0(\boldsymbol{d}\cdot\boldsymbol{e}_{\mathrm{v}})^2 \ , \quad \mathcal{Y} = 4\tau T(\boldsymbol{d}\cdot\boldsymbol{e}_{\mathrm{v}})^4 g_{\mathrm{v}}^2/\hbar^2\kappa^2 g_{\mathrm{s}}g_{\mathrm{p}} \quad .$$

As one can easily see from the equations for a_{v} and a_{s} in (6.4.7), the increase of the total energy of the resonant and Raman fields can only be caused by Raman interaction. Indeed, multiplying the first of them by a_{v} and the second by a_{s} and adding the results, we obtain

$$\frac{d(a_{\mathrm{v}}^2 + a_{\mathrm{s}}^2)}{d\zeta} = 2(a_{\mathrm{s}}^2a_{\mathrm{p}}^2 - a_{\mathrm{v}}^2)\frac{\eta}{\eta_0} \quad .$$

On the other hand, parametric interaction described by expressions that are proportional to $\cos\theta$ only affects the redistribution of energy between fields with frequency ω_{v} and ω_{s}. If the difference $a_{\mathrm{s}}^2 - a_{\mathrm{v}}^2$ is positive at the boundary ($z = 0$), it remains positive for any $z>0$ since it follows from the first two equations of (6.4.7) that

$$\frac{d(a_{\mathrm{s}}^2 - a_{\mathrm{v}}^2)}{dz} = 2(\boldsymbol{d}\cdot\boldsymbol{e}_{\mathrm{v}})^2 g_{\mathrm{v}}(a_{\mathrm{s}}^2a_{\mathrm{p}}^2 + 2a_{\mathrm{v}}a_{\mathrm{s}}a_{\mathrm{p}} \cos\theta + a_{\mathrm{v}}^2)\eta \geq 0 \tag{6.4.8}$$

(in a non-inverted medium we have always $\eta>0$).

The first integral

$$a_{\mathrm{s}}^2 + a_{\mathrm{p}}^2 = a_{\mathrm{s}0}^2 + a_{\mathrm{p}0}^2 \tag{6.4.9}$$

can be found without difficulty from the second and third equations of (6.4.7); it enables one to exclude the equation for a_p from further analysis. Taking into account the phase locking which occurs under the boundary conditions,

$$a_{\mathrm{p}0}\ ,\quad a_{\mathrm{s}0} \gg a_{\mathrm{v}0} \approx 0\ , \tag{6.4.10}$$

makes it possible to exclude at the first step also the equation for θ. This is justified since for $a_\mathrm{v} \to 0$, the phase difference θ varies along the z axis much more rapidly than the quantities a_v and a_s, and it approaches a region close to the value $\theta_1 = \pi$, which is stable when $\delta k = 0$. Under the initial conditions (6.4.10) relation (6.4.8) ensures that $a_\mathrm{s}^2 - a_\mathrm{v}^2$ preserves its sign as a function of z. Hence, in the space a_v, a_s, θ the plane $\theta = \theta_1$ will be stable for any z. Therefore, it can in a first approximation be assumed that $\theta = \pi$ holds in the first two equations of (6.4.7). Using (6.4.9) one can write them in the simple form

$$\begin{aligned}\frac{da_\mathrm{v}}{d\zeta} &= \left(a_\mathrm{s}\sqrt{a_{\mathrm{p}0}^2 + a_{\mathrm{s}0}^2 - a_\mathrm{s}^2} - a_\mathrm{v}\right)\frac{\eta}{\eta_0}\ ,\\ \frac{da_\mathrm{s}}{d\zeta} &= \sqrt{a_{\mathrm{p}0}^2 + a_{\mathrm{s}0}^2 - a_\mathrm{s}^2}\left(a_\mathrm{s}\sqrt{a_{\mathrm{p}0}^2 + a_{\mathrm{s}0}^2 - a_\mathrm{s}^2} - a_\mathrm{v}\right)\frac{\eta}{\eta_0}\ .\end{aligned} \tag{6.4.11}$$

When the spatial synchronization is complete ($\delta k = 0$), these equations are exact.

In phase space the above system of equations has a line of equilibrium states with nonvanishing values of the fields a_{vlim} and a_{slim}, which is represented by

$$a_{\mathrm{vlim}} = a_{\mathrm{slim}}\sqrt{a_{\mathrm{p}0}^2 + a_{\mathrm{s}0}^2 - a_{\mathrm{slim}}^2}\ . \tag{6.4.12}$$

When the quantities a_v and a_s satisfy condition (6.4.12) then it follows from the last equation of (6.4.7) that the population difference is equal to η_0. In other words, the regime of parametric self-induced transparency is achieved if $\delta k = 0$. When the fields at the input are such that $\theta = \pi$ and $a_{\mathrm{v}0} = a_{\mathrm{s}0}a_{\mathrm{p}0}$, this regime can also establish itself over the whole thickness of the nonlinear medium.

Let us determine the dependence of the limiting value of the resonant field $a_\mathrm{v}|_{z=\infty} = a_{\mathrm{vlim}}$ on the magnitudes of fields at the boundary. If the first equation of (6.4.11) is divided by the second and the result is integrated, we obtain the phase trajectory of system (6.4.11):

$$a_\mathrm{v} = a_{\mathrm{v}0} + \arcsin\frac{a_\mathrm{s}}{\sqrt{a_{\mathrm{p}0}^2 + a_{\mathrm{s}0}^2}} - \arcsin\frac{a_{\mathrm{s}0}}{\sqrt{a_{\mathrm{p}0}^2 + a_{\mathrm{s}0}^2}}\ . \tag{6.4.13}$$

When $z \to \infty$, the representing point moving along this trajectory should reach a location whose coordinates satisfy condition (6.4.12). Hence one can easily obtain an equation for a_{vlim}:

$$\frac{2a_{\mathrm{vlim}}}{a_{\mathrm{p0}}^2 + a_{\mathrm{s0}}^2} = \sin 2\left(a_{\mathrm{vlim}} - a_{\mathrm{av0}} + \arcsin\frac{a_{\mathrm{s0}}}{\sqrt{a_{\mathrm{p0}}^2 + a_{\mathrm{s0}}^2}}\right) \quad . \tag{6.4.14}$$

From the practical point of view, the parametric generation of the difference frequency for low values of the Stokes component field at the boundary is of great interest

$$a_{\mathrm{v0}} = 0 \ , \quad a_{\mathrm{s0}} \ll a_{\mathrm{p0}} \quad . \tag{6.4.15}$$

Note, however, that it is impossible to assume $a_{\mathrm{s0}} = 0$ since in this case the fields cannot increase:

$$\left.\frac{da_{\mathrm{v}}}{d\zeta}\right|_{\zeta=0} = \left.\frac{da_{\mathrm{s}}}{d\zeta}\right|_{z=0} = 0 \quad .$$

In real experiments the role of the input signal at the frequency ω_{s} is played by the spontaneous emission of the Stokes component by the initial section of the nonlinear medium.

As follows from (6.4.12, 14, 15), the value $a_{\mathrm{p0}} = a_{\mathrm{p\,cr}} = 1$ corresponds under these conditions to a sharp increase of the limiting magnitude of the resonant field (Fig. 6.6). Its generation, however, is also possible when $a_{\mathrm{p0}} < 1$ since

$$a_{\mathrm{v\,lim}} \approx a_{\mathrm{s0}} a_{\mathrm{p0}} / (1 + a_{\mathrm{s0}}^2 - a_{\mathrm{p0}}^2) \quad .$$

Certainly, this expression is only applicable if $a_{\mathrm{p0}}^2 \ll 1 + a_{\mathrm{s0}}^2$.

It must be recalled that when $a_{\mathrm{p0}} > 1$, in the approximation of the constant pumping field, the amplitude of the field with the frequency ω_{v} tends to infinity.

In order to estimate the quantity $a_{\mathrm{v\,lim}}$ when $a_{\mathrm{p\,cr}}$ is slightly exceeded, let us assume that $a_{\mathrm{v\,lim}}$ is equal to $\tilde{a}_{\mathrm{v}}$, the coordinate of the maximum of curve (6.4.14):

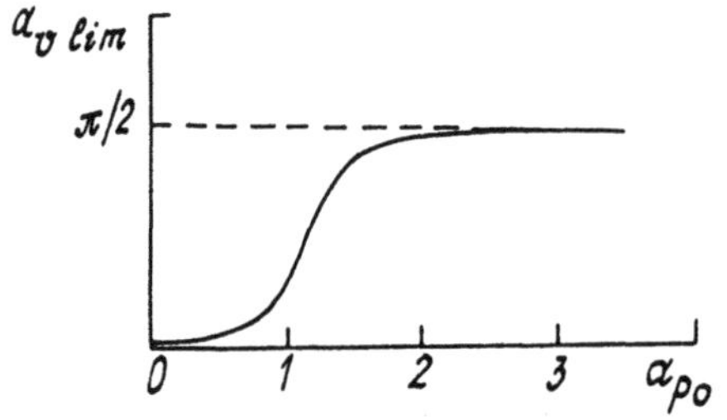

Fig. 6.6. The maximal possible value of the resonance frequency field as a function of the pumping field boundary value for $a_{\mathrm{s0}} \ll 1$

$$\tilde{a}_{\mathrm{v}} = 0.5(a_{\mathrm{p}0}^2 + a_{s0}^2) \quad . \tag{6.4.16}$$

In the general case the equation $a_{\mathrm{v\,lim}} = \tilde{a}_{\mathrm{v}}$ is valid if $a_{\mathrm{p}0}$ and $s_{\mathrm{s}0}$ are such that

$$a_{\mathrm{p}0}^2 + a_{\mathrm{s}0}^2 + 2\arcsin\frac{a_{\mathrm{s}0}}{\sqrt{a_{\mathrm{p}0}^2 + a_{\mathrm{s}0}^2}} - 2a_{\mathrm{v}0} = \frac{\pi}{2} \quad . \tag{6.4.17}$$

Using (6.4.17) and boundary conditions (6.4.15) we have

$$a_{\mathrm{v\,lim}} \approx -a_{\mathrm{p}0}^{-1} + \sqrt{a_{\mathrm{p}0}^{-2} - a_{\mathrm{p}0}^2 + \frac{\pi}{2}} \quad .$$

One can see from this formula that

$$a_{\mathrm{v\,lim}} = \tilde{a}_{\mathrm{v}} = \frac{\pi}{4}$$

can be reached when the pumping field amplitude is still less than $a_{\mathrm{v}0} = \sqrt{\pi/2} \approx 1.25$, i.e., when the critical value $a_{\mathrm{v\,cr}} = 1$ is only slightly exceeded. At the same time, the input value of the Stokes component, which is necessary for the amplitude to reach the quantity $a_{\mathrm{p\,cr}}$, is very small.

To conclude the study in the case $\delta k = 0$, we have to determine the length ζ_1 corresponding to a certain specified value of the resonant field $a_{\mathrm{v}1}$ being attained. This length is

$$\begin{aligned}\zeta_1 = 2\int\limits_{a_{\mathrm{v}0}}^{a_{\mathrm{v}1}} & \frac{da_{\mathrm{v}}}{(a_{\mathrm{p}0}^2 + a_{\mathrm{s}0}^2)\sin 2\left(a_{\mathrm{v}} - a_{\mathrm{v}0} + \arcsin\frac{a_{\mathrm{s}0}}{\sqrt{a_{\mathrm{p}0}^2+a_{\mathrm{s}0}^2}}\right) - 2a_{\mathrm{v}}} \\ & + \frac{\mathcal{Y}}{4}\left\{a_{\mathrm{p}0}^2 - a_{\mathrm{s}0}^2 + 2(a_{\mathrm{v}0}^2 - a_{\mathrm{v}1}^2) - (a_{\mathrm{p}0}^2 + a_{\mathrm{s}0}^2)\right. \\ & \left.\times\ \cos 2\left(a_{\mathrm{v}1} - a_{\mathrm{v}0} + \arcsin\frac{a_{\mathrm{s}0}}{\sqrt{a_{\mathrm{p}0}^2 + a_{\mathrm{s}0}^2}}\right)\right\}.\end{aligned} \tag{6.4.18}$$

Consider the limits of applicability of solutions obtained in the previous section when $\delta k \neq 0$. It can be easily shown in a manner similar to that of Sect. 6.1 that if

$$(\delta k)^2 \ll f_{\min}^2 \quad , \tag{6.4.19}$$

where $f_{\min}$ is the minimum of the factor near $\sin\theta$ in the fourth equation of (6.4.7), the phase difference θ approaches the value $\theta = \pi + \tilde{\theta}$, where

$$|\tilde{\theta}| \leq |\delta k| f_{\min}^{-1} \ll 1 \quad . \tag{6.4.20}$$

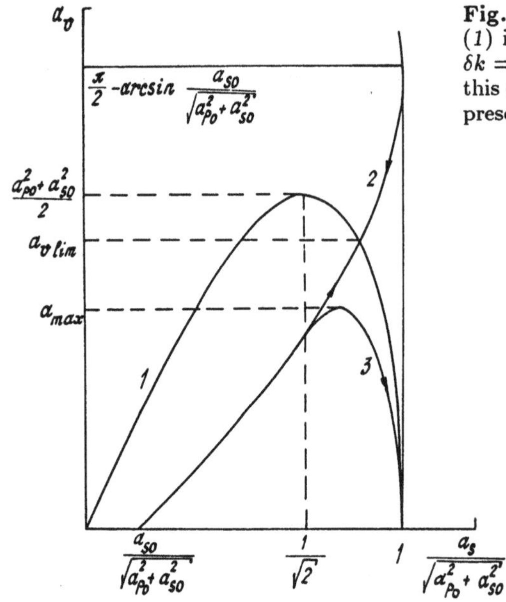

Fig. 6.7. The phase plane for $a_{p0} > a_{p\,cr} = 1$. (*1*) is the equilibrium states' line (6.4.12) for $\delta k = 0$, (*2*) is the phase trajectory (6.4.13) in this case and (*3*) is the phase trajectory in the presence of a small wave detuning

Then $\cos\theta \approx -(1 - \tilde{\theta}^2/2)$, and according to (6.4.7),

$$\frac{da_v}{d\zeta} = \left[a_s\sqrt{a_{p0}^2 + a_{s0}^2 - a_s^2}\left(1 - \frac{\tilde{\theta}^2}{2}\right) - a_v\right]\frac{\eta}{\eta_0}\ , \ \frac{da_s}{d\zeta}$$
$$= \sqrt{a_{p0}^2 + a_{s0}^2 - a_s^2}\left[a_s\sqrt{a_{p0}^2 + a_{s0}^2 - a_s^2} - a_v\left(1 - \frac{\tilde{\theta}^2}{2}\right)\right]\frac{\eta}{\eta_0}\ . \qquad (6.4.21)$$

Obviously, when (6.4.19) is valid, in the region far from the curve (6.4.12) in the plane (a_v, a_s), the quantity $\tilde{\theta}^2/2$ can be neglected, and the phase trajectories can be assumed to coincide with those obtained for $\delta k = 0$. When the representing point approaches curve (6.4.12), one has to use (6.4.21), instead of (6.4.11). It follows from (6.4.21) that for $\delta k \neq 0$, curve (6.4.12) of the equilibrium states is replaced by a single stable state of equilibrium, namely, a point with coordinates $a_v = 0$, $a_s = \sqrt{a_{p0}^2 + a_{s0}^2}$, see Fig. 6.7.

Thus, the wave detuning is responsible for the incomplete parametric self-induced transparency and for the field of the Stokes frequency to approaching the same value which it would assume in the absence of parametric interaction (and, therefore, in the absence of generation of the resonant field). The spatial behavior of the fields a_s and a_v during the motion of the representing point near curve (6.4.12) can be approximately determined if it is taken into account that $\eta \approx \eta_0$ holds in this region (parametric self-induced transparency is still present, although it is incomplete), and that (6.4.12) is valid (up to a certain accuracy). The integration of the equation for a_s from (6.4.21) under these assumptions results in

$$a_s^2 = \frac{a_{s2}^2(a_{p0}^2 + a_{s0}^2)\exp[(a_{s0}^2 + a_{p0}^2)\tilde{\theta}^2(\zeta - \zeta_2)]}{a_{p2}^2 + a_{s2}^2\exp[(a_{s0}^2 + a_{p0}^2)\tilde{\theta}^2(\zeta - \zeta_2)]} \quad . \tag{6.4.22}$$

Substituting this expression into (6.4.12), we obtain the amplitude of the field of the difference frequency:

$$a_v^2 \to \frac{a_{s2}a_{p2}(a_{p0}^2 + a_{s0}^2)\exp[\frac{1}{2}(a_{p0}^2 + a_{s0}^2)\tilde{\theta}^2(\zeta - \zeta_2)]}{a_{p2}^2 + a_{s2}^2\exp[(a_{p0}^2 + a_{s0}^2)\tilde{\theta}^2(\zeta - \zeta_2)]} \quad . \tag{6.4.23}$$

Equations (6.4.22, 23) contain the values of fields at a certain "initial" point of motion near the curve (6.4.12): a_{s2} and $a_{p2} = \sqrt{a_{p0}^2 + a_{s0}^2 - a_{s2}^2}$, as well as coordinate ζ_2 at which the representing point coincides with this "initial" point. Apparently, one can choose as the latter a point corresponding to the zero value of the derivative $da_v/d\zeta$ in the first equation of (6.4.21). At this point

$$\left| a_s\sqrt{a_{p0}^2 + a_{s0}^2 - a_s^2} - a_v \right| \approx \frac{a_v}{2}\tilde{\theta}^2 \quad . \tag{6.4.24}$$

If the quantity a_s from (6.4.13) is substituted into this formula, it becomes possible to compute $a_{max} < a_{v\,lim}$, which is the maximal value of the resonant field. Assuming that a_{v1} in (6.4.18) is equal to this maximal value a_{max}, one can find the conversion length ζ_{max}, i.e., the length over which the quantity a_{max} is reached. Further variation (attenuation) of the resonant field is described approximately by (6.4.23) in which ζ_{max} is substituted in place of ζ_2, and a_{s2} is related to a_{max} through (6.4.13). Figure 6.8 illustrates the qualitative dependence of a_s and a_v on coordinate z.

It is of interest to note that the increase of the Stokes component accompanying the motion of the representing point to the equilibrium state along the

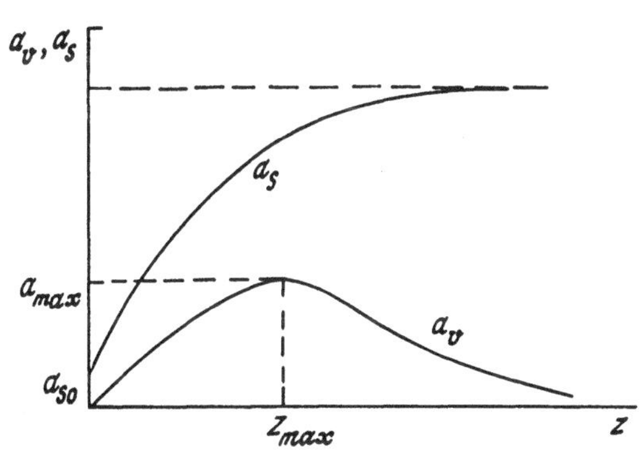

Fig. 6.8. The qualitative coordinate dependence of the field amplitudes a_s and a_v under condition (6.4.19)

curve (6.4.12) turns out to be slower by a factor of approximately $\tilde{\theta}^{-2}$ than it is during the ordinary Raman scattering with no parametric interaction. This can be easily demonstrated if (6.4.22) is compared with the solution of the second equation in (6.4.7) in which the term with $\cos\theta$ is discarded. The effect of parametric interaction on SRS becomes small when $\tilde{\theta} \geq 1$, i.e., when the values of wave detuning satisfy inequalities inverse with respect to (6.4.20).

6.4.2 Generation of the Difference Frequency During SRS in Gases

In all experiments that involve the observation of the difference frequency generation during SRS known at present, a gas (H_2) in an electrostatic field (EF) is used as the nonlinear medium. Therefore, it would be of interest to consider this process in gases. In such a case it is necessary to take into account the broadening of the vibrational transition line, as well as the distribution of molecular orientations.

The nonuniform broadening (in rarefied gases) which is a consequence of the Doppler shift of the transition frequency can be allowed for if T is replaced by T' (the corresponding expression is given in Sect. 6.1). For compressed gases the quantity T' should be interpreted as the transition line width that has been formed in the presence of collisions (in this case the broadening is uniform).

The spread of molecular orientations can lead to more drastic (qualitative) changes. In this case the parameter

$$\mathcal{F} = \frac{\langle (\boldsymbol{d} \cdot \boldsymbol{e}_{\text{v}})\kappa \rangle}{\sqrt{\langle (\boldsymbol{d} \cdot \boldsymbol{e}_{\text{v}})^2 \rangle \langle \kappa^2 \rangle}}$$

introduced in Sect. 4.1 plays an important role. When $\mathcal{F} = 1$, the qualitative picture of the process is the same as described above. In many gaseous media $\mathcal{F} < 1$ holds. In this case it is easily seen, that there is no stationary state with C_{s} and C_{v} having nonzero values. In this respect the effect of the spread of orientations is similar to that of wave detuning (as in the process of the second harmonic generation considered in Sect. 6.1).

If the parameter $\mathcal{F}$ is small ($\mathcal{F} \lesssim 0.3$) or the wave detuning large, system (6.4.2) can be solved by the method of successive approximations, with the influence of the resonant field C_{v} on the pumping field C_{p} and the Stokes component C_{s} being regarded as insignificant.

In the zeroth approximation it will be assumed that $C_{\text{v}}(z) = 0$. Then the last two equations of (6.4.2) make it possible (see [6.33]) to determine the variation of the pumping field intensity $\mathcal{J}_{\text{p}} = |C_{\text{p}}(z)|^2$ and the intensity of the Stokes component $\mathcal{J}_{\text{s}} = |C_{\text{s}}(z)|^2$ with z:

$$\mathcal{J}_{\text{s}}(z) = \mathcal{J}_{\text{s}0} \frac{1+M}{1+M\exp(2\mathcal{K}z)} \exp(2\mathcal{K}z) \; ,$$
$$\mathcal{J}_{\text{p}}(z) = \mathcal{J}_{\text{p}0} \frac{1+M}{1+M\exp(2\mathcal{K}z)} \; . \tag{6.4.25}$$

In these expressions $\mathcal{J}_{s0} = \mathcal{J}_s(z = 0)$, $\mathcal{J}_{p0} = \mathcal{J}_p(z = 0)$ are the quantities proportional to the energy of the Stokes component and the pumping field at the input, $M = g_p \mathcal{J}_{s0}/g_s \mathcal{J}_0$ is a quantity proportional to the ratio of the number of the Stokes component quanta to that of the quanta of the pumping field at the input and

$$\mathcal{K} = g_s \langle \kappa^2 \rangle \left(\mathcal{J}_{p0} + \frac{g_p}{g_s} \mathcal{J}_{s0} \right)$$

is the Stokes component amplification coefficient. Substitution of (6.4.25) into the equation for C_v of system (6.4.2) and integration under the condition that $C_v(z = 0) = 0$ results in

$$|C_v(z)|^2 = \frac{\pi^2}{4} \frac{g_v^2}{g_s g_p} \frac{\langle (\boldsymbol{d} \cdot \boldsymbol{e}_v) \kappa \rangle^2}{\langle \kappa^2 \rangle^2} |f(z)|^2 \quad . \tag{6.4.26}$$

The function $f(z)$ that appears in (6.4.26) is expressed through the Gaussian hypergeometric function ${}_2\mathcal{F}_1$:

$$f(z) = \frac{2}{\pi} (M \, \mathrm{e}^{2\mathcal{K}z})^{-b} \left[\frac{\xi^{(\mathrm{i}\delta + b + 1)/2}}{\mathrm{i}\delta + b + 1} \right.$$
$$\left. \times \, {}_2\mathcal{F}_1 \left(\frac{\mathrm{i}\delta + b + 1}{2} \; ; \; \frac{\mathrm{i}\delta + b + 1}{2} \; ; \; \frac{\mathrm{i}\delta + b + 3}{2} \; ; \; \xi \right) \right]_{\xi = M/(1+M)}^{\xi = M/(\mathrm{e}^{-2\mathcal{K}z} + M)} \quad . \tag{6.4.27}$$

Here the notations

$$\delta = (\delta k)/\mathcal{K} \; ; \quad b = g_v \langle (\boldsymbol{d} \cdot \boldsymbol{e}_v)^2 \rangle \mathcal{K}^{-1} = \alpha/\mathcal{K}$$

for the dimensionless wave detuning and the IR absorption coefficient have been used. If there are losses at the frequency ω_v that are not related to the above vibrational transition, the corresponding absorption coefficient is added to α.

Let us analyse the function $|f(z)|^2$. When $M \exp(\mathcal{K}z) \ll 1$ (the intensity of the Stokes component at the input and the interaction length are small), this function behaves as

$$|f(z)|^2 \approx \frac{4 g_s g_p \langle \kappa^2 \rangle^2}{\pi^2} \mathcal{J}_{p0} \mathcal{J}_{s0} \frac{(\mathrm{e}^{\mathcal{K}z} - \mathrm{e}^{-\alpha z})^2 + 4\mathrm{e}^{(\mathcal{K} - \alpha)z} \sin^2(\delta k z/2)}{(\delta k)^2 + (\mathcal{K} + \alpha)^2} \quad . \tag{6.4.28}$$

According to (6.4.28), there are beats between the curves

$$\frac{4M(\mathrm{e}^{\mathcal{K}z} - \mathrm{e}^{-\alpha z})^2}{\pi^2[(1 + b)^2 + \delta^2]} \quad \text{and} \quad \frac{4M(\mathrm{e}^{\mathcal{K}z} + \mathrm{e}^{-\alpha z})^2}{\pi^2[(1 + b)^2 + \delta^2]} \quad .$$

In the case the intensity of resonant IR radiation is by a factor of $1 + \delta^2(1+b)^{-2}$ smaller than in the presence of exact spatial synchronization. As the interaction length increases, the contribution from beats diminishes.

If z is large, the asymptotic behavior of the function $|f(z)|^2$ is different for $b<1$ and $b>1$. When $b<1$, we have

$$f(z) \approx 2[M^b e^{2\alpha z}(\cosh \delta\pi + \cos b\pi)]^{-1} \quad . \tag{6.4.29}$$

When $b>1$, the function $|f(z)|^2$ decreases for large z as $\exp(-2\mathcal{K}z)$ and does not depend on δ.

From (6.4.29) and (6.4.28) it is seen that the intensity of resonant radiation reaches its maximum for $\delta k = 0$ and the negative effect of detuning can practically be eliminated if a sufficiently intense pumping field is directed into the gas so that

$$\mathcal{K}>|\delta k| \quad . \tag{6.4.30}$$

The realization of this condition implies the phase locking of interacting waves. Since the solution of (6.4.27) depends continuously on the parameter δ, the spatial behaviour of $\mathcal{J}_{\text{v}}$ is nearly the same as for $\delta k = 0$. Therefore, assumed below that $\delta k = 0$. Then (6.4.27) can be expressed through simpler functions. For $\alpha \gg \mathcal{K}$

$$|f(z)|^2 \approx \frac{4M\mathcal{K}^2 \exp(2\mathcal{K}z)}{\pi^2\alpha^2[1 + M\exp(2\mathcal{K}z)]^2} \tag{6.4.31}$$

holds. At $M\exp(2\mathcal{K}z) = 1$ this function has a maximum with the value $\mathcal{K}^2(\pi\alpha)^{-2} \ll 1$. If the inequality

$$\mathcal{K} \gg \alpha \quad (b \ll 1) \quad , \tag{6.4.32}$$

is valid, the $f(z)$ lies between the limits

$$f_1(z) = \frac{2}{\pi}[\arctan(\sqrt{M}\, e^{\mathcal{K}z}) - \arctan\sqrt{M}] \quad \text{and} \tag{6.4.33}$$

$$f_2(z) = e^{-\alpha z} f_1(z) \quad . \tag{6.4.34}$$

For its growing branch $f(z)$ is closer to $f_1(z)$, having reached its maximum, it falls off as $f_2(z)$. Under the condition (6.4.32), the maximum of $|f(z)|^2$ is located between 1 and $(\pi b/2)^{2b}$ that is less than 1 but tends to 1 when $b \to 0$. For $b = 0.1$ the quantity $|f(z)|^2$ becomes maximal with $|f(z)|^2_{\max} \approx 0.67$. Thus, $|f(z)|^2$ never exceeds unity, although it can approach 1. Therefore, the factor in (6.4.26) near $|f(z)|^2$ gives the limiting intensity of resonant radiation, which can be obtained in a substance with given parameters:

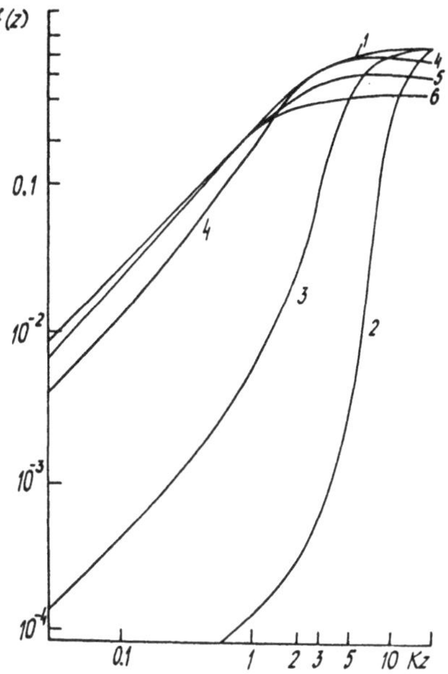

Fig. 6.9. The dependence of the resonance field amplitude on $\mathcal{K}z$ for $b \ll 1$ and $\delta = 0$. Curves *1–6* correspond to values $\ln M = -\mathcal{K}z, -20, -10, -3, -1, 0$

$$\mathcal{J}_\mathrm{v} = C^2_{\mathrm{v\,lim}} = \frac{\pi^2 g_\mathrm{v}^2 \langle (\boldsymbol{d} \cdot \boldsymbol{e}_\mathrm{v})\kappa \rangle^2}{4 g_\mathrm{s} g_\mathrm{p} \langle \kappa^2 \rangle^2} \quad . \tag{6.4.35}$$

Let us now formulate the requirements for the interaction length and the relationship between the input intensities of the pumping radiation and the Stokes component [under the assumption that (6.4.30, 32) are valid with sufficient accuracy]. As follows from (6.4.33, 34), $|f(z)|^2$ increases with growing z (as long as $z < \alpha^{-1}$). For fixed $\mathcal{K}z$ the maximal intensity of IR radiation given by

$$|f(z)|^2 = \frac{\mathcal{J}_{\mathrm{v\,max}}}{\mathcal{J}_{\mathrm{v\,lim}}} = \frac{4}{\pi^2}\left[\arctan\left(\sinh\frac{\mathcal{K}z}{2}\right)\right]^2 \tag{6.4.36}$$

is reached when the interaction length and the input values $\mathcal{J}_{\mathrm{s0}}$ and $\mathcal{J}_{\mathrm{p0}}$ satisfy the condition

$$\frac{g_\mathrm{p}\mathcal{J}_{\mathrm{s0}}}{g_\mathrm{s}\mathcal{J}_{\mathrm{p0}}} = M = \exp(-\mathcal{K}z) \quad . \tag{6.4.37}$$

In this case, when $\mathcal{K}z < 1$, it can be approximately assumed that $f(z) = \mathcal{K}z/\pi$. Since for $\mathcal{K}z = 5$ we have $f(z) = 0.9$, a further increase of z produces little effect on the resonant field generation. According to Fig. 6.9, the influence of the ratio $\mathcal{J}_{\mathrm{s0}}/\mathcal{J}_{\mathrm{p0}}$ can for $\mathcal{K}z \to 1$ turn out to be quite significant. When $\mathcal{K}z > 10$, the magnitude of the resonant field is practically independent of the ratio $\mathcal{J}_{\mathrm{s0}}/\mathcal{J}_{\mathrm{p0}}$.

In the above, the angular brackets denote the averaging of the corresponding quantities over orientations. This procedure involves multiplication of the respective quantities by the distribution function then an integration over the angles. In our case the distribution function is

$$\frac{1}{4\pi}\exp(E_0 d_{11}/k_B T^0) \approx \frac{1}{4\pi}(1 + E_0 d_{11}/k_B T^0) \quad , \tag{6.4.38}$$

where E_0 is the strength of the applied constant field and d_{11} is the projection of the static dipole moment of a molecule on the direction of the field $\boldsymbol{E}_0$. The projection of the total dipole moment of the transition 2–1 on the direction of polarization of the field C_v is given by

$$(\boldsymbol{d}\cdot \boldsymbol{e}_v) = (\boldsymbol{d}_{12})_v + \kappa_{v0}(0)E_0 \quad . \tag{6.4.39}$$

Here $\boldsymbol{d}_{12}$ is the intrinsic dipole moment of the transition and $\kappa_{v0}(0)$ is the molecular polarizability tensor. On the basis of the distribution function (6.4.38) and the properties of isotropic tensors, the following expressions for $C_{v\,\mathrm{lim}}$ and b can be obtained:

$$\begin{aligned} C_{v\,\mathrm{lim}} &= \frac{\pi}{2}\frac{\omega_v}{\sqrt{\omega_s\omega_p}}\frac{\langle(\boldsymbol{d}\cdot\boldsymbol{e}_v)\kappa\rangle}{\langle\kappa^2\rangle} \\ &= \frac{\pi}{2}\frac{\{\langle|\kappa_{v0}\kappa|\rangle_{\mathrm{is}} + (1/k_B T^0)\langle d_{11}(d_{12})_v\kappa\rangle_{\mathrm{is}}\}}{\langle\kappa^2\rangle_{\mathrm{is}}}\frac{\omega_v}{\sqrt{\omega_s\omega_p}}E_0 \quad , \end{aligned} \tag{6.4.40}$$

$$b = \frac{\omega_v}{\omega_s}\frac{\langle(d_{12})_v^2\rangle_{\mathrm{is}} + E_0^2[\langle\kappa_{v0}^2\rangle_{\mathrm{is}} + (2/k_B T^0)\langle(d_{12})_v d_{11}\kappa_{v0}\rangle_{\mathrm{is}}]}{\langle\kappa^2\rangle_{\mathrm{is}}(\mathcal{J}_{p0} + (\omega_p/\omega_s)\mathcal{J}_{s0})} \quad . \tag{6.4.41}$$

In these formulas the symbol $\langle\,\rangle_{\mathrm{is}}$ stands for the average computed with the distribution function of the isotropic medium $1/4\pi$.

Based on (6.4.40, 41) let us formulate conditions, under which it is possible to generate a considerable power $\mathcal{J}_{v\,\mathrm{lim}}$. The quantity b is given by the ratio of the IR radiation absorption coefficient to the coefficient of amplification of the Stokes component. As shown above, it should be small ($b = \alpha/\mathcal{K} \ll 1$) indicating via (6.4.41) that the electric dipole transition 2–1 should be almost prohibited. If $\boldsymbol{d}_{12}$ corresponds to an allowed dipole transition, the condition $b \ll 1$ can only be fulfilled for such intensities of the pumping field $\mathcal{J}_{p0}$ which are accompanied by transitions into the continuous spectrum (by laser breakdown). At the same time, if

$$\langle\kappa_{v0}\kappa\rangle_{\mathrm{is}} < \langle d_{11}(d_{12})_v\kappa\rangle_{\mathrm{is}}(k_B T^0)^{-1} \quad ,$$

the presence of the static molecular dipole moment $\boldsymbol{d}_{11}$ turns out to be essentially useful for generation of the resonant field. This holds, however, only as long as $\alpha \ll \mathcal{K}$, i.e.,

$$\frac{\omega_v}{\omega_s} \frac{\langle (d_{12})_v d_{11} \kappa_{v0} \rangle_{is} E_0^2}{k_B T^0 \langle \kappa^2 \rangle_{is} \mathcal{J}_{p0}} \ll 1 \quad .$$

In [6.29], molecular hydrogen was chosen for an experimental study of the above resonant field intensity as a function of various parameters: frequency and intensity of the pumping radiation, gas pressure and the degree to which the electric dipole transition is allowed. In a hydrogen molecule there is a vibrational transition $Q_1(1)$ corresponding to SRS and characterized by a frequency $4155.21\,\mathrm{cm}^{-1}$. When the hydrogen pressure equals 1 atm then $T \approx 10^{-9}\,\mathrm{s}$ [6.34]. In hydrogen there is neither an intrinsic dipole moment of the vibrational transition d_{12} nor a constant dipole moment d_{11}. For this reason, the gas was placed in a constant electric field to allow for electric dipole transitions. The following formula [6.35]

$$(\boldsymbol{d} \cdot \boldsymbol{e}_v) = \kappa_{0v}(0) E_0 = 4.1 \times 10^{-25} E_0 \tag{6.4.42}$$

was used to calculate the matrix element of the dipole moment operator. In the above expression E_0 is in $\mathrm{kV\,cm}^{-1}$, $(\boldsymbol{d} \cdot \boldsymbol{e}_v)$ is in cgse units. The limiting intensity of resonant IR radiation is determined by

$$|C_{v\,\mathrm{lim}}|^2 = \frac{\pi^2}{4} \frac{\omega_v^2}{\omega_s \omega_p} \frac{\langle \kappa_{0v}^2 \rangle}{\langle \kappa^2 \rangle} \mathcal{F}^2 E_0^2 \quad . \tag{6.4.43}$$

For making estimates it may be assumed that the local coherence parameter $\mathcal{F}$ is equal to 1. This is based on the following considerations. According to the experimental study of polarization characteristics, $\boldsymbol{e}_s \| \boldsymbol{e}_p$, and $\boldsymbol{e}_v \| \boldsymbol{e}_0$. It has also been discovered that up to the accuracy of experimental errors ($\approx 30\%$), the angle between $\boldsymbol{e}_p$ and $\boldsymbol{e}_0$ does not affect the generation of the difference frequency field [6.29]. In other words, for two typical cases, $\boldsymbol{e}_0 \| \boldsymbol{e}_p$ and $\boldsymbol{e}_0 \perp \boldsymbol{e}_p$, the ratio of the intensities $\mathcal{J}_{vA}/\mathcal{J}_{vB}$ which is proportional to[6]

$$\mathcal{F}_A^2/\mathcal{F}_B^2 = [G^0(0,\omega_p) + \tfrac{2}{15} G^s(0,\omega_p)]^2 [G^0(0,\omega_p) - \tfrac{1}{15} G^s(0,\omega_p)]^{-2} \quad ,$$

[see (4.1.6-8)] is equal to 1 to within 30%. This implies that the contribution from the symmetric part of the polarizability tensors is small, and therefore $\mathcal{F} \approx 1$ [see Sect. 4.1.2, (1)].

Table 6.1 presents the results of numerical computations of the IR radiation intensity for three different wavelengths of the pumping field: $1.054\,\mu$m (the neodymium laser), $0.694\,\mu$m (the ruby laser) and $0.527\,\mu$m (the second harmonic of the neodymium laser). According to these data, it is in compressed

[6] The quantities $\langle \kappa_{0v}^2 \rangle$ and $\langle \kappa^2 \rangle$ do not depend on the angle between $\boldsymbol{e}_0$ and $\boldsymbol{e}_p$.

Table 6.1. λ_s is the wavelength of the first Stokes component corresponding to the pumping field with the wavelength λ_p, the $\mathcal{J}_{p\,lim}$ are the limiting (pre-breakdown [6.36]) values of the pumping field intensity for hydrogen pressure 10 atm, $|\kappa|^2$ are the values of polarizability describing the intensity of the Raman process as calculated on the basis of data on Raman scattering cross-sections in hydrogen. $\mathcal{J}_{v\,lim}$ are the limiting values of the generated resonant IR field intensity calculated from (6.4.40) and (6.4.42) without taking into account the wave detuning (for a constant electric field with $E_0 = 50\,kV/cm$), δk are the values of wave detuning for hydrogen pressure 10 atm. $\mathcal{K}$ are the Stokes component amplification coefficients for this pressure and the pumping field intensity $\mathcal{J}_{p\,lim}$, $|f(z)|^2$ the values of function (6.4.29) obtained under the assumption that the IR radiation absorption coefficient $\alpha = 0$ (which holds with a sufficient accuracy) and the $\mathcal{J}_v(\delta k)$ are the limiting values of the generated IR field intensity with the wave detuning taken into account [see (6.4.26)]

	λ_p [μm]		
	1.054	0.694	0.527
λ_s [μm]	1.875	0.976	0.675
$\mathcal{J}_{p\,lim}$ [GW/cm^2]	1.1	1.5	2.0
$\lvert\kappa\rvert^2 \times 10^{-51}$ [cgs units]	5.5	7.0	7.5
$\mathcal{J}_{v\,lim}$ [MW/cm^2]	33.6	12.3	4.5
δk [1/cm]	0.42	1.25	2.43
$\mathcal{K}$ [1/cm]	0.3	1.0	2.0
$\lvert f(z)\rvert^2 \times 10^{-3}$	2.6	6.9	3.7
$\mathcal{J}_v(\delta k)$ [MW/cm^2]	0.089	0.083	0.017

hydrogen subject to the action of the field $50\,kV\,cm^{-1}$ possible to generate resonant radiation with a wavelength of 2.41 μm and an intensity $\sim 100\,kW\,cm^{-2}$.

Resonant IR radiation of this wavelength has been recorded for experiments, using the pumping field of both ruby and neodymium lasers, and also the second harmonic of the neodymium laser. The intensity reaches its maximal value (25–$30\,kW\,cm^{-2}$) at the output of a plane capacitor inside of which an EF with a strength of $50\,kV\,cm^{-1}$ has been produced for hydrogen (at a pressure of 7 atm) pumped by the neodymium laser radiation. The order of magnitude of this intensity is close to that of the computed limiting intensity for a chamber with plane electrodes, where $E_0 = 50\,kV\,cm^{-1}$ and the hydrogen pressure is 10 atm (see Table 6.1). As follows from the theory, the dependence between $\mathcal{J}_v$ and E_0 is quadratic so that one should expect an increase in the field E_0 up to its breakdown value ($90\,kV\,cm^{-1}$) to lead to a resonant radiation intensity as high as $\approx 80\,kW\,cm^{-2}$.

The conclusion to the effect that the long wave pumping radiation is more effective (see Table 6.1) was confirmed experimentally [6.29]. It was also demonstrated that generation of the difference frequency can be made more efficient if the ratio of intensities of the Stokes component and the pumping field at the entrance into the space between the electrodes is kept close to its optimal value [see (6.4.37)]. In particular, for a pumping field with a wavelength of 0.527 μm, the resonant radiation was obtained only under the conditions of preliminary generation of the Stokes component in the electrodeless part of the hydrogen chamber. The same scheme was also used for producing the maximal intensity of resonant radiation.

The dependence of the resonant radiation intensity on gas pressure, observed in the experiment (a sharp decrease of intensity with the growth of pressure, as a result of the increase of the wave detuning), agrees with the theory, too. Such a dependence, as well as a breakdown in the gas due to EF, are the main factors that limit the possibility of raising the efficiency of the difference frequency generation during SRS in hydrogen. Decreasing the pressure we improve the wave synchronism, but at the same time we reduce the EF strengths left for work due to the lower breakdown threshold. It follows from the estimates, that the SRS thresholds in CO, CO_2 and N_2 at pressures $p \lesssim 10-30$ atm are much too high (and therefore higher pressures must be used). Therefore, in order to obtain noticeable intensities of radiation at vibrational frequencies in these gases, it is also necessary to overcome the negative effect of wave detuning. The same requirement should be satisfied for observing the difference frequency generation during SRS on rotational transitions.

6.4.3 Generation of the Difference Frequency During SRS in the Presence of a Nonuniform Electrostatic Field[7]

Two factors have been established above that are responsible for the lowering of the difference frequency (DF) generation efficiency during SRS that are specific for RPI. The first of them is established in experiments with gases described in the preceding section. It is associated with the failure to ensure the phase locking of interacting waves for the wave detunings of real gases. In principle, the wave detuning in gaseous media can be made sufficiently small if the gas, whose particles participate in the wave conversion, is added to a buffer gas with a specially chosen dispersion law of the refractive index. However, for vibrational or rotational transitions, no such buffer gas is known as yet.

The second factor that limits the efficiency of DF generation is parametric self-induced transparency caused by the reverse effect of the generated DF field on the pumping radiation and the Stokes component. Parametric self-induced transparency is most essential when there is wave synchronism. It can be observed if one employs the travelling wave scheme in experiments on DF generation on polaritons in crystals in which $\delta k = 0$ for certain directions. (In this case it is possible to assume that the dipole moments of elementary units in the crystal are induced by a very strong electric field whose order of magnitude is close to that of the intracrystalline one, while the crystal itself represents a medium consisting of oriented particles: $\mathcal{F} = 1$). The presence of parametric self-induced transparency in such situations is confirmed, for example, by the

[7] For the first observation of the difference frequency generation during SRS in a medium placed in a periodically nonuniform EF, see [6.31].

results of [6.37]. In this study, generation of radiation at polariton frequencies was accompanied by a considerable decrease of the SRS Stokes component amplification coefficient (see Sect. 6.4.1). If wave synchronism is ensured by means of the buffer gas, parametric self-induced transparency can also affect the efficiency of DF generation during SRS on electron and vibrational transitions in gaseous media.

In the following, a method based on the use of a nonuniform EF [6.38] for neutralizing the negative influence of the wave detuning and parametric self-induced transparency will be considered. For further analysis it would be convenient to introduce the explicit form of the dependence of EF on the coordinate $E_0 = \mathcal{E}_0 u(z)$ and the normalized quantities:

$$
\begin{aligned}
&c_j = C_j|C_{\mathrm{p}0}|^{-1} \ ; \quad \xi = g_{\mathrm{s}}\langle\kappa^2\rangle|C_{\mathrm{p}0}|^2 z \ , \\
&\delta = \delta k g_{\mathrm{s}}^{-1}|C_{\mathrm{p}0}|^{-2}\langle\kappa^2\rangle^{-1} \\
&q_{\mathrm{p,v}} = \frac{\omega_{\mathrm{p,v}}^2 k_{\mathrm{s}}}{\omega_{\mathrm{s}}^2 k_{\mathrm{p,v}}} \ ; \quad \mathcal{F} = \frac{\langle\kappa_{\mathrm{v}0}(0)\kappa\rangle}{\sqrt{\langle\kappa^2\rangle\langle\kappa_{\mathrm{v}0}^2\rangle}} \ , a_0 = \sqrt{\frac{\langle\kappa_{\mathrm{v}0}^2(0)\rangle}{\langle\kappa^2\rangle}}\frac{C_0}{|C_{\mathrm{p}0}|} \ ; \qquad (6.4.44)
\end{aligned}
$$

(here $C_{\mathrm{p}0}$ is the pumping field amplitude at the entrance into the medium). Then, allowing for the fact that in (6.4.39), $(\boldsymbol{d}_{12})_{\mathrm{v}} = 0$, one can write the system of equations (6.4.2) as

$$
\begin{aligned}
&c_{\mathrm{v}}' = -q_{\mathrm{v}}(a_0^2u^2c_{\mathrm{v}} + \mathcal{F}a_0u\,\mathrm{e}^{\mathrm{i}\delta\xi}c_0c_{\mathrm{s}}^*) \\
&c_{\mathrm{s}}' = c_{\mathrm{s}}|c_{\mathrm{p}}^2| + \mathcal{F}a_0u\,\mathrm{e}^{\mathrm{i}\delta\xi}c_{\mathrm{p}}c_{\mathrm{v}}^* \ ; \\
&c_{\mathrm{p}}^{*\prime} = -q_{\mathrm{p}}(\mathcal{F}a_0u\,\mathrm{e}^{\mathrm{i}\delta\xi}c_{\mathrm{v}}^*c_{\mathrm{s}}^* + c_{\mathrm{p}}^*|c_{\mathrm{s}}|^2) \ . \qquad (6.4.45)
\end{aligned}
$$

In these formulas, primes denote differentiation with respect to ξ. For a uniform EF $u = 1$ must be assumed in (6.4.45). If EF is governed by a sinusoidal law, $u = \sin \Omega\xi$. Our present discussion will be limited to the case of a periodically nonuniform (PNU) electrostatic field; systems with large and small values of wave detuning will be considered separately.

Systems with Large Wave Detuning. Assume that $|\delta| \gg 1$, $\gamma = q_{\mathrm{v}}a_0^2$, and the "frequency" Ω of the periodically nonuniform EF is sufficiently close to δ[8], so that $|\delta| \gg |\delta - \Omega|$. Then it is possible to use the averaging method according to which the solution (6.4.45) consists of slowly varying ($\bar{c}_j$) and rapidly varying ($\tilde{c}_j$) components. In the first approximation the equations

[8] In the following, $\delta > 0$ will be supposed. And if $\delta < 0$, the quantity δ should be replaced by $-\delta$ everywhere.

$$\bar{c}_{\mathrm{v}}' = -q_{\mathrm{v}}\left[\frac{a_0^2}{2}\bar{c}_{\mathrm{v}} + \mathrm{i}\frac{\mathcal{F}a_0}{2}\,\mathrm{e}^{\mathrm{i}(\delta-\Omega)\xi}\bar{c}_{\mathrm{s}}^*\bar{c}_{\mathrm{p}}\right] \; ,$$

$$\bar{c}_{\mathrm{s}}' = \mathrm{i}\frac{\mathcal{F}a_0}{2}\,\mathrm{e}^{\mathrm{i}(\delta-\Omega)\xi}\bar{c}_{\mathrm{v}}^*\bar{c}_{\mathrm{p}} + \bar{c}_{\mathrm{s}}|\bar{c}_{\mathrm{p}}|^2 \; ,$$

$$\bar{c}_{\mathrm{p}}^{*\prime} = -q_{\mathrm{p}}\left[\mathrm{i}\frac{\mathcal{F}a_0}{2}\,\mathrm{e}^{\mathrm{i}(\delta-\Omega)\xi}\bar{c}_{\mathrm{v}}^*\bar{c}^{*_{\mathrm{s}}} + \bar{c}_{\mathrm{p}}^*|\bar{c}_{\mathrm{s}}|^2\right] \tag{6.4.46}$$

are valid for the slowly varying components, while the rapidly varying ones are given by

$$\tilde{c}_{\mathrm{v}} = q_{\mathrm{v}}\left[\frac{a_0^2\,\sin 2\Omega\xi}{4\Omega}\bar{c}_{\mathrm{v}} + \frac{\mathcal{F}a_0}{2(\Omega+\delta)}\mathrm{e}^{\mathrm{i}(\Omega+\delta)\xi}\bar{c}_{\mathrm{p}}\bar{c}_{\mathrm{s}}^*\right] \; ,$$

$$\tilde{c}_{\mathrm{s}} = -\frac{\mathcal{F}a_0}{2(\Omega+\delta)}\mathrm{e}^{\mathrm{i}(\Omega+\delta)\xi}\bar{c}_{\mathrm{p}}\bar{c}_{\mathrm{v}}^* \; ,$$

$$\tilde{c}_{\mathrm{p}}^* = q_{\mathrm{p}}\frac{\mathcal{F}a_0}{2(\Omega+\delta)}\mathrm{e}^{\mathrm{i}(\Omega+\delta)\xi}\bar{c}_{\mathrm{s}}^*\bar{c}_{\mathrm{v}}^* \quad . \tag{6.4.47}$$

The boundary conditions for $\bar{c}_j$ are determined by relations

$$c_j|_{\xi=0} = c_{j0} = \bar{c}_{j0} + \tilde{c}_{j0} \quad .$$

Usually, at the entrance there is no field of the frequency $\omega_{\mathrm{v}}(c_{\mathrm{v}0} = 0)$, and the average flux of the pumping radiation photons is much greater than the average flux of photons at the Stokes frequency $(g_{\mathrm{s}}|C_{\mathrm{p}0}|^2 \gg g_{\mathrm{p}}|C_{\mathrm{s}0}|^2))$; in this case

$$\bar{c}_{\mathrm{p}0} \simeq 1 \; , \quad \bar{c}_{\mathrm{s}0} \simeq \frac{c_{\mathrm{s}0}}{1 + q_{\mathrm{v}}(\mathcal{F}^2 a_0^2/4(\Omega+\delta)^2)} \; ,$$

$$\bar{c}_{\mathrm{v}0} = -q_{\mathrm{v}}\frac{\mathcal{F}a_0}{2(\Omega+\delta)}\bar{c}_{\mathrm{s}0}^* \quad . \tag{6.4.48}$$

Obviously, when $|\Omega + \delta| \gg 1$, the boundary value for the slowly varying component at the Stokes frequency practically coincides with $c_{\mathrm{s}0}$, and $|\bar{c}_{\mathrm{v}0}| \ll |c_{\mathrm{s}0}|$. Since at the beginning of DF generation the slowly varying components of the fields at frequencies ω_{v} and ω_{s} are small, the relative variation of $\bar{c}_{\mathrm{p}}$ is also insignificant and the system (6.4.46) can be solved at this stage by applying the constant pumping field approximation, when it is assumed that in the first two equations $\bar{c}_{\mathrm{p}} = 1$. This procedure results in

$$\bar{c}_{\mathrm{v}} = \frac{[(p_2 + \frac{\gamma}{2})\bar{c}_{\mathrm{v}0} + \mathrm{i}q_{\mathrm{v}}\frac{\mathcal{F}a_0}{2}\bar{c}_{\mathrm{s}0}^*]\mathrm{e}^{p_1\xi} - [(p_1 + \frac{\gamma}{2})\bar{c}_{\mathrm{v}0} + \mathrm{i}q_{\mathrm{v}}\frac{\mathcal{F}a_0}{2}\bar{c}_{\mathrm{s}0}^*]\mathrm{e}^{p_2\xi}}{p_2 - p_1} \; , \tag{6.4.49}$$

where $p_{1,2}$ are the roots of the characteristic equation

$$p^2-\left[1-\frac{\gamma}{2}+\mathrm{i}(\delta-\Omega)\right]p-\frac{\gamma}{2}\left[1-\frac{\mathcal{F}^2}{2}+\mathrm{i}(\delta-\Omega)\right]=0 \quad . \tag{6.4.50}$$

If

$$\mathcal{F}^2\gamma \ll \left(1+\frac{\gamma}{2}\right)^2+(\delta-\Omega)^2 \quad , \tag{6.4.51}$$

the approximate expression

$$p_{1,2}\approx \left\{\begin{array}{l} 1-\dfrac{\mathcal{F}^2\gamma(1+\frac{\gamma}{2})}{4[(1+\frac{\gamma}{2})^2+(\delta-\Omega)^2]}+\mathrm{i}(\delta-\Omega)\left(1+\dfrac{\mathcal{F}^2\gamma}{4[(1+\frac{\gamma}{2})^2+(\delta-\Omega)^2]}\right) \\ -\dfrac{\gamma}{2}+\dfrac{\mathcal{F}^2\gamma(1+\frac{\gamma}{2})}{4[(1+\frac{\gamma}{2})^2+(\delta-\Omega)^2]}-\mathrm{i}(\delta-\Omega)\dfrac{\mathcal{F}^2\gamma}{4[(1+\frac{\gamma}{2})^2+(\delta-\Omega)^2]} \end{array}\right. \tag{6.4.52}$$

can be conveniently used. Inequality (6.4.51) is valid with a sufficient accuracy for any values of parameters a_0, $\mathcal{F}$ and $(\delta-\Omega)$, except the region where $\delta\to\Omega$, $\gamma\to 2$ and $\mathcal{F}\to 1$ (it must be recalled that $\mathcal{F}\leq 1$). Even in this case, however, the roots $p_{1,2}$ computed from (6.4.52) are quite close to their values obtained directly from (6.4.50) $[p_{1,2}\approx\pm 1/\sqrt{2}]$. Since $\mathrm{Re}\{p_1\}-\mathrm{Re}\{p_2\}>1$, the dependence of c_v on z rapidly becomes

$$\overline{c}_\mathrm{v}\approx \mathrm{i}\frac{q_\mathrm{v}\mathcal{F}a_0\overline{c}^*_{\mathrm{s}0}}{2(p_2-p_1)}\mathrm{e}^{p_1\xi} \quad . \tag{6.4.53}$$

Here the term $[p_2+(\gamma/2)]\overline{c}_{\mathrm{v}0}$ has been neglected in comparison with $\mathrm{i}q_\mathrm{v}(\mathcal{F}a_0/2)c^*_{\mathrm{s}0}$ [see (6.4.48, 49, 52)]. If (6.4.53) is applicable,

$$\overline{c}^*_\mathrm{s}\approx \mathrm{i}\overline{c}_\mathrm{v}\frac{2p_1+\gamma}{q_\mathrm{v}\mathcal{F}a_0}\mathrm{e}^{-\mathrm{i}(\delta-\Omega)\xi} \quad . \tag{6.4.54}$$

One can easily see from (6.4.47) that $|\tilde{c}_\mathrm{v}|\ll|\overline{c}_\mathrm{v}|$, $|\tilde{c}_\mathrm{s}|\ll|\overline{c}_\mathrm{s}|$ and, therefore, it is possible, on the basis of (6.4.53, 54), to determine quite precisely the field intensities at the frequencies $\omega_\mathrm{k}(\mathcal{J}_\mathrm{k}\sim|C_\mathrm{v}|^2)$ and $\omega_\mathrm{c}(\mathcal{J}_\mathrm{s}\sim|C_\mathrm{s}|^2)$.

Thus, in the case of PNU EF, as with a uniform EF, a constant ratio of field intensities at the frequencies ω_v and ω_s is established at the first stage [see (6.4.54)]. Because $\mathrm{Re}\{p_1\}<1$, the increase of these fields in space also occurs more slowly than the increase of the Stokes component during ordinary SRS, i.e., the field c_v generated in the parametric process exercises its reverse influence on the Stokes component increment. At the same time, the phase locking of interacting waves for PNU EF has a different character: the quantity $\theta=\delta\xi+\varphi_\mathrm{s}+\varphi_\mathrm{v}-\varphi_\mathrm{p}$, instead of being constant, depends on a coordinate according to

$$\theta = \Omega\xi - \frac{\pi}{2} + \arg\left(p_1 + \frac{\gamma}{2}\right) \quad . \tag{6.4.55}$$

Subsequently, when $|\overline{c}_{\rm s}|^2$ becomes of the order of $|\overline{c}_{\rm p}|^2 \approx 1$, the pumping field should no longer be regarded as constant. In order to obtain information on this stage of the process, note that the substitution

$$\overline{c}_{\rm p} \rightarrow c_{\rm p} \ , \quad \overline{c}_{\rm s} \rightarrow c_{\rm s} \ , \quad -{\rm i}\overline{c}_{\rm v} \rightarrow c_{\rm v} \tag{6.4.56}$$

reduces (6.4.46) to a form coinciding with that of equations for a uniform EF [the system (6.4.45) where $u = 1$] in which the quantities $\mathcal{F}$, a_0 and δ should be replaced, respectively, by $\mathcal{F}/\sqrt{2}$, $a_0/\sqrt{2}$ and $(\delta - \Omega)$. We can therefore use the results, valid for the uniform field. First of all, it must be said that, since $\mathcal{F}/\sqrt{2}<1$, the intensity of the field at the frequency $\omega_{\rm v}$ approaches zero when $z \rightarrow \infty$, even if $\delta = \Omega = 0$ [in the case of PNU EF there is no stable state of the system (6.4.45) for $\mathcal{F} = 1$ associated with parametric self-induced transparency]. If under these conditions [for $\Omega = \delta$] $\gamma/2 \ll 1$ [the absorption at the frequency $\omega_{\rm v}$ is small], we have

$$|\overline{c}_{\rm v}|^2 \approx \frac{q_{\rm v}^2 \mathcal{F}^2 a_0^2}{4q_{\rm p}} X^2(\xi) \quad . \tag{6.4.57}$$

The function $X(\xi)$ is bounded by the limits

$$X_1(\xi) = \arctan\{\sqrt{M}\exp[|\overline{c}_{\rm p0}|^2(1+M)\xi]\} - \arctan\sqrt{M}$$

and

$$X_2 = X_1(\xi)\exp\left(-\frac{\gamma}{2}\xi\right) \ ,$$

where $M = q_{\rm p}(|\overline{c}_{\rm s0}|^2/|\overline{c}_{\rm p0}|^2)$. The quantity $|c_{\rm v}|^2$ reaches its maximum

$$|c_{\rm v}|^2_{\max} \approx |\overline{c}_{\rm v}|^2_{\max} \approx \frac{\pi^2 q_{\rm v}^2 a_0^2 \mathcal{F}^2}{16 q_{\rm p}} \tag{6.4.58}$$

when $\xi_{\max} \sim (1\text{–}3)\overline{\xi}_1$; here

$$\overline{\xi}_1 = \frac{|c_{\rm p0}|^2}{1+M}\left|\ln\frac{1}{\sqrt{M}}\right| \quad .$$

For a large value of absorption at the frequency $\omega_{\rm v}$ and for $\Omega \sim \delta$ in the first of equations (6.4.46), the derivative $\overline{c}_{\rm v}'$ can be neglected. The relationship between $\overline{c}_{\rm v}$ and $\overline{c}_{\rm s}^*$, $\overline{c}_{\rm p}$ thus found, can be substituted into two remaining equations in order to compute the dependence of the pumping field and the Stokes component of SRS on ξ. Knowing this dependence one can easily find that

$$\bar{c}_{\mathrm{v}} \approx -\mathrm{i}\frac{\mathcal{F}}{a_0}\mathrm{e}^{\mathrm{i}(\delta-\Omega)\xi}\bar{c}^*_{\mathrm{s}0}\bar{c}_{\mathrm{p}0} \times \frac{(1+M)\exp\{(1+M)|\bar{c}_{\mathrm{p}0}|^2[1-(\mathcal{F}^2/2)]\xi\}}{1+M\exp\{2(1+M)|\bar{c}_{\mathrm{p}0}|^2[1-(\mathcal{F}^2/2)]\xi\}} \quad . \tag{6.4.59}$$

In this case the maximal value of $|c_{\mathrm{v}}|^2$ equal to

$$|c_{\mathrm{v}}|^2_{\max} = |\bar{c}_{\mathrm{v}}|^2_{\max} \approx \frac{\mathcal{F}^2}{4q_{\mathrm{p}}a_0^2}|\bar{c}_{\mathrm{p}0}|^4(1+M)^2 \quad , \tag{6.4.60}$$

is reached when

$$\xi_{\max} \approx \frac{\bar{\xi}}{[1-(\mathcal{F}^2/2)]} \quad .$$

It is also possible that the wave detuning δ and the frequency of PNU EF Ω differ substantially from each other ($|\delta-\Omega|\gg\gamma/2,\ 1$). The expression for $\bar{c}_{\mathrm{v}}$

$$\bar{c}_{\mathrm{v}} = -\mathrm{i}\frac{q_{\mathrm{v}}\mathcal{F}a_0}{2(\delta-\Omega)}\bar{\bar{c}}^*_{\mathrm{s}0}\bar{\bar{c}}_{\mathrm{p}0} \times \left\{\mathrm{e}^{(-\gamma/2)\xi} - \mathrm{e}^{\mathrm{i}(\delta-\Omega)\xi}\frac{(1+M)\exp[(1+M)|\bar{\bar{c}}_{\mathrm{p}0}|^2\xi]}{1+M\exp[2(1+M)|\bar{\bar{c}}_{\mathrm{p}0}|^2\xi]}\right\} \tag{6.4.61}$$

can be found by applying the averaging method for the second time, namely, averaging the system (6.4.46) over intervals that are much greater than $|\delta-\Omega|^{-1}$ but less than 1 and $2/\gamma$. If $|\delta-\Omega|\gg1$ and $|c_{\mathrm{s}0}|\ll|c_{\mathrm{p}0}|=1$ the boundary values for the slowy varying components that are selected by the second averaging, namely the $\bar{\bar{c}}_{\mathrm{s}0}$ and $\bar{\bar{c}}_{\mathrm{p}0}$ which occur in (6.4.61), are nearly equal to $c_{\mathrm{s}0}$ and $c_{\mathrm{p}0}$. Making use of (6.4.61) it is possible to demonstrate that if $M\ll1$ the maximal value of $|c_{\mathrm{v}}|^2$ is

$$|c_{\mathrm{v}}|^2_{\max}\simeq|\bar{c}|^2_{\max} \approx \frac{\mathcal{F}^2q_{\mathrm{v}}^2a_0^2|\bar{c}_{\mathrm{p}0}|^4}{16q_{\mathrm{p}}(\delta-\Omega)^2}(1+M)^2 \quad . \tag{6.4.62}$$

As in the previous cases, it is nearly reached a point where the flux densities of the pumping field photons and the photons of the Stokes component become equal:

$$\xi_{\max} \approx \bar{\xi} \quad .$$

The following conclusions can be drawn from (6.4.58, 60, 62): (i) as long as $|\delta-\Omega|\gg\max\{\pi^{-1};\gamma/2\}$, the decrease of $|\delta-\Omega|$ (i.e., the "frequency" of the periodic nonuniformity of EF approaching the wave detuning) leads to the increase of the maximal intensity $\mathcal{J}_{\mathrm{v\,max}}$ of radiation at the frequency ω_{v}. However, this intensity is little affected when the quantity $|\delta - \Omega|$ decreases from the level $\max\{\pi^{-1}; \gamma/2\}$ down to zero. (ii) If $|\delta - \Omega| < \min\{\pi^{-1}; \gamma/2\}$, this intensity increases as the ratio $\mathcal{E}_0^2/|C_{\mathrm{p}0}|^2$ tends to the optimal value

$$\left(\frac{\mathcal{E}_0^2}{|C_{\mathrm{p}0}|^2}\right)_{\mathrm{opt}} \approx \frac{2}{\pi}\frac{\omega_{\mathrm{s}}^2 k_{\mathrm{v}}}{\omega_{\mathrm{v}}^2 k_{\mathrm{s}}}\frac{\langle\kappa^2\rangle}{\langle\kappa_{\mathrm{v}0}^2(0)\rangle} \quad , \tag{6.4.63}$$

for which

$$|C_{\mathrm{v}}|_{\mathrm{opt}}^2 \approx \frac{\pi\omega_{\mathrm{v}}^2 k_{\mathrm{p}}}{8\omega_{\mathrm{p}}^2 k_{\mathrm{v}}}\frac{\langle\kappa_{\mathrm{v}0}(0)\kappa\rangle^2}{\langle\kappa_{\mathrm{v}0}^2(0)\rangle\langle\kappa^2\rangle}|C_{\mathrm{p}0}|^2 \tag{6.4.64}$$

[under the conditions of exact snychronism in a uniform EF the quantity $(\mathcal{E}_0^2|C_{\mathrm{p}0}|^{-2})_{\mathrm{opt}}$ is less by a factor of two, and $|C_{\mathrm{v}}|_{\mathrm{opt}}^2$ is greater by a factor of two].

A similar result concerning the effect of $|\Omega - \delta|$ on the intensity $\mathcal{J}_{\mathrm{v}}(z)$ at a certain distance z can be obtained from the analysis of (6.4.52, 53). In this case there is also an optimal ratio $\mathcal{E}_0^2|C_{\mathrm{p}0}|^{-2}$ the value of which, however, is determined not only by the parameters of matter and the frequencies of fields, but also by the coordinate z.

Systems with Small Wave Detuning. As has been done previously, we first conduct approximate calculations of c_{v} using the fact that some of the parameters, Ω, γ or $|c_{\mathrm{p}0}|^2(1+M) \simeq 1$ are much greater than the rest. After that, setting the approximate expressions for c_{v} equal to each other, we find the optimal ratio $\mathcal{E}_0|C_{\mathrm{p}0}|^{-1}$, the greatest attainable intensity $|C_{\mathrm{v}}|_{\mathrm{opt}}^2$ and the limits of variation of the quantity Ω.

For a sufficiently small period of EF nonuniformity (i.e., if $\Omega \gg \gamma$, $|\delta|$, 1), the averaging method can be applied. The resulting solution

$$\begin{aligned} c_{\mathrm{v}} = &-\frac{q_{\mathrm{v}}\mathcal{F}a_0}{\Omega}\bar{c}_{\mathrm{s}0}^*\bar{c}_{\mathrm{p}0}\left[\left(1+\frac{\gamma}{4\Omega}\sin 2\Omega\xi\right)\exp\left(-\frac{\gamma}{2}\xi\right)\right. \\ &\left.-\frac{(1+M)\exp[(1+M)|\bar{c}_{\mathrm{p}0}|^2\xi]}{1+M\exp[2(1+M)|\bar{c}_{\mathrm{p}0}|^2\xi]}\mathrm{e}^{\mathrm{i}\delta\xi}\cos\Omega\xi\right] \end{aligned} \tag{6.4.65}$$

makes it possible to demonstrate that if $M \ll 1$, the maximal value

$$|c_{\mathrm{v}}|_{\mathrm{max}}^2 \approx \frac{q_{\mathrm{v}}^2\mathcal{F}^2 a_0^2}{4q_{\mathrm{p}}\Omega^2}|c_{\mathrm{p}0}|^4(1+M)^2 \tag{6.4.66}$$

is reached near the points where $\xi = \bar{\xi}$ and $|\cos\Omega\xi| = 1$.

In order to increase the efficiency of DF generation, it is according to (6.4.65, 66) necessary to make the spatial "frequency" of the periodically nonuniform EF smaller. When Ω becomes less than $\gamma/2$ or $|c_{p0}|^2(1+M)$, the solution (6.4.65) is not applicable. Under these conditions, the expressions for c_v can be obtained if one neglects the terms with c_v in the second and third equations (6.4.45) and substitutes their solution into the first equation. Carrying out the approximate integration, we arrive for $\gamma/2 \gg \Omega \gg \delta$ and $|c_{p0}|^2(1+M) \approx 1$, at

$$c_v \simeq -\frac{2\mathcal{F}}{a_0} \frac{c_{s0}^* c_{p0}(1+M)\exp[(1+M)|c_{p0}|^2\xi]}{1+M\exp[2(1+M)|c_{p0}|^2\xi]} \, e^{i\delta\xi} \sin \Omega\xi \tag{6.4.67}$$

and for $|c_{p0}|^2(1+M) \gg \Omega \gg \gamma/2, |\delta|$, we have

$$c_v \approx -\frac{q_v \mathcal{F} a_0 \, e^{-i\varphi_{s0}}}{\sqrt{q_p}|c_{p0}|^2}[\arctan(\sqrt{M}\exp[(1+M)|c_{p0}|^2\xi]) - \arctan\sqrt{M}]$$

$$\times \begin{cases} \sin \Omega\xi \; e^{i\delta\xi} & \text{for} \quad \xi<\bar{\xi} \\ \sin \Omega\bar{\xi} \; \exp[i\delta\bar{\xi} - \frac{\gamma}{2}(\xi-\bar{\xi}) + \frac{\gamma}{4\Omega}(\sin 2\Omega\xi - \sin 2\Omega\bar{\xi})] & \\ & \text{for} \quad \xi>\bar{\xi} \end{cases} \quad . \tag{6.4.68}$$

In both cases the quantity $|c_v|^2$ reaches its maximum when $|\sin \Omega\xi| = 1$ and at the same time $\xi \approx \bar{\xi}$. According to (6.4.67),

$$|c_v|^2_{\max} \simeq \mathcal{F}^2 \frac{|c_{p0}|^4(1+M)^2}{q_p a_0^2} \tag{6.4.69}$$

and from (6.4.68) we obtain

$$|c_v|^2_{\max} \simeq \frac{\pi^2 q_v^2 \mathcal{F}^2 a_0^2}{4 q_p} \quad . \tag{6.4.70}$$

Considering (6.4.66, 68–70) and taking into account the necessary requirement $|\sin \Omega\xi| = 1$ for reaching the maximal value, one can easily see that the "frequency" of EF nonuniformity should satisfy the condition

$$\begin{aligned} \Omega_{\min} &= \frac{\pi}{2}\bar{\xi}^{-1} < \Omega < \Omega_{\max} \\ &= \max\{\pi^{-1}; \gamma/2\} \approx |\Omega_{\max} \pm \delta| \quad . \end{aligned} \tag{6.4.71}$$

The optimal ratio $\mathcal{E}_0^2|C_{p0}|^{-2}$ is determined, as before, by (6.4.71) and the quantity $|C_v|^2_{\rm opt}$ is four times greater than in (6.4.71).

The ratio of intensities at the frequency ω_v, produced in a periodically nonuniform ($\mathcal{J}_\mathrm{v}^\mathrm{pn}$) and a uniform ($\mathcal{J}_\mathrm{v}^\mathrm{u}$) electric field, makes it possible to estimate the effect of application of PNU EF. When the length of the chamber is small ($\bar{\xi}>\xi$) and the wave detuning is substantial, this ratio is equal to

$$\begin{aligned} R &= \frac{\mathcal{J}_\mathrm{v}^\mathrm{pn}}{\mathcal{J}_\mathrm{v}^\mathrm{u}} \simeq \frac{|p_2^\mathrm{u}-p_1^\mathrm{u}|^2}{4|p_2^\mathrm{pn}-p_1^\mathrm{pn}|^2}\exp[2\xi\,\mathrm{Re}\{p_1^\mathrm{pn}-p_1^\mathrm{u}\}] \\ &\approx \frac{(1+\gamma)^2+\delta^2}{4[(1+\frac{\gamma}{2})^2+(\delta-\Omega)^2]}\exp\left\{2\gamma\mathcal{F}^2\xi\right. \\ &\quad\times\left.\left[\frac{1+\gamma}{(1+\gamma)^2+\delta^2}-\frac{1+\gamma/2}{4[(1+\gamma/2)^2+(\delta-\Omega)^2]}\right]\right\} \quad . \end{aligned} \tag{6.4.72}$$

If $\gamma\xi\ll1$, the greatest gain is obtained when $\Omega=\delta$; it is due to compensation for the negative effect of the wave detuning and is equal to $\delta^2/4$ for small γ. Therefore, under these conditions PNU EF can be applied up to the values of reduced wave detuning $|\delta|=2$. And if $\mathcal{F}^2\gamma\xi$ is much greater than 1, the ratio R can exceed 1 if

$$4\frac{1+\gamma}{(1+\gamma)^2+\delta^2}>\frac{1+\gamma/2}{(1+\gamma/2)^2+(\delta-\Omega)^2} \quad .$$

When $(1+\gamma)(3+\gamma)>\delta^2$, this inequality is only valid for $\delta-\Omega\neq0$. In this case the gain produced by using PNU EF is associated with the decrease of the reverse effect of the DF wave on the development of SRS.

Let us now compare the maximal intensities that can be attained in PNU and uniform EF. For large values of wave detuning one can estimate the ratio $R=(\mathcal{J}_\mathrm{v}^\mathrm{pn})_\mathrm{max}/(\mathcal{J}_\mathrm{v}^\mathrm{u})_\mathrm{max}$ by applying (6.4.58, 60, 62) and similar expressions for a uniform field. In this situation, in systems with $\gamma\gg1$ the use of PNU EF leads to a gain when $|\delta|>\gamma$ (since this is equivalent to $R\simeq\delta^2/\gamma^2>1$). If $\gamma\ll1$, for $|\delta|>2/\pi$ the ratio $R=\pi^2\delta^2/4>1$, too. The necessary requirement in both cases is that $|\delta-\Omega|\to0$.

As follows from (6.4.66, 69, 70) and the equations for a uniform EF, when the wave detuning and the dipole moment of the resonant transition are small ($|\delta|$, $\gamma\ll1$), the maximal intensity that can be reached in PNU EF is the same as in a uniform one. And if $\gamma\gg|\delta|$, 1 and $\mathcal{F}\neq1$, the application of PNU EF results in the four-fold increase of the maximal intensity $\mathcal{J}_\mathrm{v}$, which is explained by the effective decrease of absorption at the frequency ω_v. Of special interest is the regime when $\gamma\gg|\delta|$, 1 and $\mathcal{F}=1$, when in a uniform EF the effect of the field C_v on SRS is particularly great. A quite considerable gain can be produced in this case (for $M\ll1$) if PNU EF is employed:

$$R\approx\frac{|c_{\mathrm{p}0}|^2(1+M)^2}{M}\simeq\frac{1}{M}\gg1 \quad . \tag{6.4.73}$$

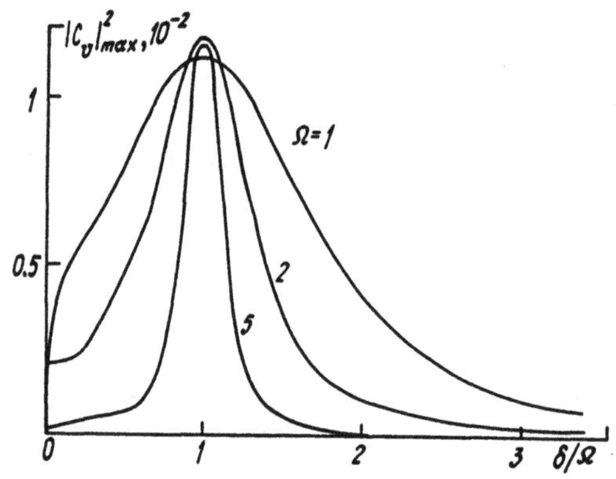

Fig. 6.10. The dependence of $|c_v|^2_{max}$ on δ/Ω for $\gamma = 0.05$, calculated for $\Omega = 1, 2, 5$

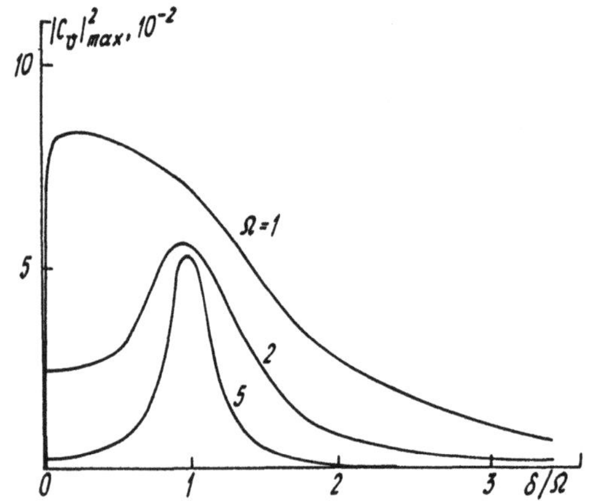

Fig. 6.11. The dependence of $|c_v|^2_{max}$ on δ/Ω for $\gamma = 0.5$, calculated for $\Omega = 1, 2, 5$

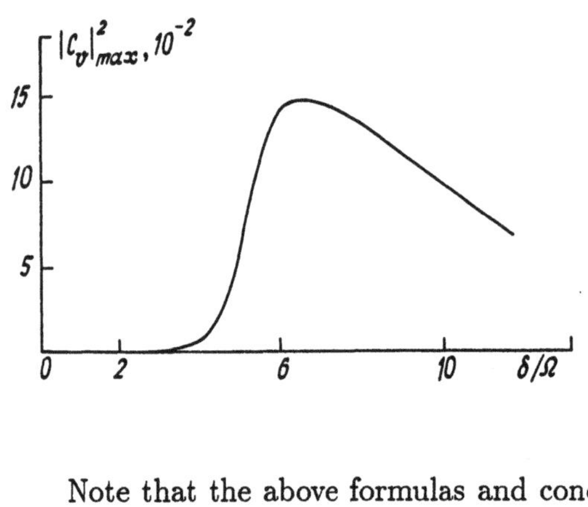

Fig. 6.12. The dependence of $|c_v|^2_{max}$ on δ/Ω for $\gamma = 10$ and a small wave detuning $\delta = 0.1$

Note that the above formulas and conclusions are based on approximate solutions of (6.4.45). Therefore, it makes sense to consider some results of the study of DF generation that have been obtained through the numerical solution of these equations by means of a computer. Figures 6.10–12 depict the quantity $|c_v|^2_{max}$ as a function of the ratio δ/Ω for various γ. The values q_p

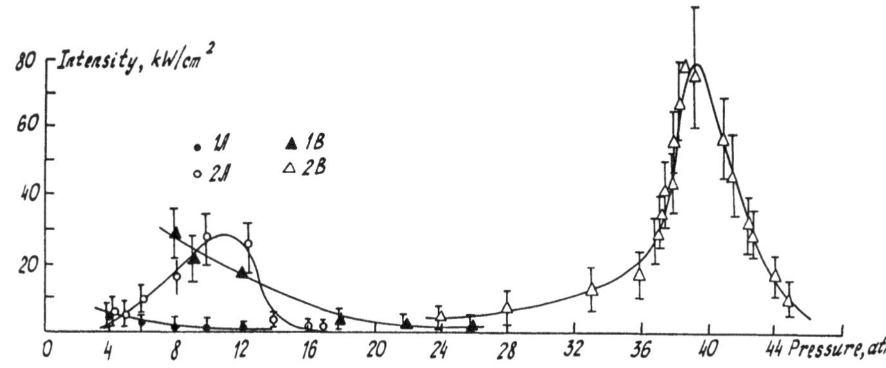

Fig. 6.13. The dependence of the resonant IR radiation intensity ($\lambda_v = 2.41\,\mu$m) on hydrogen pressure. The pumping field with the wavelength $0.694\,\mu$m (**A**) and 1.054 (**B**) as collimated by a lens into a chamber containing hydrogen. The focal region of the lens contains electrodes that produced (*1*) uniform or (*2*) periodically nonuniform EF with a strength of 40 kV/cm. The length of the electrode system is 20 cm and the period of EF nonuniformity 4 cm. The beam diameter of the pumping field in the focal region of the lens is 0.9 mm The pumping field pulse duration is 3×10^{-8} s and its intensity in the focal region, $(0.5 \pm 0.1) \times 10^9$ W/cm^2, the Raman conversion coefficient is $\mathcal{J}_s/\mathcal{J}_{p0} = 0.15$

and q_v correspond to the pumping field with a wavelength $\lambda_p = 1.054\,\mu$m of the vibrational transition $Q_1(1)$ in molecular hydrogen, with $\mathcal{F} = 1$. In accordance with the results of an analytical study, when $\gamma \ll 1$, the dependence of $|c_v|^2_{max}$ on δ/Ω reaches its maximum if the synchronization length coincides with the half-period of EF nonuniformity (Fig. 6.10). From Figs. 6.11 and 6.12 one can see that for larger values of γ (0.5 and 10) and small values of wave detuning, the maximum of the intensity of DF radiation occurs when the "frequency" of nonuniformity Ω differs from the wave detuning δ. Note that if $\gamma < 1$, this maximum is shifted towards the interval where $\delta/\Omega < 1$, its magnitude approaching that obtained in the case of a uniform EF. If $\gamma > 1$, the maximum is in the interval where $\delta/\Omega > 1$. If the pumping field cannot be regarded as constant and the reverse effect of DF radiation on SRS is essential, these results cannot be obtained in closed form. On the basis of numerical computations it is also possible to determine the optimal ratio $(\delta/\Omega)_{opt}$ for which the quantity $|c_v|^2_{max}$ reaches its maximum [for instance, for systems with $\gamma = 10$, $\delta = 0.1$ (Fig. 6.12) we have $(\delta/\Omega)_{opt} \approx 6$].

For experimental studies of DF generation in PNU EF, compressed hydrogen was used. Figure 6.13 presents the intensity of resonant radiation as a function of hydrogen pressure for the pumping field with the wavelengths $0.694\,\mu$m and $1.054\,\mu$m. For the sake of comparison the similar dependence for a uniform EF is given on the same figure. As follows from experiment, the intensities of the resonant field as a functions of pressure in a uniform EF are monotone, whereas in PNU EF they are characterized by a sharp maximums at pressures ≈ 12 atm and 38 atm (see Fig. 6.13). It is exactly for these values of the pressure that the synchronization length $L_s = \pi/\delta k$ coincides with the half-period of EF nonuniformity.

As follows from Fig. 6.13, the maximal intensity of the DF field (for EF strength of 40 kV/cm) is equal to $(75 \pm 20) \times 10^3$ W/cm^2. This corresponds to a coefficient of conversion of the pumping radiation into DF of the order 1.2×10^{-4}. If one uses (6.4.57) to estimate the conversion coefficient, one obtains 1.3×10^{-3} [$\gamma = 0.035$; $X(\xi) = 0.56$]. Apparently, such a discrepancy between the experimental and theoretical values of the conversion coefficient is associated with the fact that in the computations, the actual distribution of the pumping field intensity in space and time was not taken into account. Since the effect produced by it is the same for a uniform and PNU EF, the difference between the theoretical and experimental data for the gain R in DF radiation intensity turned out to be much less. The quantity R is bound by the limits 20–80 for the pumping radiation of a ruby laser (the spread essentially being due to irregular distribution of its intensity over the beam cross-section) and is greater than 10^2 for the pumping field produced by the neodymium laser, when pressures are such that $\Omega = \delta$. An attempt to estimate the gain in the latter case more accurately encounters difficulties, since the intensity of the resonant field generated in a uniform EF at pressures ~ 40 atm is below the registration threshold. If the gain is evaluated by means of the formula $R = \pi^2\delta^2/4$, the values 40 and 120 for the ruby and the neodymium lasers, respectively, are obtained.

Note that in the mentioned experiments γ did not exceed 0.06. Therefore, the observed gain in the DF intensity can be interpreted as a result of compensation of the wave detuning effect. The necessary requirement for observing the phenomenon of overcoming the parametric self-induced transparency is that γ should reach a value of the order of unity, and the wave detuning should be small. Although this requirement, as a rule, is not met when DF is generated on rotational transitions in gases, it is possible to fulfill it rather easily in the case of transitions in which high-lying electron levels participate. The condition for γ can be also satisfied for DF generation during SRS on vibrational transitions. For instance, if special measures are taken in constructing the system of electrodes, one can increase the breakdown EF strength in hydrogen up to ~ 230 kV/cm, the gap between the electrodes being equal to 0.35 cm [6.39]. This corresponds to an increase of γ up to 1.1. For the same pumping field the limiting value of γ for CO_2 is greater by a factor of six. If a mixture of the SRS-active gas and a gas characterized by a high breakdown threshold (for instance, SF_6 [6.39]) is used, the quantity γ can also be increased somewhat. Therefore, the main task for the experimental study of the influence of PNU EF on the efficiency of DF generation on vibrational transitions in gases under the conditions of parametric self-induced transparency is to guarantee that the wave detuning is small.

In crystals γ is large and the wave detuning is small for certain directions and polarization states of interacting fields. It is possible to produce PNU EF inside crystals, for example, in ferroelectrics during the formation of the domain structure. For this reason, it would certainly be of interest to conduct experiments on DF generation accompanying SRS on polaritons in ferroelectrics in the presence of domains. Similar studies in liquid crystals whose parameters are relatively easily controlled by external fields [6.40] would also be worthwhile.

As well as this, we would like to mention papers devoted to DF and sum frequency generation in media subject to the action of a magnetic field [6.41, 42]. Since there is an analogy between the electric and magnetic fields (see Sect. 1.1), the nonuniformity of a magnetic field should in such experiments lead to results resembling the ones discussed above.

The authors of [6.43] have demonstrated that, by creating non-uniform EF, one can reach an efficiency of resonant generation of the second harmonic in gases close to 100%. In this case the mechanism leading to such an increase in the efficiency is the same as during generation of difference frequency on the basis of SRS: the elimination of the negative influence of parametric self-induced transparency and wave detuning. An analogous method can obviously be applied to the three-wave sum frequency generation in crystals with impurities and metal vapors which yielding small conversion coefficients in experiments conducted with an uniform field [6.44, 45].

7. Four-Photon Resonant Parametric Interactions (RPI)

This chapter is devoted to four-photon resonant parametric processes which are now more and more widely used in IR and visible and UV frequency ranges to produce coherent radiation, including such with tunable frequency. Much attention is paid to explaining the role of the generated wave phase locking and confinement, of parametric self-induced transparency and of the saturation of resonant transition populations. Based on the developed theory and on especially designed experiments, the problem of the SRS anti-Stokes component generation mechanism is clarified. The nature of the dynamics of biharmonic pumping SRS that is used in the frequency conversion of weak (sub-threshold) radiation is illuminated. The process of frequency increase and decrease under two-photon absorption of the pumping radiation are also discussed. In the last section the possibility for the use of four-photon parametric processes for soft excitation of a two-photon laser is discussed.

Four-photon RPIs may arise when the wave frequencies satisfy two second-order resonance conditions

$$\omega_\alpha + \omega_\beta = \omega_\gamma + \omega_\delta = \omega_{21} + \nu \quad ,$$

see Fig. 7.1. Apparently, this process was observed for the first time when in 1963, the anti-Stokes SRS component (ASRS) that was produced by a four-photon interaction degenerate with respect to frequency was discovered [7.1]. The corresponding level scheme is depicted in Fig. 7.1d, for the pumping field frequency $\omega_1 = \omega_3$. At first, both experimental and theoretical studies were devoted mainly to this phenomenon, detected in many liquids, solids and gases [7.2–8].

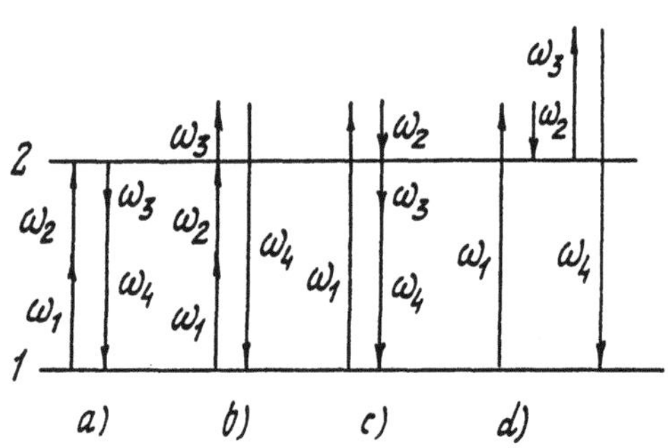

Fig. 7.1. Resonant four-photon parametric processes on the basis of the pumping field TPA and SRS; ω_1 and ω_2 are the frequencies of the pumping field (or of the pumping field and its Stokes SRS component), ω_3 is the trigger field frequency, ω_4 is the generated field frequency

Relying on processes as illustrated in Fig. 7.1c the authors of [7.9] considered, for the first time, the possibility of frequency transformation. In [7.10], the third harmonic generation during the propagation of radiation through a medium which is resonant for the second harmonic was studied theoretically and the maximal transformation coefficient was determined.

After electron SRS and TPA had been detected in metal vapors [7.11–16], a new stage in the research of four-photon RPIs began. This led to the discovery of a whole series of parametric processes [7.13–19], including anti-Stokes SRS [7.13]. The main cause for the interest in four-photon RPIs is at present that they make it possible to obtain frequency-tunable coherent radiation sources. If they are triggered with tunable lasers (for instance, liquid ones) then their operation ranges cover the infrared (IR), ultraviolet (UV) and vacuum ultraviolet (VUV) spectral intervals.

Although there was no doubt about the fundamental possibility of constructing tunable sources on the basis of the above processes[1] they were first observed in metal vapors. So far, experiments on four-photon interactions associated with the generation of the Stokes component of electron SRS in potassium and cesium vapors [7.17, 22] and with the TPA of the pumping field in vapors of rubidium, strontium and sodium [7.16, 18, 19] have been conducted. The authors of [7.17] report the generation of tunable radiation in the range 2–5.4 μm, and those of [7.18] in the VUV range.

In [7.19], the addition of the frequencies of a CO_2-laser and of a dye laser in Na vapor was achieved. Thus the possibility of using the four-photon processes for visualization of infrared radiation was demonstrated. It should be noted that the authors of [7.19] obtained quite a high coefficient of transformation of IR radiation into the sum frequency wave, namely, 58% in terms of the number of photons.

The visualization of an image illuminated by radiation with $\lambda = 2.94\,\mu$m into the violet spectral region was accomplished in [7.23].

References [7.13, 19, 22, 23] established the importance of a detailed theoretical and experimental study of four-photon RPIs. Until recently, however, the state of the art in this field provided no possibility for the development of ideas about specific features that are common to all four-photon RPIs and distinguish them from similar nonresonant phenomena. In the absence of such ideas it is naturally impossible to work out methods for the design of optimal regimes for various devices in which such processes are employed.

It is worth mentioning that the peculiarities of four-photon interactions were discovered as early as 1965 [7.2], when effective ASRS was observed along the pumping beam axis, i.e., in the direction characterized by a considerable degree of detuning of the interacting waves' phase velocities. A typical feature of this phenomenon is that axial radiation may occur before the radiation in the

[1] See [7.20] which discuss the feasibility of using, for this purpose, the process depicted in Fig. 7.1a and excited in the resonator, and also [7.21] where processes shown in Fig. 7.1a, b are under consideration.

direction of the wave synchronism [7.2, 5, 6, 24]. In [7.25] it was attempted to explain the possibility of effective generation of parametrically coupled waves outside the directions of linear synchronism by radiation arising from thin filaments which are formed during self-focussing of light in matter.

However, in further experiments it was established that axial radiation is not necessarily connected with self-focussing. Axial radiation was observed in the absence of self-focussing in various liquids, mixtures of liquids [7.6] and in gaseous hydrogen [7.8]. Besides, as has been ascertained experimentally, the nondegenerate four-photon RPIs based on the SRS pumping field (see Fig. 7.1c, 7.1d), like ASRS, have components that propagate outside the synchronization direction [7.26]. Their behavior cannot be explained by self-focussing. All this gives evidence for the necessity to improve the theory of four-photon interactions.

Usually four-photon RPIs were considered (see, e.g., [7.9, 21, 27]) in the same way as nonresonant parametric interactions of waves, such as the third harmonic generation in a transparent medium [7.28, 29][2]. Within this approach, polarization of matter at the sum or difference frequency $\omega_\delta = \omega_\alpha + \omega_\beta - \omega_\gamma$ (see Fig. 7.1) is assumed to be proportional to $\chi E_\alpha E_\beta E_\gamma$ [7.21], where χ is the corresponding nonlinear susceptibility of matter. It is also supposed that the only change produced by resonance concerns the magnitude of χ, and waves only interact effectively over the length $(\delta k)^{-1}$ which is determined by the detuning of wave vectors.

Such a description cannot be applied to real experiments (with large radiation flux densities up to 1 GW/cm^2 and considerable thicknesses of the lasing substance). This is due to the fact that the four-photon RPIs, as the three-photon RPIs discussed in the preceding chapter, proceed in a manner which essentially differs from that of nonresonant interactions. This has been demonstrated in [7.24, 31, 32]. The properties of these processes turned out to permit an explanation from a unified point of view, on the basis of phase locking of interacting waves. In particular, the phase locking may result in energy transformation from one wave into the other that occurs in collinear beams and over lengths that are much greater than the synchronization interval. This is exactly what guarantees the possibility of intense radiation outside the directions of synchronism. The limited nature of the phase locking region, on the other hand, leads in general to a coordinate dependence of the amplitudes of interacting waves, which is quite different from that for the corresponding nonresonant interactions.

The theory of ASRS that has been developed further than any other (see [7.30]), was constructed under the assumption of a constant pumping field. This does not allow to take into account the limited nature of the spatial

[2] References [7.10, 30], in which ASRS and the third harmonic generation in a two-photon absorbing medium [7.10] were analyzed by taking into account not only parametric but also Raman (two-photon) processes, constitute an exception.

region where phase locking occurs and, therefore, leaves many peculiarities of ASRS unexplained.

The authors of [7.32] used another RPI, the four-photon interaction that arises in the presence of biharmonic pumping between its components and their two Stokes components. Such a process had been studied earlier for the purposes of nonlinear spectroscopy when the Raman interaction of fields is negligible and the conditions of linear synchronization are fulfilled (this had been ensured by choosing the corresponding geometry of interacting beams) [7.33, 34]. In contrast to that, [7.32] reports the generation of Stokes components during biharmonic pumping observed in collinear beams for a considerable value of wave detuning. It has been demonstrated that the phase locking may in this case also lead to parametric interaction providing the main contribution to the generation of the Stokes component by the less intense component of the pumping field. As in ASRS, the characteristic interaction length is not determined by the magnitude of the wave detuning, but by the scale of transformation of the strong (i.e., exceeding the SRS threshold) component of the pumping field into its Stokes component. As a result of this, the intensity of the weak pumping field component which is necessary to produce its Stokes component, can be lowered by several orders of magnitude [7.32].

Another peculiar feature of RPI under consideration is that both Raman and parametric processes have the same direction, i.e., they transfer the pumping field energy into Stokes components ("Raman-parametric generation" [7.32]). In the remaining four-photon interactions the wave amplified by the parametric process is always weakened by the Raman or two-photon process (see Sects. 7.3, 4).

Because of the above properties of Raman-parametric generation, it is a promising tool for transforming the radiation of low intensity tunable lasers into the IR range. In schemes with a constant ray geometry, the poor sensitivity in respect to wave detuning makes it possible to obtain broad frequency tuning. This was demonstrated for the first time in [7.32]. Later on, using the above method the continuously tuning radiation at the 4–25 μm range was obtained by the frequency conversion of the parametric light oscillator radiation. In [7.71] the radiation at the 16 μm range was obtained with the help of Raman-parametric conversion of a CO_2-laser radiation on the rotational transition $j = 0 \rightarrow j = 2$ of parahydrogen.

The experimental and theoretical studies presented in [7.32] also lead to the conclusion that "the lowering of the SRS threshold" during irradiation of matter, both by the pumping field and by its second harmonic, observed as early as 1968 [7.35], is caused by Raman-parametric generation.

The possibility of such a lowering of the SRS threshold exists not only for SRS of the additional field [7.32, 35], but as well for its two-photon absorption. A similar method can be used to produce a tunable Stokes component of electron SRS that lies in the IR spectral range.

The following analysis of four-photon RPIs begins by considering the anti-Stokes SRS.

7.1 Anti-Stokes Stimulated Raman Scattering

7.1.1 Specific Features of ASRS

In noninverted media, the anti-Stokes SRS (ASRS) results from parametric interaction of the pumping radiation and the Stokes component of SRS. ASRS was the first phenomenon of four-photon interaction between the light waves discovered experimentally [7.1], and a number of further papers (see, for instance, [7.30, 36, 37]) are devoted to its theoretical study. ASRS is the only resonant parametric process which is discussed by *Bloembergen* [7.30]. His analysis, however, as well as that of subsequent theoretical papers [7.36, 37] was based on the approximation of the constant pumping field. Although it is possible to establish certain important features of ASRS such as the absence of generation in the direction characterized by the zero value of δk, and some other regularities of behavior of cone ASRS rings[3] by applying this approach, it cannot explain the nature of the dependence of the anti-Stokes component energy (especially in the case of the axial component, i.e., the one propagating along the pumping beam) on the input energy of the pumping field even when the transformation of the latter into the anti-Stokes component is weak.

A number of peculiar features of ASRS that could not be interpreted from the theoretical point of view have been discovered in the numerous experiments. Here are the main ones.

1) In addition to cone scattering which satisfies the conditions of linear wave synchronization, the axial component has been detected. In most experiments this component, for which the wave detuning is considerable, appeared before the cone one [7.5]. The initial explanation of this fact by the effect of self-focussing was disproved by the observation of the axial ASRS in gases and liquids, where there was no self-focussing [7.6].

2) The behavior of the anti-Stokes component energy as a function of the thickness of the scattering substance and the pumping field energy differs a lot from the behavior of the Stokes component, although the increments for both components computed under the assumption of the constant pumping field are the same [7.30].

3) The behavior of the energies of the cone and axial components is also different [7.6]. This difference turned out to be so large that the authors of [7.6] put forward the hypothesis that the natures of axial and cone ASRS are not the same.

It will be shown below that the above peculiarities of ASRS can be explained by the influence of the changing phase difference of interacting fields caused by the opposed actions of Stokes and anti-Stokes components on the

[3] That is, of the maxima of the anti-Stokes component intensity that occur in directions for which δk is close to zero [7.30].

pumping radiation [7.24]. The decisive role is due to the fact that the phase locking of generated fields (or, the nonlinear synchronization) takes place over a length which is much greater than the linear synchronization length. For the application of the approximation of a constant pumping field it is not sufficient that the coefficient of the pumping field energy transformation into the anti-Stokes component is small. Because of the energy transfer into the Stokes component, which always occurs when the thickness of the lasing substance exceeds a certain value, the phase locking is broken (while in the approximation of the constant pumping field the phase locking is present in all cases). As a result of this, the dependence of the energy W_a of the anti-Stokes component on the pumping field energy becomes nonmonotonic to a degree determined by the magnitude of wave detuning δk. If this phenomenon is taken into account, all differences in the behavior of the axial and cone ASRS components and SRS Stokes component can be explained.

The breaking of phase locking also leads to an unusual relationship between W_a and the length of the cell containing the nonlinear substance, namely, in a certain range of pumping field strengths the quantity W_a diminishes as the cell length increases. This effect was predicted and observed in [7.24].

7.1.2 Basic Equations

Assume that three waves, the pumping field, its Stokes and anti-Stokes components (with frequencies, respectively, equal to ω_p, ω_s and ω_a) propagate in a Raman-active medium. The resonance conditions for the frequencies of fields and matter are

$$\omega_{21} = \omega_p - \omega_s = \omega_a - \omega_p \quad . \tag{7.1.1}$$

The derivation of equations for amplitudes and phases of interacting fields and polarization of matter will be based on the following assumptions:

1. each electromagnetic wave

   $$\boldsymbol{E}(\omega_j) = \boldsymbol{e}_j \mathcal{A}_j \exp\{\mathrm{i}[\omega_j t - \boldsymbol{k}_j \boldsymbol{r} - \varphi_j(\boldsymbol{r})]\} + \mathrm{c.c.}$$

   is characterized by linear polarization, and all three vectors of wave polarization $\boldsymbol{e}_j$ ($j = \mathrm{p, s, a}$) lie in the same plane (in the experimental study [7.24] these conditions were fulfilled);
2. the interaction is quasistationary, which corresponds to the regime of lasers operating without mode synchronization;
3. the nonresonant part of nonlinear polarization can be neglected (in gases it is less than the resonant one by 4–5 orders of magnitude [7.38];
4. saturation of populations is insignificant (in experiments with a pulse duration $\approx 10^{-8}$ s and intensity less than $10\,\mathrm{GW/cm^2}$ $\eta \approx \eta_0$).

Under these assumptions the methods presented in the first two chapters make it possible to obtain expressions for projections of polarization components on the field directions. When these are substituted into (1.1.3) and the real and imaginary parts are separated, the system of equations for dimensionless real amplitudes of fields $a_j = \mathcal{A}_j/\mathcal{A}_{p0}$ (where $\mathcal{A}_{p0}$ denotes $\mathcal{A}_p$ for $z=0$) and the phase difference $\theta = 2\varphi_p - \varphi_s - \varphi_a - \delta\boldsymbol{k}\cdot\boldsymbol{r}$ (the wave detuning $\delta\boldsymbol{k} = \boldsymbol{k}_a + \boldsymbol{k}_s - 2\boldsymbol{k}_p$) can be written as

$$\begin{aligned}
\frac{\partial a_p}{\partial\zeta} &= (\kappa_1 a_a^2 - a_s^2)a_p \cos(\alpha_p - \alpha_a) \quad , \\
\frac{\partial a_s}{\partial\zeta} &= g_s(a_s + \kappa_2 a_a \cos\theta)a_p^2 \cos(\alpha_s - \alpha_a) \quad , \\
\frac{\partial a_a}{\partial\zeta} &= -g_a(\kappa_1 a_a + \kappa_2 a_s \cos\theta)a_p^2 \quad , \\
\frac{\partial\theta}{\partial\zeta} &= \kappa_2\left[g_a\frac{a_s}{a_a} - g_s\frac{a_a}{a_s}\cos(\alpha_s - \alpha_a)\right]a_p^2 \sin\theta + \delta \quad .
\end{aligned} \tag{7.1.2}$$

Here $\alpha_{p,a,s}$ are the angles which are formed by $k_{p,a,s}$ with the z axis,

$$\begin{aligned}
\zeta &= z\frac{2\pi\omega_p^2 T\eta_0 N}{\hbar c^2 k_p \cos\alpha_a}\langle\kappa_s^2\rangle\mathcal{A}_{p0}^2 \quad ; \\
\delta &= (\delta\boldsymbol{k})_a\left(\frac{2\pi\omega_p^2 T\eta_0 N}{\hbar c^2 k_p}\langle\kappa_s^2\rangle\mathcal{A}_{p0}^2\right)^{-1} \quad .
\end{aligned}$$

In these formulas $(\delta\boldsymbol{k})_a$ is the projection of $\delta\boldsymbol{k}$ on $\boldsymbol{k}_a$;

$$\begin{aligned}
g_{s,a} &= \frac{\omega_{s,a}^2 k_p}{\omega_p^2 k_{s,a}} \quad ; \quad \kappa_1 = \frac{\langle\kappa_a^2\rangle}{\langle\kappa_s^2\rangle} \quad ; \quad \kappa_2 = \frac{\langle\kappa_s\kappa_a\rangle}{\langle\kappa_s^2\rangle} \quad , \\
\kappa_s &= \hbar^{-1}\sum_m\left(\frac{(\boldsymbol{d}_{m1}\boldsymbol{e}_c)(\boldsymbol{d}_{2m}\boldsymbol{e}_p)}{\omega_{m2}+\omega_p} + \frac{(\boldsymbol{d}_{1m}\boldsymbol{e}_p)(\boldsymbol{d}_{2m}\boldsymbol{e}_s)}{\omega_{m2}-\omega_s}\right) \quad ,
\end{aligned}$$

the quantity κ_a is obtained from κ_s by the change of indices $s \to p$, $p \to a$.

Note that the system (7.1.2) describes both the cone and the axial component of ASRS, only $(\delta\boldsymbol{k})_a$ and the magnitudes of α_p, α_s and α_a being different. The main parameter, the variation of which results in the fact that the behavior of these components is not the same, is the quantity δ.

7.1.3 Spatial Distribution of the Anti-Stokes Component

Reference [7.30] contains a detailed study of the solution of (7.1.2) in a constant pumping field. This approach, however, does not correspond to the experimental conditions under which a considerable transfer of the pumping field energy into at least one (the Stokes) scattering component occurs. Let us solve the system (7.1.2), taking into account the influence of the latter on the pumping field for a sufficiently large value of detuning δ and a small value of the anti-Stokes component, so that its reverse effect on the pumping field can be neglected. The solution of the first two equations (7.1.2) in which $a_{\mathrm{a}} \equiv 0$ (6.4.25) approximately describes the behavior of the pumping field and the Stokes component of SRS:

$$a_{\mathrm{p}}^2 = a_{\mathrm{p0}}^2 \frac{1+M}{1+M\exp\gamma\zeta} \quad , \quad a_{\mathrm{s}}^2 = a_{\mathrm{s0}}^2 \frac{(1+M)\exp\gamma\zeta}{1+M\exp\gamma\zeta} \quad , \quad \text{where}$$

$$M = a_{\mathrm{s0}}^2/g_{\mathrm{s}}a_{\mathrm{p0}}^2 \quad , \quad \gamma = 2g_{\mathrm{s}}a_{\mathrm{p0}}^2(1+M) \quad ,$$
$$a_{\mathrm{p0}} = a_{\mathrm{p}}|_{z=0} \quad , \quad a_{\mathrm{s0}} = a_{\mathrm{s}}|_{z=0} \quad .$$

Substitute these formulas into the equation for the complex amplitude:

$$\frac{\partial[a_{\mathrm{a}}\exp(-\mathrm{i}\varphi_{\mathrm{a}})]}{\partial\zeta} = -g_{\mathrm{a}}a_{\mathrm{p}}^2\{\kappa_1[a_{\mathrm{a}}\exp(-\mathrm{i}\varphi_{\mathrm{a}})] + \kappa_2 a_{\mathrm{s}}\exp(\mathrm{i}\delta\zeta)\} \quad .$$

Solving the above equation we find that the behavior of the anti-Stokes component amplitude is determined by the Gaussian hypergeometric function ${}_2\mathcal{F}_1$:

$$\begin{aligned} a_{\mathrm{a}}\,\mathrm{e}^{-\mathrm{i}\varphi_{\mathrm{a}}} = a_{\mathrm{s0}} & \frac{g_{\mathrm{a}}\kappa_2}{2g_{\mathrm{s}}} \frac{(1+M)^{(-1/2)+\mathrm{i}\tilde{\delta}}(M+\mathrm{e}^{-\gamma\zeta})^R}{(1-\mathrm{i}\tilde{\delta})M^{0.5+R+\mathrm{i}\tilde{\delta}}} \left\{\left(\frac{1+M}{1+M\,\mathrm{e}^{\gamma\zeta}}\right)^{1-\mathrm{i}\tilde{\delta}}\right. \\ & \times {}_2\mathcal{F}_1\left(\frac{1}{2}-R-\mathrm{i}\tilde{\delta};\, 1-\mathrm{i}\tilde{\delta};\, 2-\mathrm{i}\tilde{\delta};\, \frac{1}{1+M\,\mathrm{e}^{\gamma\zeta}}\right) \\ & \left. - {}_2\mathcal{F}_1\left(\frac{1}{2}-R-\mathrm{i}\tilde{\delta};\, 1-\mathrm{i}\tilde{\delta};\, 2-\mathrm{i}\tilde{\delta};\, \frac{1}{1+M}\right)\right\} \quad , \end{aligned} \tag{7.1.3}$$

where $\tilde{\delta} = \delta/2(1+M)$, $R = g_{\mathrm{a}}\kappa_1/2g_{\mathrm{s}}$.

It follows from this that for distances at which a substantial part of the pumping field energy is transferred into the Stokes component, the function $a_{\mathrm{a}}(z)$ is nonmonotonic (see, for instance, the curve $\delta = -3$ on Fig. 7.2). The behavior of the SRS anti-Stokes component is strongly influenced by the reverse effect of the Stokes component on the pumping field. Finally it leads to the breakdown of the phase locking:

$$\lim_{z\to\infty} \theta = \delta\boldsymbol{k}\cdot\boldsymbol{r} \quad .$$

As a result of this, a stationary value of the anti-Stokes component intensity is reached:

$$\begin{aligned} \lim_{z\to\infty} a_{\mathrm{a}}^2 &= a_{\mathrm{p0}}^2 \frac{g_{\mathrm{a}}^2\kappa_2^2}{g_{\mathrm{s}}(1+R^2)} \frac{\pi\delta/2}{\sinh(\pi\delta/2)} \prod_{m=0}^{\infty}\left\{1+\frac{\delta^2}{(1+R+2m)^2}\right\}^{-1} \\ &= \mathrm{const} \quad . \end{aligned} \tag{7.1.4}$$

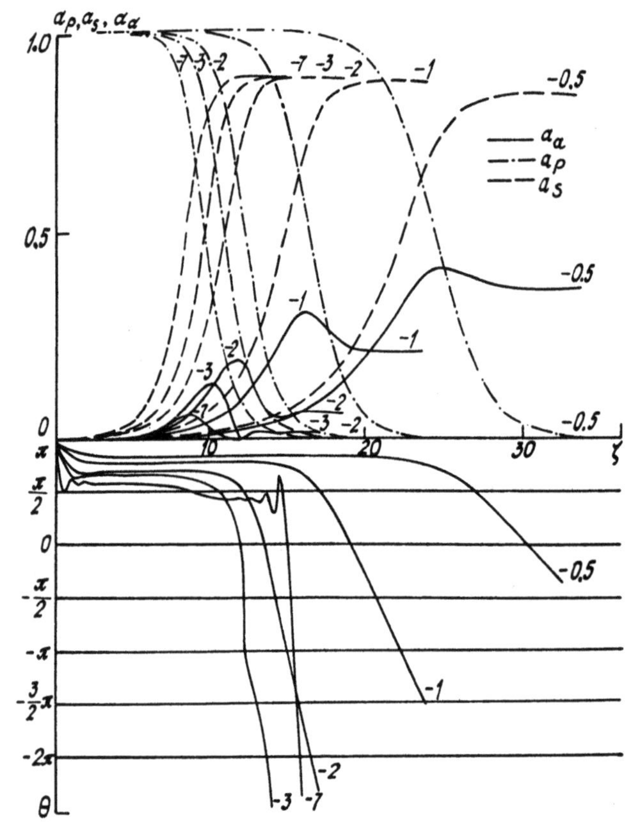

Fig. 7.2. Dependences of a_a, a_s, a_p, θ on ζ for $\delta = -7, -3, -2, -1, -0.5$, obtained for the value of the ratio $\omega_{21}/\omega_p = 0.22$ and the boundary values $a_a(\zeta = 0) = 10^{-6}$, $a_s(\zeta = 0) = 10^{-3}$, $a_p(\zeta = 0) = 1$, $\theta(\zeta = 0) = \pi$

In (7.1.4), the ratio of the numbers of the Stokes component photons to that of the photons of the pumping field at the entrance has been neglected in comparison with unity ($M \ll 1$).

As the input energy of the pumping field increases, the limiting value of the amplitude a_a becomes of the same order of magnitude as the amplitudes of the pumping field and the Stokes component, and therefore the solution (7.1.4) is inapplicable when $\pi\delta/2 < 1$, i.e., when

$$\mathcal{A}_{p0} > \frac{\hbar c^2 k_p^2 (\delta \boldsymbol{k})_a}{4\omega_p^2 T \eta_0 N \langle \kappa_s^2 \rangle} \quad .$$

In this case, in order to describe the anti-Stokes component, it is necessary to account for the opposed effect it has on the pumping field and the Stokes component. Although it does not appear possible to obtain a closed-form solution of the system (7.1.2), one can see from it that the qualitative pattern of behavior of $a_a(z)$ is also preserved at large values of the pumping field (small δ). Indeed, according to the last equation (7.1.2), and inequality $a_p^2 a_s / a_a \gg 1$ valid at the initial stage of interaction, the phase difference is

$$\theta \approx \pi + \arcsin \frac{\delta}{\kappa_2 \left(g_a \frac{a_s}{a_a} - g_s \frac{a_a}{a_s} \right) a_p^2} \tag{7.1.5}$$

(for large values of the pumping field $\theta \approx \pi$). Because the parametric energy transfer from the pumping field and the Stokes component into the anti-Stokes one [$a_s + a_p \to a_a$, see the second term in the third equation (7.1.2)] exceeds the purely Raman transformation of the anti-Stokes component into the pumping field ($a_a \to a_p$), the amplitude of the anti-Stokes component increases. Further, the diminishing a_p causes the phase variation in such a way that the rate of growth of a_a decreases. The result of this is that both the rate of transformation $a_p \to a_s$ and the velocity with which the phase moves away from π become greater. When the difference between θ and π is such that the processes $a_s + a_p \to a_a$ and $a_a \to a_p$ are balanced, the quantity a_a reaches its maximum and then decreases. Since a_s continues to grow, the pumping field is finally lowered to a very small value, and the waves practically no longer interact. A stationary value of a_a and $\theta \approx \delta \boldsymbol{k} \cdot \boldsymbol{r}$ is built up.

The numerical solution of (7.1.2) confirms the above qualitative picture. Figure 7.2 displays the curves $a_a(\zeta)$ for different values of $(\delta \boldsymbol{k})_a$ for a constant pumping field $\mathcal{J}_{p0}$[4].

Note one consequence that follows from plots on Fig. 7.2. According to the results of the calculations of the anti-Stokes wave increment in the approximation of a constant pumping field, for a certain $\delta = \delta_m$ the initial rate of growth of $a_a(\zeta)$ is maximal. However, from Fig. 7.2 it is seen that the decrease of $|\delta|$ (i.e., the increase of the pumping field or the decrease of the detuning $|\delta \boldsymbol{k}|$) is accompanied by the increase of the quantity $\lim_{z \to \infty} a_a$. Therefore, for a fixed interaction length L the pumping field for which the anti-Stokes component has the maximal intensity may correspond to the value of $|\delta| < |\delta_m|$.

7.1.4 Energy Characteristics of ASRS

For a fixed length L of wave interaction the solution for a_a (Fig. 7.2) makes it possible to determine the output energy of the anti-Stokes component as a function of the pumping field input energy. Obviously, this function will be nonmonotonic (Fig. 7.3). First, the increase of $\mathcal{J}_{p0}$ will lead to a local low-value maximum of the anti-Stokes component intensity $\mathcal{J}_a$, associated with the maximum of the corresponding curves $a_a(z, \delta)$ passing through the exit of the

[4] See also Fig. 1 in [7.24], where the behavior of a_j for a constant $(\delta \boldsymbol{k})_a$ and various values of the pumping field intensity $\mathcal{J}_{p0}$ are presented.

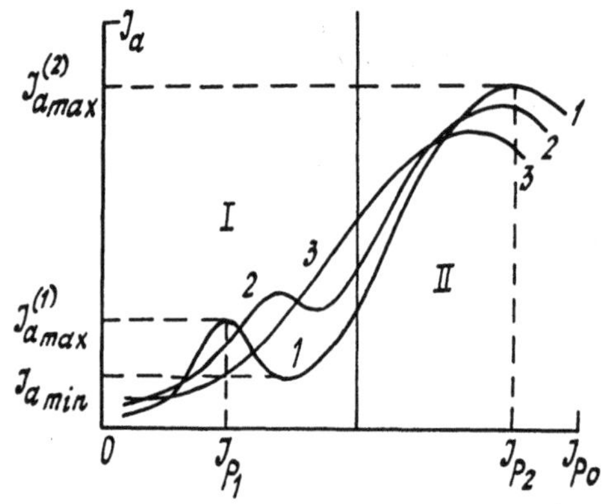

Fig. 7.3. The function $\mathcal{J}_a(\mathcal{J}_{p0})$ for various values of detuning δ. For curves 1–3 $|\delta_1|>|\delta_2|>|\delta_3|$

chamber ($z = L$) (Fig. 7.2). Then the quantity $\mathcal{J}_a(\mathcal{J}_{p0})|_{z=L}$ starts growing again. Finally, as $\mathcal{J}_{p0}$ increases, the output intensity $\mathcal{J}_a$ will diminish for the second time, since for $|\delta| < |\delta_m|$ both the growth of $a_a(z)$ and the growth of the Stokes component slow down. This fact, in particular, can explain the asymmetry of the energies of SRS Stokes components propagating in the forward and backward directions, which has been observed in experiment for a large value of the pumping field (for instance, when it is focussed into the nonlinear substance) [7.39]. Because for reverse waves $(\delta \boldsymbol{k})_a \sim k$ holds and as a consequence no phase locking occurs at realistic intensities $\mathcal{J}_{p0}$ and there is no anti-Stokes wave.

In [7.40] a detailed study can be found of the competition between the Stokes and anti-Stokes SRS components which leads to a decrease of the Stokes "forward" scattering in the approximation of the constant pumping field.

Consider now the difference between the functions $\mathcal{J}_a$ $(\mathcal{J}_{p0})$ for the axial and cone ASRS.

It should be noted that in the general case, in order to compute the angular distribution, the combined solution of the equations for interacting waves that takes into account the spatial spectrum of both Stokes and anti-Stokes components is required, which is an extremely difficult task – even for a computer-based numerical analysis. For pumping field energies ensuring a clear distinction between the axial and cone ASRS it is, however, possible to establish certain qualitative peculiarities specific for the behavior of the cone component, in contrast to the axial one. To this end, we shall use the results of the calculation of the function $\mathcal{J}_a$ $(\mathcal{J}_{p0})$ for different values of $\delta \boldsymbol{k}$, presented in Fig. 7.3. The more rapid saturation and the subsequent falloff of the intensity of the cone ASRS in comparison with the axial one, which is observed in region II, is associated with the fact that in the case of the axial ASRS the dimensionless detuning $\delta = \delta_m$ is reached at lower values of the pumping field. In the initial region I, on the other hand, the decrease of $|\delta \boldsymbol{k}|$ is accompanied by the shift of the first maximum of $\mathcal{J}_a$ into the strong pumping field range. The magnitude of the "dip" on the segment of the curve $\mathcal{J}_a$ $(\mathcal{J}_{p0})$ is decreased, and for the compo-

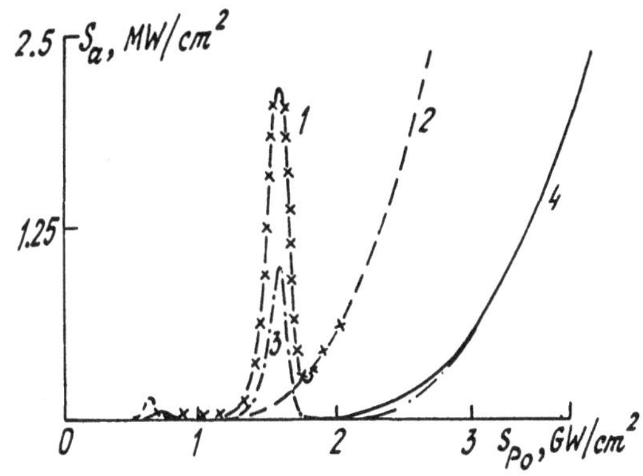

Fig. 7.4. The dependence of the power flux density of axial ASRS in the output section of the chamber on the pumping field. $S = c\sqrt{\varepsilon}|\mathcal{E}(\omega)|^2/2\pi$, $1 = 75$ atm, 5 cm, $2 = 75$ atm, 12 cm, $3 = 110$ atm, 5 cm, and $4 = 110$ atm, 12 cm (the numbers of centimeters indicate the length of chambers, the numbers of atmospheres indicate the hydrogen pressure)

nent that propagates in the direction for which $|\delta| < |\delta_m|$, it actually disappears completely. This characterizes the qualitative differences in the behavior of $\mathcal{J}_a$ $(\mathcal{J}_{p0})$ for the cone and axial ASRS.

These properties of ASRS can be explained as follows. The function $\mathcal{J}_a$ $(\mathcal{J}_{p0})$ may be nonmonotonic in region I (Fig. 7.3) when the phase locking is broken off at $z < L$ [the dependence $a_a(\zeta)$ on Fig. 7.2]. As the quantity $|\delta \boldsymbol{k}|$ decreases, the phase locking is destroyed for interaction lengths with values of $\mathcal{J}_{p0}$ exceeding those along the axial direction.

Because of this, the fall off of $\mathcal{J}_a$ after passing $\mathcal{J}_{a\,max}^{(1)}$ will also be less since the decrease of $|\delta \boldsymbol{k}|$ leads to the maximal value of $a_a(z)$ on the curves $a_a(z)$ approaching the asymptotic one.

Thus, in the initial region the behavior of $\mathcal{J}_a$ deviates most sharply from the monotonic one of the axial component of ASRS.

As follows from a similar argument in the case of a given value of detuning $|\delta \boldsymbol{k}|$, the increase of the interaction length L is accompanied by the decrease of $\mathcal{J}_{a\,max}^{(1)}$ (Fig. 7.4). At the same time the difference between $\mathcal{J}_{a\,max}^{(1)}$ and $\mathcal{J}_{a\,min}$ becomes less for sufficiently small L. For this reason, a quantity L which is optimal for observing the nonmonotonic nature of the function $\mathcal{J}_a$ $(\mathcal{J}_{p0})$ should exist.

Thus, the following qualitative peculiarities of the behavior of the ASRS axial and cone components can be formulated:

1. the intensity of the axial anti-Stokes component is a nonmonotonic function of the pumping field;
2. as the direction of the exact synchronism is approached, the deviation of the function $\mathcal{J}_a$ $(\mathcal{J}_{p0})$ from a monotonic behavior becomes less and less pronounced;
3. the increase of the interaction length should result in the decrease of the value of the maximum $\mathcal{J}_{a\,max}^{(1)}$ and its shift towards lower values of the pumping field;

4. the output intensity $\mathcal{J}_a(L)$ does not depend on the length L if $|\delta \boldsymbol{k}| =$ const and the pumping field is sufficiently strong (after the passing of the value $\mathcal{J}_{a\,min}$);
5. if $\mathcal{J}_p < \mathcal{J}_{p2}$, the variation of the Stokes component intensity $\mathcal{J}_s$ ($\mathcal{J}_{p0}$) is monotonic for any $|\delta \boldsymbol{k}|$.

All the above results are obtained from (7.1.2) which describe the interaction of plane waves. It must be taken into account that even in the plane wave of the pumping field, a complex angular distribution of the scattering components arises [7.41]. The situation becomes even more complicated if one takes into account the actual distribution of intensity over the cross-section of the pumping field beam, as well as the fact that its pulse cannot be described as a rectangular function of time. All this makes the observation of the above specific features difficult. For instance, if the integral energy of the anti-Stokes component W_a is measured, the nonrectangular nature of the spatial and temporal distribution of the pumping field would lead to a more smooth dependence $W_a(W_{p0})$ than the dependence between the intensities depicted in Fig. 7.3. According to the results of numerical computations, when the cross-section and the temporal distribution of the pumping field beam have a Gaussian shape, the dip on the curves $W_a(W_{p0})$ may be replaced by a "plateau", the height of which decreases with the increasing L.

Further, as the pumping field increases, the intensity of the Stokes scattering forming an angle with the beam axis, increases as well. Because of phase locking, more and more components with vairous values of $\delta \boldsymbol{k}$ will make contributions to the intensity $\mathcal{J}_a$ in a certain fixed direction, and this will lead to "smearing" of the ASRS angular spectrum.

7.1.5 The Experimental Analysis of Energy Characteristics

The experiments [7.24, 42] were devoted mainly to the observation of the above peculiarities of ASRS. Hydrogen was used as the Raman-active substance. The role of the pumping field belonged to the second harmonic of a single-mode neodymium glass laser (the equipment is presented schematically in Fig. 3 of [7.42]). The distribution of the intensity over the beam cross-section for the pumping radiation, characterized by a duration of 3×10^{-8} s and energy up to 0.3 J, was close to Gaussian.

Experiments were conducted with various chamber lengths, hydrogen pressures and focal lengths of the lenses used to collect the radiation into the chamber. Figure 7.5 depicts the typical dependences of the energy of the axial, cone and intermediate ASRS parts on the pumping field energy. One can see that the behavior of the axial and cone components is in qualitative agreement with the theory. The cone component increases monotonically and becomes saturated before the axial one. The order of magnitude of the pumping field intensity at which this saturation occurs ($S \approx 3\,\mathrm{GW/cm^2}$) is the same as that of the computed value (Fig. 7.3). The axial part of ASRS as a function of the pumping field intensity clearly exhibits a dip (Fig. 7.5). The intensity of the pumping

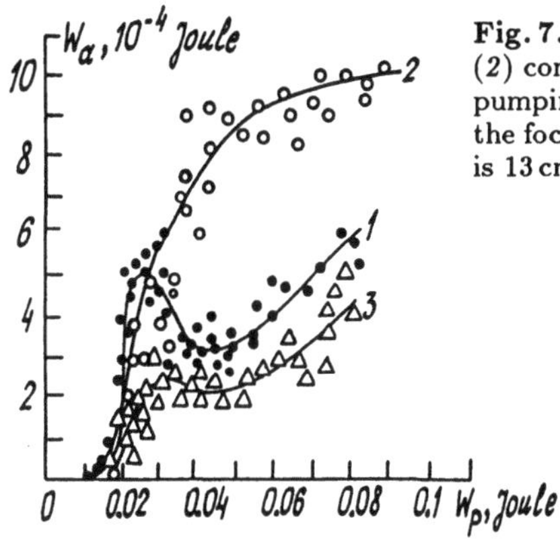

Fig. 7.5. The dependence of the energy of the (*1*) axial, (*2*) cone and (*3*) intermediate ASRS components on the pumping field energy. The hydrogen pressure is 75 atm, the focal region length ≈4 cm, the length of the chamber is 13 cm

field corresponding to the extremum of the axial component intensity is equal to 1.5–2 GW/cm^2 and is close to the computed quantity.

Consider the behavior of the anti-Stokes component energy in the region between the axis and the synchronization cone (curve 3 on Fig. 7.5). In this region ASRS is characterized (in the pumping field range used in the given experiment) by a much lower intensity than the intensity of the axial and cone components. Although curve 3 on Fig. 7.5 is similar to curve 1, the dip on it is less pronounced. This can be explained by the natural assumption that the main contribution to ASRS intensity in this region stems from plane waves with the wave detuning $|\delta \boldsymbol{k}|$ being greater than for the cone ASRS but less than for the axial one (see Fig. 7.3).

The energy of the ASRS axial component as a function of the pumping field energy for various values of the chamber length is given in Fig. 7.6. In correspondence with the results of the previous section, the energy of the anti-Stokes component turned out to be less for a larger interaction length.

The authors of [7.42] also studied the dependence of the SRS Stokes component, emitted into the same solid angles in which ASRS was analyzed, on the pumping field energy. Figure 7.7 presents the data obtained under the same conditions as the curves on Fig. 7.5. In the pumping field energy variation range 0.01–0.06 J in which curve 1 on Fig. 7.5 has a dip, both the axial and cone components of Stokes radiation increase monotonically. The intensity of the axial Stokes component becomes saturated and decreases because of the appearance of the second-order Stokes component at the pumping field energies exceeding 0.07 J, i.e., outside the pumping field energy range corresponding to the typical peculiarities of the ASRS axial component behavior.

On Fig. 7.8, one can see photographs of ASRS angular distribution obtained with the diameter of the pumping field beam in the chamber being 0.08 mm. They show that the threshold is reached first for axial ASRS, then the cone component appears; as the pumping field increases further, the radiation fills the whole solid angle inside the cone which matches the angle of synchronism.

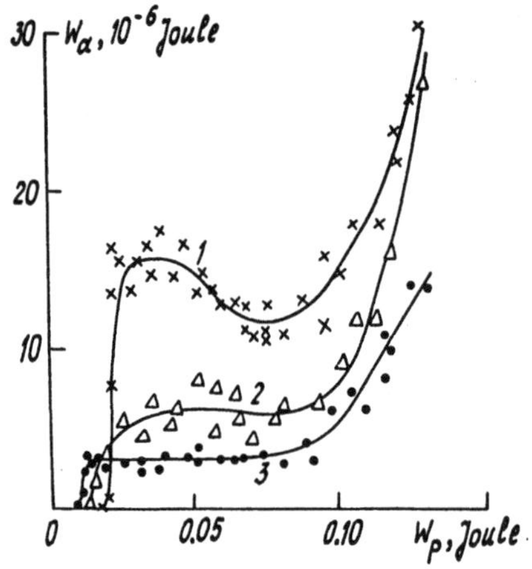

Fig. 7.6. The dependence of the energy of the axial component on the pumping field energy for various values of pressure and chamber length. *1* = 75 atm, 5 cm, *2* = 75 atm, 12 cm, and *3* = 110 atm, 12 cm

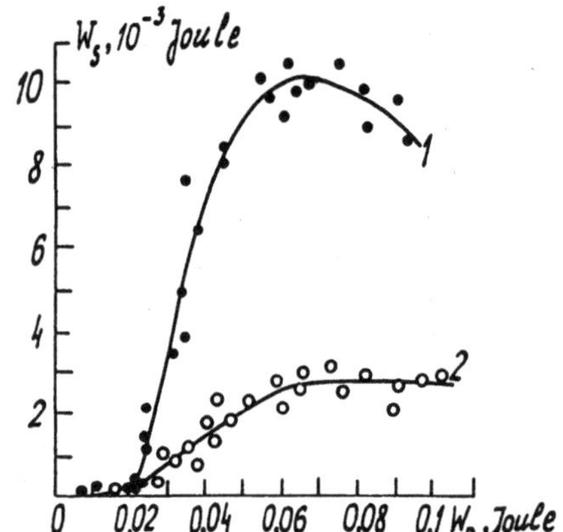

Fig. 7.7. The dependence of the energy of the SRS Stokes component on the pumping field energy. *1* = the "axial" part of the SRS Stokes component and corresponds to curve *1* on Fig. 7.5, *2* = the "cone" part of the SRS Stokes component and corresponds to curve *2* on Fig. 7.5

When the diameter of the pumping field beam is 0.3 mm, the cone component of ASRS emerges first. Thus, the differences between the thresholds of the axial and cone ASRS can be explained by the dependence of these thresholds on the pumping field beam diameter. When the latter is increased, the interaction length for the cone SRS components becomes greater, and the pumping field for which the cone ASRS reaches a recordable value diminishes.

Note that no focussing was observed in the experiments. Besides, according to the estimates of the focal length of a lens produced by SRS [7.43, 44] (see Chap. 8), even for the longest chamber among those used in [7.42], the increase of the energy density in the beam was negligible. As is well known, other causes of self-action, such as the optical Kerr effect, electrostriction and thermal self-focussing, are still less essential for hydrogen.

To conclude this section, we would like to mention publications in which other four-photon RPIs degenerate with respect to frequency (i.e., RPIs that have one of the frequencies participating in both resonances) have been stud-

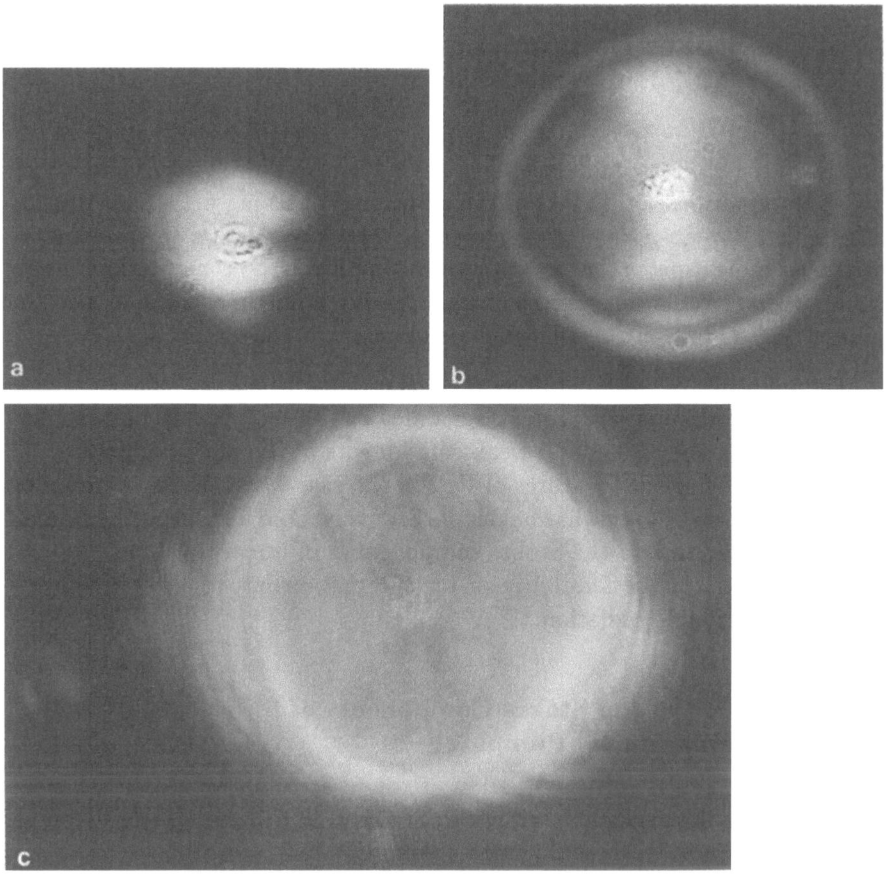

Fig. 7.8a–c. The angular distribution of the SRS anti-Stokes component for three values of the pumping field energy: (**a**) 0.028 J, (**b**) 0.049 J, (**c**) 0.11 J. The hydrogen pressure is 80 atm, the length of the chamber is 13 cm

ied. The authors of [7.72] have determined the conditions when the parametric mechanism is dominant in the formation of SRS second Stokes component and demonstrated that this occurs in gaseous media. In [7.73] the mutual influence of the processes associated with generation of the anti-Stokes and the second Stokes component of SRS is discussed. Reference [7.74] reports the investigation of SRS of the pumping field with the frequency $\omega_p < 2\omega_{21}$ (this process concludes the chain of successive generation of higher order Stokes components). It is shown that in this case, due to four-photon RPIs, the formation of the field at the frequency $\omega_{21} - \omega_s = 2\omega_{21} - \omega_p$ may be accompanied by the production of more photons at the frequency of the Stokes component than there was at the entrance into the medium at the pumping field frequency.

7.2 The Influence of Four-Photon RPIs on the Dynamics of the Stokes Components of SRS

The present section is devoted to the theoretical and experimental study of the dynamics of the Stokes component generation during RPI that results from SRS or two-photon absorption (TPA) of the additional pumping field. This process will be called Raman-parametric generation. The introduction to this chapter contains an indication of the possibility of using it for the construction of new tunable coherent radiation sources in the infrared range. In this connection the emphasis will be put on the following questions:

1. the influence of wave detuning on generation efficiency;
2. the dependence of the process dynamics on the relationship between frequencies of the strong and weak components of the pumping field;
3. the study of specific features of RPI in the case when the weak pumping field intensity is somewhat below the SRS threshold and its purely Raman transformation into the Stokes component is of importance;
4. the analysis of the feasibility of transformation into the low frequency range of tunable radiation.

7.2.1 Generation of the Stokes Components of SRS During Biharmonic Pumping

We shall be concerned with interaction of waves in collinear beams; consider therefore the one-dimensional interaction of fields $\boldsymbol{E} = \boldsymbol{e}_j[\mathcal{E}_j \exp(\mathrm{i}\omega_j t) +$ compl. conj.] whose frequencies are subject to a condition

$$\omega_1 - \omega_{1\mathrm{s}} = \omega_2 - \omega_{2\mathrm{s}} = \omega_{21} \quad . \tag{7.2.1}$$

For exact resonance (7.2.1) the complex polarization amplitudes at the frequencies of interacting fields with nonresonant interactions being neglected, can be found by means of the formulas of Sect. 1.3.

The substitution of the expressions for polariaztions into the Maxwell equations results in the following relations for the complex amplitudes of the pumping fields (C_1 and C_2) and of the corresponding Stokes components ($C_{1\mathrm{s}}$ and $C_{2\mathrm{s}}$):

$$\begin{aligned}
\frac{dC_1}{dz} &= -\alpha_{11}C_1|C_{1\mathrm{s}}|^2 - \alpha_{12}C_2C_{2\mathrm{s}}^*C_{1\mathrm{s}}\,\mathrm{e}^{\mathrm{i}(\delta k)z} \quad , \\
\frac{dC_{1\mathrm{s}}}{dz} &= \alpha_{21}C_2^*C_{2\mathrm{s}}C_1\,\mathrm{e}^{-\mathrm{i}(\delta k)z} + \alpha_{22}C_{1\mathrm{s}}|C_1|^2 \quad , \\
\frac{dC_2}{dz} &= -\beta_{11}C_2|C_{2\mathrm{s}}|^2 - \beta_{12}C_1C_{1\mathrm{s}}^*C_{2\mathrm{s}}\,\mathrm{e}^{-\mathrm{i}(\delta k)z} \quad , \\
\frac{dC_{2\mathrm{s}}}{dz} &= \beta_{21}C_1^*C_{1\mathrm{s}}C_2\,\mathrm{e}^{\mathrm{i}(\delta k)z} + \beta_{22}C_{2\mathrm{s}}|C_2|^2 \quad ,
\end{aligned} \tag{7.2.2}$$

where

$$\alpha_{11}=\frac{2\pi\omega_1^2\eta NT}{k_1c^2\hbar}\kappa^2(\omega_1)\quad,\quad \alpha_{12}=\frac{2\pi\omega_1^2\eta NT}{k_1c^2\hbar}\kappa(\omega_1)\kappa(\omega_2)$$
$$\alpha_{21}=\alpha_{12}\frac{\omega_{1s}^2k_1}{\omega_1^2k_{1s}}\quad,\quad \alpha_{22}=\frac{2\pi\omega_{1s}^2\eta NT}{k_{1s}c^2\hbar}\kappa^2(\omega_1)\quad. \tag{7.2.3}$$

T is the inverse half-width of the line for the transition 1–2, $\kappa^2(\omega_j)$ and $\kappa(\omega_1)\kappa(\omega_2)$ are convolutions of tensors $\kappa_{ab}(\omega_j)\kappa_{cd}(\omega_j)$ and $\kappa_{ab}(\omega_1)\kappa_{cd}(\omega_2)$ over pairs of indices ab, cd (see Sect. 4.1), formulas for β_{ij} are obtained from (7.2.3) by the substitution $2\rightleftarrows 1$ in k_j, ω_j, and

$$\delta k=(k_1-k_{1s})-(k_2-k_{2s})$$

is the wave detuning.

In the quasistationary case the equation for the population difference η is

$$\begin{aligned}\frac{d\eta}{dt}+\frac{\eta-\eta_0}{\tau}&\\ =-\frac{4\eta T}{\hbar^2}&\mathrm{Re}\{[\kappa_{ab}(\omega_1)\mathcal{E}_b(\omega_{1c})\mathcal{E}_a^*(\omega_1)+\kappa_{ab}(\omega_2)\mathcal{E}_a^*(\omega_2)\mathcal{E}_b(\omega_{2s})]\\ &\times[\kappa_{cd}(\omega_1)\mathcal{E}_c(\omega_1)\mathcal{E}_d^*(\omega_{1s})+\kappa_{cd}(\omega_{2s})\mathcal{E}_c(\omega_2)\mathcal{E}_d(\omega_{2s})]\}\quad.\end{aligned}$$

The rhs of the equation for η contains both the terms that do not depend on the phase difference θ of the fields (the Raman process) and those depending on θ (the parametric process). One can easily see that the real part of the parametric term is proportional to $\cos\theta$, the phase difference θ given by $\theta=\delta kz+(\varphi_1-\varphi_{1s})-(\varphi_2-\varphi_{2s})$, and $\mathcal{E}_b(\omega_j)=\mathcal{A}_b(\omega_j)\exp[-\mathrm{i}\{(k_jz)-\varphi_j(z)\}]$.

In our problem $\mathcal{A}(\omega_{1s})|_{z=0}\ll\mathcal{A}(\omega_1)_{z=0}$ holds. It will be shown below that $-\pi/2<\theta<\pi/2$ and therefore $\cos\theta>0$ (if $\kappa_{ab}(\omega_1)\kappa_{cd}(\omega_2)>0$). This means that the signs of the changes of the population produced both by parametric and Raman interactions are the same. In this sense the RPI under consideration differs essentially from other resonant parametric processes in which the changes of the population effected by Raman and parametric interactions have opposite signs (see Sect. 7.3). In the following only the field intensities associated with the insignificant saturation of populations will be discussed.

The Constant Pumping Fields Approximation. Assume that a system is subject to the action of a biharmonic pumping field with components C_1 and C_2. If $|C_{1s}|^2\ll|C_1|^2$ and $|C_{2s}|^2\ll|C_2|^2$, the amplitudes of the components of the biharmonic pumping field can be regarded as constant. It will be supposed further that

$$C_1|_{z=0}=C_{10}<C_2|_{z=0}=C_{20}\quad,$$

the quantity C_1 will be called the weak, and C_2 the strong component of the pumping field. Then in the approxmation $C_{1,2}$ = const, the second and the fourth equations (7.2.2) for the Stokes components of the pumping field doublet yield

$$C_{1s} = \left\{ \frac{\mathcal{K}_{11} + i(\delta k/2) + \lambda_2}{\lambda_1 - \lambda_2} e^{\lambda_1 z} + \frac{\mathcal{K}_{11} + i(\delta k/2) + \lambda_1}{\lambda_2 - \lambda_1} e^{\lambda_2 z} \right\} \times C_{1s0} \exp\left(-i\frac{\delta k}{2} z\right) + \left\{ \frac{\mathcal{K}_{12}[\exp(\lambda_1 z) - \exp(-\lambda_2 z)]}{\lambda_1 - \lambda_2} \right\} \times C_{2s0} \exp\left(-i\frac{\delta k}{2} z\right) \quad ,$$

$$C_{2s} = \left\{ \frac{\mathcal{K}_{21}[\exp(\lambda_1 z) - \exp(\lambda_2 z)]}{\lambda_1 - \lambda_2} \right\} C_{1s0} \exp\left(i\frac{\delta k}{2} z\right) + \left\{ \frac{\mathcal{K}_{22} - i(\delta k/2) - \lambda_2}{\lambda_1 - \lambda_2} e^{\lambda_1 z} + \frac{\mathcal{K}_{22} - i(\delta k/2) - \lambda_1}{\lambda_2 - \lambda_1} e^{\lambda_2 z} \right\} \times C_{2s0} \exp\left(i\frac{\delta k}{2} z\right) \quad , \tag{7.2.4}$$

where

$$\lambda_{1,2} = \frac{1}{2}(\mathcal{K}^+) \pm \sqrt{\frac{\{[(\mathcal{K}^+)^2 - (\delta k)^2]^2 + 4(\delta k)^2(\mathcal{K}^-)^2\}^{1/2} + (\mathcal{K}^+)^2 - (\delta k)^2}{2}} \pm i\sqrt{\frac{\{[(\mathcal{K}^+)^2 - (\delta k)^2]^2 + 4(\delta k)^2(\mathcal{K}^-)^2\}^{1/2} - (\mathcal{K}^+)^2 + (\delta k)^2}{2}} \tag{7.2.5}$$

and

$$\mathcal{K}^{\pm} = \mathcal{K}_{11} \pm \mathcal{K}_{22} \quad , \quad \mathcal{K}_{11} = \alpha_{22}|C_{10}|^2 \quad , \quad \mathcal{K}_{22} = \beta_{22}|C_{20}|^2 \quad , \quad \mathcal{K}_{12} = \alpha_{21}|C_{20}||C_{10}| \quad , \quad \mathcal{K}_{21} = \beta_{21}|C_{10}||C_{20}| \tag{7.2.6}$$

are the coefficients of Raman and parametric amplification at frequencies ω_{1s} and ω_{2s}[5].

Consider three limiting cases.

1) $\delta k = 0$ corresponding to linear synchronization. From (7.2.4, 5) we obtain $\lambda_1 = \mathcal{K}^+$, $\lambda_2 = 0$, i.e., the increment for the Stokes components of the pumping field doublet is equal to the sum of coefficients of their Raman am-

[5] In [7.33], the case when $\mathcal{K}_{22} \gg \mathcal{K}_{11}$ is studied.

plification produced by each component of the pumping field separately. If the intensity of the weak component of the pumping field is low, the amplification efficiency does not depend on the intensity of the weak wave and is determined by the intensity of the strong component.

2) $(\mathcal{K}^+)^2 \ll (\delta k)^2$. According to (7.2.4, 5),

$$\lambda_1 \approx \mathcal{K}_{22} + \mathrm{i}\frac{\delta k}{2} \quad , \quad \lambda_2 \approx \mathcal{K}_{11} - \mathrm{i}\frac{\delta k}{2} \quad ,$$

$$\begin{aligned} C_{1s} &\approx C_{1s0}\,\mathrm{e}^{\mathcal{K}_{11}z} \\ &\quad + \frac{\mathcal{K}_{12}}{\delta k}\exp\left(-\mathrm{i}\arctan\frac{\delta k}{\mathcal{K}^-}\right)\left(\mathrm{e}^{(\mathcal{K}_{22}+\mathrm{i}\delta k)z} - \mathrm{e}^{\mathcal{K}_{11}z}\right)C_{2s0} \quad , \\ \mathcal{A}_{2s} &\approx C_{1s0}\frac{\mathcal{K}_{21}}{\delta k} \\ &\quad \times \exp\left(-\mathrm{i}\arctan\frac{\delta k}{\mathcal{K}^-}\right)\left(\mathrm{e}^{\mathcal{K}_{22}z} - \mathrm{e}^{(\mathcal{K}_{11}+\mathrm{i}\delta k)z}\right) + C_{2s0}\,\mathrm{e}^{\mathcal{K}_{22}z} \quad . \end{aligned} \tag{7.2.7}$$

With the increase of the wave detuning the efficiency of parametric interaction diminishes as $(\delta k)^{-1}$. When $\delta k \to \infty$, the processes become decoupled, and SRS Stokes components are amplified independently, only due to Raman interaction with the corresponding pumping field.

3) $(\mathcal{K}^+)^2 \gg (\delta k)^2$. The expansion of radicands in (7.2.5) into a series and retaining the first terms result in

$$\lambda_1 \approx \mathcal{K}^+ + \mathrm{i}\frac{\delta k}{2}\frac{\mathcal{K}^-}{\mathcal{K}^+} \quad , \quad \lambda_2 \approx \frac{\mathcal{K}_{11}}{\mathcal{K}^+}\frac{(\delta k)^2}{(\mathcal{K}^+)^2} - \mathrm{i}\frac{\delta k}{2}\frac{\mathcal{K}^-}{\mathcal{K}^+} \quad . \tag{7.2.8}$$

In these expressions $\mathrm{Re}\{\lambda_1\} \gg \mathrm{Re}\{\lambda_2\}$. Equations (7.2.8) and (7.2.4) allow to obtain formulas for the field amplitudes C_{1s} and C_{2s}. The variation of amplitudes turns out to be approximately the same as in the case $\delta k = 0$. Thus, if $(\mathcal{K}^+)^2 \gg (\delta k)^2$, the behavior of interacting waves resembles their behavior under the conditions of complete synchronization. Their inequality $(\mathcal{K}^+)^2 \gg (\delta k)^2$ is equivalent to the phase locking requirement $\alpha_{11}^2 W \gg (\delta k)^2$, see below.

In the constant pumping fields approximation, a solution similar to (7.2.4) can also be obtained for the generation of a SRS anti-Stokes component with the frequency ω_{1a}[6]. Then, if $[\mathcal{K}_{22}(\omega_2,\omega_{2s}) - \mathcal{K}_{11}(\omega_1,\omega_{1a})]^2 \gg (\delta k)^2$, the increment of the anti-Stokes component of the weak pumping field approaches $\mathcal{K}_{22}(\omega_2,\omega_{2s}) - \mathcal{K}_{11}(\omega_1,\omega_{1a})$. Therefore, taking into account (7.2.8), we arrive at the conclusion that when $\mathcal{K}_{22} \gg \mathcal{K}_{11}$ (i.e., when $|C_{20}|^2 \gg |C_{10}|^2$), the increase

[6] The substitution $\mathcal{K}_{11} \to -\mathcal{K}_{11}(\omega_1,\omega_{1a})$ and $\mathcal{K}_{12} \to -\mathcal{K}_{12}(\omega_2,\omega_{1a})$ makes it possible to derive formulas for C_{2s} and C_{1a}^* from (7.2.4, 5).

of the weak pumping field Stokes and anti-Stokes components occurs at nearly the same rate. Otherwise, when the inequality $\mathcal{K}_{22}>\mathcal{K}_{11}$ is weak, the increment of the Stokes wave is greater than the increment of the anti-Stokes one.

Scattering of the Weak Component of the Pumping Field in the Presence of the Strong Component. If $|C_{20}|^2 \gg |C_{10}|^2$, the parametric terms in the third and fourth equations (7.2.2) and the Raman terms in the first and second equations can be neglected. Such an approximation is valid until the product of the intensities of the pumping field strong component and its Stokes component becomes approximately equal to the corresponding product for the weak pumping field component: $|C_2 C_{2s}^*| \gtrsim |C_1 C_{1s}^*|$. In this approximation it follows from (7.2.2) that

$$\dot{C}_1 = -\gamma \, e^{i\delta\zeta} f(\zeta) C_{1s} \quad , \quad \dot{C}_{1s} = \gamma \frac{\omega_{1s}^2}{\omega_1^2} \frac{k_1}{k_{1s}} e^{-i\delta\zeta} f(\zeta) C_1 \quad , \tag{7.2.9}$$

where the dot means the derivative with respect to ζ,

$$\delta = \frac{\delta k}{\mathcal{K}_{22}} \quad , \quad \gamma = \frac{\omega_1^2}{\omega_{2s}^2} \frac{k_{2s}}{k_1} \frac{\kappa^2(\omega_1)}{\kappa^2(\omega_2)} \left| \frac{C_{2s0}}{C_{20}} \right| \quad ,$$

$$\zeta = \mathcal{K}_{22} z \quad , \quad f(\zeta) = e^{\zeta} \left(1 + \left| \frac{C_{2s0}}{C_{20}} \right|^2 \frac{\omega_2^2 k_{2s}}{\omega_{2s}^2 k_2} e^{2\zeta} \right)^{-1} \quad . \tag{7.2.10}$$

1) The case of the zero wave detuning ($\delta k = 0$). According to (7.2.9),

$$\begin{aligned} C_{1s} &= C_{2s0} \cos\left[R \arctan b(e^{\zeta} - 1)\right] \\ &\quad + \frac{\omega_{1s}}{\omega_1} \frac{k_1^{1/2}}{k_{1s}^{1/2}} C_{10} \sin\left[R \arctan b(e^{\zeta} - 1)\right] \quad , \\ C_1 &= -\frac{\omega_1}{\omega_{1c}} \frac{k_{1s}^{1/2}}{k_1^{1/2}} C_{1s0} \sin\left[R \arctan b(e^{\zeta} - 1)\right] \\ &\quad + C_{10} \cos\left[R \arctan b(e^{\zeta} - 1)\right] \quad , \end{aligned} \tag{7.2.11}$$

where

$$R = \frac{\omega_{1s}\omega_1}{\omega_{2s}\omega_2} \sqrt{\frac{k_{2s}k_2}{k_{1s}k_1} \frac{\kappa^2(\omega_1)}{\kappa^2(\omega_2)}} \quad , \quad b = \left| \frac{C_{2s0}}{C_{20}} \right|^2 \frac{\omega_2^2 k_{2s}}{\omega_{2s}^2 k_2} \quad . \tag{7.2.12}$$

When $\omega_1 \approx \omega_2$ ($R \approx 1$), (7.2.11) yield:

$$C_{1s} = C_{1s0}\frac{1}{\sqrt{1+b^2\,e^{2\zeta}}} + \frac{\omega_{1s}}{\omega_1}\sqrt{\frac{k_1}{k_{1s}}}C_{10}\frac{b(e^{\zeta}-1)}{\sqrt{1+b^2\,e^{\zeta}}} \quad ,$$

$$C_1 = -\frac{\omega_1}{\omega_{1s}}\sqrt{\frac{k_1}{k_{1s}}}C_{1s0}\frac{b(e^{\zeta}-1)}{\sqrt{1+b\,e^{2\zeta}}} + C_{10}\frac{1}{\sqrt{1+b^2\,e^{2\zeta}}} \quad . \qquad (7.2.13)$$

If $R \approx 1$, it follows from (7.2.13) that for $z \gtrsim \mathcal{K}_{22}^{-1}$, the increase of the Stokes components of the weak and strong pumping fields occurs at the same rate. In terms of photon numbers the asymptotic ($z \rightarrow \infty$) transformation of the weak pumping field into the Stokes component constitutes 100% (see curve 2 on Fig. 7.9a).

When $R<1\,(\omega_1<\omega_2)$, the transformation of the weak component is less efficient than that of the strong one (see curve 1 on Fig. 7.9a). The limiting transformation coefficient in terms of photon numbers is

$$\beta_{\mathrm{ph}} = \lim_{z\rightarrow\infty}\frac{|C_{1s}|^2\omega_1}{|C_{10}|^2\omega_{1s}} = \sin^2\left(R\frac{\pi}{2}\right) \quad . \qquad (7.2.14)$$

For $R>1\,(\omega_1>\omega_2)$ the weak pumping field is transformed into its Stokes component faster than the strong one. When

$$z_{\max} = \frac{1}{\mathcal{K}_{22}}\ln\left(\frac{1}{b}\tan\frac{\pi}{2R}\right) \quad , \qquad (7.2.15)$$

the intensity of the Stokes wave reaches its maximum ($\beta_{\mathrm{ph\,max}} = 100\%$, see Fig. 7.9a, curves 3–5). For example, for $R=2$,

$$C_{1s} = C_{1s0}\frac{1-b^2\,e^{2\zeta}}{1+b^2\,e^{2\zeta}} + \frac{\omega_{1s}}{\omega_1}\sqrt{\frac{k_1}{k_{1s}}}C_{10}\frac{2b(e^{\zeta}-1)}{1+b^2\,e^{2\zeta}} \quad ,$$

$$C_1 = C_{1s0}\frac{\omega_1}{\omega_{1s}}\sqrt{\frac{k_{1s}}{k_1}}\frac{2b(e^{\zeta}-1)}{1+b^2\,e^{2\zeta}} + C_{10}\frac{1-b^2\,e^{2\zeta}}{1+b^2\,e^{2\zeta}} \qquad (7.2.16)$$

and the maximal transformation takes place at a point, where approximately 50% (in terms of the photon numbers) of the strong component of the pumping field has been transformed into its Stokes component.

2) The case of arbitrary wave detuning δk. Two second-order differential equations for determining the complex amplitudes C_{1s} and C_1

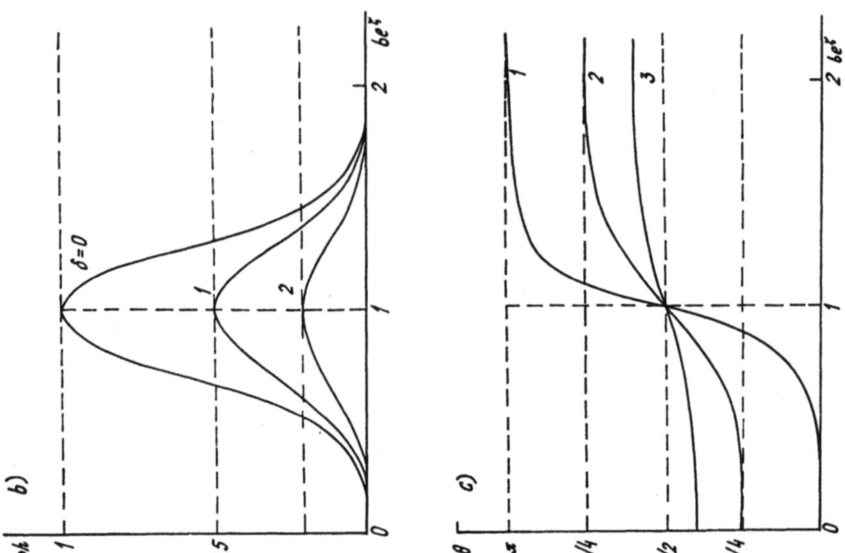

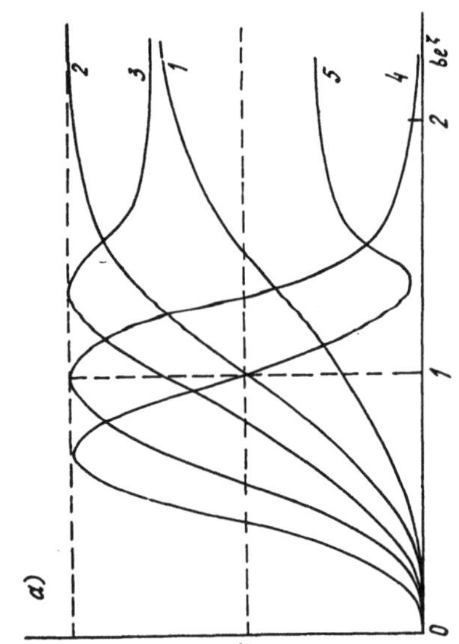

Fig. 7.9. **(a)** The coordinate dependence of the coefficient of the pumping field transformation [or (for a fixed length of the nonlinear medium) the dependence of the coefficient mentioned above on the strong pumping field intensity] for $\delta = 0$. Curves *1–5* correspond to values $R<1$, $R = 1$, $1<R<2$, $R = 2$, $2<R<3$. **(b)** The dependence of interaction efficiency on the value of reduced wave detuning δ. **(c)** The phase difference θ as a function of the dimensionless coordinate ζ for various values of reduced wave detuning δ. Curves *1*, *2*, *3* correspond to $\delta_1 \ll 1$, $\delta_2 > \delta_1$ and $\delta_3 > \delta_2$

$$\ddot{C}_{1s} + \left(\mathrm{i}\delta - \frac{1-b^2\,\mathrm{e}^{2\zeta}}{1+b^2\,\mathrm{e}^{2\zeta}}\right)\dot{C}_{1s} + R^2\frac{b^2\,\mathrm{e}^{2\zeta}}{(1+b^2\,\mathrm{e}^{2\zeta})^2}C_{1s} = 0 \quad ,$$
$$\ddot{C}_{1} - \left(\mathrm{i}\delta + \frac{1-b^2\,\mathrm{e}^{2\zeta}}{1+b^2\,\mathrm{e}^{2\zeta}}\right)\dot{C}_{1} + R^2\frac{b^2\,\mathrm{e}^{2\zeta}}{(1+b^2\,\mathrm{e}^{2\zeta})^2}C_{1} = 0 \tag{7.2.17}$$

can be easily obtained from (7.2.9).

In general it is not possible to obtain the solution of (7.2.17) in closed form. In order to clarify the role of wave detuning, consider a special case $R = 2$, when (7.2.17) represent equations in total differentials. Their solution is

$$
\begin{aligned}
C_{1s} = & \frac{\omega_{1s}}{\omega_1}\sqrt{\frac{k_1}{k_{1s}}}C_{10}\frac{1+\mathrm{i}\delta}{1+\delta^2}\frac{2b\{\exp[(1-\mathrm{i}\delta)\zeta]-1\}}{1+b^2\,\mathrm{e}^{2\zeta}} \\
& + C_{1s0}\left(1-b^2\frac{(1-\mathrm{i}\delta)^2\mathrm{e}^{2\zeta}}{1+\delta^2}\right)\frac{1}{1+b^2\,\mathrm{e}^{2\zeta}} \\
C_1 = & -\frac{\omega_1}{\omega_{1s}}\sqrt{\frac{k_{1s}}{k_1}}C_{1s0}\frac{1-\mathrm{i}\delta}{1+\delta^2}\frac{2b(\mathrm{e}^{(1-\mathrm{i}\delta)\zeta}-1)}{1+b^2\,\mathrm{e}^{2\zeta}} \\
& + C_{10}\left(1-b^2\frac{(1-\mathrm{i}\delta)^2\,\mathrm{e}^{2\zeta}}{1+\delta^2}\right)\frac{1}{1+b^2\,\mathrm{e}^{2\zeta}} \quad . \qquad (7.2.18)
\end{aligned}
$$

Let us analyze solutions (7.2.18) for generation of the Stokes component ω_{1s} when $C_{1s0} \approx 0$. Isolating slowly varying real amplitudes and phases of fields, we arrive at

$$
\begin{aligned}
C_{1s} \approx & \frac{\omega_{1s}}{\omega_1}\sqrt{\frac{k_1}{k_{1s}}}\frac{\mathcal{A}_{10}}{\sqrt{1+\delta^2}}\frac{2b\,\mathrm{e}^{\zeta}}{1+b^2\,\mathrm{e}^{2\zeta}}\exp\{\mathrm{i}(\delta\zeta-\arctan\delta)\} \quad , \\
C_1 \approx & \mathcal{A}_{10}\frac{\sqrt{[1+\delta^2-b^2\,\mathrm{e}^{2\zeta}(1-\delta^2)]^2+4\delta^2b^4\,\mathrm{e}^{4\zeta}}}{(1+b^2\,\mathrm{e}^{2\zeta})(1+\delta^2)} \\
& \times\exp\left\{\mathrm{i}\arctan\frac{2\delta^2\,\mathrm{e}^{2\zeta}}{(1+\delta^2)-b^2\,\mathrm{e}^{2\zeta}(1-\delta^2)}\right\} \quad . \qquad (7.2.19)
\end{aligned}
$$

Thus, in the presence of wave detuning the efficiency of parametric energy transformation into the Stokes component of the weak pumping field decreases as $(1+\delta^2)^{-1}$. The transformation coefficient becomes maximal and equal to $(1+\delta^2)^{-1}\cdot 100\%$ (in terms of photon numbers) at the same point as in the case $\delta k = 0$ (see Fig. 7.9b). If $\delta^2 \ll 1$, the efficiency of transformation remains practically equal to that characteristic of complete synchronization ($\delta = 0$). Until $|\delta k| \ll \mathcal{K}_{22}$ and $b\,\mathrm{e}^{\zeta} < 1$ the parametric interaction depends so weakly on conditions of linear synchronization because of phase locking. In order to illustrate this, let us determine the phase difference θ from (7.2.19):

$$
\begin{aligned}
\theta & = \delta kz + (\varphi_1 - \varphi_{1s}) - (\varphi_2 - \varphi_{2s}) \\
& = \arctan\delta + \arctan\frac{2\delta b^2\,\mathrm{e}^{2\zeta}}{1+\delta^2-b^2\,\mathrm{e}^{2\zeta}(1-\delta^2)} \quad . \qquad (7.2.20)
\end{aligned}
$$

Figure 7.9c depicts the quantity θ as a function of the coordinate ζ for various δ. When $\delta \ll 1$ (curve 1), the phase difference $\theta_0 \approx 0$ is formed in the region of constant pumping fields, and the behavior of the system is similar to its behavior under the conditions of complete synchronization. As δ increases, the phase difference θ_0 approaches the value $\pi/2$ (curves 2, 3 on Fig. 7.9c), parametric interaction becomes less efficient, coherent interaction of fields ends, and their variation is caused only by the Raman process.

The anti-Stokes component of the weak pumping field as a function of δ can be obtained from (7.2.18) by making the substitution of subcripts $1\text{s} \to 1$, $1 \to 1\text{a}$ and setting $C_{1\text{a}0}$ equal to zero. This function is similar to that representing its Stokes component, the only difference being that for $\delta \ll 1$ the phase difference $\theta = \pi$ is formed which approaches the value $\pi/2$ as δ increases. Note, however, that, as has already been mentioned above, the efficiency of the transformation of the weak pumping field into the Stokes and anti-Stokes regions is the same only if $C_{10} \ll C_{20}$.

Equations for Real Amplitudes and Phase Difference in the General Case. For an arbitrary relationship between boundary conditions, the solution of the system (7.2.2) does not appear to be possible. In general, the analysis of equations for slowly varying real amplitudes $\mathcal{A}_j = C_j \exp[\mathrm{i}\varphi_j(z)]$ and the phase difference $\theta = (\delta k)z + (\varphi_1 - \varphi_{1\text{s}}) - (\varphi_2 - \varphi_{2\text{s}})$ enables one to formulate certain regularities of the dynamics of parametric interactions. For the above process the equations are

$$\begin{aligned}
\frac{d\mathcal{A}_1}{d\zeta_1} &= -\mathcal{A}_{1\text{s}}(\mathcal{A}_1\mathcal{A}_{1\text{s}} + s\mathcal{A}_2\mathcal{A}_{2\text{s}} \cos\theta) \quad , \\
\frac{d\mathcal{A}_{1\text{s}}}{d\zeta_1} &= g_{1\text{s}}\mathcal{A}_1(\mathcal{A}_1\mathcal{A}_{1\text{s}} + s\mathcal{A}_2\mathcal{A}_{2\text{s}} \cos\theta) \quad , \\
\frac{d\mathcal{A}_2}{d\zeta_1} &= -g_2\mathcal{A}_{2\text{s}}(\mathcal{A}_1\mathcal{A}_{1\text{s}} \cos\theta + s\mathcal{A}_2\mathcal{A}_{2\text{s}}) \quad , \\
\frac{d\mathcal{A}_{2\text{s}}}{d\zeta_1} &= g_{2\text{s}}\mathcal{A}_2(\mathcal{A}_1\mathcal{A}_{1\text{s}} \cos\theta + s\mathcal{A}_2\mathcal{A}_{2\text{s}}) \quad , \\
\frac{d\theta}{d\zeta_1} &= \frac{\delta k}{\alpha_{11}} + W(\mathcal{A}_1, \mathcal{A}_{1\text{s}}, \mathcal{A}_2, \mathcal{A}_{2\text{s}}) \sin\theta \quad ,
\end{aligned} \tag{7.2.21}$$

where

$$W = \mathcal{A}_1\mathcal{A}_{1\text{s}}\left(\frac{\mathcal{A}_{2\text{s}}}{\mathcal{A}_2}g_2 - \frac{\mathcal{A}_2}{\mathcal{A}_{2\text{s}}}g_{2\text{s}}\right) + s\mathcal{A}_2\mathcal{A}_{2\text{s}}\left(\frac{\mathcal{A}_{1\text{s}}}{\mathcal{A}_1} - g_{1\text{s}}\frac{\mathcal{A}_1}{\mathcal{A}_{1\text{s}}}\right) \quad . \tag{7.2.22}$$

In these formulas the following notations are introduced:

$$\begin{aligned}
&\zeta_1 = \alpha_{11} z \quad , \quad s = \kappa(\omega_2)/\kappa(\omega_1) \quad , \\
&g_{1\text{s}} = \frac{\omega_{1\text{s}}^2}{\omega_1^2}\frac{k_1}{k_{1\text{s}}} \quad , \quad g_2 = \frac{\omega_2^2}{\omega_1^2}\frac{k_1}{k_2}s \quad , \quad g_{2\text{s}} = \frac{\omega_{2\text{s}}^2}{\omega_1^2}\frac{k_1}{k_{2\text{s}}}s \quad .
\end{aligned} \tag{7.2.23}$$

It has been demonstrated in Chap. 6 that; if the initial stage of field interaction corresponds to the inequality $\alpha_{11}^2 W^2 \gg (\delta k)^2$, in the region where

$$|\alpha_{11} W(\mathcal{A}_{1s}, \mathcal{A}_1, \mathcal{A}_{2s}, \mathcal{A}_2)| > |\delta k| \quad ; \tag{7.2.24}$$

the phase difference is stable (because of phase locking of the interacting waves)[7], and the solution of the system (7.2.21) can be found approximately, by assuming that $\delta k = 0$ and $\theta = 0$ if $\kappa(\omega_1)\kappa(\omega_2) > 0$. And the region of field values itself, in which this solution is valid, is determined when the solution of the system (7.2.21) for $\theta = 0$ is substituted into the expression for W (7.2.22).

Thus we arrive at the following formula for the length of transformation of the weak pumping field into the Stokes component

$$L = \frac{1}{2\alpha_{11}} \int_{\mathcal{J}_{1s0}}^{\mathcal{J}_{1s}} \frac{d\mathcal{J}_{1s}}{\mathcal{J}_{1s}(M_1 - \mathcal{J}_{1s}) + s\sqrt{\mathcal{J}_2 \mathcal{J}_{2s} g_{1s} \mathcal{J}_{1s}(M_1 - \mathcal{J}_{1s})}} \quad , \tag{7.2.25}$$

where

$$\mathcal{J}_j = \mathcal{A}_j^2 = |C_j|^2 \quad ,$$

$$\mathcal{J}_{2s} = g_2^{-1} M_2 \sin^2 \left\{ \arcsin \frac{g_2 \mathcal{J}_{2s0}}{M_2} + \sqrt{\frac{g_{2s}}{g_{1s}}} \left[\arcsin \sqrt{\frac{\mathcal{J}_{1s}}{M_1}} - \arcsin \sqrt{\frac{\mathcal{J}_{1s0}}{M_1}} \right] \right\} \quad , \tag{7.2.26}$$

$$\mathcal{J}_2 = g_{2s}^{-1} M_2 \sin^2 \left\{ \arcsin \frac{g_2 \mathcal{J}_{20}}{M_2} + g_2 \sqrt{\frac{g_{2s}}{g_{1s}}} \times \left[\arcsin \sqrt{\frac{\mathcal{J}_{1s}}{M_1}} - \arcsin \sqrt{\frac{\mathcal{J}_{1s0}}{M_1}} \right] \right\} \quad , \tag{7.2.27}$$

$$M_1 = g_{1s} \mathcal{J}_1 + \mathcal{J}_{1s} = g_{1s} \mathcal{J}_{10} + \mathcal{J}_{1s0} \quad , \tag{7.2.28}$$

$$M_2 = g_{2s} \mathcal{J}_2 + g_2 \mathcal{J}_{2s} = g_{2s} \mathcal{J}_{20} + g_2 \mathcal{J}_{2s0} \quad . \tag{7.2.29}$$

This solution is approximately valid in the region where

$$W = g_{1s}^{-1/2} \sqrt{\mathcal{J}_{1s}(M_1 - \mathcal{J}_{1s})} \left(g_2 \sqrt{\frac{\mathcal{J}_{2s}}{\mathcal{J}_2}} - g_{2s} \sqrt{\frac{\mathcal{J}_2}{\mathcal{J}_{2s}}} \right) + s\sqrt{\mathcal{J}_2 \mathcal{J}_{2s} g_{2s}} \frac{2\mathcal{J}_{1s} - M_1}{\sqrt{\mathcal{J}_{1s}(M_1 - \mathcal{J}_{1s})}} > \frac{\delta k}{\alpha_{11}} \quad . \tag{7.2.30}$$

[7] If at the medium boundary the amplitudes of the three fields are much greater than the amplitude of the fourth one, condition (7.2.24) is always satisfied in the boundary region, see (7.2.22).

The phase difference θ in this region is close to zero. In the denominator of the integrand in (7.2.25), the first term describes Raman intreaction of fields, and the second their parametric interaction. If the boundary value of the weak pumping field intensity is a little bit below its threshold value, the main contribution to the intensity of the Stokes wave $\mathcal{J}_{1s}$ in the beginning of interaction comes from its parametric form. As $\mathcal{J}_{1s}$ increases, Raman interaction, which prevails over the parametric one after phase locking is "broken down", can also become effective.

If at the boundary of the medium the condition $\alpha_{11}^2 W^2 \gg (\delta k)^2$ (or $\delta \ll 1$) is not satisfied, it is impossible to describe the behavior of the system under the assumption that $\delta k = 0$ and $\theta = 0$. In this case the initial phase difference θ is formed in the region $0 < \theta_0 < \pi/2$ (when the anti-Stokes component is generated in the region $-\pi/2 < \theta_0 < \pi/2$), and the system (7.2.21) can only be solved numerically.

7.2.2 The Effect of Strong Pumping TPA on Weak Pumping SRS

Scattering of a Weak Pumping Field in the Presence of a Strong Pumping Field. Consider the generation of the Stokes component of the weak pumping field with a frequency ω_1 during TPA of the strong pumping field with a frequency ω_2:

$$2\omega_2 = \omega_1 - \omega_{1s} \approx \omega_{21} \quad . \tag{7.2.31}$$

This process is described by

$$\begin{aligned} \frac{dC_1}{dz} &= -\alpha_{11} C_1 |C_{1s}|^2 - \alpha_{12} C_2^2 C_{1s}\, \mathrm{e}^{\mathrm{i}(\delta k)z} \quad , \\ \frac{dC_{1s}}{dz} &= \alpha_{21} C_2^{*2} C_1\, \mathrm{e}^{-\mathrm{i}(\delta k)z} + \alpha_{22} C_{1s} |C_1|^2 \quad , \\ \frac{dC_2}{dz} &= -\beta_{11} |C_2|^2 C_2 - \beta_{12} C_1 C_{1s}^* C_2^*\, \mathrm{e}^{-\mathrm{i}(\delta k)z} \quad , \end{aligned} \tag{7.2.32}$$

where the notations of (7.2.3) have been used, and $\delta k = k_1 + k_{1s} - 2k_2$.

When $|C_{20}|^2 \gg |C_{10}|^2$, both Raman $C_1 \to C_{1s}$ and parametric transformations C_1, $C_{1s} \to C_2$ can be neglected. Then the system (7.2.32) yields:

$$\dot{C}_1 = -\gamma_1 \frac{\mathrm{e}^{\mathrm{i}\delta\zeta}}{1+2\zeta} C_{1s} \quad , \quad \dot{C}_{1s} = \gamma_1 \frac{\omega_{1s}^2 k_1}{\omega_1^2 k_{1s}} \frac{\mathrm{e}^{-\mathrm{i}\delta\zeta}}{1+2\zeta} C_1 \quad , \tag{7.2.33}$$

where

$$\gamma_1 = \frac{\omega_1^2}{\omega_2^2} \frac{k_2}{k_1} \frac{\kappa^2(\omega_1)}{\kappa^2(\omega_2)} \quad .$$

For zero value of detuning the solution of the system (7.2.33) is given by

$$C_{1s} = C_{1s0} \cos\left[\frac{k_1^{1/2}}{k_{1s}^{1/2}}\frac{\omega_{1s}}{\omega_1}\gamma_1 \ln\sqrt{1+2\zeta}\right]$$

$$+ C_{10}\frac{\omega_{1s}}{\omega_1}\sqrt{\frac{k_1}{k_{1s}}} \sin\left[\frac{k_1^{1/2}}{k_{1s}^{1/2}}\frac{\omega_{1s}}{\omega_1}\gamma_1 \ln\sqrt{1+2\zeta}\right] \quad ,$$

$$C_1 = C_{1s0}\frac{\omega_1}{\omega_{1s}}\sqrt{\frac{k_{1s}}{k_1}} \sin\left[\frac{k_1^{1/2}}{k_{1s}^{1/2}}\frac{\omega_{1s}}{\omega_1}\gamma_1 \ln\sqrt{1+2\zeta}\right]$$

$$+ C_{10} \cos\left[\frac{k_1^{1/2}}{k_{1s}^{1/2}}\frac{\omega_{1s}}{\omega_1}\gamma_1 \ln\sqrt{1+2\zeta}\right] \quad . \tag{7.2.34}$$

Formulas (7.2.34) are approximately valid until the strong pumping field is partially adsorbed and its order of magnitude becomes equal to that of the boundary value of the weak pumping field. According to them, if $\mathcal{J}_{1s0} \ll \mathcal{J}_{10}$, the length

$$z_{\max} = \frac{1}{2\beta_{22}}\frac{1}{\mathcal{J}_{20}}\left(\exp\frac{\pi\omega_1}{\omega_{1s}\gamma_1} - 1\right) \tag{7.2.35}$$

corresponds to the complete transformation of the weak pumping field into the Stokes component. As follows from (7.2.35), when

$$\frac{\omega_1\omega_{1s}}{\omega_2^2}\frac{k_2}{\sqrt{k_1 k_{1s}}}\frac{\kappa^2(\omega_1)}{\kappa^2(\omega_2)} = \pi \quad ,$$

the decrease through absorption of the strong pumping field by a factor of e should occur over the length of the active medium in order to ensure the complete transformation in the case of linear synchronism ($\delta k = 0$).

The Analysis of Equations for Slowly Varying Real Amplitudes and Phase Difference. Equations for slowly varying amplitudes and phase difference $\theta = -2\varphi_2 + \varphi_1 - \varphi_{1s} + \delta k$ have the form

$$\frac{d\mathcal{A}_1}{d\zeta_1} = -\mathcal{A}_{1s}(\mathcal{A}_1\mathcal{A}_{1s} + s\mathcal{A}_2^2 \cos\theta) \quad ,$$

$$\frac{d\mathcal{A}_{1s}}{d\zeta_1} = g_{1s}\mathcal{A}_1(\mathcal{A}_1\mathcal{A}_{1s} + s\mathcal{A}_2^2 \cos\theta) \quad ,$$

$$\frac{d\mathcal{A}_2}{d\zeta_1} = -g_2\mathcal{A}_2(s\mathcal{A}_2^2 + \mathcal{A}_1\mathcal{A}_{1s} \cos\theta) \quad ,$$

$$\frac{d\theta}{d\zeta_1} = \frac{\delta k}{\alpha_{11}} + W(\mathcal{A}_1, \mathcal{A}_{1s}, \mathcal{A}_2)\sin\theta \quad , \tag{7.2.36}$$

where

$$W(\mathcal{A}_1, \mathcal{A}_{1s}, \mathcal{A}_2) = 2g_2\mathcal{A}_1\mathcal{A}_{1s} + s\mathcal{A}_2^2\left(\frac{\mathcal{A}_{1s}}{\mathcal{A}_1} - g_{1s}\frac{\mathcal{A}_1}{\mathcal{A}_{1s}}\right) \quad . \tag{7.2.37}$$

Formulas (7.2.36, 37) are based on the notations of (7.2.23). If at the boundary of the medium $\mathcal{J}_{1s0} \ll \mathcal{J}_{10}$, we have

$$\alpha_{11}^2 W^2|_{z=0} \gg (\delta k)^2$$

and interacting fields exhibit phase locking for which $\theta = 0$ [if $\kappa(\omega_2)\kappa(\omega_1) > 0$]. As the quantity $\mathcal{A}_{1s}$ increases and $\mathcal{A}_2$, $\mathcal{A}_1$ diminish, the magnitude of W decreases and for $|\alpha_{11}W| \approx \delta k$ the phase locking is broken down after that θ approaches $\pi/2$ and parametric interaction becomes less efficient. The behavior of the system in the region of phase locking can be described in the same way as when $\delta k = 0$. It follows from (7.2.36) that the given transformation $\mathcal{J}_1 \to \tilde{\mathcal{J}}_{1s}$ is ensured over a length

$$L = \frac{1}{2\alpha_{11}} \int_{\mathcal{J}_{1s0}}^{\tilde{\mathcal{J}}_{1s}} \frac{d\mathcal{J}_{1s}}{\mathcal{J}_{1s}(M_1 - \mathcal{J}_{1s}) + s\mathcal{J}_2\sqrt{g_{1s}\mathcal{J}_{1s}(M_1 - \mathcal{J}_{1s})}} \quad . \tag{7.2.38}$$

In this expression

$$\mathcal{J}_2 = \mathcal{J}_{20} \exp\left\{\frac{2g_2}{g_{1s}}\left(\arcsin\sqrt{\frac{\mathcal{J}_{1s0}}{M_1}} - \arcsin\sqrt{\frac{\mathcal{J}_{1s}}{M_1}}\right)\right\} \quad . \tag{7.2.39}$$

Solution (7.2.38) is valid for those $\mathcal{J}_{1s}$ that satisfy the inequality

$$|\alpha_{11}W| = \alpha_{11}\left|\frac{2g_2}{g_{1s}}\sqrt{g_{1s}\mathcal{J}_{1s}(M_1 - \mathcal{J}_{1s})} + s\mathcal{J}_2\frac{2\mathcal{J}_{1s} - M_1}{\sqrt{g_{1s}\mathcal{J}_{1s}(M_1 - \mathcal{J}_{1s})}}\right| > |\delta k| \quad . \tag{7.2.40}$$

As an example, consider the possibility of producing radiation in potassium vapors through the generation of a SRS Stokes component by low intensity radiation from a dye laser, tunable in the range 410–460 nm (Fig. 7.10, the Stokes component will be tuned in the range 3–15 μm) in the presence of TPA of radiation from a high intensity source with $\lambda_2 = 951\,\mathrm{n}\mu$. In the approximation based on the assumption that levels $4P$ and $5P$ participate in two-photon transitions, the quantities $\kappa(\omega_2)$ and $\kappa(\omega_1)$ are equal, respectively, to 2.2×10^{-23} cgse and 1.2×10^{-22} cgse (for $\lambda_1 = 455\,\mathrm{n}\mu$, $\lambda_{1s} = 9\,\mu\mathrm{n}$). For a pressure of 15 torr, $\alpha_{11} = 2 \times 10^{-5}$ cgse, $S_{20} = 2.5 \times 10^8\,\mathrm{W/cm^2}$, $S_{10} = 2.5 \times 10^7\,\mathrm{W/cm^2}$, $I_{1s0} = 10^{-6} I_{10}$, $\delta k < 1.25\,\mathrm{cm^{-1}}$, Raman-parametric generation in the range defined by (7.2.40) ensures the coefficient of transformation

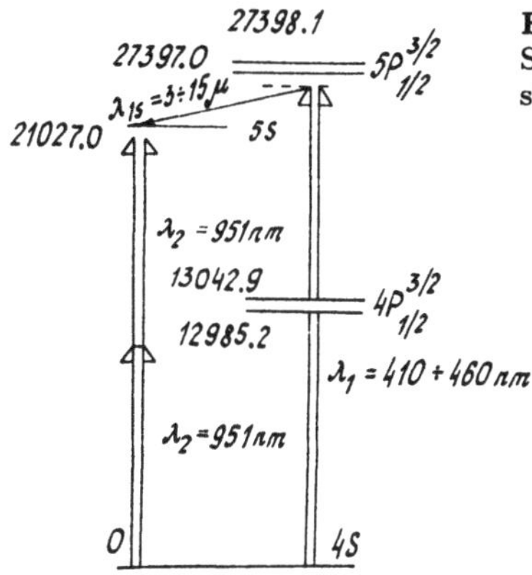

Fig. 7.10. The diagram of the weak pumping field SRS Stokes component generation in the presence of the strong pumping field TPA in potassium vapors

$I_1 \to I_{1s} \approx 30\%$ in terms of photon numbers[8], the transformation length being equal to 15 cm. Under the conditions of confocal focussing the realization of the process requires laser powers $I_{20} \approx 100\,\mathrm{kW}$, $I_{10} \approx 3\,\mathrm{kW}$. If there is no radiation with $\lambda_2 = 951\,\mathrm{n}\mu$, a similar value of the transformation coefficient is obtained over the length (600 cm) of the active region of matter.

7.2.3 Discussion of Experimental Results

The authors of [7.32] have conducted an experimental study of the generation of SRS Stokes components in compressed hydrogen in the presence of isolated and biharmonic pumping fields. The second harmonic of a frequency-stabilized single-mode neodymium laser ($\lambda_2 = 530\,\mathrm{nm}$) was used as the strong pumping field $\mathcal{J}_2$. The pumping field $\mathcal{J}_1$ was produced by a SRS Stokes component ($\lambda_1 = 630\,\mathrm{nm}$) of radiation with the wave length $\lambda_2 = 530\,\mathrm{nm}$ obtained in a deuterium chamber. The radiation with $\lambda_1 = 630\,\mathrm{nm}$ and the nontransformed portion of the second harmonic were focussed into a chamber with hydrogen, the pressure in the chamber being $\approx 75\,\mathrm{atm}$. Light filters were applied to vary the ratio of intensities $\mathcal{J}_{10}$ and $\mathcal{J}_{20}$ of the pumping fields at the entrance. The power of the Stokes component $I_{1s}\,(\lambda_{1s} = 850\,\mathrm{nm})$ was recorded by a calorimeter with the maximal sensitivity $I^{\mathrm{r}} \approx 0.1\,\mathrm{W}$.

The ratio $I_{10}(I^{\mathrm{r}})/I_{10}^{\mathrm{th}}$, where I_{10}^{th} is the threshold for SRS in the absence of the strong pumping field and $I_{10}(I^{\mathrm{r}})$ is the input power sufficient for generation of I_{1s} for a given sensitivity of the receiver I^{r}, was measured as a function of the quantity $I_{20}/I_{20}^{\mathrm{th}}$. At the wave length λ_2 the SRS threshold was $I_{20}^{\mathrm{th}} = 50\,\mathrm{kW}$, and at the wave length λ_1, $I_{10}^{\mathrm{th}} \approx 5\,\mathrm{kW}$[9].

[8] The solution of the transcendental equation (7.2.40) and the value of the integral (7.2.38) were obtained by a computer.

[9] The frequency dependence of SRS cross-sections cannot explain the difference between the threshold intensities, which is apparently associated with the somewhat different geometry of the strong and weak pumping field beams under the experimental conditions.

As I_{20} increased up to 150 kW, the value of I_{10}, necessary for ensuring the inequality $I_{1s} \gtrsim 0.1$ W, diminished by a factor of approximately 5×10^3 as compared to I_{10}^{th}.

In order to establish whether the experimental results agree with the theory, the lowering of the threshold by SRS described by the quantity I_{10}/I_{10}^{th} was computed as a function of the excess of the strong pumping field intensity over the threshold intensity. The computation was based on results of Sect. 7.2.1. The recorded value of the Stokes component power was $I_{1s}^{r} \lesssim (0.2 - 0.3)\, I_{10}$; in this range it can be assumed that

$$\theta = \theta_0 = \arctan \frac{\delta k}{\beta_{22} \mathcal{J}_{20}} = \text{const.} \quad .$$

Solving (7.2.21), we obtain in this case the following form of dependence I_{10}/I_{10}^{th} on I_{20}/I_{20}^{th} for a given value of I_{1s}^{r}:

$$f = \frac{\omega_1}{\omega_{1s}} \frac{I_{1s}^{r}}{I_{20}^{th}} \sin^{-2} \{R \cos\theta_0 \arctan(b\, e^{\beta_{22} \mathcal{J}_{20} L})\} \quad . \tag{7.2.41}$$

In correspondence with experimental conditions $TN = 4.5 \times 10^{11}$ cgse, $\kappa^2(\omega_1) \approx \kappa^2(\omega_2) \approx 0.68 \times 10^{-50}$ cgse, $\delta k = 5.6\,\text{cm}^{-1}$ [7.45–47], $L = 5$ cm, $S_{10}^{th} \approx 5.7 \times 10^8\,\text{W/cm}^2$, $S_{20}^{th} = 4.4 \times 10^8\,\text{W/cm}^2$, while the input intensities of Stokes components were I_{1s0}, $I_{2s0} \approx 10^{-10}\, I_{10}$, and I_{20}. The dispersal of the energy of the initial generator pulses that also led to dispersal of I_{10} ($\lambda_1 = 630$ nm) was such that the power of the recorded signal with the wave length $\lambda_{1s} = 850$ nm varied from 0.1 W to 5 W. Figure 7.11 depicts the respective theoretical curves representing functions $-\lg(I_{10}/I_{10}^{th})$ for recorded powers $I_{1s}^{r} \approx 10^{-3}\, I_{20}^{th} \approx 5$ W and $I_{1s}^{r} \approx 2 \times 10^{-5}\, I_{20}^{th} \approx 0.1$ W. Experimental points lie between the computed curves. In the case under consideration $R = 0.8$ [see (7.2.12)], and for the asymptotic transformation coefficient in terms of photon numbers which is due to parametric interaction, we have $\beta_{ph} \approx 0.9$, see (7.2.14) and Fig. 7.9a. The maximal transformation coefficient obtained in the experiment was $\approx 50\%$.

In order to verify the fact that the Stokes component of the weak pumping field appears in the presence of the strong one due to a parametric process, an experiment was conducted with the waves λ_1 and λ_2 directed into chambers facing each other. Under these conditions $|\delta k|^{-1} \sim \lambda_1$, and no phase locking is possible. The SRS Stokes component from radiation with $\lambda = 530$ nm in H_2 ($\lambda_1 = 680$ nm) propagating in the direction opposite to that of the pumping radiation, was used as the weak component of the pumping field I_1, while the role of the pumping field strong component belonged, as before, to the second harmonic of the neodymium laser with $\lambda_2 = 530$ nm. When the values of I_{10} were the same as in the experiment in which the pumping fields propagated in one direction, no Stokes component I_{1s} was observed even for $I_{20}/I_{20}^{th} \approx 30$.

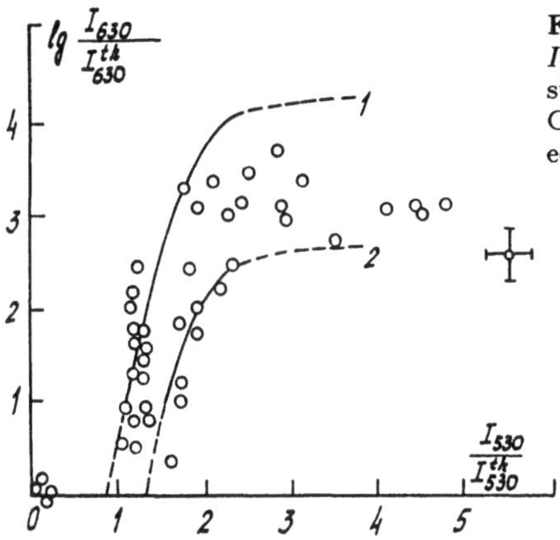

Fig. 7.11. The decrease of the threshold power I^{th}_{10} depending on the excess of the power of the strong pumping field I_{20} over the SRS threshold. Curves *1* and *2* are obtained for values of $(I^{\mathrm{r}}_{1\mathrm{s}})_{\mathrm{min}}$ equal to 0.1 W and 5 W, respectively

According to theoretical results, the Stokes radiation at the frequency $\omega_{1\mathrm{s}}$ can be a nonmonotonic function of a coordinate (or of the strong pumping field intensity for a fixed cell length) if $R>1$ (see Fig. 7.9a). In the case of hydrogen this corresponds to the low frequency strong pumping field. Then the dependence of the Stokes component intensity $I_{1\mathrm{s}}$ on the intensity I_2 for a fixed value of the weak pumping field intensity I_1 should be characterized by a sharp maximum. This was verified in an experiment using the first and second harmonics of a neodymium laser ($\lambda_1 = 530\,\mathrm{nm}$, $\lambda_2 = 1060\,\mathrm{nm}$) as the strong and weak pumping fields. The SRS threshold was $I^{\mathrm{th}}_{10} \approx 30\,\mathrm{kW}$, $I^{\mathrm{th}}_{20} = 250\,\mathrm{kW}$, the effective interaction region length $L \approx 10\,\mathrm{cm}$ and the wave detuning $\delta k \approx 9.6\,\mathrm{cm}^{-1}$ [7.47]. The range of variation of the weak pumping field was $(0.4–0.6)I^{\mathrm{th}}_{10}$. The experimental results are depicted in Fig. 7.12.

As follows from the statistical pattern of many flares on Fig. 7.12, there is a sufficiently well pronounced maximum near the value $I_{20} \approx 2I^{\mathrm{th}}_{20}$ of the strong pumping field intensity, where the coefficient of its transformation into the Stokes component is 20–35% (in terms of photon numbers). Figure 7.12 also shows the theoretical curve obtained for these conditions and $I_{10} \approx 0.5\,I^{\mathrm{th}}_{10}$ on the basis of (7.2.11), i.e., without taking into account Raman terms in polarization at the weak pumping field frequency (curve 1). The position of the maximum on this curve corresponds to the coefficient of transformation $I_2 \rightarrow I_{2\mathrm{s}}$ equal to 28%, which is close to its position observed in the experiment (see curve 2 on Fig. 7.12).

The very fact of the presence of a sharp maximum gives evidence of the determining role of RPI in the generation of the Stokes component of the weak pumping field, even when $I_{10} \approx 0.5\,I^{\mathrm{th}}_{10}$. For such values of I_{10}, the behavior of the anti-Stokes component intensity $I_{1\mathrm{a}}$ (curve 3 on Fig. 7.12) differs essentially from that of the Stokes one. In accordance with the theoretical results (Sect. 7.2.1), in the same range of intensities I_{20}, the anti-Stokes component is substantially weaker than the Stokes component. Apparently, the six-photon interaction with the simultaneous participation of the Stokes and anti-Stokes

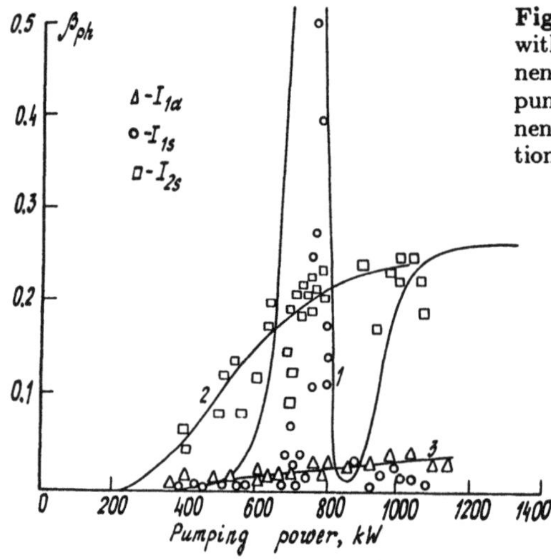

Fig. 7.12. The coefficients of transformation with respect to photons for Stokes components of the strong (I_{2s}) and weak (I_{1s}) pumping fields and the anti-Stokes component (I_{1a}) of the weak pumping field as functions of the strong pumping field intensity I_2

waves from the weak pumping field also plays a certain role. From qualitative point considerations it is clear that, because of the transfer of I_{10} into I_{1s}, this interaction should lead to a still faster decrease of the rate of the anti-Stokes component growth in comparison with that of the Stokes one.

Since the dependence of the above processes on the value of wave detuning is weak, it is possible to obtain the transformation of a broad-band or tunable weak pumping field. Reference [7.32] describes an experiment on transformation of a dye laser radiation into Stokes and anti-Stokes regions. The radiation from the dye laser (using rhodamine 6 G), together with the second harmonic of a neodymium laser, was directed into a hydrogen chamber. At the chamber exit the spectra were recorded by an STE-1 spectrograph and are depicted on Fig. 7.13. The spectrograms (a) and (b) were obtained, respectively, in the absence and in the presence of the strong pumping field SRS. When there was no radiation with $\lambda_2 = 530\,\mathrm{nm}$, the SRS threshold for $\lambda_1 = 575\,\mathrm{nm}$ ($\delta\lambda = 20\,\mathrm{nm}$) was 500 kW. The spectrogram (b) was produced when $I_1 \approx 10\,\mathrm{kW}$, $I_2 \approx 500\,\mathrm{kW}$ ($I_2^{\mathrm{th}} \approx 50\,\mathrm{kW}$).

The spectrum at the hydrogen chamber exit for the spectral width of rhodamine radiation $\delta\lambda = 2\,\mathrm{nm}$ ($I_1 \approx 10\,\mathrm{kW}$) is shown on Fig. 7.13c. Figure 7.13d presents the same spectrum for a dye laser radiation tuned by means of a Fabry-Perot resonator. As the frequency of the rhodamine laser varies in the range 562–585 nm, the anti-Stokes and Stokes waves were tuned in the ranges 455–469 nm and 737–773 nm. In this experiment the efficiency of transformation of the weak pumping field was of the order of 50% (in terms of energy).

Thus, a typical example of resonant RPIs associated with nonlinear synchronization is provided by the mixed Raman-parametric generation of Stokes components of the biharmonic pumping field. The coherent interaction length is determined not by the wave detuning, but by the length corresponding to substantial Raman transformation or two-photon absorption of the strong com-

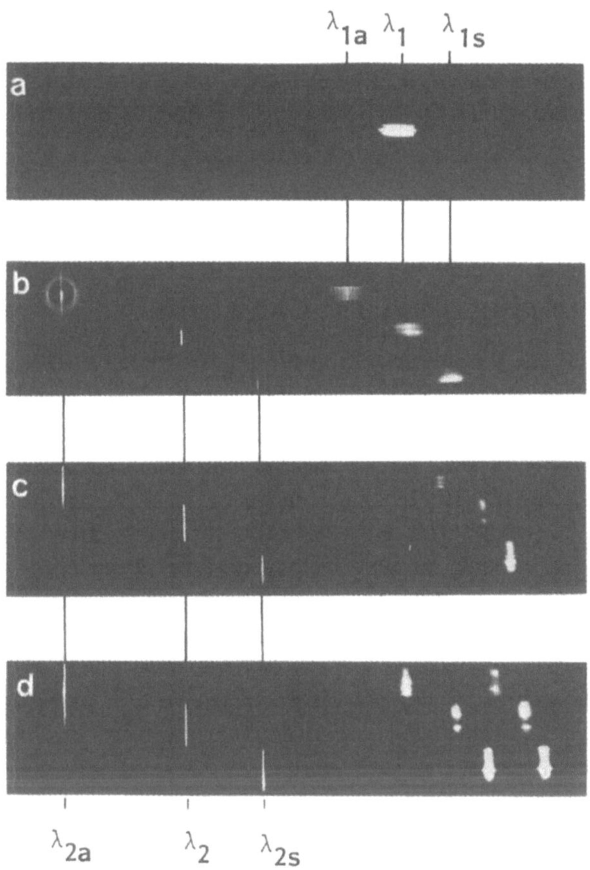

Fig. 7.13. Spectrograms of radiation in the output section of a hydrogen chamber irradiated by the second harmonic of a neodymium laser and a broad-band or narrow-band tunable rhodamine laser

ponent of the pumping field. The cause of this is the phase locking of waves generated during four-photon interaction. The above experimental results on the generation of Stokes and anti-Stokes components in hydrogen, where the lengths of coherent parametric interaction satisfied the inequality $L \gg (\delta k)^{-1}$, confirm the importance of this phenomenon.

Since it is possible to realize a highly efficient ($\approx 50\%$) transformation of low intensity tunable radiation into the low frequency range, there is a feasibility of constructing tunable generators of IR radiation on the basis of the processes considered in this section. Because of the existence of conditions, under which the coordinate dependence of the Stokes component corresponds to a monotonic increase (see Fig. 7.9a), Raman-parametric generation can be produced in waveguide devices, such as fiber light guides.

Note in conclusion that when several waves propagate in matter, the RPI of the components of the pumping field with their Stokes components can affect the dynamics of nonlinear interactions. For instance, in cesium vapors generation of many Stokes components of the electron SRS due to RPI was observed [7.48].

7.3 Radiation Frequency Transformation in Four-Photon RPIs Based on Pumping Field TPA and SRS

7.3.1 Introductory Remarks and Basic Equations

As has been demonstrated by experiments [7.17, 19], there is a possibility of efficient generation of sum and difference frequencies of field and matter by means of four-photon resonant processes. In this connection it is of interest to consider them in detail from the point of view of obtaining new sources of coherent radiation, including those with tunable frequency. One of such processes, namely, Raman-parametric generation for which parametric interaction leads to excitation of matter, has been investigated from their point of view in Sect. 7.2.

The present section deals with frequency converters based on four-photon processes, in which parametric interactions of waves bring the atoms of the matter back into their ground states. In this case there is competition between Raman and parametric processes. For the successful construction of such frequency converters, it is necessary to know their possible operation modes as well as the optimal conditions of generation. The effect of the lasing transition levels' population saturation on the energy transformation coefficient is also of interest. In [7.21] it was stated that the value of the latter is limited by saturation.

In the following we shall analyze these questions regarding four-photon RPI's on the basis of the pumping field TPA and SRS (Fig. 7.1). It will be demonstrated, in particular, that in both situations there is an optimal relationship between the pumping field intensity and the intensity of the trigger field corresponding to the maximum of the pumping field – the generated field transformation coefficient. Because of the peculiarities of four-photon interactions caused by the phase locking of the generated fields, there also exist modes for which this coefficient is not limited by population saturation due to SRS or TPA; it only increases the transformation length.

Let us describe the RPI's under study by equations for real amplitudes $\mathcal{A}_j$ $(j = \alpha, \beta, \gamma, \delta)$ and the phase difference for interacting waves $\theta = \varphi_\alpha + \varphi_\beta - \varphi_\gamma - \varphi_\delta + z\delta k$

$$
\begin{aligned}
\frac{d\mathcal{A}_\alpha}{dz} &= -g_\alpha(\langle\kappa_1^2\rangle\mathcal{A}_\beta^2\mathcal{A}_\alpha + \langle\kappa_1\kappa_2\rangle\mathcal{A}_\beta\mathcal{A}_\gamma\mathcal{A}_\delta \cos\theta)\eta \quad ,\\
\frac{d\mathcal{A}_\gamma}{dz} &= -g_\gamma(\langle\kappa_2^2\rangle\mathcal{A}_\delta^2\mathcal{A}_\gamma + \langle\kappa_1\kappa_2\rangle\mathcal{A}_\alpha\mathcal{A}_\beta\mathcal{A}_\delta \cos\theta)\eta \quad ,\\
\frac{d\theta}{dz} &= \delta k + \langle\kappa_1\kappa_2\rangle\left[\mathcal{A}_\alpha\mathcal{A}_\delta\left(g_\gamma\frac{\mathcal{A}_\delta}{\mathcal{A}_\gamma} + g_\delta\frac{\mathcal{A}_\gamma}{\mathcal{A}_\delta}\right)\right.\\
&\quad \left. + \mathcal{A}_\delta\mathcal{A}_\gamma\left(g_\alpha\frac{\mathcal{A}_\beta}{\mathcal{A}_\alpha} + g_\beta\frac{\mathcal{A}_\alpha}{\mathcal{A}_\beta}\right)\right]\eta \sin\theta \quad ,\\
\eta &= \eta_0 f_\eta = \eta_0[1 + 4\hbar^{-2}\tau T(\langle\kappa_1^2\rangle\mathcal{A}_\alpha^2\mathcal{A}_\beta^2\\
&\quad + 2\langle\kappa_1\kappa_2\rangle\mathcal{A}_\alpha\mathcal{A}_\beta\mathcal{A}_\gamma\mathcal{A}_\delta \cos\theta + \langle\kappa_2^2\rangle\mathcal{A}_\gamma^2\mathcal{A}_\delta^2)]^{-1} \quad . \qquad (7.3.1)
\end{aligned}
$$

The subsitution of subscripts $\alpha\rightleftarrows\beta$, $\gamma\rightleftarrows\delta$ in the first two equations (7.3.1) lead to the equations for $\mathcal{A}_\beta$ and $\mathcal{A}_\delta$. The following notations

$$\delta k = k_\alpha + k_\beta - k_\gamma - k_\delta \quad , \quad g_j = 2\pi N\omega_j^2 T\hbar^{-1}c^{-2}k_j^{-1} \quad ;$$

have been used; where κ_1 and κ_2 are the polarizabilities associated, with the resonances $\omega_\alpha + \omega_\beta = \omega_{21}$ and $\omega_\gamma + \omega_\delta = \omega_{21}$, respectively. The averaging of the quantities κ_1^2, κ_2^2 and $\kappa_1\kappa_2$ over the molecular orientations in a rarefied gas or vapor is performed in accordance with the formulas of Sect. 4.1. The value of T is determined by the mean velocity of molecules $\overline{v}$:

$$T = \sqrt{\pi}\overline{v}^{-1}|k_\alpha + k_\beta|^{-1} = \sqrt{\pi}\overline{v}^{-1}|k_\gamma + k_\delta|^{-1}$$

(interaction of waves with a small wave detuning $|\delta k| \ll |k_j|$ is considered). If the sublevels of states 1 and 2 corresponding to degeneration with respect to the magnetic quantum number have identical population (see Sect. 4.1), the last equation of (7.3.1) can be used for gases. For media consisting of fixed molecules with the same orientation it must be assumed that $T = T_{21}$, and the angular brackets should be omitted.

The following processes are described by (7.3.1).

1) Resonant four-photon parametric interactions on the basis of the pumping field TPA with the amplification (Fig. 7.1a) or the absorption (Fig. 7.1b) of the trigger field with the frequency ω_3. The process accompanied by the trigger field amplification can also be called the generation of the difference frequency, and that occurring with the absorption of the trigger field, generation of the sum frequency. In order to obtain equations describing these phenomena, one has to suppose that in (7.3.1), $\alpha = 1$, $\beta = 2$, $\gamma = \pm 3$ (the plus sign is for the generation of the difference frequency and the minus for the sum frequency generation) and $\delta = 4$. The pumping field degenerate in respect to frequency will be considered below. In this case the pumping field frequency $\omega_\mathrm{p} = \omega_1 = \omega_2$, the frequencies of the trigger and the generated fields are $\omega_\mathrm{d} = \omega_3$, $\omega_\mathrm{g} = \omega_4$.

2) Parametric processes on the basis of the SRS Stokes component generation are accompanied either by amplification (Fig. 7.1c) or absorption (Fig. 7.1d) of the trigger field. Equations describing these processes are obtained if it is assumed that in (7.3.1)

$$\omega_\alpha = \omega_p \quad , \; \omega_\beta = -\omega_s \quad , \; \omega_\gamma = \pm\omega_d \quad , \; \omega_\delta = \omega_g \quad .$$

Note that generation of the SRS anti-Stokes component studied in Sect. 7.1 represents a special case of RPI on the basis of SRS corresponding to the frequencies of the trigger and the pumping fields being equal: $\omega_d = \omega_p$.

Under general conditions no exact solution of (7.3.1) can be found. Therefore, their approximate solution will be sought by making use of the following considerations.

At the point $z = 0$, the coefficient near $\sin\theta$ in the third equation (7.3.1) [as before, it will be denoted by $W(\mathcal{A}_j)$] becomes infinite for any of the above processes, since at this point the amplitude of the generated field $\mathcal{A}_g = 0$. According to the results of Sects. 7.1, 2, when $W>\delta k$, the phase locking of generated fields occurs and the phase difference θ turns out to be approximately equal to

$$\theta \approx \pi + \delta k/W \tag{7.3.2}$$

(it is supposed that $\langle\kappa_1\kappa_2\rangle>0$).

As the generated field intensity increases and the intensity of the pumping field decreases, the coefficient W diminishes, and for certain values of the field amplitudes $\tilde{\mathcal{A}}_p$, $\tilde{\mathcal{A}}_g$, $\tilde{\mathcal{A}}_d$ that satisfy the condition $W(\tilde{\mathcal{A}}_j) = \delta k$, the phase locking is broken down and the generated field begins to decrease.

We have to analyze the solution of (7.3.1) in the range of values $W>\delta k$ since it is exactly in this range that the field $\mathcal{A}_g$ is generated. It is possible to show in a way similar to that of Sect. 7.2 that, because of the fact that at the initial stage (in the region where the variation of $\mathcal{A}_p$, $\mathcal{A}_d$ is small) the rate of growth of $\mathcal{A}_g$ is maximal when $\delta k = 0$, and also on the basis of the study of (7.3.1) near the boundary of the phase locking area, the solution of (7.3.1) for $|\delta k|^2/|W|^2 \ll 1$ is close to their solution in the case of the exact synchronization $\delta k = 0$[10]. We shall use this solution to estimate $\mathcal{A}_g$ over the whole region of phase locking, while the region itself will be determined approximately, by substituting into the equations $W = \delta k$ the solutions of (7.3.1) obtained for $\delta k = 0$.

[10] One can easily demonstrate that this statement is wrong only when there is a degeneration $\omega_1 = \omega_3$ which happens, for instance, during SRS. Then the variation of the pumping field is described by the sum of the first two equations (7.3.1) which contains no parametric term (see Sect. 7.1). A typical feature of such processes is the absence of generation in the direction of exact wave synchronization.

Let us add a few remarks about the competition between the various processes depicted in Fig. 7.1. When matter is irradiated by the pumping and trigger fields, simultaneous generation of the difference (Fig. 7.1a,c) and the sum (Fig. 7.1b,d) frequencies is possible. Their competition can only be considered on the basis of solving the problem of six-photon interaction. We shall study the generation of the difference and sum frequency under the assumption that only one of these processes occurs. Such a situation takes place, for example, in the presence of a substance, that absorbs the corresponding frequency ω_g, or in the presence of a difference in values of the wave detuning δk, sufficient for the phase locking region in one of these processes to be much larger than in the other.

7.3.2 Generation of the Difference Frequency During TPA

In accordance with the above program, let us discuss the case of exact phase synchronization: $\delta k = 0$. Later the role of mismatch of the wave phase velocities will be analyzed.

As can be easily shown by using the third equation of (7.3.1), for $\langle\kappa_1\kappa_2\rangle>0$ the plane $\theta = \pi$ is stable. If $\langle\kappa_1\kappa_2\rangle<0$, the stability exists for the plane $\theta = 0$, and the equations for the field amplitudes remain unchanged. It will be supposed that $\langle\kappa_1\kappa_2\rangle>0$. Setting $\theta = \pi$ and introducing notations

$$\begin{aligned} &\zeta = g_p\eta_0\langle\kappa_1^2\rangle z = Gz \quad , \quad s = \langle\kappa_1\kappa_2\rangle\langle\kappa_1^2\rangle^{-1} \quad , \\ &q_d = s\omega_d/\omega_p \quad , \quad q_g = s\omega_g/\omega_p \quad , \end{aligned} \tag{7.3.3}$$

we rewrite (7.3.1) for the amplitudes of the pumping field $\mathcal{A}_p = \mathcal{A}_\alpha = \mathcal{A}_\beta$, the trigger field $\mathcal{A}_d = \mathcal{A}_\gamma$ and the generated field $\mathcal{A}_g = \mathcal{A}_\delta$ in the form

$$\begin{aligned} \frac{d\mathcal{A}_p}{d\zeta} &= -\mathcal{A}_p(\mathcal{A}_p^2 - s\mathcal{A}_d\mathcal{A}_g)f_\eta \quad , \\ \frac{d\mathcal{A}_d}{d\zeta} &= q_d\mathcal{A}_g(\mathcal{A}_p^2 - s\mathcal{A}_d\mathcal{A}_g)f_\eta , \\ \frac{d\mathcal{A}_g}{d\zeta} &= q_g\mathcal{A}_d(\mathcal{A}_p^2 - s\mathcal{A}_d\mathcal{A}_g)f_\eta \quad , \end{aligned} \tag{7.3.4}$$

where f_η is determined from the last equation of (7.3.1). When (7.3.4) were derived, it had been assumed that

$$\mathcal{F} = \langle\kappa_1\kappa_2\rangle\langle\kappa_1^2\rangle^{-1/2}\langle\kappa_2^2\rangle^{-1/2} = 1 \quad . \tag{7.3.5}$$

In the following we shall discuss the effect of the deviation of $\mathcal{F}$ from unity.

The first integrals of system (7.3.4) under the initial conditions

$$\mathcal{A}_p|_{\zeta=0} = \mathcal{A}_{p0} \quad , \quad \mathcal{A}_d|_{\zeta=0} = \mathcal{A}_{d0} \quad , \quad \mathcal{A}_g|_{\zeta=0} = 0$$

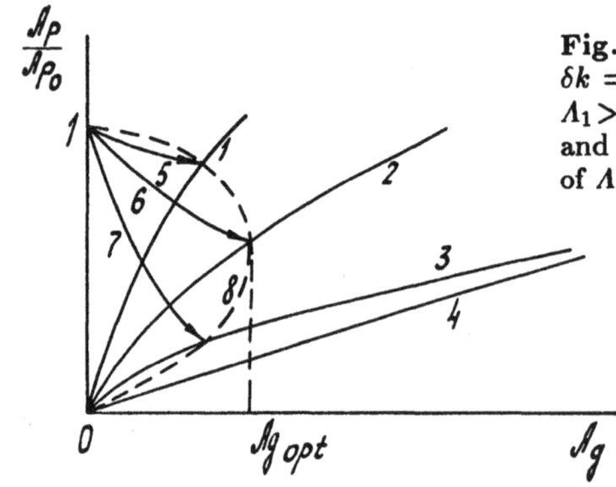

Fig. 7.14. The difference frequency generation for $\delta k = 0$; 1–4 are the equilibrium state curves for $\Lambda_1 > \Lambda_2 > \Lambda_3 > \Lambda_4 = 0$, respectively ($\Lambda = \mathcal{J}_{d0}/\mathcal{J}_{p0}$) and *5*, *6*, *7* the integral curves for the same values of Λ_1, Λ_2, Λ_3, *8* is the locus of equilibrium states

are

$$\mathcal{A}_d^2 = \mathcal{A}_{d0}^2 + (q_d/q_g)\mathcal{A}_g^2 \quad , \tag{7.3.6}$$

$$\mathcal{A}_g = \frac{1}{2}\sqrt{\frac{q_g}{q_d}}\mathcal{A}_{d0}\left[\left(\frac{\mathcal{A}_{p0}}{\mathcal{A}_p}\right)^{\sqrt{q_g q_d}} - \left(\frac{\mathcal{A}_p}{\mathcal{A}_{p0}}\right)^{\sqrt{q_q q_d}}\right]. \tag{7.3.7}$$

Equation (7.3.6) is valid for any values of δk. Making use of (7.3.4) it can easily be shown that the equilibrium state

$$\overline{\mathcal{A}}_p^2 = s\overline{\mathcal{A}}_d\overline{\mathcal{A}}_g \tag{7.3.8}$$

is stable. The substitution of (7.3.6, 7) into (7.3.8) results in an equation for the equilibrium amplitude of the pumping field:

$$\overline{\mathcal{J}}_p - m\mathcal{J}_{d0}\left[\left(\frac{\mathcal{J}_{p0}}{\overline{\mathcal{J}}_p}\right)^{\sqrt{q_g q_d}} - \left(\frac{\overline{\mathcal{J}}_p}{\mathcal{J}_{p0}}\right)^{\sqrt{q_g q_d}}\right] = 0 \quad , \tag{7.3.9}$$

where

$$\overline{\mathcal{J}}_p = \overline{\mathcal{A}}_p^2 \quad , \quad \mathcal{J}_{d0} = \mathcal{A}_{d0}^2 \quad , \quad m = 0.25\, s(\omega_g/\omega_d)^{1/2} \quad .$$

The solution of (7.3.9) combined with (7.3.6, 7) determines the values of the equilibrium amplitudes $\overline{\mathcal{A}}_d$ and $\overline{\mathcal{A}}_g$.

Figure 7.14 depicts the integral curves (7.3.7) and the curves of the equilibrium states (7.3.8) for various ratios of the boundary values of the trigger and the pumping fields

$$\Lambda = \mathcal{A}_{d0}^2/\mathcal{A}_{p0}^2 \quad .$$

This figure demonstrates that the limiting value of the generated field is small both for low and high values of Λ. If parameter Λ is small, the amplification is slow and results in the equilibrium state for the pumping field attained after strong TPA. Large Λ gives rise to an effective inverse parametric transformation of the amplified fields into the pumping field. It leads both to small attenuation of the pumping field and an insignificant increase of the trigger field, while the equilibrium state is established at a low value of the generated field [see (7.3.8)]. Thus, there is an optimal ratio Λ_{opt} corresponding to the maximal amplification of fields.

The optimal relationship between the boundary values $\mathcal{J}_{\text{d0}}$ and $\mathcal{J}_{\text{p0}}$ can be found if one solves the problem of the maximum of the function $\overline{\mathcal{J}}_{\text{g}} = f(\mathcal{J}_{\text{p0}}, \mathcal{J}_{\text{d0}})$ which is defined implicitly by the system of equations (7.3.8) and (7.3.7). The final form of the equation for the optimal relationship between $\mathcal{J}_{\text{d0}}$ and $\mathcal{J}_{\text{p0}}$ is

$$\begin{aligned}&\mathcal{J}_{\text{p0}}(2\sqrt{q_{\text{g}}q_{\text{d}}}m\mathcal{J}_{\text{d0}})^{\sqrt{q_{\text{g}}q_{\text{d}}}+1}\\&\quad = m\mathcal{J}_{\text{d0}}\left[\mathcal{J}_{\text{p0}}^{2\sqrt{q_{\text{g}}q_{\text{d}}}} - (2\sqrt{q_{\text{g}}q_{\text{d}}}m\mathcal{J}_{\text{d0}})^{2\sqrt{q_{\text{g}}q_{\text{d}}}}\right] \quad . \qquad (7.3.10)\end{aligned}$$

In general it appears impossible to obtain a closed-form expression for the transformation coefficient. As an example, consider the optimal transformation regime under special conditions when $\sqrt{q_{\text{g}}q_{\text{d}}} = 1$ [for instance, $\sqrt{q_{\text{g}}q_{\text{d}}} = 1$ for $\omega_{\text{g}}/\omega_{\text{d}} = 7$ and $s = 4/\sqrt{7} \approx 1.5$]. It follows from (7.3.6–8) that

$$\begin{aligned}&\overline{\mathcal{J}}_{\text{p}} = \mathcal{J}_{\text{p0}}\left(1 + \frac{\mathcal{J}_{\text{p0}}}{m\mathcal{J}_{\text{d0}}}\right)^{-1/2} \quad , \quad \overline{\mathcal{J}}_{\text{g}} = \frac{1}{4}\frac{\omega_{\text{g}}}{\omega_{\text{d}}}\frac{\overline{\mathcal{J}}_{\text{d0}}(\mathcal{J}_{\text{p0}} - \overline{\mathcal{J}}_{\text{p}})^2}{\mathcal{J}_{\text{p0}}\overline{\mathcal{J}}_{\text{p}}} \quad ,\\&\overline{\mathcal{J}}_{\text{d}} = \frac{1}{4}\frac{\mathcal{J}_{\text{d0}}(\mathcal{J}_{\text{p0}} + \overline{\mathcal{J}}_{\text{p}})}{\mathcal{J}_{\text{p0}}\overline{\mathcal{J}}_{\text{p}}} \quad . \qquad (7.3.11)\end{aligned}$$

According to (7.3.11), if $\Lambda_{\text{opt}} = 0.21\,\text{m}^{-1}$, the optimal transformation regime takes place, with the optimal transformation coefficient with respect to the trigger field energy being

$$\overline{\beta}_{\text{d}}^{\text{opt}} = \overline{\mathcal{J}}_{\text{g}}^{\text{opt}}/\mathcal{J}_{\text{d0}} \approx 0.2\omega_{\text{g}}/\omega_{\text{d}} \quad , \qquad (7.3.12)$$

and with respect to the energy of the pumping field

$$\overline{\beta}_{\text{p}}^{\text{opt}} = \overline{\mathcal{J}}_{\text{p}}^{\text{opt}}/\mathcal{J}_{\text{p0}} \approx 0.17\text{s}^{-1}\sqrt{\omega_{\text{g}}/\omega_{\text{d}}} \quad . \qquad (7.3.13)$$

When $\omega_{\text{g}}/\omega_{\text{d}} = 7$, the quantities $\overline{\beta}_{\text{d}}^{\text{opt}}$ and $\overline{\beta}_{\text{p}}^{\text{opt}}$ are equal to 1.4 and 0.31, respectively.

7.3.3 Generation of the Sum Frequency During TPA

As in the case of the difference frequency generation, we consider first the situation corresponding to the exact phase synchronization between the interacting waves. The equation for the phase of system (7.3.1) shows that for $z = 0$ the value $\theta = \pi$ is set. As the field $\mathcal{A}_g$ increases, however, the negative term proportional to $\mathcal{A}_g/\mathcal{A}_d$ in the coefficient W near $\sin\theta$ may exceed the sum of the remaining terms in this coefficient, which would result in a change of the phase difference from the value $\theta = \pi$ to $\theta = 0$; the quantity θ having passed the value $\pi/2$ will mark the beginning of the inverse parametric transformation of the field $\mathcal{A}_g$ into $\mathcal{A}_d$.

In order to obtain the rigorous solution of the problem, one has to solve the whole system of equations of (7.3.1) simultaneously. It does not appear possible to do that (analytically). In accordance with Sect. 7.3.2 we shall seek an approximate solution of the system (7.3.1) assuming $\theta = \pi$ and then making use of this solution to determine the values of fields for which $W = 0$.

Under the condition $\theta = \pi$, the equations for the field amplitudes are given by (7.3.4); in the second equation of this system the sign of the rhs has to be changed. The first integrals and the equations corresponding to the stable equilibrium states are

$$\mathcal{J}_d = \mathcal{J}_{d0} - \frac{\omega_d}{\omega_g}\mathcal{J}_g \quad , \tag{7.3.14}$$

$$\mathcal{J}_g = \frac{\omega_g}{\omega_d}\mathcal{J}_{d0}\sin^2\ln\left(\frac{\mathcal{J}_{p0}}{\mathcal{J}_p}\right)^{\sqrt{q_g q_d}/2} \quad , \tag{7.3.15}$$

$$\overline{\mathcal{J}}_p = s\left[\overline{\mathcal{J}}_g\left(\mathcal{J}_{d0} - \frac{\omega_d}{\omega_g}\overline{\mathcal{J}}_g\right)\right]^{1/2} \quad . \tag{7.3.16}$$

The substitution of (7.3.14, 15) into the equation for the phase difference in (7.3.1) results in the following expression for the coefficient W near $\sin\theta$:

$$W = g_p\langle\kappa_1\kappa_2\rangle\eta_0\sqrt{\frac{\omega_g}{\omega_d}}\left[\sin\ln\left(\frac{\mathcal{J}_{p0}}{\mathcal{J}_p}\right)^{\sqrt{q_g q_d}} + \frac{\sqrt{q_g q_d}}{2m}\frac{\mathcal{J}_p}{\mathcal{J}_{p0}}\cot\ln\left(\frac{\mathcal{J}_{p0}}{\mathcal{J}_p}\right)^{\sqrt{q_g q_d}}\right]\mathcal{J}_{d0}f_\eta \quad . \tag{7.3.17}$$

As the quantity $\ln(\mathcal{J}_p/\mathcal{J}_p)^{\sqrt{q_g q_d}}$ varies from 0 to π, the variation of the function W occurs within the limits $(+\infty, -\infty)$, the function passing through zero in the interval specified by

$$\frac{\pi}{2} < \ln(\mathcal{J}_{p0}/\mathcal{J}_p)^{\sqrt{q_g q_d}} < \pi \quad .$$

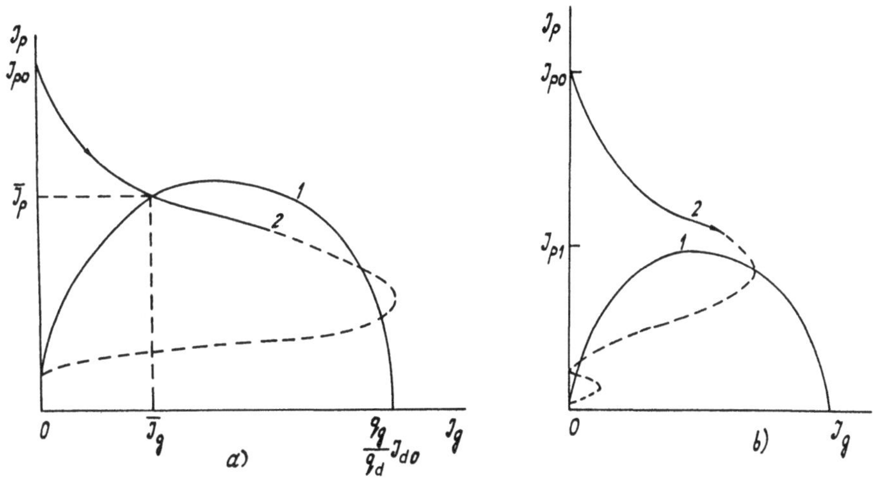

Fig. 7.15a,b. The sum frequency generation for $\delta k = 0$. Figures **(a)** and **(b)** correspond to modes *1* and *2* ($\Lambda_1 > \Lambda_2$); *1* are the equilibrium state curves, *2* the integral curves, $\overline{\mathcal{J}}_p$ and $\overline{\mathcal{J}}_g$ are intensities of the pumping and generated fields in the state of equilibrium. (——) depicts a part of the integral curve for which $W>0$

When $W>0$, there are two different transformation modes that can be easily identified if the integral curves (7.3.15) and the curves of the stationary states (7.3.16) are constructed for different values of $\Lambda = \mathcal{J}_{d0}/\mathcal{J}_{p0}$ (Fig. 7.15).

1) The first mode is realized under the condition

$$2m\mathcal{J}_{d0} > \mathcal{J}_{p0} \exp(-\pi/2\sqrt{q_g q_d}) \quad . \tag{7.3.18}$$

In this case interacting fields reach a steady state, and the course of the above process is similar to that of the difference frequency generation. The equilibrium value of the pumping field can be found from

$$\overline{\mathcal{J}}_p - 2m\mathcal{J}_{d0} \sin \ln(\mathcal{J}_{p0}/\overline{\mathcal{J}}_p)^{\sqrt{q_g q_d}} = 0 \quad . \tag{7.3.19}$$

According to (7.3.15, 16), the transformation coefficients with respect to the energy of the trigger field

$$\beta_d = \overline{\mathcal{J}}_g/\mathcal{J}_{d0} = \omega_g/2\omega_d \quad , \tag{7.3.20}$$

and with respect to the total energy of the trigger and the pumping fields

$$\beta = \frac{\overline{\mathcal{J}}_g}{\mathcal{J}_{d0} + \mathcal{J}_{p0}} = \frac{\omega_g}{2\omega_d} \frac{\exp(-\pi/2\sqrt{q_g q_d})}{2m + \exp(-\pi/2\sqrt{q_g q_d})} \quad , \tag{7.3.21}$$

become maximal at a certain value of Λ

$$\Lambda_{\mathrm{opt}} = \exp(-\pi/2\sqrt{q_{\mathrm{g}} q_{\mathrm{d}}})/2m \quad . \tag{7.3.22}$$

2) When

$$2m\mathcal{J}_{\mathrm{d0}} < \mathcal{J}_{\mathrm{p0}} \exp(-\pi/2\sqrt{q_{\mathrm{g}} q_{\mathrm{d}}}) \quad , \tag{7.3.23}$$

the condition $W = 0$ is attained before the interacting fields reach the equilibrium state. The maximal value of the generated field for this mode is not less than it is at the point $W(\tilde{\mathcal{J}}_{\mathrm{g}}, \tilde{\mathcal{J}}_{\mathrm{p}}, \tilde{\mathcal{J}}_{\mathrm{d}}) = 0$. In spite of the solution (7.3.15) being valid only up to this point, a qualitative analysis of the further behavior of interacting fields is not difficult. When the phase difference approaches the value $\pi/2$ from below, the generated field $\mathcal{J}_{\mathrm{g}}$ continues to increase although at a smaller rate. Reaching of the maximum at the point $\theta = \pi/2$ the reverse process starts, namely, the decrease of $\mathcal{J}_{\mathrm{g}}$ and the increase of $\mathcal{J}_{\mathrm{d}}$. This process represents a mixed Raman-parametric generation of $\mathcal{J}_{\mathrm{d}}$ as the Stokes component of the wave $\mathcal{J}_{\mathrm{g}}$ during TPA of the pumping field (see Sect. 7.2). Figure 7.15 gives a qualitative picture of the field interaction dynamics.

In the case of a strong inequality (7.3.23) (weak trigger fields), there is no significant inverse parametric transformation of the generated wave into the pumping field, and therefore one can describe interaction assuming that the pumping field is affected only by TPA. Raman transformation of the generated field into the triggered one can be neglected. Then the behavior of $\mathcal{J}_{\mathrm{g}}(z)$ and $\mathcal{J}_{\mathrm{d}}(z)$ is characterized by (7.2.34), where $\mathcal{J}_{\mathrm{g}} = |C_1|^2$ and $\mathcal{J}_{\mathrm{d}} = |C_{1\mathrm{s}}|^2$.

7.3.4 The Effect of Wave Detuning

In the above, the model with $\delta k = 0$ was considered. We have demonstrated in the frame of this model that the maximal transformation coefficients can be obtained if the interaction length is sufficient for producing fields that are almost stationary[11]. Usually in real media $\delta k \neq 0$, and it may turn out that the length of linear synchronization $(\delta k)^{-1}$ is less than the distance over which the stationary state would have been reached for $\delta k = 0$. As has been said earlier, under certain conditions the coherent interaction of light waves in resonant four-photon processes can take place over lengths that are greater than the length of linear synchronization. Because of this, it is in the cases under consideration (when $\delta k \neq 0$) also possible to obtain fields approaching the stationary values (7.3.9, 19).

Indeed, if $W > |\delta k|$, phase locking of interacting fields occurs in the system; if $W|_{z=0} \gg |\delta k|$, the system rapidly reaches a state with $\theta \approx \pi + \delta k/W$, and the process can be approximately studied by assuming at the first stage that $\delta k = 0$. If one uses (7.3.17) for $W(\mathcal{J}_{\mathrm{p}})$, it becomes possible to determine the

[11] The mode 2 for generation of the sum frequency is an exception.

conditions for the sum frequency generation under which the phase locking would ensure the formation of fields that are close to the stationary.

For the difference frequency generation the substitution of $\mathcal{A}_g$ and $\mathcal{A}_d$ from (7.3.6, 7) into the third equation of (7.3.1) leads to a formula for $W(\mathcal{J}_p)$:

$$W(\mathcal{J}_p) = L_{TPA}^{-1} f_\eta \left(\frac{1-y^2}{y} m\Lambda + \sqrt[4]{q_g q_d} \frac{1+y^2}{1-y^2} y^{-1/\sqrt{q_g q_d}} \right) \quad . \tag{7.3.24}$$

Here L_{TPA} is the TPA characteristic length,

$$L_{TPA} = \left(\frac{4\pi T \eta_0 N \omega_p}{\hbar c} \langle \kappa_1^2 \rangle \mathcal{J}_{p0} \right)^{-1} \quad , \quad y = \left(\frac{\mathcal{J}_p}{\mathcal{J}_{p0}} \right)^{\sqrt{q_g q_d}} \quad .$$

At the input section ($z = 0$) the function (7.3.24), as well as (7.3.17), becomes infinite. Therefore, there is always phase locking at the initial stage of interaction. Depending on the value of δk and the initial field intensities, a breakdown of the phase locking can occur both before and after the state which is at equilibrium when $\delta k = 0$ is reached. The pumping field intensity $\mathcal{J}_p^{br}$ corresponding to the breakdown is determined approximately from:

$$W(\mathcal{J}_p^{br}) = |\delta k| \quad .$$

It is obvious that if $W(\overline{\mathcal{J}}_p) > |\delta k|$, no breakdown of the phase locking occurs before reaching the stationary state.

As an example, consider a situation where $\sqrt{q_g q_d} = 1$. In the case of the difference frequency generation, the equation for $W(\overline{\mathcal{J}}_p)$ would contain the boundary values of intensity $\mathcal{J}_{p0}$ and $\mathcal{J}_{d0}$ if one makes use of (7.3.11, 24); then the range of values $\mathcal{J}_{p0}$ and $\mathcal{J}_{d0}$ in which the phase locking ensures the formation of fields approaching the equilibrium ones (7.3.11) is specified by the inequality:

$$W(\overline{\mathcal{J}}_p) = 2L_{TPA}^{-1} [m\Lambda(1 + m\Lambda)]^{1/2} \gtrsim |\delta k| \quad . \tag{7.3.25}$$

After the fields reach the values close to equilibrium fields they will decay slowly. Indeed, taking into account the fact that in this region $\theta \approx \pi + \delta k / W(\overline{\mathcal{J}}_p)$, as well as the first equation of (7.3.1), we can easily obtain the law governing the variation of the pumping field:

$$\mathcal{J}_p \approx \overline{\mathcal{J}}_p \left[1 + \left(\frac{\delta k}{W} \right)^2 \frac{z}{2\overline{L}_{TPA}} \right]^{-1} \quad , \tag{7.3.26}$$

i.e., its characteristic attenuation length is greater by a factor of $(W/\delta k)^2$ than the length of two-photon absorption

$$\overline{L}_{TPA} = \left(\frac{4\pi T \eta_0 N \omega_p}{\hbar c} \langle \kappa_1^2 \rangle \overline{\mathcal{J}}_p \right)^{-1} \quad .$$

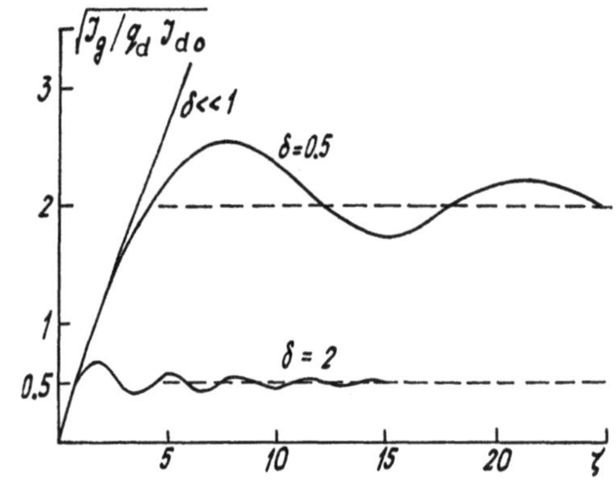

Fig. 7.16. The behavior of the difference frequency field for $\mathcal{J}_{d0} \ll \mathcal{J}_{p0}$

As follows from (7.3.26) and (7.3.1), for $\mathcal{A}_g \approx \overline{\mathcal{A}}_g$, $\mathcal{A}_d \approx \overline{\mathcal{A}}_d$, both generated and trigger fields will also decay, although somewhat slower than the pumping field. The behavior of fields during the sum frequency generation in 1 mode is similar.

When the values of wave detuning $|\delta k|$ are large, the input field intensities are low, and instead of (7.3.25) a strong reverse inequality is valid the breakdown of phase locking occurs far from the equilibrium state. Under these conditions the energy transfer from the generated and trigger fields into the pumping field, as well as Raman transformation of the generated field into the trigger field, can be neglected. In order to illustrate the effect of the wave detuning on RPI dynamics, it is possible to use the complex form of the corresponding equations for the fields, as has been done in Sect. 7.2. These equations permit a solution for a certain value of the parameter

$$R = \frac{\omega_d \omega_g}{\omega_p^2} \sqrt{\frac{k_p^2}{k_d k_g} \frac{\kappa_1^2}{\kappa_2^2}} = 2$$

[see (7.2.19)] and an arbitrary wave detuning δk.

Figure 7.16 depicts the behavior of the function $\mathcal{J}_g(z)$ during difference frequency generation for three values of dimensionless wave detuning δ. As in nonresonant parametric processes, the quantity $\mathcal{J}_g$ reaches its maximum at distances $z \approx \pi/\delta k$. Unlike these processes, however, in our case the amplitude modulation coefficient for the generated field is small. When the coordinate increases, $\mathcal{J}_g$ approaches a constant value which is close to the maximal quantity that decreases with the increase of δk.

Thus, in practice the difference frequency generation does not depend critically on the interaction length; it is sufficient that the latter is greater than $\pi/\delta k$. Therefore, it is for strong enough pumping fields not necessary to ensure a careful control of the synchronization conditions in experiments.

To conclude this section, note that when the parameter F differs somewhat from unity [see (7.3.5)], the equations for the field amplitudes are changed in the same way as when the phase difference θ deviates from the value π as a result of incomplete synchronism. Therefore, for $1-\mathcal{F}\ll 1$, the solution of (7.3.1) is to vary slightly in the region far away from the stationary solutions of (7.3.4). When the fields approach their equilibrium values, the difference between $\mathcal{F}$ and unity will lead to their attenuation even when $\delta k = 0$. [A similar phenomenon takes place during resonant frequency doubling in gases (see Sect. 6.2). In both cases its nature is the same].

7.3.5 Transformation Length and Effect of Population Saturation

Let us define the transformation length L as the distance over which a given value of the transformation coefficient (or a given value of $\mathcal{J}_g$, for instance, $\mathcal{J}_g = 0.9\overline{\mathcal{J}}_g$) is reached. According to the first equation of (7.3.4),

$$L = \frac{\hbar c^2 k_p}{4\pi T N \eta_0 \omega_p^2 \langle \kappa_1^2 \rangle} \int\limits_{\mathcal{J}_{p1}}^{\mathcal{J}_{p0}} \frac{d\mathcal{J}_p}{\mathcal{J}_p(\mathcal{J}_p - s\sqrt{\mathcal{J}_g \mathcal{J}_d}) f_\eta(\mathcal{J}_p, \mathcal{J}_g, \mathcal{J}_d)} \quad , \tag{7.3.27}$$

where $\mathcal{J}_{p1}$ corresponds to a given value of $\mathcal{J}_g$. On the basis of the integrals of motion (7.3.6, 7) or (7.3.14, 15) for generation of the difference and the sum frequencies, respectively, it is possible to calculate the integral in (7.3.27). When the sum frequency in mode 2 is generated, the quantity $\mathcal{J}_{p1}$ should be set equal to the value of $\mathcal{J}_p$ at the point $W = 0$.

Consider, for instance, the difference frequency generation in a medium consisting of (fixed) oriented molecules or in a dense gaseous medium. Then

$$f_\eta = [1 + \mathcal{J}_{sat}^{-2}(\mathcal{J}_p - s\sqrt{\mathcal{J}_g \mathcal{J}_d})^2]^{-1} \quad , \tag{7.3.28}$$

where

$$\mathcal{J}_{sat}^{1/2} = (\hbar / 2\langle \kappa_1 \rangle \sqrt{\tau T})^{1/2} \tag{7.3.29}$$

is the saturating field amplitude. For the special case $q_g q_d = 1$, the substitution of (7.3.6–8) into (7.3.27) and integration results in

$$\begin{aligned} L = L_1 + L_2 = {} & \frac{L_{TPA}}{2\sqrt{m\Lambda(1+m\Lambda)}} \ln \frac{(\mathcal{J}_{p0} - \overline{\mathcal{J}}_p)(\mathcal{J}_{p1} + \overline{\mathcal{J}}_p)}{(\mathcal{J}_{p0} + \overline{\mathcal{J}}_p)(\mathcal{J}_{p1} - \overline{\mathcal{J}}_p)} \\ & + L_{TPA} m\Lambda \frac{\mathcal{J}_{p0}^2(\mathcal{J}_{p0} - \mathcal{J}_{p1})(\mathcal{J}_{p1}\mathcal{J}_{p0} - \overline{\mathcal{J}}_p^2)}{\mathcal{J}_{sat}^2 \overline{\mathcal{J}}_p^2 \mathcal{J}_{p1}} \quad . \end{aligned} \tag{7.3.30}$$

The quantity $\overline{\mathcal{J}}_p$ appearing in this formula is determined from (7.3.11). The first term (L_1) in (7.3.30) is the transformation length in a medium without saturation, the second (L_2) describes the increase of the transformation length due to population saturation. If $\mathcal{J}_{p0} \gg \mathcal{J}_{sat}$, the contribution of L_2 into the transformation length is essential.

It is obvious that in rarefied gases the effect of population saturation on four-photon interaction is qualitatively the same as in the above case. For gases with the Doppler line width, however, (7.3.27) makes it possible to calculate the transformation length only when the fields are not stronger than the saturating ones and $f_\eta \approx 1$, the transformation length under these conditions coinciding with L_1.

It has been shown above that if $\delta k \neq 0$, the fields close to $\overline{\mathcal{J}}_p$, $\overline{\mathcal{J}}_g$, $\overline{\mathcal{J}}_d$ are reached in the presence of phase locking. In this situation the transformation length is also described by (7.3.27) or (7.3.30). Therefore, one can say that if there is phase locking, the population saturation, instead of changing the transformation coefficient, only increases the transformation length. The latter may turn out to be much greater than the linear synchronization length.

According to [7.21], the transformation coefficient is limited because of the saturation effect. This statement is valid only when the breakdown of phase locking occurs long before the stationary state is reached. In that case the local effectiveness of interaction is decreased by the population saturation, and transformation takes place just over the linear synchronization length.

Consider the choice of the optimal regime of the sum and difference frequency generation in vapors and gases. First, we shall formulate conditions necessary for the realization of such a regime.

For generation of the difference and sum frequencies, respectively, the input values of the trigger and the pumping fields $\mathcal{A}_{p0}$, $\mathcal{A}_{d0}$ should satisfy (7.3.10, 22). The fields $\mathcal{A}_{d0}$ and $\mathcal{A}_{p0}$ must be sufficient to ensure the phase locking up to the value of the generated field close to the equilibrium one in the presence of exact synchronism ($\delta k = 0$). The necessary condition for this is that $W(\mathcal{J}_{d0}, \mathcal{J}_{p0}, \overline{\mathcal{J}}_p) > \delta k$, where W is determined by (7.3.24) and (7.3.17) and $\overline{\mathcal{J}}_p = \overline{\mathcal{A}}_p^2$ is the solution of (7.3.9, 19).

If the intensity of one of the fields $\mathcal{A}_{d0}$ and $\mathcal{A}_{p0}$ is not sufficient for the conditions of the optimal ratio $\mathcal{A}_{d0}/\mathcal{A}_{p0}$ and the phase locking to be satisfied simultaneously, it is desirable to increase the intensity of the second field up to a level for which $W \approx \delta k$.

As the first example, consider the possibility of the difference frequency generation in Hg vapors by the radiation of a neodymium laser with the wave length $1.075\,\mu$m [7.49] and its fourth harmonic; the latter corresponds to two-photon resonance between the ground $6s^2S$ and $8s^1S$ states ($\hbar\omega_{21} = 74404.6\,\mathrm{cm}^{-1}$ [7.50]). In order to obtain estimates, let us assume that $\langle \kappa_1^2 \rangle \approx 10^{-46}\,\mathrm{cm}^6$, $s = 1.5$; then the solution (7.3.11) is valid. For a pressure of 10 Torr ($t^0 = 200°$C, $N \approx 2.6 \times 10^{17}\,\mathrm{cm}^{-3}$ [7.51]), and $\tau = 5 \times 10^{-9}$ s, the quantities T and G are equal, respectively, to 2×10^{-10} s and 4×10^{-6} cgse, see (7.3.3). If $\delta k \approx 0.2\,\mathrm{cm}^{-1}$ and the input pumping field power is $75\,\mathrm{MW/cm}^2$, both the optimal generation regime and the phase locking condition (7.3.25) take place for the trigger field power $15\,\mathrm{MW/cm}^2$. The transformation coeffi-

Table 7.1. Transformation coefficients, transformation length and phase locking breakdown length calculated for difference frequency generation in neon for various values of $\Lambda = \mathcal{J}_{d0}/\mathcal{J}_{p0}$

$\Lambda = \mathcal{J}_{d0}/\mathcal{J}_{p0}$	$\overline{\beta}_d$ [%]	$\overline{\beta}_p$ [%]	β_d [%]	β_p [%]	L [cm]	L_{br} [cm]
0.167	184	31	38	6.5	0.96	0.21
0.66	41.8	27.5	38	24	0.77	0.77
1.33	11.5	15	10	13.5	0.40	1.85
2	6.5	12	5.8	10.8	0.25	6.25

cient with respect to the pumping field (7.3.13) is equal to $\beta_p = \overline{\beta}_p^{opt} \approx 0.28$, and with respect to the trigger field $\beta_d = \overline{\beta}_d^{opt} \approx 1.25$. Note that in this case the pumping field intensity is by a factor of 2.2 less than the intensity of the saturating field.

Consider the sum frequency generation in Mg produced by the radiation of two dye lasers [7.21] (for $\lambda = 4597$ Å TPA occurs between $3s^{2'}S_0$ and $4s'S$ states). If $\omega_g/\omega_d = 10$, the estimate of the quantities $\langle\kappa_1^2\rangle$ and $\langle\kappa_2^2\rangle$, taking into account the contribution into TPA and SRS of states $3p'P^o$ and $4p'P^o$ results, respectively, in the values 1.4×10^{-45} and 2×10^{-46} cm^6; then $s = 3.8$, $\sqrt{q_g q_d} = 2.7$. For the vapor pressure 10 Torr, $\tau = 5 \times 10^{-9}$ s, we obtain $T = 8 \times 10^{-11}$ s, $G = 3.6 \times 10^{-6}$ cgse. When $\delta k = 0.3\,\text{cm}^{-1}$, $S_{p0} = 25\,\text{MW/cm}^2$ and $S_{d0} = 2.7\,\text{MW/cm}^2$, the first transformation regime is realized (see Sect. 7.3.3), and due to phase locking a state close to equilibrium is reached in which the transformation coefficient with respect to the total energy of the trigger and the pumping fields is equal to $\overline{\beta} = 0.31$; $\overline{\beta}_d = 3.5$. The transformation length corresponding to $\beta = 0.9\overline{\beta}$ is 15 cm.

In conclusion, let us evaluate the possibility of the difference frequency generation in neon ($2p^{6'}S \rightarrow 5p''$ [1 1/2]) irradiated by the ninth harmonic of the neodymium laser and the laser using a rhodamine-6 G solution (in order to satisfy the two-photon resonance condition, it is necessary to tune the neodymium laser to the wave number 32 cm^{-1}). Let us set $N = 10^{20}\,\text{cm}^{-3}$, $T = 10^{-11}$ s, $\langle\kappa_1^2\rangle \approx 10^{-48}\,\text{cm}^6$, $s = 1.65$. In this case $q_g q_d = 1$, and solution (7.3.11) is valid. Table 7.1 presents the results of the calculation of the transformation coefficients $\overline{\beta}_p$ and $\overline{\beta}_d$ for $\delta k = 0$ and β_d and β_p for $\delta k = 5\,\text{cm}^{-1}$, the pumping field power equal to 375 MW/cm^2, the transformation length L (the length over which the transformation coefficient $\beta_p = 0.9\overline{\beta}_p$ is reached), and the length L_{br} of the phase locking breakdown for various values of $\Lambda = \mathcal{J}_{d0}/\mathcal{J}_{p0}$. According to these data, when $\Lambda = 0.167$, the stationary value $\mathcal{J}_g$ is maximal. However, it is not attained, because the phase locking breaks down too early. For $\delta k = 5\,\text{cm}^{-1}$ the optimal regime corresponds to the value $\Lambda = 0.66$. As Λ increases further, the transformation coefficient diminishes and phase locking allows to obtain coherent interaction over lengths that are much greater than

$(\delta k)^{-1}$. Note that under the conditions of such a regime the transformation coefficient is nearly independent of the length of the lasing gas chamber.

Thus, the last example demonstrates the feasibility of constructing a tunable source in the range 650–656 Å. For $\Lambda = 0.66$, its excitation in the scheme with confocal focussing [7.52] requires the power of the pumping field to be 1.6 kW and of the trigger field, 6 kW.

7.3.6 Four-Photon RPI's Based on SRS of the Pumping Field

In the following we shall consider generation of the field under the assumption that

$$\begin{aligned} \mathcal{A}_{\rm g}|_{z=0} &= \mathcal{A}_{\rm g0} \quad , \quad \mathcal{A}_{\rm p}|_{z=0} = \mathcal{A}_0 \quad , \\ \mathcal{A}_{\rm d}|_{z=0} &= \mathcal{A}_{\rm d0} \quad , \quad \mathcal{A}_{\rm s}|_{z=0} = \mathcal{A}_{\rm s0} \end{aligned} \tag{7.3.31}$$

hold at the boundary of the medium. For practical applications two situations are of interest:

a) There are not only the pumping and trigger fields at the input, but the Stokes component of SRS as well, and $\mathcal{A}_{\rm g0} = 0$. This situation can be realized if one first obtains SRS of the pumping field. Then the latter, together with the Stokes component of the scattered field, is directed into a cell containing the same substance which is also irradiated by a source with a frequency $\omega_{\rm d}$.

b) Both the field $\mathcal{A}_{\rm g}$ and the field $\mathcal{A}_{\rm s}$ are generated in the same cell: $\mathcal{A}_{\rm s0}$, $\mathcal{A}_{\rm g0} \ll \mathcal{A}_{\rm p0}, \mathcal{A}_{\rm d0}$.

The coefficient near $\sin\theta$ in the second to last equation of (7.3.1) will be denoted by W. As a rule, in the above two cases at the point $z = 0$ W is much greater than the wave detuning δk. For (b), it is obvious that for sufficiently strong pumping and the trigger fields the condition $W \gg \delta k$ can be assumed to be valid since

$$\frac{\mathcal{A}_{\rm s0}}{\mathcal{A}_{\rm g0}} \sim \exp\frac{\hbar\omega_{21}}{k_{\rm B}T^0} \gg 1 \quad .$$

We have demonstrated above that if $|W| > |\delta k|$, phase locking of interacting fields occurs, and θ reaches values close to

$$\theta = \pi + \delta k / W \tag{7.3.32}$$

(under the condition that $\langle \kappa_1 \kappa_2 \rangle > 0$). As the fields $\mathcal{A}_{\rm s}$ and $\mathcal{A}_{\rm g}$ increase, the coefficient W diminishes, and for a certain value of the fields $\overline{\mathcal{A}}_{\rm p}$, $\overline{\mathcal{A}}_{\rm s}$, $\overline{\mathcal{A}}_{\rm d}$ and $\overline{\mathcal{A}}_{\rm g}$ such that

$$W(\overline{\mathcal{A}}_j) = \delta k \quad , \tag{7.3.33}$$

one observes the breakdown of phase locking and attenuation of the field $\mathcal{A}_{\rm g}$.

We must consider the solution of (7.3.1) in the range $W > \delta k$ since it is here that the effective generation of the field $\mathcal{A}_g$ takes place. In a way similar to that of Sect. 7.3.1 it is possible to show that the solution of (7.3.1) for $(\delta k/W)^2 \ll 1$ is close to their solution in the presence of exact synchronizaton $\delta k = 0$. We shall use this solution to estimate $\mathcal{A}_g$ over the whole range of phase locking, while the range itself will be determined approximately, by substituting for $\delta k = 0$ the solutions of (7.3.1) into (7.3.33).

Let us simplify (7.3.1) somewhat in the case of exact wave synchronization. Assume $\langle\kappa_1\kappa_2\rangle > 0$ to hold. This means that the phase difference $\theta = \pi$ is stable in the region $W > 0$ (when $\langle\kappa_1\kappa_2\rangle < 0$, the stability condition holds for the phase difference $\theta = 0$, and the equations for the field amplitudes given below remain the same). Introducing notations

$$\begin{aligned}
&\zeta = g_p\langle\kappa_1^2\rangle z = Gz \quad , \quad s = \langle\kappa_1\kappa_2\rangle\langle\kappa_1^2\rangle^{-1} \quad , \\
&q_s = \omega_s/\omega_p \quad , \qquad q_{d,g} = s\omega_{d,g}/\omega_p \quad ,
\end{aligned} \tag{7.3.34}$$

we shall write equations for the amplitudes of interacting fields in the form

$$\begin{aligned}
\frac{d\mathcal{A}_p}{d\zeta} &= -\mathcal{A}_s(\mathcal{A}_p\mathcal{A}_s - s\mathcal{A}_d\mathcal{A}_g) \quad , \\
\frac{d\mathcal{A}_s}{d\zeta} &= q_s\mathcal{A}_p(\mathcal{A}_p\mathcal{A}_s - s\mathcal{A}_d\mathcal{A}_g) \quad , \\
\frac{d\mathcal{A}_d}{d\zeta} &= \pm q_d\mathcal{A}_g(\mathcal{A}_p\mathcal{A}_s - s\mathcal{A}_d\mathcal{A}_g) \quad , \\
\frac{d\mathcal{A}_g}{d\zeta} &= q_g\mathcal{A}_d(\mathcal{A}_p\mathcal{A}_s - s\mathcal{A}_d\mathcal{A}_g) \quad .
\end{aligned} \tag{7.3.35}$$

The above is based on the assumption $\mathcal{F} = 1$, see (7.3.5), and also Sect. 4.1.

7.3.7 Generation of the Difference Frequency During SRS

Integration of (7.3.35) results in the following formulas containing the squares of the amplitudes of interacting fields ($j = \mathrm{p}, d, s, \mathrm{g}$):

$$\mathcal{J}_s + q_s\mathcal{J}_p = \mathcal{J}_{s0} + q_s\mathcal{J}_{p0} = M_1 \quad , \tag{7.3.36}$$

$$\mathcal{J}_d + \frac{\omega_d}{\omega_g}\mathcal{J}_g = \mathcal{J}_{d0} + \frac{\omega_d}{\omega_g}\mathcal{J}_{g0} = M_2 \quad , \tag{7.3.37}$$

$$\mathcal{J}_d = M_2\cosh^2[u(\xi) - u(\xi_0)] \quad , \tag{7.3.38}$$

$$\mathcal{J}_g = \frac{\omega_g}{\omega_d}M_2\sinh^2[u(\xi) - u(\xi_0)] \quad , \tag{7.3.39}$$

where

$$u(\xi)=\left(\frac{q_{\rm g}q_{\rm d}}{q_{\rm s}}\right)^{1/2}\arccos\xi^{1/2}\quad ;\quad \xi=\frac{q_{\rm s}\mathcal{J}_{\rm p}}{M_1}(0<\xi<\xi_0)\quad . \tag{7.3.40}$$

According to the analysis of (7.3.35), the state of equilibrium

$$\mathcal{A}_{\rm p}\mathcal{A}_{\rm s}-s\mathcal{A}_{\rm g}\mathcal{A}_{\rm d}=0 \tag{7.3.41}$$

is stable, i.e., when $z\to\infty$, the interacting fields approach a steady state. The pumping field intensity in this state is determined by[12]

$$\sqrt{\bar{\xi}(1-\bar{\xi})}=\frac{s}{2}\sqrt{\frac{q_{\rm g}}{q_{\rm d}q_{\rm s}}}\Lambda\sinh 2[u(\bar{\xi})-u(\xi_0)]\quad . \tag{7.3.42}$$

Here

$$\bar{\xi}=q_{\rm s}\bar{\mathcal{J}}_{\rm p}/M_1\quad ,\quad \Lambda=q_{\rm s}\mathcal{J}_{\rm d0}/M_1\quad . \tag{7.3.43}$$

Since, as we have mentioned above, the phase difference $\theta=\pi$ is stable just for $W(\mathcal{A}_{\rm p},\mathcal{A}_{\rm s},\mathcal{A}_{\rm d},\mathcal{A}_{\rm g})>0$,(7.3.35) correctly describe the variation of fields along the z direction only in this region. It can be shown, however (see [7.53]) that the coefficient W is positive for all values of fields that satisfy (7.3.36–39) and (7.3.42). In other words, these formulas are valid for any boundary values of fields and all z such that $0\leq z$ (under the assumption that $\mathcal{F}=1$).

The substitution of the value $\bar{\xi}$ in (7.3.36–38) enables one to determine the intensities of the Stokes, trigger and generated fields in the state of equilibrium. Let us now find the maximum of the function $\mathcal{J}_{\rm g}=f(\mathcal{J}_{\rm d0},\mathcal{J}_{\rm p0})$ which is defined implicitly by (7.3.38, 39, 42). If $\Lambda=\Lambda_{\rm opt}$, the latter being specified by

$$\begin{aligned}&(1-4s^2q_{\rm g}^2\Lambda_{\rm opt}^2q_{\rm s}^{-1})^{1/2}\\&=s\left(\frac{q_{\rm g}}{q_{\rm d}q_{\rm s}}\right)^{1/2}\sinh 2\left[u\left(0.5-\frac{sq_{\rm g}}{q_{\rm s}}\Lambda_{\rm opt}\right)-u(\xi_0)\right]\quad ,\end{aligned} \tag{7.3.44}$$

then the value of the generated field in the steady state is maximal. Thus, there is a certain value of the trigger field for any input value of the pumping field, such that the transformation coefficient $\bar{\beta}_{\rm p}=\bar{\mathcal{J}}_{\rm g}/\mathcal{J}_{\rm p0}$ is maximal.

For the wave detuning δk differing from zero, the behavior of the generated field is described approximately [up to the accuracy of terms of the order $(\delta k)^2/W^2$] by solution (7.3.38) only for

$$W(\xi,\Lambda)>\delta k\quad , \tag{7.3.45}$$

[12] We would like to recall that the bar above a symbol denotes the steady state values of the corresponding quantities.

see above. It is obvious that the optimal transformation conditions (7.3.44) can be realized if the phase locking ensures field values close to the stationary ones; the necessary condition for this is $W(\overline{\xi}, \Lambda) > \delta k$. Otherwise, there may be a considerable difference between the actual transformation coefficient and the optimal one.

It should be noted that the optimal ratio of the input fields $\mathcal{J}_{d0}/\mathcal{J}_{p0}$ also exists in the case when inequality (7.3.45) is not valid. Indeed, $\lim_{\Lambda \to 0} \mathcal{J}_g = 0$; on the other hand, for $\Lambda \to \infty$ one can demonstrate [7.53] that the condition $W = |\delta k|$ holds when $\xi \to \xi_0$, i.e., the point of the phase locking breakdown corresponds to the pumping field value close to $\mathcal{J}_{p0}$. According to (7.3.39), the quantity $\mathcal{J}_g$ at this point is also close to zero.

7.3.8 Generation of the Sum Frequency During SRS

The result of the integration of the system (7.3.35) when the rhs of the third equation is negative, is that the intensities of interacting fields satisfy the relationships (7.3.36–39) in which the substitution sinh → sin, cosh → cos must be performed. [All formulas obtained from the preceding section by means of this substitution will be referred to as having the corresponding number supplemented by a letter "*a*", for instance, (7.3.39a), (7.3.42a)]. As in the case of the difference frequency generation, these expressions are valid when $W > 0$.

Unlike the difference frequency generation, in the process under consideration, even under the conditions of exact synchronization, the fields can reach the equilibrium state only in a certain region of values of Λ. In order to find this region, one has to construct the integral curves (7.3.38a, 39a), as well as the curves of steady states. Through substitution of (7.3.36a, 39a) into (7.3.41), the latter can be expressed as

$$\overline{\xi}^2 - \overline{\xi} + s^2(\Lambda - q_s q_d q_g^{-1}\overline{\psi})\overline{\psi} = 0 \quad , \tag{7.3.46}$$

where $\overline{\psi} = \overline{\mathcal{J}}_g/M$; the quantities s, ξ, Λ and M are defined, as before, by (7.3.34, 40). According to the result of the construction [7.53], the decisive role is played by the parameter

$$q = \left(\frac{q_g q_d}{q_s}\right)^{1/2} = s\left(\frac{\omega_g \omega_d}{\omega_p \omega_s}\right)^{1/2} \quad . \tag{7.3.47}$$

If $q < 1$, it follows from the comparison of the slopes of the integral curves and the curves of steady states for $\mathcal{J}_p \to \mathcal{J}_{p0}$ that the equilibrium state can be reached when the input fields are governed by the relationship

$$\Lambda = \frac{q_s \mathcal{J}_{d0}}{q_s \mathcal{J}_{p0} + \mathcal{J}_{s0}} < \frac{\omega_s}{s^2 \omega_g} \quad . \tag{7.3.48}$$

We must recall that the second necessary condition for reaching the steady state is, as before

$$W(\overline{\xi},\Lambda)>0 \quad . \tag{7.3.49}$$

When the inequalities (7.3.48, 49) are satisfied simultaneously, the magnitude of the generated field is determined by the solution of (7.3.42a) and (7.3.39a).

If $q>1$, it is possible to demonstrate that $W(\xi,\Lambda)<0$, and the fields cannot reach their equilibrium values.

To conclude this section, we would like to note that, as in the case of the difference frequency generation, for $\delta k\neq 0$ the conditions (7.3.49) and $W=0$ must be replaced by $W(\overline{\xi},\Lambda)>\delta k$ and $W(\overline{\xi},\Lambda)=\delta k$, respectively. It can be shown in a similar manner to that of the preceding section, that for the above process there is always an optimal value of Λ for which the transformation coefficient $\beta_{\rm p}=\mathcal{J}_{\rm g}/\mathcal{J}_{\rm p0}$ is maximal.

7.3.9 Discussion

One of the results of Chap. 6 is the demonstration of the fact that in three-photon RPI's in the presence of exact synchronization for $\mathcal{F}=1$, the fields reach a steady state for which the population difference returns to its equilibrium value. According to the above, the four-photon processes differ in this respect from the three-photon ones. That is, they include such interactions for which there are even for $\delta k=0$ and $\mathcal{F}=1$ no steady states for generated fields. For certain relationships between the field values at the boundary of a nonlinear medium, this situation takes place in the case of sum frequency generation on the basis of the pumping field TPA (see Sect. 7.2). When the sum frequency is generated on the basis of the pumping field SRS, there are regimes for certain relations between the field frequencies and SRS cross sections in which interacting fields never reach an equilibrium state for any ratios of the trigger field and the pumping one.

From the point of view of making the transformation coefficients maximal, it is reasonable to obtain regimes with fields approaching equilibrium. Obviously, under these conditions the transformation coefficient $\beta_{\rm p}=\mathcal{J}_{\rm g}/\mathcal{J}_{\rm p0}$ is maximal and practically independent of the accuracy with which the length of the lasing substance containing cell is specified.

In the general case of the sum frequency generation the ratio of the field boundary values and the cell length must be selected more carefully than for generation of the difference frequency. As far as the latter process is concerned, only the validity of (7.3.45) is required to obtain a regime that is noncritical with respect to the thickness of the lasing substance and corresponds to the maximal transformation coefficient. This can be frequently achieved by the matching of the boundary values of the pumping and the trigger fields. For $q<1$ the inequality (7.3.48) limits the possibility of such a matching for generation of the sum frequency.

As an example, let us calculate the optimal regime parameters for the transformation of radiation of two dye lasers in the IR and UV ranges in sodium vapors.

The experimental generation of the difference frequency was performed in [7.17]. It is possible to obtain [7.54] for $\omega_p = 4.7 \times 10^{15}\,s^{-1}$, $\omega_d = 3.15 \times 10^{15}\,s^{-1}$, $\omega_s = 0.73 \times 10^{15}\,s^{-1}$, $\omega_g = 0.80 \times 10^{15}\,s^{-1}$, under the assumption that only $5p^2P^o$ and $6p^2P^o$ states participate in two-photon transitions, the quantity $\langle\kappa_1^2\rangle \approx 2 \times 10^{-45}$ cgse and $s \approx 0.4$. As follows from the solution of (7.3.44)[13], $\Lambda_{opt} = 1.65$, $\overline{\beta}_p^{opt} = 0.027$ and $\overline{\beta}_d^{opt} = \overline{\mathcal{J}}_g^{opt}/\mathcal{J}_{d0} = 0.072$. For $\delta k = 0.3\ cm^{-1}$, the vapor pressure equal to 15 Torr, $\mathcal{J}_{p0} = 5 \times 10^4$ cgse, the solution of (7.3.45) yields $\mathcal{J}_p^{br} < \overline{\mathcal{J}}_p$. The phase locking under these conditions would ensure that the fields reach a state close to the stationary one, for instance, $\mathcal{J}_g = 0.9\overline{\mathcal{J}}_g$.

The transformation length can be determined from

$$L = \frac{\hbar c^2 k_p}{4\pi T N \eta_0 \omega_p^2 \langle\kappa_1^2\rangle} \int\limits_{\mathcal{J}_{p1}}^{\mathcal{J}_{p0}} \frac{d\mathcal{J}_p}{\mathcal{J}_p(\sqrt{\mathcal{J}_p\mathcal{J}_s} - s\sqrt{\mathcal{J}_g\mathcal{J}_d})} \ , \tag{7.3.50}$$

where the quantities $\mathcal{J}_s$, $\mathcal{J}_g$, $\mathcal{J}_d$ should be replaced by their expressions containing the pumping field intensity $\mathcal{J}_p$. $\mathcal{J}_{p1}$ corresponds to $0.9\overline{\mathcal{J}}_p$ or to the value of $\mathcal{J}_g$ corresponding to the breakdown of phase locking, depending on whether the interacting fields do or do not reach the values close to the stationary ones. In the above example the transformation length computed on the basis of (7.3.50) is 6 cm. For confocal focussing [7.52], the required laser radiation power is $I_p = 1.5$ kW, $I_d = 3.1$ kW.

If the sum frequency generation is excited by radiation with $\lambda_d = 985$ nm, then under the same conditions, tunable radiation can be generated in sodium vapors near $\lambda_g = 319$ nm, with the $7p^2P^o$ state now making the main contribution to $\langle\kappa_2^2\rangle$. Setting $s = 1$, we obtain $q = 1.82 > 1$, see (7.3.47). Since $s^2\omega_d < \omega_p$, the interacting fields do not reach a steady state for any values of $\Lambda = \mathcal{J}_{d0}/\mathcal{J}_{p0}$, even when $\delta k = 0$. If $\mathcal{J}_{p0} = 2 \times 10^5$ cgse, the remaining parameters have the same values and $\Lambda_{opt} \approx 0.2$, the quantity $\mathcal{J}_p^{br} \approx 0.85\,\mathcal{J}_{p0}$ and the transformation coefficients $\beta_d = 0.66$, $\beta_p = 0.13$ are reached over a length $L = 15$ cm.

Let us consider a resonant four-photon process on the basis of SRS in gases, for example, in hydrogen ($\omega_{21} = 7.9 \times 10^{14}\,s^{-1}$) irradiated by neodymium ($\lambda_p = 1060$ nm) and CO_2 ($\lambda_d = 10600$ nm) lasers. When the pressure in the chamber is equal to 75 atm, we have $T = 3.5 \times 10^{-10}$ s and $\langle\kappa_1^2\rangle \approx 2 \times 10^{-50}$ cgse and we assume that for the difference frequency generation $s = 0.5$ and for generation of the sum frequency $s = 1$.

[13] The solution of the transcendental equations and the calculation of the transformation length (7.3.50) were done numerically.

In the case of difference frequency generation ($\lambda_g = 3270\,\mathrm{nm}$) we obtain $\Lambda_{opt} = 2.1$, $\overline{\beta}_d^{opt} \approx 0.07$ and $\overline{\beta}_p^{opt} \approx 0.14$. When $\delta k = 3\,\mathrm{cm}^{-1}$ and $\mathcal{J}_{p0} = 7.5 \times 10^5$ cgse, the transformation length corresponding to $\mathcal{J}_g = 0.9\overline{\mathcal{J}}_g$ is 0.85 cm.

In the case of the generation of the sum frequency ($\lambda_g = 1940\,\mathrm{nm}$), $q = 0.314<1$, and for $\Lambda_{opt} = 0.34$ the interacting fields can reach a state close to the stationary one for which $\beta_p = 0.9\overline{\beta}_p^{opt} = 0.19$ and $\beta_d = 0.9\overline{\beta}_d^{opt} = 0.56$. As follows form (7.3.50), under the assumption that $\delta k = 3\,\mathrm{cm}^{-1}$ and $\mathcal{J}_{p0} = 2 \times 10^6$ cgse, the corresponding transformation length is ≈ 1 cm. Under the conditions of confocal focussing, the power of a CO_2 laser should be $\approx$ 225 kW and $\approx$66 kW for generation of the difference and the sum frequencies, respectively.

Let us demonstrate that in the third equation of (7.3.1), the coefficient in front of $\sin\theta$ does not change sign until the fields reach a steady state, whatever the boundary conditions (7.3.31) are. To this end, W will be expressed through the reduced pumping field intensity ξ and the ratio of the boundary values of fields Λ [see (7.3.43)] by making use of (7.3.36–39):

$$W(\xi,\Lambda) = GM_1\left\{\frac{s}{2}\sqrt{\frac{q_g}{q_s q_d}}\Lambda\frac{1-2\xi}{\sqrt{\xi(1-\xi)}}\sinh 2[u(\xi)-u(\xi_0)]\right.$$
$$\left.\times\sqrt{\xi(1-\xi)q_g q_d/q_s}\,2\coth 2[u(\xi)-u(\xi_0)]\right\} \quad . \tag{7.3.51}$$

At the point of phase locking ($\xi = \xi_0$) the function (7.3.51) tends to infinity, and in the interval $0\le\xi\le\xi_0$ [$\max\xi_0 = 1$, see (7.3.40)] it has a single minimum, with the possibility of the latter taking on negative values only for $1/2<\xi<\xi_0$. Note that within this range

$$\partial W/\partial\xi < 0 \quad . \tag{7.3.52}$$

Further, we shall compute the values W and W'_ξ in the steady state:

$$W(\overline{\xi},\Lambda) = GM_1\left\{1-2\overline{\xi}+s\frac{q_g}{q_s}\Lambda\cosh 2[u(\overline{\xi})-u(\xi_0)]\right\} \quad , \tag{7.3.53}$$

$$\left.\frac{\partial W}{\partial\xi}\right|_{\xi=\overline{\xi}} = \frac{2\sqrt{q_g q_d}GM_1(\Lambda^2-q_s^2/s^2q_g^2)}{\sqrt{q_s}\Lambda^2\sinh^2 2[u(\overline{\xi})-u(\xi_0)]} \quad . \tag{7.3.54}$$

According to (7.3.54) and (7.3.53), when $\Lambda = q_s/sq_g$, the stationary value of W is reached at the point $\xi = \overline{\xi}_1$. It corresponds to the minimum. At this point we have

$$W(\overline{\xi}_1)>0 \quad . \tag{7.3.55}$$

Owing to (7.3.52, 55), for $\Lambda<q_s/sq_g$ the quantity $W(\xi,\Lambda)$ is positive in the interval $1/2\leq\xi\leq\xi_0$ and, therefore, for all ξ.

When $\Lambda>q_s/sq_g$, the minimal value of W can be negative, although, as follows from (7.3.53), in this case $W(\bar{\xi})>0$, and according to (7.3.54)

$$\left.\frac{\partial W}{\partial\xi}\right|_{\xi=\bar{\xi}}>0 \quad .$$

In other words, $W(\xi)>0$ for all $\xi\geq\bar{\xi}$.

Thus, in the range $\bar{\xi}\leq\xi\leq\xi_0$ for any values of Λ the function $W(\xi,\Lambda)$ is positive and, therefore, in the case of the difference frequency generation, (7.3.35) describe the variation of amplitudes of interacting fields for any boundary values $\mathcal{A}_{p0}$, $\mathcal{A}_{d0}$, $\mathcal{A}_{s0}$.

7.4 On Soft Excitation of Stimulated Two-Photon Radiation

Let us discuss another possibility of the application of four-photon interactions, which is associated with the problem of a two-photon laser (TPL) excitation. The ways of constructing such a laser using stimulated two-photon radiation (STPR) were suggested more than two decades ago [7.55–60]. The prospects for TPL are quite promising, its main advantage over the conventional laser being the possibility of frequency tuning in a broad range.

As follows from the design of the TPL single-mode model presented in [7.57, 58], it can be realized in principle. At the same time, the construction of such an instrument faces some difficulties. The main problem is associated with the necessity to obtain a higher density of inverted particles than in ordinary lasers and with the large values of the inversion rate for which quasistationary generation is possible.

The second problem arises because one has to find a method of excitation that would allow the tuning of TPL radiation frequencies from one flash to another. Indeed, in order to excite a TPL, a certain threshold intensity of the external source at the frequencies ω_1 or ω_2 (their sum is equal to the transition frequency) is required, the switching on of the source being followed by the system emitting the field at both frequencies ω_1 and ω_2 [7.57]. In other words, this laser represents an oscillator circuit with a hard mode of excitation. Therefore, if the frequency of the trigger field is fixed, the TPL frequencies can be tuned only during generation. Since the high rate and density of inversion which are necessary for the laser operation are only possible in the pulse regime, this is impracticable.

In order to be able to tune TPL frequencies between flashes, we have to achieve the soft mode of excitation, i.e., to obtain conditions under which the radiation at the frequencies ω_1 and ω_2 starts from the noise level. The authors

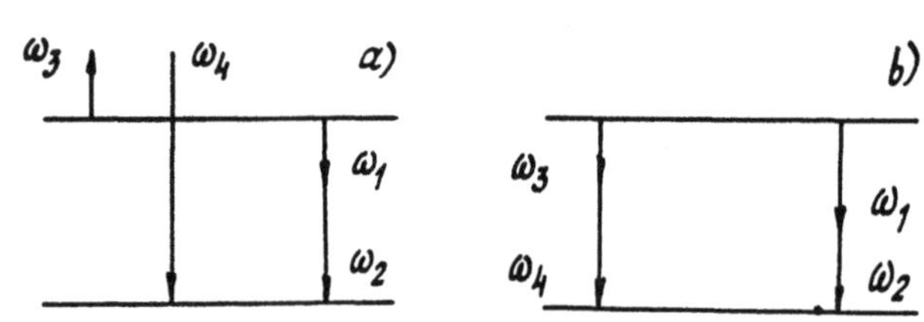

Fig. 7.17a,b. The frequency diagrams of four-photon parametric processes in an inverted medium

of [7.59, 60] suggest for this purpose, the three-photon parametric interaction of fields $E(\omega_1)$, $E(\omega_2)$ with the field at the resonant frequency $\omega_3 = \omega_{21}$ (see Sect. 6.1), the latter being either generated in the system or supplied from the outside. It was established, however, that the efficiency of this method of soft TPL excitation is limited to quite a narrow range of the electrodynamic parameters of the system and the characteristics of matter. In particular, it is required that the ratio of cross sections for single and two-photon transitions between lasing levels has a certain value. No substance with such specific parameters is known.

At present another possibility of constructing TPL with the soft mode of excitation has appeared. It is based on producing inversion in forbidden transitions of iodine [7.61] and oxygen [7.62, 63] atoms during photodissociation. This applies first of all to the transition ${}^2P^{\rm o}_{1/2} - {}^2P^{\rm o}_{3/2}$ in iodine. Because of the high density of excited particles and the pumping rate (which is determined by the rate of photodissociation), as well as the small line width of this transition, the use of iodine vapors as the lasing substance in TPL is quite promising. An important factor here is the possibility of obtaining the soft mode of TPL excitation on the above transition by employing four-photon parametric processes [7.70], the frequency diagram of which is given in Fig. 7.17.

All the properties of TPL that are of interest to us can be studied on the basis of the following simple model:

1. it will be assumed that there are plane single-mode resonators at the frequencies ω_1 and ω_2 with the same qualities,
2. the fields $E(\omega_3)$ and $E(\omega_4)$ will be regarded as given, and the duration of their action will be supposed to be much greater than T_{21}.

If the field $E(\omega_3)$ is produced in a separate volume by means of SRS of the field $E(\omega_4)$ in the noninverted TPL lasing substance for the diagram depicted in Fig. 7.17a, and by means of STPR in the inverted substance for the diagram in Fig. 7.17b, one can obtain a situation close to the above.

The equations that describe the self-consistent variation of the population difference for lasing levels, the amplitudes and phases of the fields $E(\omega_1)$ and $E(\omega_2)$, have the form

$$\frac{d\mathcal{A}_j^2}{dt} + \frac{\mathcal{A}_j^2}{\tau_r}$$
$$= \frac{4\pi T\Phi\omega_j N}{\hbar k_j c}\eta[G(\omega_1,\omega_1)\mathcal{A}_2^2\mathcal{A}_1^2 + G(\omega_1,\omega_3)\mathcal{A}_1\mathcal{A}_2\mathcal{A}_3\mathcal{A}_4 \cos\theta] \quad ,$$

$$\frac{d\eta}{dt} + \frac{\eta-\eta_0}{\tau} = -\eta T\hbar^{-2}[G(\omega_1,\omega_1)\mathcal{A}_1^2\mathcal{A}_2^2 + G(\omega_3,\omega_3)\mathcal{A}_3^2\mathcal{A}_4^2 + 2G(\omega_1,\omega_3)\mathcal{A}_1\mathcal{A}_2\mathcal{A}_3\mathcal{A}_4 \cos\theta],$$

$$\frac{d\theta}{dt} = c(\delta k) - \frac{2\pi T\Phi N G(\omega_1,\omega_3)}{\hbar c}\left[\mathcal{A}_3\mathcal{A}_4\left(\frac{\mathcal{A}_2\omega_1^2}{\mathcal{A}_1 k_1} + \frac{\mathcal{A}_1\omega_2^2}{\mathcal{A}_2 k_2}\right) + \mathcal{A}_1\mathcal{A}_2\left(\frac{\mathcal{A}_4\omega_3^2}{\mathcal{A}_3 k_3} \mp \frac{\mathcal{A}_3\omega_4^2}{\mathcal{A}_4 k_4}\right)\right]\sin\theta \quad . \tag{7.4.1}$$

In these formulas $j = 1, 2$ and τ_r is the field attenuation time in the resonator, Φ is the coefficient of filling the resonator with the lasing substance, τ is the time of relaxation of the population difference to its equilibrium value in the presence of the excitation source [7.57], T is the inverse half-width of the lasing transition line and $G(\omega_\alpha, \omega_\beta)$ are convolutions of products of polarizability tensors $\kappa_{ab}^*(\omega_\alpha)$, $\kappa_{cd}(\omega_\beta)$ [see Sect. 4.1]

$$\theta = (\delta k)ct - (\varphi_3 \mp \varphi_4) + \varphi_1 + \varphi_2 \quad ; \quad \delta k = k_1 + k_2 - (k_3 \mp k_4) \quad ;$$
$$\eta = \sigma_{22} - \sigma_{11} \quad ; \quad \nu = \omega_3 \mp \omega_4 - \omega_{21} = \omega_1 + \omega_2 - \omega_{21} = 0 \quad .$$

In the presence of exact wave synchronization ($\delta k = 0$) the value of the phase difference $\theta = \theta_0 = 0$ is stable. If $\delta k \neq 0$, but the phase locking condition $W > (\delta k)c$ [where W is the coefficient near $\sin\theta$ in the last equation of (7.4.1)] is fulfilled, it can be assumed that $\theta \approx 0$. Even in the absence of this condition we can set $\theta \approx 0$ for the time interval such that $t_{\text{synchr}} \lesssim 1/(\delta k)c$, which corresponds to the approximate realization of the linear synchronism regime. Therefore, it will be supposed that in (7.4.1), $\theta = 0$. Introducing the notation

$$\mathcal{J} = \frac{\omega_2 k_1^{1/2}}{\omega_1 k_2^{1/2}}\mathcal{A}_1^2 = \frac{\omega_1 k_2^{1/2}}{\omega_2 k_1^{1/2}}\mathcal{A}_2^2 \quad , \tag{7.4.2}$$

one can easily cut the number of equations down to two:

$$\frac{d\mathcal{J}}{dt} + \frac{\mathcal{J}}{\tau_r} = g\mathcal{J}\eta[G(\omega_1,\omega_1)\mathcal{J} + G(\omega_1,\omega_3)\mathcal{J}_0] \quad ,$$
$$\frac{d\eta}{dt} + \frac{\eta-\eta_0}{\tau} = -\frac{1}{\mathcal{J}_{\text{sat}}^2\tau}\eta[\mathcal{J} + s\mathcal{J}_0]^2, \tag{7.4.3}$$

where

$$\mathcal{J}_0 = \mathcal{A}_3\mathcal{A}_4 \quad , \quad g = \frac{4\pi T\Phi\omega_1\omega_2 N}{\hbar c(k_1 k_2)^{1/2}} \quad ,$$

$$s = \frac{G(\omega_1,\omega_3)}{G(\omega_1,\omega_1)} \quad , \quad \mathcal{J}_{\rm sat} = \frac{\hbar G^{-1/2}(\omega_3,\omega_3)}{(T\tau)^{1/2}} \quad . \tag{7.4.4}$$

Formulas (7.4.3) have been derived for $G(\omega_1,\omega_3) = G^{1/2}(\omega_1,\omega_1)\cdot G^{1/2}(\omega_3,\omega_3)$, i.e., for $\mathcal{F} = 1$ (see Sect. 4.1).

In dimensionless variables

$$x = \mathcal{J}/\mathcal{J}_{\rm sat} \quad , \quad y = gG(\omega_1,\omega_1)\alpha\mathcal{J}_{\rm sat}\eta \quad ,$$
$$\alpha = \tau_{\rm r}/\tau \quad , \quad \beta = s\mathcal{J}_0/\mathcal{J}_{\rm sat} \quad , \tag{7.4.5}$$

these equations can be written as

$$\alpha\dot{x} = -x + yx(x+\beta) \quad , \quad \dot{y} = y_0 - y - y(x+\beta)^2 \quad , \tag{7.4.6}$$

where the dot means differentiation with respect to time $t_{\rm n} = t/\tau$ and $y_0 = y(\eta = \eta_0)$.

On the basis of the qualitative study of (7.4.6), we arrive at the following results.

If the conditions of TPL excitation are not fulfilled, there is one stable state of equilibrium for system (7.4.6):

$$x_1 = 0 \quad , \quad y_1 = y_0(1+\beta)^{-2} \quad . \tag{7.4.7}$$

The following conditions of TPL excitation are formulated with respect to the values of the reduced population difference y_0 and the trigger fields β:

$$y_0 \geq 2 \quad ,$$
$$\frac{y_0}{2} - \sqrt{\frac{y_0^2}{4} - 1} = \beta_{\rm min} < \beta < \beta_{\rm max} = \frac{y_0}{2} + \sqrt{\frac{y_0^2}{4} - 1} \quad . \tag{7.4.8}$$

If these are met, the state (7.4.7) becomes a saddle point, and the only state of equilibrium which is real from the physical point of view and is characterized by nonzero values of generated fields emerges:

$$x_2 = \frac{y_0}{2} + \sqrt{\frac{y_0^2}{4} - 1} - \beta \quad ; \quad y_2 = \frac{y_0}{2} - \sqrt{\frac{y_0^2}{4} - 1} \quad . \tag{7.4.9}$$

The existence of the upper limit in (7.4.8) is associated with the fact that the trigger fields that are much too strong, sharply decrease the population difference for the lasing levels of the medium, and this prevents the excitation of radiation at the frequencies ω_1 and ω_2. As the equilibrium population difference y_0 increases, the region of possible values of β expands.

If

$$\alpha<y_2^2(1-y_2^2)(1+y_2^2)^{-1} \quad , \tag{7.4.10}$$

and $\beta<\beta_1 = y_2^{-1} + \alpha(1+y_2^2)y_2^{-3}$, the state (7.4.9) is unstable, and there is a stable limiting cycle around it; if $\beta>\beta_1$, this state is stable. Thus, when the trigger fields are low, the four-photon parametric interaction leads to amplitude-modulated radiation at the frequencies ω_1 and ω_2; the increase of β is accompanied by both the modulation depth and the strength of generated fields approaching zero.

Under the condition

$$\alpha>y_2^2(1-y_2^2)(1+y_2^2)^{-1} \quad , \tag{7.4.11}$$

the equilibrium values (7.4.9) become stable immediately after the trigger fields exceed the minimal quantity $\beta>\beta_{\min}$, see (7.4.8).

Consider now the possibility of STPR following the switching off of the trigger fields.

One of the results of [7.57] is that if the reduced energy of fields in the steady state satisfies the inequality

$$x_2^2+1>\alpha^{-1} \quad , \tag{7.4.12}$$

for the hard mode of TPL, then excitation generation continues during the whole time of the action of the pumping field. One can easily show on the basis of (7.4.8, 9) that condition (7.4.12) can be fulfilled only on the presence of (7.4.11); for $\beta = \beta_{\min}$, inequality (7.4.12) holds when

$$y_0^2>\alpha^{-1}+3 \quad . \tag{7.4.13a}$$

If $\tau_r = \tau(\alpha = 1)$, the quantity y_0 defined by (7.4.13a) coincides with the minimal value $y_0 = 2$ necessary for STPR excitation, see (7.4.8).

In the case of the validity of (7.4.10), there are no values of β for which condition (7.4.12) can be fulfilled.

Thus, two essentially different regimes of TPL operation are possible:

1. Generation occurs only when the trigger fields are switched on.
2. STPR at the frequencies ω_1 and ω_2 also continues after the switching off of these fields, as long as there is a pump which inverts the population of lasing levels. The necessary requirement for this is that the parameters of the active medium and the resonator satisfy inequality (7.4.13a).

In rarefied gases this can be achieved very easily in the case of the generation of the degenerate frequencies $\omega_1 = \omega_2 = \omega$ in a standing wave resonator.

In this situation the peculiar features of STPR are associated with the fact that in counter propagating waves, there is a component with the natural width in the two-photon radiation line viewed against the background of the Doppler contour [7.64].

Consider the influence of this circumstance on TPL dynamics. Let us find polarization at the frequency ω which is determined by the formulas of Sect. 4.1. The averaging of $\boldsymbol{P}(\omega)$ with the Maxwell velocity distribution and the neglect of spatial harmonics of polarization leads to an expression for the product $\langle \boldsymbol{P}(\omega)\rangle \boldsymbol{e}_1 \mathcal{A}_1$ which appears in the rhs of (7.4.1):

$$\langle \boldsymbol{P}(\omega)\rangle \boldsymbol{e}_1 \mathcal{A}_1 = \frac{iN\eta}{\hbar}\left[G(\omega_1,\omega_2)\left(\frac{\sqrt{\pi}}{2k_1\overline{v}} + 2T_{21}\right)\mathcal{A}_1^4 + G(\omega_1,\omega_3)\frac{\sqrt{\pi}}{(k_3 \mp k_4)\overline{v}}\mathcal{A}_1^2\mathcal{A}_3\mathcal{A}_4 \cos\theta\right] \quad . \tag{7.4.14}$$

The first term in (7.4.14) describing two-photon radiation contains a component inversely proportional to the uniform width of the lasing transition line. In the second, parametric term, there is only the Doppler line width $(k_3 \mp k_4)\overline{v}/\sqrt{\pi}$, where $\overline{v}$ is the average velocity of molecules, and the minus sign corresponds to the TPL excitation mode depicted in Fig. 7.17a. The term proportional to T_{21} is absent in the parametric part of polarization for an obvious reason: no parametric interaction is possible between counter waves since the wave detuning δk in this case is of the order of k.

On the basis of calculations similar to the above we arrive at the following conditions for soft TPL excitation in mode 2 [instead of (7.4.13a)]:

$$y_0^2 > \frac{\sqrt{\pi}}{4k_1\overline{v}T_{21}}(\alpha^{-1} - 1) + 4 \quad . \tag{7.4.13b}$$

If $\sqrt{\pi}/2k_1\overline{v}T_{21} \ll 1$, in other words, if the line broadening is Doppler, quasistationary generation can be obtained more easily in counter waves than in the travelling wave resonator.

Consider the possibility of the construction and soft excitation of TPL on $^2P^{\mathrm{o}}_{1/2} \rightarrow {}^2P^{\mathrm{o}}_{3/2}$ in iodine.

The iodine atoms in the excited state $^2P^{\mathrm{o}}_{1/2}$ are produced during photodissociation of molecules of RI type, where R is one of the radicals CF_3, iso-C_3F_7, $n - C_3F_7$. Many authors studied this inversion mechanism in connection with the investigation of the iodine laser (see, for instance, [7.61, 65, 66]). In our estimates given below we shall use the values of parameters of the lasing substance obtained in these studies.

According to [7.57], in a three-level system the population difference η and the relaxation time τ are related to the transition probabilities through

$$\eta_0 = \frac{W_{13}W_{32} - W_{21}(W_{31}+W_{32})}{W_{13}W_{32} + W_{21}(W_{31}+W_{32})} \quad ,$$

$$\tau = \frac{W_{31}+W_{32}}{W_{13}W_{32} + W_{21}(W_{31}+W_{32})} \quad . \tag{7.4.15}$$

In the case under consideration, the quantity W_{13} describes the probability of transition from the iodine ground state into the RI states; W_{31} and W_{32} are the probabilities of photodissociation of RI leading to the formation of $I(^2P^o_{3/2})$ and $I(^2P^o_{1/2})$, respectively and W_{21} is the probability of dissociation of the level $^2P^o_{1/2}$ caused by chemical reactions and collisions. As follows from the analysis of the basic processes that accompany photolysis of RI molecules [7.66], $W_{32} \gg W_{31}$, W_{13}. Therefore, it can be assumed that

$$\eta_0 \approx \frac{W_{13} - W_{21}}{W_{13} + W_{21}} \quad , \quad \tau \approx (W_{13} + W_{21})^{-1} \quad . \tag{7.4.16}$$

The quantity W_{13} can be determined through the rate r of reaction $R + I \rightarrow RI$:

$$W_{13} \approx rN\eta_0 \quad . \tag{7.4.17}$$

In [7.66], $r \approx 10^{-10} - 10^{-11}\,\mathrm{cm^3\,s^{-1}}$ was obtained. In our case the quantity η_0 describes quantum photodissociation yield; usually $\eta_0 < 0.2$, so that $\tau \approx W_{13}^{-1}/2$, see (7.4.16).

Let us now perform numerical estimates.

When the buffer gas pressure is 0.4 atm, the lasing transition line width is determined by collisions and constitutes $0.06\,\mathrm{cm^{-1}}$ ($T^{-1} = 1.80 \times 10^9\,\mathrm{s^{-1}}$) [7.67]. Assume that $\omega_1 = \omega_2 = \omega \approx 0.72 \times 10^{15}\,\mathrm{s^{-1}}$, $\phi = 0.5$, $\tau_r \approx 5 \times 10^{-8}\,\mathrm{s}$, $s = 1$ and $G(\omega, \omega) = 1.4 \times 10^{-48}\,\mathrm{cm^6}$ [7.68]. If the excited iodine density $\eta_0 N \approx 10^{17}\,\mathrm{cm^{-3}}$ (this density exists in an iodine laser with the partial pressure of RI being equal to 0.085 atm and $\eta_0 \approx 0.04$), the quantity $\tau \approx 5 \times 10^{-8}$ $-5 \times 10^{-7}\,\mathrm{s}$.

For such values of parameters $y_0 \approx 2.6 - 0.85$, and according to (7.4.13a), when $y_0 = 2.6$, there is a possibility of soft TPL excitation and mode 2. This value of y_0 corresponds to the reduced strength of the driving fields such that $0.4 < \beta < 2.5$. For $\beta = 0.4$ ($S_0 \approx 0.4\, S_{\mathrm{sat}} \simeq 6 \times 10^7\,\mathrm{W/cm^2}$), the quasistationary generation power obtained from one cubic centimeter of the active medium is $\approx 120\,\mathrm{kW}$, and the width of the pulse leading edge $t_{\mathrm{pulse}} \simeq 10^{-8}\,\mathrm{s}$.

When τ reaches its maximal value, the quantity $y_0 \approx 0.85$, and the density of the excited particles of $10^{17}\,\mathrm{cm^{-3}}$ is not sufficient for TPL excitation. If N is increased up to $2.5 \times 10^{17}\,\mathrm{cm^{-3}}$, we obtain $y_0 \approx 2.2$, and there is a possibility of TPL excitation, although this time in mode 1.

The iodine laser can work in the modulated quality regime [7.69] in which the number of excited particles can be much greater than in the quasistationary regime: $N \approx (3\text{–}6) \times 10^{17}\,\mathrm{cm^{-3}}$ [7.68]. If in (7.4.1) we set $\eta = \eta_0$, the condition of TPL excitation in this case follows immediately:

$$\mathcal{J}_0 > \frac{\hbar c\sqrt{k_1 k_2}}{2\pi\omega_1\omega_2 T\tau_r \Phi N\eta_0 G(\omega_1,\omega_3)} \quad . \tag{7.4.18}$$

For the same values of the remaining parameters inequality (7.4.18) yields that it is possible to excite TPL when the trigger field intensity is of the order of $(2-4)\times 10^7\,\mathrm{W/cm^2}$.

Thus, the excitation of a tunable TPL on the transition ${}^2P^{\mathrm{o}}_{1/2} \rightarrow {}^2P^{\mathrm{o}}_{3/2}$ in iodine is apparently feasible.

In conclusion we would like to mention a fundamentally new possibility of realization of multiphoton emission based on the effect of removal of selection rules in a strong light field. An appropriate material for the observation of this phenomenon is the active medium of CO- and CO_2-lasers [7.75]. Among recent studies on the problem of TPL let us mention the papers [7.76, 77].

Instead of a conclusion of Chap. 7 we shall limit ourselves to some comments.

As follows from the results presented in this chapter (Figs. 7.2 and 16 and the analysis of (7.3.25, 45) for $\Lambda = \Lambda_{\mathrm{opt}}$), as well as from [7.72, 74] the RPI based on the pumping TPA considerably differs from the RPI based on SRS, where the Stokes component at the input is zero or small. In the first ones the length of the confinement of the generated wave phase near the value attained in the result of its locking at the input practically coincides with the length of linear synchronism ($\sim\pi|\delta k|^{-1}$). In the processes based on SRS the length of phase confinement is much longer than the length of spatial synchronism. This is connected with the difference in the behavior of the medium nonlinear polarization that is responsible for parametric interaction. In the case of the pumping TPA it is proportional to $\mathcal{A}_{\mathrm{p}}^2$ and is diminished beginning from the input into the medium; under SRS in the pumping channel the corresponding nonlinear polarization proportional to $\mathcal{A}_{\mathrm{p}}\mathcal{A}_{\mathrm{s}}$ turns out to rise exponentially for a sufficiently long distance.

The above-mentioned also concerns the RPI of ultrashort light pulses (UP), the duration of which is much shorter than the typical relaxation times τ and T (see Chap. 1). UP RPI are interesting not only from the point of view of the frequency conversion of such pulses, but may also be considered as a possibility for light pulse formation (particularly as one of the ways of further pulse shortening without any notable energy losses) [7.78–88].

Recently the stationary resonant and quasi-resonant four-photon parametric interactions have been studied extensively, see, e.g., [7.89–115] most of which are of experimental character. The earlier works were discussed in [7.116], and subsequently – in connection with the problems of coherent UV radiation generation and IR radiation visualization – in [7.117–120]. The prominent feature of such investigations is that they increase the variety of applicable pumping sources and nonlinear media. Besides the originally used alkali metal vapors first the ones of the alkali-earth metals, cadmium, and mercury, and the noble gases were established as alternatives to be subsequently supplemented by some molecular gases (ozone, carbon monoxide, nitrogen) and, finally, by the vapors of more complex compounds (organic dyes).

8. Self-Action of Light Beams Caused by Resonant Interaction with the Medium*

One of the most universal manifestations of third-order nonlinear optical interactions in general and resonant interactions in particular is the formation of the so-called nonlinear (or intensity-dependent) refractive index.

This phenomenon, in turn, causes many interesting nonlinear optical effects such as self-focussing, self-defocussing and self-bending of laser beams, four-wave mixing, soliton pulse propagation in optical fibers, recently discovered optical switching at so-called nonlinear interfaces and optical bistability based on mutual self-action of counter-propagating laser beams, etc. Since the resonant nonlinear interactions greatly enhance the nonlinear refractive index, it is of special interest to us to look into the theory and feasible applications of the self-action effects based on it. In this chapter we will consider in detail stationary resonant self-action effects such as self-(de)focussing and self-bending effects, as well as optical switching and bistability at nonlinear interfaces and in counter-propagating laser beams.

8.1 Specific Features and Threshold Characteristics of Self-Focussing in an Absorbing Medium

Resonant interaction of an intense radiation with a medium is not only the cause of nonlinear light absorption and frequency transformations, but also of the appearance of a nonlinear correction ε_{nl} to the dielectric constant which results from the change of the medium state. Nonlinear polarizability leads to nonlinear refraction, which in turn can result under certain conditions in the self-action of the electromagnetic wave, in particular, in such well-studied phenomena as self-focussing and defocussing of narrow light beams.

In [8.1–6] and a number of others (see reviews [8.7–9]), the conditions for light self-focussing and self-defocussing, as well as various effects that arise in the case of ordinary nonlinearity mechanisms (for instance, Kerr-effect, striction, thermal processes), have been considered; we are not going to treat their results in detail here.

The significance of the conditions of resonant self-action follows from the fact that in this situation, there are a number of peculiar features of the dynamics of the process in space which are mainly associated with its competition with light absorption during self-focusing or self-bending [8.10].

It was demonstrated in Chap. 2 that the nonlinear dielectric constant $\varepsilon_{\mathrm{nl}}$ corresponding to resonant absorption may be large. It would be natural to assume that this can be accompanied by self-action. The authors of [8.11, 12], for instance, pointed out the possibility of self-focussing attributed to the change of quantum states of atoms or molecules in the medium. The nonlinear propagation in a two-level system was considered in [8.13], however, not all factors responsible for the formation of $\varepsilon_{\mathrm{nl}}$ were accounted for. The regions of possible observation of resonant focussing in absorbing media were determined in [8.14]. Factors such as "repolarization" and the Stark effect were also taken into account [8.15].

Apparently, this type of self-focussing was seen in resonant absorption of laser radiation by potassium vapor in [8.16], where, for intense radiation with a broad spectrum, the narrowing of the beam for some spectral components and its broadening for others was observed. Pulse self-focussing in the "travelling focus" regime was reported in [8.17]. Cw self-trapping in sodium vapor was observed in [8.18]. Most recent observations of new cw self-action effects attributed to resonant nonlinearities in sodium vapor include optical cw bistability based on cross-self-trapping [8.19] and cw self-bending of asymmetric beams [8.20].

When the quasi-stationary resonant field is sufficiently intense for saturation of the transition, the absorption can be neglected and the phenomena that may arise here are similar to self-action in a transparent medium. Not only absorption, but also nonlinear polarizability become saturated, which leads, as in [8.18], to the formation of a nonlinear waveguide oscillating in space. Here we are interested mainly in the effects that correspond to intensities below the saturating ones [8.14] (hence we will neglect the effect of pseudo-self-focussing attributed strictly to nonlinear absorption [8.21]). The role of absorption under these conditions is as follows. On the one hand, absorption causes the presence of nonlinearity $\varepsilon_{\mathrm{nl}}$ and, consequently, the self-focussing itself. However, on the other hand decreasing the total intensity of the beam along its axis of propagation suppresses these effects. One can assume that in the first approximation, the nonlinear correction to the dielectric constant is proportional to the light intensity, while the absorption is linear. The important issue is the regions of the existence of specific effects in the space of parameters of matter and field (in particular, the frequency detuning from resonance).

First, it is necessary to study the peculiar features of self-action and its threshold characteristics in the presence of absorption in the general case, i.e., regardless of the possible connection between absorption and nonlinearity of polarizability. An experimental analysis of the threshold intensity for self-focussing in absorbing media for a finite cell length can be found in [8.22, 23] (the theoretical analysis in [8.22, 23] was based on finding small corrections to the solution obtained in the absence of absorption). It is essential, however, to find the threshold values of intensity and self-focussing length in an unbounded absorbing medium, taking into account substantial variation of self-focussing dynamics in the focal region as the parameters approach the threshold ones. In particular, under the threshold conditions, the field intensity $\mathcal{J}$ in the collapse zone in an absorbing medium increases to $(z_0 - z)^{-3/2}$, where the quantity $(z_0 - z)$ is the distance from the collapse point [8.14], whereas ordinarily, $\mathcal{J} \propto (z_0 - z)^{-1}$ [8.7].

8.1.1 The Equation for the Beam Radius

In a parabolic approximation [8.7] for linear absorption which is characterized by the wave attenuation length L_a (in terms of power) and when $\varepsilon_{nl} = \varepsilon_2 \mathcal{J}$ (where $\mathcal{J}$ is the field intensity), the nonlinear equations for a beam with cylindrical symmetry which propagates along the z-axis can be written as

$$2\frac{\partial \psi}{\partial z} + \left(\frac{\partial \psi}{\partial \varrho}\right)^2 = (k_0^2 \sqrt{\mathcal{J}})^{-1} \left(\frac{\partial^2 \sqrt{\mathcal{J}}}{\partial \varrho^2} + \frac{1}{\varrho}\frac{\partial \sqrt{\mathcal{J}}}{\partial \varrho}\right) + \frac{\varepsilon_2}{\varepsilon_0}\mathcal{J} \quad ;$$

$$\frac{\partial \mathcal{J}}{\partial z} + \frac{\partial \psi}{\partial \varrho}\frac{\partial \mathcal{J}}{\partial \varrho} + \mathcal{J}\left(\frac{\partial^2 \psi}{\partial \varrho^2} + \frac{1}{\varrho}\frac{\partial \psi}{\partial \varrho}\right) = -\frac{\mathcal{J}}{L_a} \quad ,$$

where k_0 is the wave number in an unperturbed medium, ψ is the eikonal and ϱ is the transverse coordinate.

Assuming, as usual, that the beam has a Gaussian shape $\mathcal{J} \propto \exp[-\varrho^2/b^2(z)]$, where $b(z)$ is the beam radius and the eikonal is quadratic with respect to its cross-section, i.e., that $\partial\psi/\partial\varrho \propto \varrho$, one can easily obtain from this system the following equation for the dynamics of the beam radius $b(z)$:

$$f^3 \frac{d^2 f}{dz^2} = L_d^{-2} - R_{nl}^{-2} \exp(-z/L_a) \quad . \tag{8.1.1}$$

Here, $f = b/b_0$ is the dimensionless beam radius (b_0 being the radius of the beam at the input section), $L_d = k_0 b_0^2$ is the beam diffraction length, and

$$R_{nl}^2 = \frac{\varepsilon_0 b_0^2}{\varepsilon_2 \mathcal{J}_{in}} = \frac{\varepsilon_0 L_d}{\varepsilon_2 k \mathcal{J}_{in}} \quad . \tag{8.1.2}$$

In this formula, $\mathcal{J}_{in}$ is the field intensity at the center of the beam in the input section. The quantity R_{nl} is the self-focussing length of an initially nonconvergent beam in a transparent nonlinear medium, obtained in the approximation of geometrical optics (if $\varepsilon_2 > 0$).

Note that when $\varepsilon_2 > 0$ (i.e., when $R_{nl}^2 > 0$), there is a possibility of self-focussing both in an absorbing and amplifying medium. If $\varepsilon_2 < 0$ ($R_{nl}^2 < 0$), self-defocussing always takes place.

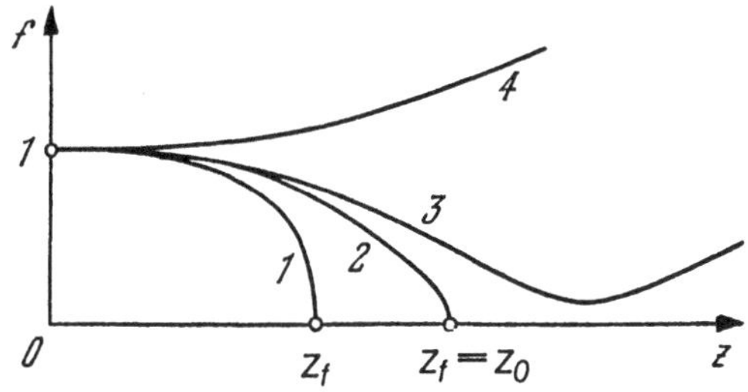

Fig. 8.1. Dependence of the relative beam radius f on coordinate z

All solutions of (8.1.1) can be classified into two groups: those that approach zero for a certain value of $z = z_f > 0$ (Fig. 8.1, curves 1, 2), and those that never take on zero values if $z > 0$ (curves 3, 4). The first type of solution corresponds to self-focussing. Let us obtain the relations that determine those parameters of the beam and the medium (in particular, the incident beam intensity) for which the second type of solution changes into the first one, i.e., the relations that specify the self-focussing threshold.

First, we shall prove the unconditional validity of the inequality:

$$z_f \leq z_0(L_\mathrm{d}, L_\mathrm{a}, R_\mathrm{nl}), \tag{8.1.3}$$

where z_0 is the value of z at which the rhs of (8.1.1) goes to zero:

$$z_0 = L_\mathrm{a} \ln(L_\mathrm{d}/R_\mathrm{nl})^2 \quad . \tag{8.1.4}$$

Inequality (8.1.3) follows from the fact that for $z > z_0$, the solutions of (8.1.1) cannot assume zero and vice versa, for $z < z_0$, there are always some solutions that go to zero. Indeed, in order to obtain solutions of (8.1.1) in a small vicinity of an arbitrary point $z_1 > z_0$, we solve an equation

$$\bar{f}^3 \bar{f}'' = B \quad , \qquad \text{where}$$

$$B = \text{const} = L_\mathrm{d}^{-2} - R_\mathrm{nl}^{-2} \exp(-z_1/L_\mathrm{a}) > 0 \quad ,$$

and arrive at the result that near z_1, any solution of (8.1.1) can be presented as

$$f = \bar{f} + 0(\bar{f}) \quad , \qquad \text{where}$$

$$\bar{f}^2 = \frac{1}{C_1}[B + C_1^2(z + C_2)^2] \quad ,$$

in other words, $f(z_1) \neq 0$. On the other hand, it is possible to show in a similar way that any point $z_2<z_0$ corresponds to certain solutions that vanish at this point. All these solutions have the form

$$f \propto (z_2 - z)^{1/2} + 0(\sqrt{z_2 - z})$$

in a small vicinity of the point z_2. Finally, there are solutions that go to zero for $z_3 = z_0$. In a small vicinity of the point z_0, their behavior is described by

$$f \propto (z_0 - z)^{3/4} + 0[(z_0 - z)^{3/4}] \quad .$$

This proves the validity of inequality (8.1.3).

It follows from the above that the threshold is represented by

$$(z_f)_{\rm th} = (z_0)_{\rm th} \tag{8.1.5}$$

which enables one to derive threshold relations. If it were possible to deduce the general form of the solution of (8.1.1) that vanishes for $z = z_0$ (i.e., a solution that contains only one arbitrary constant) then taking into consideration two boundary conditions at the point $z = 0$ one would be able to obtain yield formulas that determine the quantity z_0, in addition to the above constant. Hence, taking (8.1.4) into account, we would also obtain an equation connecting the threshold intensity $\mathcal{J}_{\rm th}$, the attenuation length $L_{\rm a}$ and the wave parameter of the beam $L_{\rm d}$. But, it is impossible to find the exact solution of (8.1.1) and one should look for an approximate solution. When attenuation is weak ($L_{\rm a} \gg L_{\rm d}$) or, on the contrary, strong ($L_{\rm a} \ll L_{\rm d}$), the form of the solution $f_{\rm th}(z)$ of (8.1.1) and, therefore, the quantity $(z_0)_{\rm th}$, can be calculated quite accurately.

8.1.2 The Threshold for Weak Attenuation

Consider the case of weak attenuation ($L_{\rm a} \gg L_{\rm d}$). To this end, let us first determine the behavior of the "threshold" solution $f_{\rm th}(z)$ of (8.1.1) near the collapse point $(z_f)_{\rm th} = z_0$. Linearization of rhs of (8.1.1) in the vicinity of $z = z_0$ results in the equation

$$f^3 f'' = -(z_0 - z)/L_{\rm a} L_{\rm d}^2 \tag{8.1.6}$$

for the function $f(z)$ in this region. The particular solution of (8.1.6) that goes to zero for $z = z_0$ is

$$f_1(z) = \frac{2}{3^{1/4}} \left[\frac{z_0 - z}{(L_{\rm a} L_{\rm d}^2)^{1/3}} \right]^{3/4} \quad . \tag{8.1.7}$$

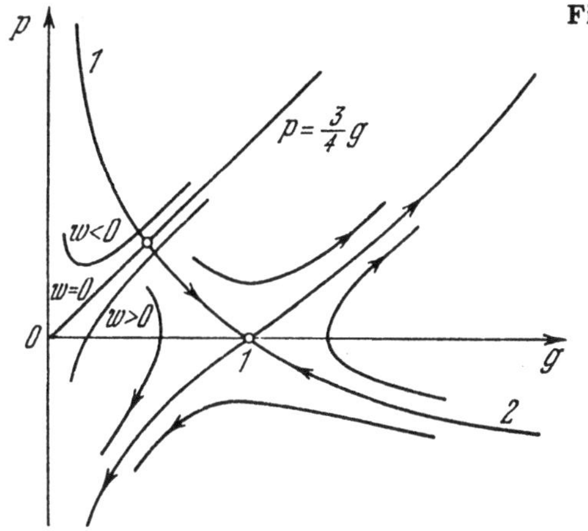

Fig. 8.2. Phase plane for (8.1.9)

Changing variables in (8.1.6), i.e.,

$$\zeta = -\ln\left[\frac{z_0 - z}{(L_a L_d^2)^{1/3}}\right] \quad ; \quad g(\zeta) = f(z)/f_1(z) \quad , \tag{8.1.8}$$

we obtain a homogeneous equation for $g(\zeta)$:

$$16g_\zeta'' - 8g_\zeta' - 3(g - g^{-3}) = 0 \quad . \tag{8.1.9}$$

Introducing the notation $p = dg/d\zeta$, we arrive at the "phase" equation

$$dp/dg = (16p)^{-1}[3(g - g^{-3}) + 8p] \quad . \tag{8.1.10}$$

The singular (saddle) points of (8.1.10) are $p = 0$, $g = \pm 1$ (see Fig. 8.2). Positive solutions of (8.1.6) that exist for $z<z_0$ and vanish when $z \to z_0$ (i.e., exactly the threshold solutions) correspond to a separatrix that enters the singular point $p = 0$, $g = 1$ when $\zeta \to +\infty$ (curves 1 and 2 on Fig. 8.2). For further analysis, it is necessary to find, at least approximately, the shape of this separatrix. According to (8.1.10), if $g \to 0$,

$$p = \frac{\sqrt{3}}{4}[g^{-1} + 0(g^{-1})] \quad ;$$

and for $g \to \infty$,

$$p = -\tfrac{1}{4}[g + 0(g)] \quad .$$

On the other hand, in the vicinity of the saddle point,

$$p = \tfrac{1}{4}(\sqrt{13} - 1)[(1 - g) + 0(1 - g)] \quad .$$

This enables one to construct an approximate solution $p(g)$ for the separatrix, for instance, as

$$p \equiv \frac{dg}{d\zeta} \approx \frac{\sqrt{3}}{4}\frac{1-g^2}{g}\frac{1+qg^2}{1+q\sqrt{3}g^2} \quad ;$$
$$q = \frac{2\sqrt{3}-\sqrt{13}+1}{\sqrt{3}(\sqrt{13}-3)} \quad . \tag{8.1.11}$$

This solution satisfies the above equations near the boundaries and at the center of the region. Integration of (8.1.11) and taking (8.1.8) into account leads to the formula for all threshold solutions $f_{\rm th}(z)$ of (8.1.6):

$$\left(1-\frac{f_{\rm th}^2}{f_1^2}\right)\left(1+q\frac{f_{\rm th}^2}{f_1^2}\right)^{(\sqrt{13}-3)/2} \approx C(z_0-z)^{(\sqrt{13}-1)/4} \quad , \tag{8.1.12}$$

where $C=$ const. For instance, near the singular point, i.e., for $|1-g| \ll 1$, we have

$$f_{\rm th} \approx f_1 + C(z_0-z)^{(\sqrt{13}+2)/4} \quad . \tag{8.1.13}$$

In the case of weak attenuation, the condition $L_{\rm a} \gg L_{\rm d}$ also ensures [see below, (8.1.15)] the validity of inequality $L_{\rm a} \gg z_0$. A consequence of this is that everywhere on the interval $(0, z_0)$, (8.1.6) practically coincides with the initial equation (8.1.1). Therefore, the function $f(z)$ determined by (8.1.12) is in the interval $(0, z_0)$ also close to the threshold solution of (8.1.1). Now, imposing the boundary conditions[1]

$$f_{\rm th}(z=0)=1 \quad , \quad f'_{\rm th}(z=0)=0 \tag{8.1.14}$$

on the solution (8.1.12) and eliminating the quantity C from the resulting equations, we obtain

$$(z_0/L_{\rm a})_{\rm th} \approx (1.24\Gamma)^{2/3} \approx (\tfrac{5}{4}\Gamma)^{2/3} \quad , \quad \text{where}$$
$$\Gamma = L_{\rm d}/L_{\rm a} \quad . \tag{8.1.15}$$

It is useful to recall that in an absorbing medium, z_0 is the maximum possible self-focussing length under any initial conditions. Thus, this phenomenon differs from the self-focussing in a transparent medium by that the threshold power corresponds not to the focussing at infinity, but at a final distance z_0.

[1] For the time being, we are only dealing with the case of a plane incident phase front, when $f'(0)=0$.

Note that on the basis of the known form of the separatrix (curves 1, 2 on Fig. 8.2) and the boundary conditions, one can determine the quantity $(z_0)_{\text{th}}$ without calculating the complete solution (8.1.12) for $f(z)$. Indeed, let us formulate the boundary conditions for f in the general case of a wave which is either convergent or divergent at the input plane:

$$f(z=0)=1 \quad , \quad f'(z=0)=f_0'=-R_{\text{in}}^{-1} \quad . \tag{8.1.16}$$

R_{in} is the radius of the incident wave front curvature at the input; it will be assumed that for a convergent wave, $R_{\text{in}}>0$. According to (8.1.7, 8), we have

$$(f_1 g)_{z=0}=1 \quad , \quad \frac{d}{dz}(f_1 g)_{z=0}=-R_{\text{in}}^{-1} \quad .$$

In terms of the variables $Z=z_0(L_{\text{a}}L_{\text{d}}^2)^{-1/3}$ and $w=-(L_{\text{a}}L_{\text{d}}^2)^{1/3}R_{\text{in}}^{-1}$, we obtain

$$\frac{2}{3^{1/4}}gZ^{3/4}=1 \quad , \quad \frac{3}{4}g+Z\frac{dg}{dZ}=\frac{3^{1/4}}{2}wZ^{1/4} \quad . \tag{8.1.17}$$

Making use of the first equation (8.1.17) to express Z through g and substituting it into the rhs of the second equation, taking into account the fact that by definition (8.1.8)

$$Z\frac{dg}{dZ}=-\frac{dg}{d\zeta}=-p \quad ,$$

we arrive at the relationship between $p(z=0)$ and $g(z=0)$ attributed to the boundary conditions (8.1.16):

$$p=\tfrac{3}{4}[g-(\tfrac{2}{3})^{2/3}wg^{-1/3}] \quad . \tag{8.1.18}$$

Intersection of the curve (8.1.18) with the separatrix (8.1.11) yields the quantity $g(z=0)$ (see Fig. 8.2). Together with the condition $f_1 g_{z=0}=1$ and definition (8.1.7) this leads immediately to the determination of z_0. For a plane wave $[f'(0)=0,\ w=0]$, (8.1.18) corresponds to a straight line $p=3/4g$. The point of its intersection with the separatrix (8.1.11) can be found from a biquadratic equation with respect to g, the solution of which is $g_{z=0}\approx 0.59$. This gives rise to the coefficient close to Γ in (8.1.15).

On the basis of (8.1.15) and the definition (8.1.4), the following value for $(R_{\text{nl}})_{\text{th}}$ in the case $\Gamma\ll 1$ can be obtained:

$$\frac{(R_{\text{nl}})_{\text{th}}^2}{L_{\text{a}}^2}=\Gamma^2\exp\left[-\left(\frac{5}{4}\Gamma\right)^{2/3}\right]\approx\Gamma^2\left[1-\left(\frac{5}{4}\Gamma\right)^{2/3}\right] \quad . \tag{8.1.19}$$

It is obvious that when $\Gamma \to 0$, the threshold value of R_{nl} approaches the corresponding quantity for the transparent medium $(R_{nl})_{th} = L_d$. From (8.1.2, 19) it is possible to derive an expression for the threshold intensity at the center of the incident beam

$$\begin{aligned}\mathcal{J}_{th} &= \frac{\varepsilon_0 L}{\varepsilon_2 k (R_{nl}^2)_{th}} \\ &= \frac{\varepsilon_0 \Gamma^{-1}}{\varepsilon_2 k L_a} \exp\left[\left(\frac{5}{4}\Gamma\right)^{2/3}\right] \approx \frac{\varepsilon_0 \Gamma^{-1}}{\varepsilon_2 k L_a}\left[1 + \left(\frac{5}{4}\Gamma\right)^{2/3}\right] \quad . \end{aligned} \tag{8.1.20}$$

8.1.3 The Threshold for Strong and Intermediate Attenuation

Now consider the case of "strong" attenuation: $\Gamma = L_d/L_a \gg 1$ (for example, for $b_0 = 1\,\mathrm{mm}$, $L_a = 10\,\mathrm{cm}$ and $k = 10^5\,\mathrm{cm}^{-1}$, we have $\Gamma = 10^2$). For the limiting focal length $(z_0)_{th}$, the inequalities $L_a \ll z_0 \ll L_d$ are valid [see below, (8.1.26)]; because of this, on the interval $(0, z_0)$, the geometrical optics approximation can be applied. The term L_d^{-2} in (8.1.1) is insignificant [the quantity L_d will be taken into account only in specifiying z_0 through (8.1.4)]; and it is possible to analyze the propagation with diffraction being neglected until the new diffraction length $(L_d)_{eff} = kb^2 = L_d f^2$ becomes equal to the characteristic dimension L_a. This will happen when $f^2 \sim L_a/L_d = \Gamma^{-1}$. Such an approximate approach shows that at the beginning (i.e., on the interval of the order of attenuation length) the beam size remains almost constant, i.e., $f \approx 1$, and only the second derivative $f''(z)$ exhibits essential variation diminishing in this range down to zero as

$$f'' \approx -R_{nl}^{-2} \exp(-z/L_a) \quad .$$

After that, the function $f(z)$ decreases nearly linearly with a constant slope given by

$$f' \approx \int_0^\infty f''\,dz = -L_a R_{nl}^{-2} \quad .$$

For small values of $f \lesssim \Gamma^{-1/2}$, the behavior of $f(z)$ is described by (8.1.13); over the length of the order L_a the function f diminishes down to zero. Thus, in this case, the ray structure is determined by a relatively thin entrance layer $z \sim L_a$ which plays the role of a phase corrector or a thin lens. Its focal distance is

$$z_f \approx -(f')^{-1} \approx R_{nl}^2 L_a^{-1} \quad . \tag{8.1.21}$$

From (8.1.4, 5), we obtain an equation for $(R_{nl})_{th}$

$$(R_{nl}^2)_{th} L_a^{-2} = \ln[L_d^2 (R_{nl})_{th}^{-2}] \tag{8.1.22}$$

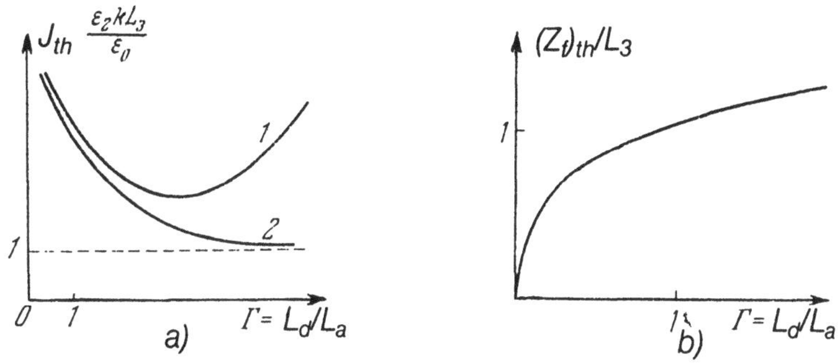

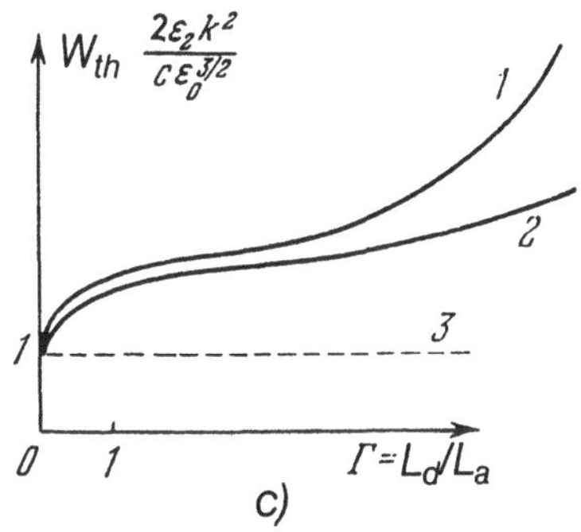

Fig. 8.3a–c. Threshold field intensity in the input section $\mathcal{J}_{\rm th}$ **(a)**, the limiting distance to the collapse point $z_0 = z_{\rm f\,th}$ **(b)**, the threshold beam power in the input section $W_{\rm th}$ **(c)**, as functions of the parameter Γ. (*1*) self-focusing with a collapse, (*2*) "weak" self-focusing

whose solution has the form

$$(R_{\rm nl})^2_{\rm th} L_{\rm a}^{-2} = \ln[\Gamma^2/\ln(\Gamma^2/\ln\ldots)] \quad . \tag{8.1.23}$$

Since $\Gamma^2 \gg 1$, the last formula yields in the first approximation

$$(R_{\rm nl})^2_{\rm th} L_{\rm a}^{-2} \approx \ln \Gamma^2 \quad . \tag{8.1.24}$$

Equations (8.1.2, 24) lead to the following expression for the threshold intensity of the beam:

$$\mathcal{J}_{\rm th} \approx \frac{\varepsilon_0}{\varepsilon_2 k L_{\rm a}} \frac{\Gamma}{\ln \Gamma^2} \quad . \tag{8.1.25}$$

Equation (8.1.4) yields the maximal focal length corresponding to a given Γ:

$$z_0 L_{\rm a}^{-1} = \ln(\Gamma^2/\ln \Gamma^2). \tag{8.1.26}$$

Thus, when $\Gamma \to \infty$, the ratio of the focal length $(z_{\rm f})_{\rm th}$ to the attenuation length $L_{\rm a}$ also becomes infinite.

A more detailed calculation makes it possible to obtain the quantity $\mathcal{J}_{\rm th}$ also at some points where $\Gamma \sim 1$. For instance, for $\Gamma = 3$, $\mathcal{J}_{\rm th} \approx 2\varepsilon_0/\varepsilon_2 k L_{\rm a}$. Combining this result with (8.1.20, 25) for $\mathcal{J}_{\rm th}$ for the near-boundary regions, one can construct an estimate for the interpolation of the function $\mathcal{J}_{\rm th}(\Gamma)$ over the whole domain of its definition (see Fig. 8.3a, curve 1):

$$\mathcal{J}_{\rm th}\frac{\varepsilon_2 k L_{\rm a}}{\varepsilon_0}\approx\frac{1}{\Gamma}\left\{1+\frac{2\Gamma^2}{3\ln\left[1+\frac{2}{3}\left(\frac{4}{5}\Gamma^2\right)^{2/3}\right]}\right\}. \qquad (8.1.27)$$

This result, together with definitions (8.1.4, 2) enables us to calculate the limiting focal length (Fig. 8.3b):

$$z_0 L_{\rm a}^{-1}=\ln\left\{1+\frac{2\Gamma^2}{3\ln\left[1+\frac{2}{3}\left(\frac{4}{5}\Gamma^2\right)^{2/3}\right]}\right\} \quad . \qquad (8.1.28)$$

Obviously, when $\Gamma\ll 1$ and $\Gamma\gg 1$, (8.1.28) turns into (8.1.15, 26), respectively.

According to (8.1.27), in an absorbing medium, the threshold intensity becomes minimal at a certain optimal value of Γ; this fact corresponds to the optimal beam width $b_0=(L_{\rm d}/k)^{1/2}$ for a given attenuation length $L_{\rm a}$. The optimal Γ is

$$\Gamma_{\rm opt}\approx 3 \quad ; \quad [L_{\rm d}\approx 3L_{\rm a}; b_{\rm opt}\approx(3L_{\rm a}/k)^{1/2}] \quad , \qquad (8.1.29)$$

and the minimal value of the threshold intensity is

$$(\mathcal{J}_{\rm th})_{\rm min}\approx 2\varepsilon_0/\varepsilon_2 k L_{\rm a} \quad ; \quad z_{\rm c}=z_0=2L_3 \quad . \qquad (8.1.30)$$

The total intensity of a Gaussian beam is given by

$$W=\frac{c\sqrt{\varepsilon_0}}{2\pi}\int\mathcal{J}ds=\frac{c\sqrt{\varepsilon_0}}{2}b_0^2\mathcal{J}_{\rm in}=\frac{c\sqrt{\varepsilon_0}}{2}\frac{\mathcal{J}_{\rm in}L_{\rm d}}{k} \quad , \qquad (8.1.31)$$

where the integration is performed over the cross-section of the beam. This equation, combined with (8.1.27), allows us to calculate the total threshold power of the beam (Fig. 8.3c, curve 1):

$$W_{\rm th}\frac{2}{c\sqrt{\varepsilon_0}}=\mathcal{J}_{\rm th}\frac{L_{\rm d}}{k}=\frac{\varepsilon_0}{\varepsilon_2 k^2}\left\{1+\frac{2\Gamma^2}{3\ln\left[1+\frac{2}{3}\left(\frac{4}{5}\Gamma^2\right)^{2/3}\right]}\right\} \quad . \qquad (8.1.32)$$

When attenuation is small ($\Gamma\to 0$), (8.1.32) turns into a well-known result, namely, the value of the threshold power for self-focussing in a transparent medium $W_{\rm th}=(2\varepsilon_2 k^2)^{-1}c\varepsilon_0^{3/2}$, which does not depend on the beam radius (curve 3, Fig. 8.3c).

Note another typical specific feature of self-focussing in the presence of absorption. The calculation of the power of the wave $W_{\rm f}$, which reaches the collapse point $z_{\rm f}$ under the threshold conditions (i.e., when $z_{\rm f}=z_0$) leads together with (8.1.4) to

$$W_{\rm f}=W_{\rm th}\exp(-z_{\rm f}/L_{\rm a})=W_{\rm th}(R_{\rm nl}/L_{\rm d})^2 \quad .$$

From this and from (8.1.2, 31), we obtain

$$W_{\mathrm{f}} = \frac{1}{2}\frac{c\varepsilon_0^{3/2}}{\varepsilon_2 k^2} = W_{\mathrm{cr}} \quad . \tag{8.1.33}$$

Thus, in the case of the threshold, it is always the same power W_{f} that reaches the focussing point. This power depends neither on its input value nor on the beam radius nor on attenuation, and equals the critical power of self-focussing in a transparent medium.

8.2 The "Weak" Self-Focussing and Self-Defocussing of a Gaussian Beam in an Absorbing Medium

In an absorbing medium there is a possibility of peculiar "weak" self-focussing [8.14]. In its presence the beam, because of the decrease of its power, reaches the constriction point without collapse and diverges, as in a linear medium (this should be observed for the values of the power below its threshold collapse value).

Consider a situation where the input intensity of the beam is less than the threshold one but greater than the critical self-focussing intensity in a transparent medium:

$$\mathcal{J}_{\mathrm{th}} > \mathcal{J} > \mathcal{J}_{\mathrm{cr}} = \frac{\varepsilon_0}{\varepsilon_2 k L} = \frac{\varepsilon_0}{\varepsilon_2 k L_{\mathrm{a}}}\Gamma^{-1} \tag{8.2.1}$$

In this case, the solution of (8.1.1) does not vanish anywhere, although the beam initially narrows down and diverges afterwards (curve 3 on Fig. 8.1)[2].

We shall call this phenomenon the "weak" self-focussing; it is essentially similar to the external self-focussing [8.24] since it can be regarded as a focussing by a finite nonlinear layer (with a thickness $\sim L_{\mathrm{a}}$). If $L_{\mathrm{d}} \gg L_{\mathrm{a}}$, it is possible to interpret this process as a focussing by a thin lens (see Sect. 8.1.3) with a focal length z_{f} given by (8.1.21). Equation (8.1.1) describes the behavior of the beam at sufficiently large distances ($z_0 \gg L_{\mathrm{a}}$); the exponential term in the rhs of this equation should be set equal to zero. For a plane phase front in the input section of the lens, its solution is

$$f^2(z) \approx \left(1 - \frac{z}{z_{\mathrm{f}}}\right)^2 + \frac{z^2}{L_{\mathrm{d}}^2} \quad . \tag{8.2.2}$$

[2] Note that the solution corresponding to $\mathcal{J}_{\mathrm{cr}}$ defined by (8.2.1) diverges from the very beginning according to a cubic law (curve 4 on Fig. 8.1). This follows directly from (8.1.1) for $R_{\mathrm{nl}}^2 = L_{\mathrm{d}}^2$ and $f'(0) = 0$.

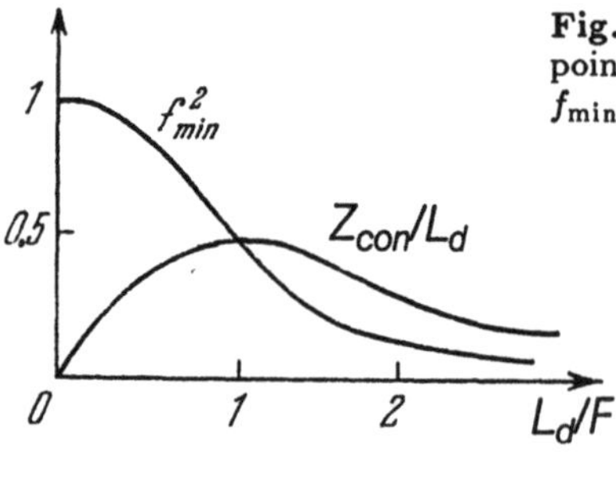

Fig.8.4. Dependence of the distance to constriction point z_{con} and the minimal relative radius of the beam f_{min} with respect to L_{d} on the focal length F

In [8.24] it has been demonstrated that a natural assumption for the threshold value of the weak self-focussing intensity is that at this value, the constriction point [i.e., the point where $f(z)$ is minimum; $f'(z) = 0$] is at the maximum distance from the origin of the coordinate system. Hence, the threshold requirement is

$$z_{\text{f}} = L_{\text{d}} \tag{8.2.3}$$

[in this case, $f^2_{\text{min}} = 0.5$ and $z(f_{\text{min}}) = L_{\text{d}}/2$].

Let us illustrate this with an example of an ordinary linear lens that focusses a thin Gaussian light beam under the condition that diffraction is essential, i.e., when the beam radius b_0 is such that the lens' focal length F is of the same order of magnitude as the diffraction length of the beam $L_{\text{d}} = kb_0^2$. If the lens is sufficiently thin, and the beam has a plane front in the input section, then the behavior of the beam is in the space behind the lens also described by (8.2.2) where $z_{\text{f}} = F$. Equation (8.2.2) enables one to easily calculate the coordinate of the constriction point z_{con} as well as the relative radius at this point f_{min}:

$$\frac{z_{\text{con}}}{L_{\text{d}}} = \frac{L_{\text{d}}}{F}\left(\frac{L_{\text{d}}^2}{F^2}+1\right)^{-1} \quad ; \quad f^2_{\text{min}} = \left(\frac{L_{\text{d}}^2}{F^2}+1\right)^{-1} \quad . \tag{8.2.4}$$

Figure 8.4 depicts the quantities $z_{\text{con}}/L_{\text{d}}$ and f^2_{min} versus of the ratio L_{d}/F. It shows, in particular, that in the beginning (when $F \gg L_{\text{d}}$), the constriction point is hardly noticeable and is located near the lens. Then, with the increase of the lens power F^{-1}, the radius at the constriction point decreases and becomes less than the radius at the input section. Meanwhile the constriction point itself moves further and further from the lens, its coordinate reaches its maximal value $z_{\text{con}} = L_{\text{d}}/2$ exactly at $L_{\text{d}} = F$. After this, the distance between the lens and the constriction point begins to diminish and approaches the value F as $F \to 0$. When this distance is maximal, the transverse beam diameter is equal to $b_0/\sqrt{2}$ and the constriction point becomes quite pronounced. Therefore, the lens' focal length F (such that $F = L_{\text{d}}$) is distinguished from the physics point of view and can naturally be regarded as the threshold length for the observation of focussing. Since this holds for any thin lens, including a nonlinear one, by choosing the equation $z_f = L_d$ and taking into account (8.1.21), we obtain

$$(R_{nl}^2)_{th} = L_a L_d, \tag{8.2.5}$$

or, on the basis of (8.1.2)

$$\mathcal{J}_{th} = \mathcal{J}_\varepsilon \equiv \varepsilon_0/\varepsilon_2 k L_a \quad . \tag{8.2.6}$$

The comparison of this result with (8.1.25) yields that the threshold value for weak self-focussing (curve 2 on Fig. 8.3a) is for $L_d \gg L_a$ much less than the corresponding quantity for self-focussing with a collapse. It is possible to demonstrate that if $L_d < L_a$, these two thresholds almost coincide with each other and with the focussing threshold in a transparent medium (8.2.1). Equations (8.1.31, 8.2.6) determine the total threshold power W of the beam (curve 2 on Fig. 8.3.c).

Let us consider a medium with negative nonlinearity ($\varepsilon_2<0$, $R_{nl}^2<0$) where self-defocussing of the beam occurs, and estimate its angular divergence. According to (8.1.1), the initial monotonic expansion of the beam [under the condition that $f'(0) = 0$] is followed by its propagation in the far region ($z \gg L_d$) with a constant divergence angle, as in a linear medium if $z \gg L_a$.

In a similar way to that discussed in Sect. 1.3, by limiting, for the sake of simplicity, our present discussion to the situation where $L_d \gg L_a$ and the intensity is not very high ($|R_{nl}^2| > L_a$), we arrive at a conclusion that the nonlinear medium acts in this case as a thin defocussing lens with a negative focal length z_f given by (8.1.21), where $R_{nl}^2<0$. Therefore, for $z \gg L_a$, the beam dynamics are described by (8.2.2) with $z_f < 0$. A consequence of this, and also of (8.1.2, 21) is that for $z^2 \gg |z_f|^2$, the total divergence angle of the beam φ_∞ determined by the equation $\tan(\varphi_\infty/2) = b_0 f'(\infty)$ is equal to

$$\varphi_\infty = \varphi_d[1 + (L_d L_a/R_{nl}^2)^2]^{1/2} = \varphi_d[1 + (\mathcal{J}/\mathcal{J}_\varepsilon)^2]^{1/2} \quad . \tag{8.2.7}$$

in this formula, $\varphi_d = 2(kb_0)^{-1} = 2b_0/L_d$ is the beam diffraction divergence angle corresponding to the value b_0 of the input section radius in a linear medium. It is worth noting that under the conditions of weak self-focussing ($\varepsilon_2>0$), the divergence of the beam after it passes the constriction point and expands beyond it, is the same (if $z^2 \gg z_f^2$).

Thus, in both cases the detection of the effect or the measurement of the nonlinearity can be performed by measuring the beam divergence angle in the far region. However, one should take into account that in these two cases the behavior of the beam near the entrance point ($0<z<|z_f|$) is quite different. (The effect of self-focussing with a collapse is excluded here, since the solution of (8.1.1) with $R_{nl}^2>0$, which vanishes for a certain $z = z_f$, cannot be extrapolated if the values of the function f are to remain real. The analysis based on (8.1.1) cannot resolve the question of the formation of a nonlinear channel).

As follows from (8.2.7), under the conditions similar to those for observation of weak self-focussing ($\mathcal{J} > |\mathcal{J}_\varepsilon|$), the nonlinear divergence of the beam is considerably higher than the one due to diffraction. Thus, according to (8.1.30, 8.2.6), in order to observe the self-action of any type in an absorbing medium, the intensity of the wave should be greater than a certain threshold quantity $\mathcal{J}_\varepsilon \sim \varepsilon_0/(|\varepsilon_2|kL_\mathrm{a})$ characteristic of the given medium. The physical meaning of this condition is rather simple. Indeed, from the complete expression for the complex dielectric constant $\varepsilon = \varepsilon_0 + \mathrm{i}\varepsilon' + \varepsilon_\mathrm{nl}$ where $\varepsilon' = \varepsilon_0/kL_\mathrm{a}$ and $\varepsilon_\mathrm{nl} = \varepsilon_2\mathcal{J}$, we obtain that for $\mathcal{J} = \mathcal{J}_\varepsilon$ the relation $|\varepsilon_\mathrm{nl}| = \varepsilon_0/kL_\mathrm{a} = \varepsilon'$ holds. Hence, the self-action in an absorbing medium can only be observed if the absolute value of the nonlinear correction to ε_0 exceeds that of the imaginary part of ε.

8.3 Self-Bending of Trajectories of Asymmetric Light Beams in Nonlinear Media

When the intensity distribution over the cross-section of a light beam in a nonlinear medium is asymmetric, the so called self-bending occurs. In the presence of self-focussing (i.e., the symmetric distribution), the nonlinear refraction tends to collect all rays at the center of the beam. On the other hand, for an asymmetric intensity distribution, a considerable portion of the beam will bend as a whole into the direction of maximal refraction. This phenomenon was predicted in [8.10] and its experimental observation was done by observing the damage tracks in a crystal [8.25]. The dynamics of this effect was studied in [8.26]; a very strong self-bending of beams was observed in glasses [8.27] and during two-photon absorption [8.28]. The recent research includes a proposal to use this effect for resonatorless optical bistability [8.76], experimental indication of a weak self-deflection in liquid crystal [8.75], calculation of the effect in thin nonlinear layers [8.101, 102] and observation of the effect in CS_2 [8.103]. Most recently, the first cw observation of self-bending was done using sodium vapor [8.20]; large deflection angle (up to eight diffraction widths) was recorded using a beam with the total power less than 200 mW.

Consider a plane-phase-front light beam with an asymmetric intensity distribution over its cross-section, which enters a nonlinear medium with a given nonlinearity $n(\mathcal{J}) = \varepsilon^{1/2}(\mathcal{J})$. In the approximation of geometrical optics at each point x of the beam cross-section, the light ray should turn around a certain axis in such a way that the radius of curvature R of the trajectory for a given ray is governed by the well-known relation

$$R^{-1} = n^{-1}(\boldsymbol{l}\,\mathrm{grad}\,n) \quad , \tag{8.3.1}$$

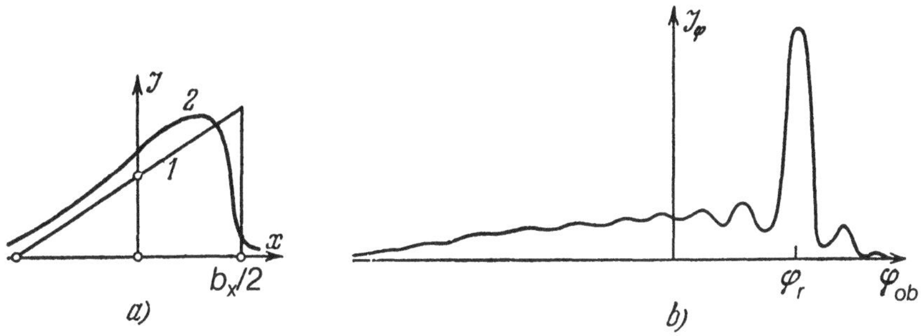

Fig. 8.5a,b. The light intensity distribution (**a**) over the cross-section of an asymmetric beam x and (**b**) over the angle of outgoing radiation φ_{ob}

where $\boldsymbol{l}$ is the unit vector along the ray's principal normal and n is the refractive index of the medium which depends in our case on the field intensity $\mathcal{J}$ at the beam point. The phenomenon of rotation in its purest form arises when in any fixed cross-section of the beam, all its rays rotate around the same axis characteristic for the given cross-section. Then it can be easily shown that

$$\operatorname{grad} n = -\boldsymbol{l}\, dn/dR \quad . \tag{8.3.2}$$

Integration of (8.3.1) and taking (8.3.2) into account leads to the following formula for the refractive index n as a function of the distance R from the center of rotation:

$$n(R) = \mathrm{const}/R \quad , \tag{8.3.3}$$

and the angle of the beam deflection φ_{r} is

$$\varphi_{\mathrm{r}} = \int \frac{ds}{R} \quad . \tag{8.3.4}$$

In this equation, s is the distance from the entrance in the direction of the beam.

Equation (8.3.3) allows one to obtain the field distribution function $\mathcal{J}(R)$ over the beam cross-section on the basis of some specific dependence of n on $\mathcal{J}$, for example, $n = \mathcal{J} n_2 + n_0$. Certainly, if an attempt had been made to create distribution (8.3.3) over the whole space, an infinite light intensity for $R \to 0$ would have been required. In the case of light beams with a limited cross-section, however, because of the condition $n_0 \gg n_2 \mathcal{J}$ that exists in optics, the transverse dimension of the beam is much less than the characteristic dimensions associated with nonlinearity (including the radius of curvature). Therefore (for the time being our discussion is limited to the two-dimensional light beam), when $n = n_0 + n_2 \mathcal{J}$, the optimal intensity distribution $\mathcal{J}(x)$ for the realization of (8.3.3) can be written as (see Fig. 8.5a, curve 1)

$$\mathcal{J}_{\text{opt}}(x) = \mathcal{J}_1 + \frac{1}{2}\Delta\mathcal{J}\left(1 - \frac{2x}{b_x}\right)$$
$$\left(\mathcal{J} = 0 \quad \text{for} \quad |x| > \frac{1}{2}b_x\right) \quad , \tag{8.3.5}$$

where x is the transverse coordinate of the ray increasing in the opposite direction to that of the rotation axis. The quantity b_x is the transverse beam dimension ($b_x \ll |R|$) and $\Delta\mathcal{J}$ is the parameter describing the beam asymmetry. According to (8.3.3), the radius of curvature R of the trajectory for the whole beam is

$$R = \frac{b_x n_0}{n_2 \Delta\mathcal{J}} = \frac{2b_x\varepsilon_0}{\varepsilon_2\Delta\mathcal{J}} \quad (\varepsilon_2 = 2n_2n_0) \quad . \tag{8.3.6}$$

Obviously, (8.3.3, 6) are also valid for the three-dimensional beam, the light intensity in the cross-section of which (where it differs from zero) does not depend on the coordinate in the direction of the rotation axis. If the cross-sectional shape of such a beam is a rectangle with sides b_x and b_y, and the field in it contains only the asymmetric component $\mathcal{J}_1 = 0$, the total power of the beam is equal to $W = (n_0c/8\pi)b_xb_y\Delta\mathcal{J}$ and (8.3.5) assumes the form

$$R = n_0^2cb_x^2b_y/8n_2\pi W \quad . \tag{8.3.7}$$

When $b_x = b_y = 0.5\,\text{mm}$, $W = 10\,\text{MW}$, $n_2 = 9 \times 10^{-12}$ cgse (caron bisulfide), we have $R \approx 300\,\text{cm}$. For the cell length $s = 10\,\text{cm}$, the angular deflection of such a beam from its initial direction is $\varphi_r \approx s/R \approx 2°$.

The angular deflection a beam experiences upon leaving the nonlinear medium has not only technological applications (for instance, scanning of a light beam over the angle through variation of its input power), but can also be used for studying the nonlinearity of various substances by measuring the magnitude of deflection.

Consider now in the geometrical optics approximation a beam which is not plane but focussed by an external lens and has the amplitude profile (8.3.5). As before, the trajectory radius of curvature at each point is determined by (8.3.6), where it should be assumed that

$$\frac{b_x}{b_{x0}} = \frac{b_y}{b_{y0}} = 1 - \frac{s}{F} \quad . \tag{8.3.8}$$

In this formula, b_{x0} and b_{y0} are the transverse dimensions of the beam at the point where it enters the medium and F is the distance between the point of entrance and the focus. Then (8.3.7) acquires the form

$$R = R_0(1 - s/F)^3 \quad , \tag{8.3.9}$$

where $R_0 = (n_0^2 c b_{x0}^2 b_{y0})/8n_2\pi W$ is the radius of curvature at the entrance point. Since $R = ds/d\varphi_r$, the integration of (8.3.9) results in

$$\varphi_r = \varphi_\infty[1 - (1 - s/F)^{-2}] \quad . \tag{8.3.10}$$

Here, $\varphi_\infty = -F/2R_0$ (if it is supposed that for $s = 0$, the deflection angle $\varphi_r = 0$).

The above equations are valid only in the geometrical optics approximation for the intensity profile (8.3.5). In the presence of diffraction one should observe "erosion" of the profile of the beam. In particular, it becomes symmetric, i.e., at large distance the effect is suppressed. Moreover, the fact that the profile in all experiments is to some extent bell-shaped leads to the beam self-focussing (see Fig. 8.5a, curve 2; even for an ideal cut-off at the entrance, this shape is caused by a strong diffraction erosion of the "cut-off" edge). The combined action of these processes limits the beam's maximal angular deflection (φ_{max}) in a sufficiently thick layer and decreases the power of its deflected part. Take $L_{max}^{-2} \approx |R_{nl}^{-2}| + L_d^{-2}$, R_{nl} to be the nonlinear scale unit (8.1.3) and $L_d = kb_0^2$ the diffraction length of the beam. If one assumes then in (8.3.4), the largest possible self-bending length to be $s \sim L_{max}$ it is possible to estimate the maximum deflection angle φ_{max} in the case of a plane phase front incident upon a transparent medium:

$$\varphi_{max} \approx \frac{1}{R_0}\left(\left|\frac{1}{R_{nl}}\right|^2 + \frac{1}{L^2{}_d}\right)^{-1/2} \quad . \tag{8.3.11}$$

In a strong field ($|R_{nl}^2| \ll L_d^2$),

$$\varphi_{max} \approx R_{nl}/R_0 = \frac{1}{2}\left(\frac{|\varepsilon_2|\mathcal{J}_0}{\varepsilon_0}\right)^{1/2} \quad , \tag{8.3.12}$$

and in a weak field ($L_d^2 \ll |R_{nl}^2|$),

$$\varphi_{max} \sim \frac{L_d}{R_0} \approx \frac{2\pi b_0 n_2}{\lambda_0}\Delta\mathcal{J} \approx \frac{16\pi^2 n_2 W}{\lambda_0 n_0^2 c b_0} \quad . \tag{8.3.13}$$

In order to estimate the maximum possible deflection of an initially focussed beam, one has to substitute the quantity

$$s \sim L_{max} \approx (F^{-2} + |R_{nl}^{-2}| + L_d^{-2})^{-1/2}$$

into (8.3.10). In a strong field (subject to the condition $|R_{nl}^2| \ll F^2$), for $L_d^2 \gg F^2$ we obtain the same value of φ_{max} as in the case of an unfocussed beam (8.3.12), and in the weak field where $F^2 \ll |R_{nl}^2|$,

$$\varphi_{\max} \approx 2/R_0 F^3(|R_{\rm nl}^{-2}| + L_{\rm d}^{-2})^2 \quad . \tag{8.3.14}$$

Suppose that a beam having passed through a nonlinear layer with thickness s is observed in the far region. Then the following features of the intensity distribution with respect to the observation angle $\varphi_{\rm ob}$ (see Fig. 8.5b) along the x-axis should emerge: (1) For an unfocussed incident beam whose maximum has a width of the order of the diffraction divergence characterising the whole beam. The main maximum is expected to be shifted by the nonlinear deflection angle

$$\varphi \sim \frac{1}{R_0} \min\{s, |R_{\rm nl}|, L_{\rm d}\} \quad .$$

(2) A broad pedestal should be noticed which is displaced into the opposite direction. The cause of this displacement is the strong nonlinear diffraction of the cut-off edge of the beam. The centroid of the whole distribution

$$\varphi_0 = \int \mathcal{J}_\varphi \varphi_{\rm in} d\varphi_{\rm in} \left(\int \mathcal{J}_\varphi d\varphi_{\rm in} \right)^{-1}$$

should coincide with the angle of inclination of the incident beam; this is a consequence of the law of the field momentum conservation which has been generalised in [8.29] for transparent nonlinear media without dispersion. (Thus, it is only a portion of the total power of the input beam which experiences self-bending. This portion is responsible for a sharp maximum in the angular distribution of the emerging light.)

In the presence of energy losses, especially when they are nonlinear, there may be a significant change in the average angle and the angle of maximum deflection. In particular, they could be strongly enhanced [8.27, 28]. In the case of linear attenuation, the deflection angle decreases. According to (8.3.6), for an incident plane phase front, one can write $R \equiv ds/d\varphi = R_0 \exp(s/L_{\rm a})$. Hence, by virtue of [8.30]

$$\varphi_{\rm r} = \frac{L_{\rm a}}{R_0}\left[1 - \exp\left(-\frac{s}{L_{\rm a}}\right)\right] \quad . \tag{8.3.15}$$

When both diffraction and self-focussing are taken into account, the angle of the beam deflection as it leaves the medium can be estimated to be

$$\varphi_{\rm r} \sim \frac{1}{R_0} \min\{s, L_{\rm a}, |R_{\rm nl}|, L_{\rm d}\} \quad . \tag{8.3.16}$$

In particular, under the condition of strong attenuation ($L_{\rm a} \ll |R_{\rm nl}|, s, L_{\rm d}$), we have

$$\varphi_{\rm r} = L_{\rm a}/R_0 \quad .$$

By setting this quantity equal to φ_d, we arrive at the value of threshold intensity

$$\Delta \mathcal{J}_{\mathrm{th}} \approx \frac{\varepsilon_0}{|\varepsilon_2| k L_{\mathrm{a}}}$$

which is in formal agreement with the characteristic value (8.2.6) corresponding to the threshold of "weak" self-focussing. Therefore, in this case, the real nonlinear correction ($\varepsilon_{\mathrm{nl}}$) to the dielectric constant ε_0 and the imaginary linear one (ε') are also equalised at the threshold.

Note that if the requirement for self-focussing is $\varepsilon_2>0$, for the bending of the beam trajectory, the sign of ε_2 is unimportant. It just determines the direction of deflection (when $\varepsilon_2<0$, the beam is deflected towards the region where the field intensity is minimal). For this reason, nonlinearities of any sign can be used.

8.4 Conditions for the Existence of Self-Action Caused by Resonant Absorption

The threshold characteristics of self-action obtained in Sects. 8.1–8.3 will be applied to the study of regions of self-action in the space of parameters of matter and field in the situation when both the degree of nonlinearity and the field attenuation are caused by resonant absorption [8.14].

According to the results of Chap. 2 the nonlinear correction to the dielectric constant in the presence of resonant absorption is determined by (2.2.20). In a somewhat modified form it reads

$$\varepsilon_{\mathrm{nl}} = 4\pi N \eta_0 \frac{d^2 T}{\hbar} m \frac{q_\kappa[\vartheta(1+\Delta^2) - 1 + \Delta^2] + 2\vartheta\Delta + q_\kappa^2 m \Delta}{(1+\Delta^2)[1 + 2m\vartheta + (\Delta + m q_\kappa)^2]} \quad . \tag{8.4.1}$$

In this formula the dimensionless intensity $m = (d^2T^2/h^2)\mathcal{J}$ has been introduced. The quantity d is the matrix element of the transition dipole moment and T the inverse halfwidth of the line corresponding to this transition. In addition, (8.4.1) contains the following parameters: $\Delta = T(\omega - \omega_{21})$, the dimensionless detuning between the field frequency ω and the frequency of the transition ω_{21}; N for the concentration of resonance particles of matter; η_0 as the equilibrium population difference for resonant levels (in the absence of the field) per particle; $\vartheta = 2\tau/T$. The symbol τ denotes the lifetime of a particle in its excited state; $q_\kappa = \hbar(\kappa_2 - \kappa_1)/d^2T$ is the "repolarizability" parameter for a given transition.

If the field intensity is small compared to the saturating intensity m_{sat} (with a detuning Δ given), i.e., if

$$m < m_{\mathrm{sat}}(\Delta) = (1+\Delta^2)/2\vartheta \quad , \tag{8.4.2}$$

it can be assumed that the quantity $\varepsilon_{\rm nl}$ is quadratic with respect to the field. In this case, $\varepsilon_{\rm nl} \approx \overline{\varepsilon}_2 m$, where

$$\overline{\varepsilon}_2 = 4\pi N \eta_0 \frac{d^2 T}{\hbar} \frac{\vartheta}{1+\Delta^2} \left[q_\kappa + \frac{2\Delta - q_\kappa(1-\Delta^2)/\vartheta}{1+\Delta^2} \right] \tag{8.4.3}$$

($\overline{\varepsilon}_2$ is dimensionless; the simultaneous substitution $\varepsilon_2 \to \overline{\varepsilon}_2$ and $\mathcal{J} \to m$ does not change the preceding results since the above equations contain their product $\varepsilon_2 \mathcal{J} = \overline{\varepsilon}_2 m$).

Condition (8.4.2) also makes it possible to interpret the resonant absorption as being linear and to describe it in terms of an attenuation length $L_{\rm a}$:

$$L_{\rm a} = (1+\Delta^2)\left(4\pi N \eta_0 \frac{k}{\varepsilon_0} \frac{d^2 T}{\hbar}\right)^{-1} \quad . \tag{8.4.4}$$

According to (8.4.1), when condition (8.4.2) for intensity at the center of the input beam is violated, a complex distribution of $\varepsilon_{\rm nl}$ over the cross-section develops: the quantity $\varepsilon_{\rm nl}$ is practically constant at the center and varies only near the beam edges. This leads to strong aberration. Under such conditions, the behavior of rays in a medium becomes very different from what is usually implied by the terms "self-focussing" or "self-bending". Therefore, one can suppose that condition (8.4.2) valid for the intensity $m_{\rm in}$ at the center of the incident beam, determines the upper boundary of the existence of self-action caused by resonant absorption. The lower boundary represents the threshold of weak self-focussing (8.2.6) that coincides with the defocussing threshold for negative ε_2 (see Sect. 8.2) and the threshold for self-bending for any sign of ε_2 (Sect. 8.3). Taking into account saturation when calculating this threshold cannot substantially change the value $\mathcal{J}_{\rm th}$ as defined by (8.2.6). That is because of the fact that in the case of weak focussing (as well as defocussing), both in the focussing layer and in the region of further propagation of the beam, condition (8.4.2) remains valid (this also applies to self-bending of an unfocussed beam). Thus, at the center of a symmetric Gaussian beam or at the edge of an asymmetric beam, the input power should meet the following requirements which are necessary for any of the above types of self-action to be realised:

$$m_{\rm sat}(\Delta) > m_{\rm in}(\Delta) > |m_\varepsilon| , \tag{8.4.5}$$

where the characteristic threshold quantity

$$m_\varepsilon \equiv \frac{\varepsilon_0}{\overline{\varepsilon}_2 k L_{\rm a}} = \left\{ \vartheta \left[q_\kappa \left(1 - \frac{1}{\vartheta} \frac{1-\Delta^2}{1+\Delta^2} \right) + \frac{2\Delta}{1+\Delta^2} \right] \right\}^{-1} \tag{8.4.6}$$

has been introduced. In this expression, $\overline{\varepsilon}_2$ and $L_{\rm a}$ have been substituted from (8.4.3,4). Hence, one can derive conditions for the parameters of material and field frequency that have to be satisfied in order to observe not only self-

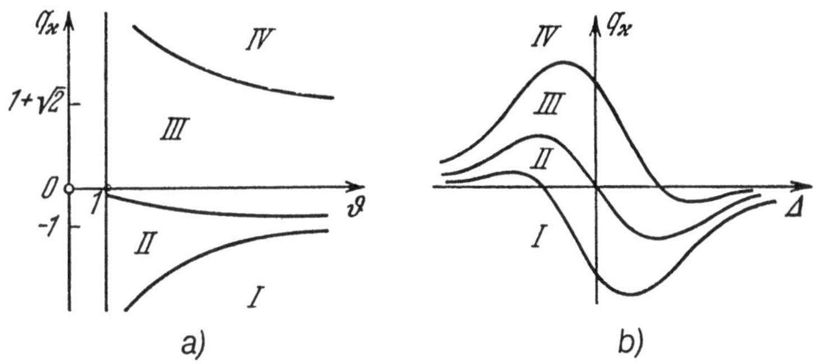

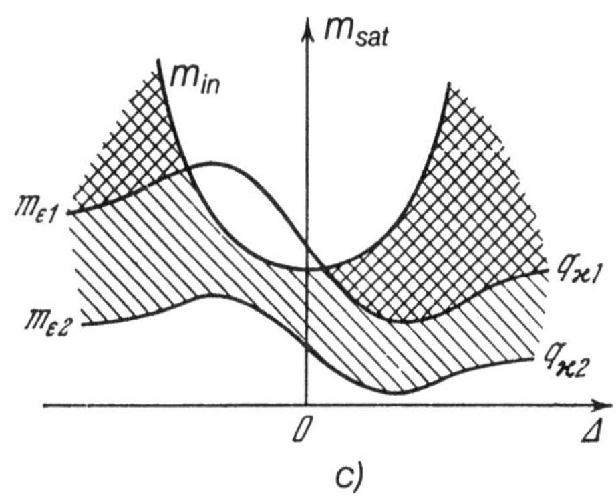

Fig. 8.6a–c. Region of existence of self-focussing (**a**) in the plane of parameters q_κ and ϑ, (**b**) in the plane of parameter q_κ and frequency detuning Δ and (**c**) in the plane of detuning Δ and the field intensity in the input section $m_{\rm in}$ for $\vartheta \gg 1$

focussing and self-defocussing, but also the self-bending of the beam. The first two cases correspond to the inequalities

$$m_{\rm sat}(\Delta) > m_\varepsilon > 0 \quad \text{and} \tag{8.4.7}$$

$$m_{\rm sat}(\Delta) > -m_\varepsilon > 0 \quad , \tag{8.4.8}$$

and the third case to

$$m_{\rm sat}(\Delta) > |m_\varepsilon| \quad . \tag{8.4.9}$$

Let us consider the conditions for self-focussing. Taking into account definition (8.4.2) and the inequality $\vartheta > 0$, let us rewrite (8.4.7) as

$$Q \equiv q_\kappa[1 - \vartheta^{-1} + \Delta^2(1 + \vartheta^{-1})] + 2\Delta > 2 \quad . \tag{8.4.10}$$

We should first point out that the region of parameters q_κ and ϑ for positive nonlinearity ($\varepsilon_2 > 0$) [i.e., $Q > 0$ (for $\eta_0 > 0$) at least in a certain range of detuning] is specified by

$$q_\kappa > -\vartheta(\vartheta^2 - 1)^{-1/2} \quad . \tag{8.4.11}$$

Figure 8.6a depicts this region as II–IV (when $\vartheta \to \infty$, $q_{\rm min} \to -1$). Near the lower edge of region II for the values of detuning close to $\Delta = [(\vartheta - 1)/(\vartheta +$

$1)]^{1/2}$, we have $\varepsilon_2>0$. In region I for any values of detuning, $\varepsilon_2<0$; it can be easily shown that in the region which is a mirror reflection of region I with respect to the ϑ-axis, $\varepsilon_2>0$ for any Δ. As follows from (8.4.10), self-focussing is possible at least in a small range of detuning values where

$$q_\kappa>\vartheta[\vartheta+1-\sqrt{2\vartheta(1+\vartheta)}](\vartheta^2-1)^{-1} \quad . \tag{8.4.12}$$

This corresponds to regions III–IV in Fig. 8.6a. When $\vartheta\to 1$, $q_{\min}\to -1/4$, and when $\vartheta\to\infty$, $q_{\min}\to 1-\sqrt{2}$; in the vicinity of the lower edge of region III,

$$Q>2 \quad \text{for} \quad \Delta=(\vartheta-1)[\sqrt{2\vartheta(\vartheta+1)}-\vartheta-1]^{-1} \quad .$$

If (8.4.10) is valid for any Δ, then there exists a region in which self-focussing is possible at any point of the detuning range[3]:

$$q_\kappa>\vartheta[\vartheta+1+\sqrt{2\vartheta(\vartheta+1)}](\vartheta^2-1)^{-1} \tag{8.4.13}$$

(region IV in Fig. 8.6a; when $\vartheta\to\infty$, $q_{\min}\to 1+\sqrt{2}$).

Thus, if $\vartheta=2\tau/T\gg 1$, it is sufficient for q_κ to exceed $1+\sqrt{2}$ to guarantee that the condition of self-focussing is satisfied for any frequency near the resonance. Then

$$Q\approx 2\Delta+q_\kappa(1+\Delta^2) \quad ,$$

and the regions where self-focussing in the plane q_κ and Δ exist assume the simplest shape. The region where $\varepsilon_2>0$ is defined by $\vartheta_\kappa>-2\Delta/(1+\Delta^2)$ (regions III and IV in Fig. 8.6b; in regions I and II, $\varepsilon_2<0$), and the region of self-focussing by

$$q_\kappa>2(1-\Delta)/(1+\Delta^2) \quad .$$

This is region IV in Fig. 8.6b where $Q>2$.

Since it is not always possible to tune the laser frequency exactly with respect to the center of the absorption line, the question of the width and location of the frequency range corresponding to self-focussing is important. In addition, because of the presence of many radiation modes or discontinuous jumps from one mode to another, it is not in all situations possible to regard the laser as a radiation source with a strictly fixed frequency. This is particularly important in our case, characterized by nonlinearity having strong dispersion within the line width. It is exactly for this reason that the materials fitting into region IV in Fig. 8.6a are of particular interest.

[3] Note that the line shape is assumed to be Lorentzian and for this reason, the accuracy of the above results is limited by those values of Δ for which the Lorentzian curve describes the real line sufficiently well.

In this connection, we would like to mention one specific feature of the range characteristic of the self-focussing threshold in a resonantly absorbing medium which is associated with the presence of linear molecular polarizabilities at the upper and lower levels $q_\kappa \propto \kappa_2 - \kappa_1$. When $q_\kappa > 0$ and the radiation frequency moves away from the resonance ($\Delta^2 \gg 1$), one can see from (8.4.6) that the threshold intensity approaches a constant value

$$m_0 = q_\kappa^{-1}(\vartheta - 1)^{-1} \tag{8.4.14}$$

or, for $\vartheta \gg 1$,

$$m_0 \approx m_\kappa = \frac{1}{q_\kappa \vartheta} = \frac{T^2 d^2}{2\hbar\tau(\kappa_2 - \kappa_1)} = \left(\frac{Td}{\hbar}\right)^2 \mathcal{J}_\kappa \quad . \tag{8.4.15}$$

According to the results of Chap. 2, for values of q_κ and ϑ that are actually possible, the intensity $\mathcal{J}_\kappa$ is rather small [note that with the increase of detuning Δ, the quantity $m_{\text{sat}}(\Delta)$ grows to $1+\Delta^2$; therefore, when $\Delta^2 \gg 1$, the inequality $m_\varepsilon < m_{\text{sat}}(\Delta)$ is valid].

Thus, if $q_\kappa > 0$, self-focussing can always be observed on the slopes of the absorption line at sufficient distances from the center. In this case, its threshold does practically not depend on detuning and is determined by the value of the characteristic intensity $\mathcal{J}_\kappa = |\mathcal{E}_\kappa|^2$. This is illustrated in Fig. 8.6c depicting the region of existence of self-focussing in the plane m_{in} (incident intensity) and detuning Δ under the condition $\vartheta \gg 1$. As follows from (8.4.6), the lower boundary of the region is in this situation defined as

$$m_\varepsilon = \frac{1+\Delta^2}{\vartheta[q_\kappa(1+\Delta^2) + 2\Delta]} \quad .$$

This formula, together with condition (8.4.2), means that the cross-hatched area corresponds to $0 < q_{\kappa 1} < 1$ and the entire shaded portion to $q_{\kappa 2} > 1$.

When $q_\kappa = 0$, the nonlinearity ε_{nl} is only due to the saturation of populations and, according to (8.4.10), self-focussing only occurs on the high-frequency side ($\Delta > 0$). For $\Delta < 0$ defocussing should take place. In this case, the threshold intensity m_ε, no longer approaches a constant value, in spite of its decrease below the saturating value [$m_{\text{sat}} \propto 1 + \Delta^2$, $m_\varepsilon \propto (1+\Delta^2)/\Delta$] as detuning increases. Instead, m_ε increases (for $\Delta^2 \gg 1$) proportional to detuning:

$$m_\varepsilon = \frac{m_{\text{sat}}}{\Delta} = \frac{1+\Delta^2}{2\vartheta\Delta} \quad . \tag{8.4.16}$$

This equation demonstrates that if $q_\kappa = 0$, the minimal value of detuning for which self-focussing is still possible is $\Delta = 1$ (see also Fig. 8.6b). The diagrams in Figs. 8.3, 6 allow to assess the possibility of observing weak or strong self-focussing for given parameters of material and to choose the optimal conditions for experiments designed to detect this phenomenon. For example, assume that parameters q_κ and ϑ correspond to a point in region III on Fig. 8.6a. Then one can find from diagrams similar to those on Fig. 8.6c the region of detuning values Δ (and, therefore, of the radiation frequencies) for which self-focussing can take place, as well as the order of the threshold power. The knowledge of the latter together with Fig. 8.3 allows one to select the beam parameters corresponding to either weak or strong self-focussing.

Consider now the region of self-defocussing and self-bending. As has been already noted in Sect. 2.2, the sign of the nonlinear correction $\varepsilon_{\mathrm{nl}}$ to polarizability varies following the simultaneous change of the signs of q_κ and Δ [this is demonstrated, in particular, by (8.4.1, 3)]. Thus, all the results obtained from (8.4.7) are valid for the region of defocussing determined by the relation (8.4.8), with the inversion of the sign near q_κ and Δ taken into account. In particular, in the plane (q_κ, ϑ) in Fig. 8.6a, the mirror reflection of region IV with respect to the ϑ-axis will play the role of the region where for any value of detuning Δ, self-defocussing should be observed. When $\vartheta \gg 1$, in the plane (q_κ, Δ) in Fig. 8.6b, self-defocussing corresponds to region I; when $q_\kappa < 1$, self-defocussing should occur on both slopes of the line for large values of detuning.

Self-bending exists in the entire region of the self-action of light. The change of the sign of the nonlinearity $\varepsilon_{\mathrm{nl}}$ alters only the sign of the effect: an asymmetric beam in the region of self-focussing will be deflected towards larger amplitudes and in the region of defocussing towards smaller amplitudes.

Let us briefly discuss the amplifying medium (the attenuation is negative, $L_{\mathrm{a}} < 0$). In the case, apparently, there is no threshold, i.e., if $\varepsilon_0 > 0$, the beam in a sufficiently thick layer should be focussed for any input intensity. It is only the sign of ε_2 which determines the occurence of either self-focussing or defocussing and, accordingly, the sign of the self-bending effect. If one takes into account the fact that now in (8.4.3) $\eta_0 < 0$, one arrives at the self-focussing condition in the form $Q < 0$. Hence, we find, for instance, that the region where self-focussing is observed for any values of detuning Δ is region I in Fig. 8.6a; the region of defocussing for any Δ is the mirror reflection of region I with respect to the ϑ-axis. In the intermediate range, some values of detuning correspond to self-focussing and others to self-defocussing. In particular, when $q_\kappa = 0$, self-focussing occurs on all the lower frequencies ($\Delta < 0$) and self-defocussing on all the upper ones ($\Delta > 0$).

The above discussion (Sects. 8.1, 2) has been limited to the spatial dynamics of self-focussing without allowing for saturation. Although in the calculation of threshold characteristics saturation plays a negligible role, when the threshold is greatly exceeded, saturation of both absorption (amplification) and nonlinearity $\varepsilon_{\mathrm{nl}}$ takes place in the vicinity of the beam collapse point as a result of an increase in intensity. Because of this, the beam no longer collapses and a quasi-waveguide channel is formed with quite a weak attenuation of the wave [8.18].

8.5 Self-Action of Light Caused by Stimulated Raman Scattering [8.31]

Let us consider self-action which is caused by one of the two-photon processes, namely, stimulated Raman scattering (SRS). As is well known [8.7, 32], in media with ordinary nonlinearity, self-focussing is frequently accompanied by SRS. Because of the redistribution of level populations, however, SRS can itself give rise to nonlinear polarizability [8.15]. According to the results of Chap. 2, it can be greater than the nonlinearity produced by the Kerr effect. For this reason, it is possible that SRS would induce self-action of light which may be viewed as the reverse process with respect to SRS stimulated by ordinary self-focussing. The conditions for observing SRS self-focussing are discussed in [8.31, 33]. The authors of [8.31] also considered self-bending of a beam with the Stokes component (SRS-refraction). An experimental discovery of the effect that was interpreted as SRS self-focussing of laser radiation can be found in [8.34] (carbon bisulfide); a number of other materials with a large SR cross-section were studied in [8.35–37]. It is interesting to note that self-focussing [8.38, 39] and self-bending [8.28] have also been observed recently for TP-absorption, the other two-photon process.

8.5.1 Formation of a Thin Lens in the Region of SRS-Transformation

In Chap. 2 we presented a derivation of the general expression for polarizability of matter during SRS as a function of fields (the pumping field E_{p} and the Stokes component E_{s}) and frequency detuning $\nu = \omega_{\mathrm{p}} - \omega_{\mathrm{s}} - \omega_{21}$, where ω_{p} and ω_{s} are the frequencies of the pumping field and the Stokes component, and ω_{21} is the frequency of the transition corresponding to SRS. In the approximation of relatively weak fields when population is far from its saturated value and with both the Stark effect and detuning being negligible ($|\nu| \ll T^{-1}$,[4], where T^{-1} is the width of the transition line), one can write the nonlinear correction to the dielectric constant as

$$\varepsilon_{\mathrm{nl}} \approx \varepsilon_4 m_{\mathrm{p}} m_{\mathrm{s}} \quad . \tag{8.5.1}$$

In this expression, m_{p} and m_{s} are dimensionless intensities of the pumping field and the Stokes component normalized with respect to the "saturating" intensity:

$$m_{\mathrm{p,s}} = \mathcal{E}^2_{\mathrm{p,s}}/\mathcal{E}^2_{\mathrm{sat}} \quad ; \quad \mathcal{E}^2_{\mathrm{sat}} = \hbar^2/2\kappa^{12}\sqrt{\tau T} \quad . \tag{8.5.2}$$

[4] This condition is valid if the Stokes component is generated from spontaneous radiation.

κ^{12} is the matrix element describing the SRS cross-section, and

$$\varepsilon_4 = 2\pi N(\kappa_2 - \kappa_1) \quad , \tag{8.5.3}$$

where κ_2, κ_1 are linear polarizabilities of states 2 and 1 and N is the particle density.

As is well known [8.40, 41], the spatial evolution of SRS with the threshold taken into account is described pretty well by

$$\frac{dm_\mathrm{p}}{dz} = -\frac{m_\mathrm{p}m_\mathrm{s}}{L_0} \quad ; \quad \frac{dm_\mathrm{s}}{dz} = \frac{m_\mathrm{p}m_\mathrm{s}}{L_0}\frac{\omega_\mathrm{s}}{\omega_\mathrm{p}} - \frac{m_\mathrm{s}}{L_\mathrm{a}} \quad . \tag{8.5.4}$$

where L_a is the attenuation length for the Stokes component and

$$L_0 = \frac{1}{k_\mathrm{p}}\frac{\sqrt{\varepsilon_0}\hbar}{2\pi N\kappa^{12}}\sqrt{\frac{\tau}{T}} \quad \left(k = \frac{\omega}{c}\right) \tag{8.5.5}$$

is the characteristic length of SRS which is inverse to the coefficient of transformation of the pumping field into the Stokes component. According to (8.5.4), for $m_{\mathrm{s\,in}} \ll m_{\mathrm{p\,in}}$ ($m_{\mathrm{s\,in}}$ and $m_{\mathrm{p\,in}}$ are the Stokes component and the pumping field intensities at the input, respectively), the characteristic length of transformation of the pumping field into the Stokes component is

$$L_\mathrm{SRS} = \left(\frac{m_\mathrm{p}}{L_0} - \frac{\omega_\mathrm{p}}{\omega_\mathrm{s}}\frac{1}{L_\mathrm{a}}\right)^{-1} = \frac{L_0}{m_\mathrm{p} - m_\mathrm{SRS}} \quad , \tag{8.5.6}$$

where m_SRS is the threshold pumping field intensity during SRS:

$$m_\mathrm{SRS} = \frac{L_0}{L_\mathrm{a}}\frac{\omega_\mathrm{p}}{\omega_\mathrm{s}} \quad . \tag{8.5.7}$$

The thickness of the layer in the material in which $\varepsilon_\mathrm{nl} \propto m_\mathrm{p}m_\mathrm{s}$ differs considerably from zero is of the same order of magnitude as the transformation length L_SRS; even when the pumping field intensity slightly exceeds its threshold value, it becomes comparatively small. For instance, in the case of benzene irradiated by a pumping field with the power flux $I \approx 30\,\mathrm{MW/cm^2}$ and having the threshold power flux value $I \approx 18\,\mathrm{MW/cm^2}$, the quantity $L_\mathrm{SRS} \approx 0.1\,\mathrm{cm}$.

In this connection, a nonlinear layer can be regarded as a phase corrector (a thin lens or a prism) formed by the pumping field and the Stokes component waves whose geometry has not yet been distorted. This enables one to use the values of the fields in the initial plane wave in the lens design. Introducing the optical thickness of the lens Δl at each point of the cross-section of the wave x

$$\Delta l(x) = \frac{1}{2\sqrt{\varepsilon_0}}\int_0^\infty \varepsilon_\mathrm{nl}(x,z)dz = \frac{\varepsilon_4}{2\sqrt{\varepsilon_0}}\int_0^\infty m_\mathrm{p}m_\mathrm{s}\,dz \quad , \tag{8.5.8}$$

we obtain directly from (8.5.4):

$$\Delta l = \frac{\varepsilon_4}{2\sqrt{\varepsilon_0}} L_0 (m_{\mathrm{p\,in}} - m_{\mathrm{p}\infty}) \quad . \tag{8.5.9}$$

In this formula, $m_{\mathrm{p}\infty}$ is the output intensity of the pumping field. Making use of (8.5.2, 3, 5) we arrive at

$$\Delta l = \frac{1}{2k_{\mathrm{p}}} \cdot \frac{\mathcal{E}^2_{\mathrm{p\,in}} - \mathcal{E}^2_{\mathrm{p}\infty}}{\mathcal{E}^2_{\kappa}} \quad , \tag{8.5.10}$$

where $\mathcal{E}^2_{\kappa} = \mathcal{J}_{\kappa} = \hbar/2\tau(\kappa_2 - \kappa_1)$ is the characteristic field of resonant repolarization.

In the absence of absorption ($L_{\mathrm{a}} = \infty$), the whole energy of the pumping field is transformed into the Stokes component, i.e., $m_{\mathrm{p}\infty} = 0$; in this case,

$$\Delta l(x) = L_0 \frac{\varepsilon_4}{2\sqrt{\varepsilon_0}} m_{\mathrm{p\,in}}(x) = \frac{1}{2k_{\mathrm{p}}} \frac{\mathcal{E}^2_{\mathrm{p\,in}}}{\mathcal{E}^2_{\kappa}} \quad . \tag{8.5.11}$$

It is worth noting that although the initial nonlinearity (8.5.1) is proportional to $\mathcal{E}^2_{\mathrm{p}}\mathcal{E}^2_{\mathrm{s}}$, in other words, to the fourth power of the field, the optical thickness of the nonlinear layer (8.5.11) only varies proportionally to the pumping field intensity. This is explained by the integral nature of the effect, see (8.5.8). The same factor is responsible for the lack of dependence of Δl either on concentration N or on the value of nonlinearity κ^{12} when there is no SRS threshold ($L_{\mathrm{a}} = \infty$). Indeed, the increase of N and κ^{12} leads not only to the increase of nonlinear polarizability [$\varepsilon_{\mathrm{nl}} \propto N(\kappa^{12})^2$, see (8.5.1, 3)], but also to the decrease in thickness of the interaction layer, $L_{\mathrm{SRS}} \propto [N(\kappa^{12})^2]^{-1}$ [see (8.5.5, 6)]. During integration, these effects compensate each other: $\Delta l \propto \varepsilon_{\mathrm{nl}} L_{\mathrm{SRS}}$. Naturally, the latter result is valid only as long as the wave geometry remains unchanged, i.e., if the quantity L_{SRS} is less than the beam diffraction length. It will be shown below that this condition is automatically fulfilled even for a small excess of pumping intensity over the threshold for SRS self-focussing.

8.5.2 The Threshold of SRS Self-Focussing and Self-Bending

If in (8.5.4) one takes into account the SRS threshold and assumes the usually fulfilled condition $m_{\mathrm{s\,in}} \ll m_{\mathrm{p\,in}}$ then one arrives at the following expression for the quantity $m_{\mathrm{p\,in}} - m_{\mathrm{p}\infty} = \Delta m$ which appears in (8.5.9) for Δl:

$$m_{\mathrm{p\,in}} = \Delta m [1 - \exp(-\Delta m / m_{\mathrm{SRS}})] \quad . \tag{8.5.12}$$

Let us first consider SRS self-focussing corresponding to the zero SRS threshold. For example, it is possible to neglect this threshold when $m_{\mathrm{p\,in}}$

$\gg m_{\mathrm{SRS}}$. This special case demonstrates most clearly the peculiar features of SRS self-focussing. From (8.5.11) it is seen that, as usual, the nonaberrational lens should correspond to the quadratic distribution of the pumping field intensity with respect to the beam cross-section radius:

$$m_{\mathrm{p\,in}}(\varrho) = m_{0\,\mathrm{in}}(1-\varrho^2/b_\varrho^2)$$
$$(m=0 \quad \text{when} \quad |\varrho| \geq b_\varrho) \quad , \tag{8.5.13}$$

where $m_{0\,\mathrm{in}}$ is the intensity of the pumping field at the center of the incident beam. The focal length of the lens is given by

$$z_{\mathrm{f}} = -\left[\frac{d^2(\Delta l)}{d\varrho^2}\right]^{-1} \quad , \tag{8.5.14}$$

or, after the substitution of (8.5.11, 13)

$$z_{\mathrm{f}} = k_{\mathrm{p}} b_\varrho^2 \frac{\mathcal{E}_\kappa^2}{\mathcal{E}_{0\,\mathrm{in}}^2} = L_{\mathrm{d}} \frac{\mathcal{E}_\kappa^2}{\mathcal{E}_{0\,\mathrm{in}}^2} \; (L_{\mathrm{d}} = k b_\varrho^2) \quad . \tag{8.5.15)}$$

Since behind the SRS-lens the medium for the beam propagation is practically linear, the whole situation can be regarded as focussing by a thin nonlinear layer ("external" self-focussing). As has been demonstrated in Sect. 8.2, the threshold power corresponds to $z_{\mathrm{f}} = L_{\mathrm{d}}$, see (8.2.3). Making use of this equation and (8.5.15) we obtain

$$\mathcal{E}_{\mathrm{p\,th}}^2 = \mathcal{E}_\kappa^2 \quad . \tag{8.5.16}$$

Let us now calculate the SRS self-focussing threshold by taking into account the threshold of SRS itself. For the quadratic distribution of the incident pumping field intensity (8.5.13), formulas (8.5.9, 10, 12, 14, 15) yield for a pumping field intensity at the center that exceeds the SRS threshold by less than a factor of 2 ($m_{\mathrm{SRS}} < m_{0\,\mathrm{in}} < 2m_{\mathrm{SRS}}$), a single nonlinear lens with a radius b_1 such that

$$b_1^2 = b_\varrho^2(1 - m_{\mathrm{SRS}}/m_{0\,\mathrm{in}}) \tag{8.5.17}$$

and a new focal length $\overline{z}_{\mathrm{f}} = 0.5 z_{\mathrm{f}}$ [with z_{f} defined by (8.5.15)] is formed. And if the intensity of the pumping field exceeds the double SRS threshold ($m_{0\,\mathrm{in}} > 2m_{\mathrm{SRS}}$) then it yields a second lens with the radius b_2 given by

$$b_2^2 = b_\varrho^2(1 - 2m_{\mathrm{SRS}}/m_{0\,\mathrm{in}}) \tag{8.5.18}$$

and the focal length z_{f} (8.5.15) appears at the center of this lens. As the pumping field intensity increases still further, the area of the first (the ring-shaped) lens decreases, while the area of the second one becomes enhanced approaching that of the pumping beam cross-section. Therefore, from the first "near" focus (at a distance $\bar{z}_f$), the power is transferred to the far one (at a distance z_f). When the major portion of energy passes through one lens, one can introduce the concept of SRS self-focussing. If only one lens (with the focal length $\bar{z}_f$) is formed, we obtain from the relation $k_p b_1^2 = \bar{z}_f$ that

$$\mathcal{E}^2_{\mathrm{p\,th}} = \mathcal{E}^2_{\mathrm{SRS}} + \tfrac{1}{2}\mathcal{E}^2_\kappa \quad . \tag{8.5.19}$$

This situation occurs if

$$\mathcal{E}^2_{\mathrm{p\,th}} > 2\mathcal{E}^2_{\mathrm{SRS}} \tag{8.5.20}$$

which is valid for $\mathcal{E}^2_\kappa < 2\mathcal{E}^2_{\mathrm{SRS}}$. If, on the other hand,

$$\mathcal{E}^2_{\mathrm{p\,th}} \gg 2\mathcal{E}^2_{\mathrm{SRS}} \tag{8.5.21}$$

(an inequality that holds when $\mathcal{E}^2_\kappa \gg 2\mathcal{E}^2_{\mathrm{SRS}}$), the whole nonlinear layer can be presented as a lens with a radius b_2 defined by (8.5.18); in this case, the quantity $\mathcal{E}^2_{\mathrm{p\,th}}$ actually coincides with (8.5.16).

Let us estimate the threshold intensities of SRS self-focussing.

The threshold power flux for SRS in benzene is $I_{\mathrm{SRS}} \approx 18\,\mathrm{MW/cm^2}$, and the quantity $\mathcal{E}^2_\kappa$ corresponds to a flux $I_\kappa \approx 1\,\mathrm{GW/cm^2}$. Therefore, in correspondence with (8.5.16, 21), the threshold for SRS self-focussing is $I_{\mathrm{th}} \approx I_\kappa \approx 1\,\mathrm{GW/cm^2}$. In hydrogen at a pressure of 30 atm, the quantity $I_{\mathrm{SRS}} \approx 12\,\mathrm{MW/cm^2}$ and $\mathcal{E}^2_\kappa$ corresponds to a flux $I_\kappa \approx 75\,\mathrm{kW/cm^2}$. In other words, the threshold is determined by (8.5.19), and the self-focussing should occur almost immediately after the SRS threshold is exceeded (one must take into account the fact that since the lifetime of a hydrogen molecule in an excited vibrational state at 30 atm is $\tau \approx 3 \times 10^{-5}\,\mathrm{s}$, in this case the pumping pulse should be sufficiently long). The lasers operating in a quasi-continuous mode make it feasible to produce light flux densities exceeding the SRS threshold in a gas, and therefore, to observe the effect under consideration.

It is worth noting that in compressed hydrogen the focal length is rather small (for $k_p = 10^5\,\mathrm{cm^{-1}}$, $b_\varrho = 0.3\,\mathrm{cm}$ we have $\bar{z}_f = 16\,\mathrm{cm}$) even when the pumping field intensity is twice as high as the SRS threshold. When $m_{0\,\mathrm{in}}/m_{\mathrm{SRS}} \approx 10$, the quantity $\bar{z}_f \approx 6\,\mathrm{cm}$. This shows that in the case of multistage transformations, it is necessary to take into account, even under ordinary conditions, the effect of SRS self-focussing.

Let us determine the requirements for applying the "thin lens" approximation. It will be assumed that the nonlinear lens thickness $\sim L_{\mathrm{SRS}}$ (8.5.6) is less than the focal length z_f as specified by (8.5.15). Hence, the condition for the input beam radius b_ϱ is

$$b_\varrho^2 > b_{\mathrm{cr}}^2 \equiv \frac{L_0}{k_p m_\kappa} = \frac{\varepsilon_0}{2\pi k_p^2}\,\frac{\hbar^2 \tau(\kappa_2 - \kappa_1)}{N(\kappa^{12})^2 T} \tag{8.5.22}$$

and for the input pumping field intensity at the center of the beam

$$m_{0\,\mathrm{in}} > m_{\mathrm{cr}} \equiv m_{\mathrm{SRS}}/(1 - b_{\mathrm{cr}}^2/b_{\varrho}^2) \quad . \tag{8.5.23}$$

For instance, for benzene, $b_{\mathrm{cr}} \sim 10^{-4}$ cm and for compressed hydrogen, $b_{\mathrm{cr}} \sim 10^{-1}$ cm, so that the inequality (8.5.22) is satisfied. Requirement (8.5.23) is considered to be obeyed everywhere except for the narrow range $m_{\mathrm{th}} < m_{\mathrm{p}} < m_{\mathrm{cr}}$ (if $m_{\mathrm{th}} < m_{\mathrm{cr}}$). Even when the input power slightly exceeds the quantity max $\{m_{\mathrm{th}}, m_{\mathrm{cr}}\}$, the SRS-lens becomes thin.

If the pumping incident intensity is much greater than the SRS threshold, a number of nearly equidistant SRS-lenses may appear. They are formed during the subsequent transformation of some Stokes components into others, higher-order ones. Such a lens system resembles a quasi-optical waveguide capable of channelling narrow light beams.

Let us now discuss the self-refraction of light in SRS [an effect similar to self-bending of asymmetric light beams in ordinary nonlinear media (see Sect. 8.3)]. In our case, this phenomenon will involve the following. If the pumping (incident) field intensity distribution is linear [see (8.3.5)] instead of quadratic [as in (8.5.13)], the SRS-lens is replaced by the "SRS-prism", and this should lead to the deflection of the Stokes component ray by an angle φ_{SRS} with respect to the initial ray. This angle is given by

$$\varphi_{\mathrm{SRS}} = d(\Delta l)/dx \quad . \tag{8.5.24}$$

The substitution of (8.3.5) into (8.5.9) results in

$$\varphi_{\mathrm{SRS}} = \frac{1}{2k_{\mathrm{p}}b_x}\frac{\mathcal{E}_{0\,\mathrm{in}}^2}{\mathcal{E}_{\kappa}^2} = \frac{1}{4}\varphi_{\mathrm{d}}\frac{\mathcal{E}_{0\,\mathrm{in}}^2}{\mathcal{E}_{\kappa}^2} \tag{8.5.25}$$

for $m_{\mathrm{SRS}} = 0$.

When one takes the SRS threshold into account, one arrives at a conclusion that for $m_{\mathrm{SRS}} < m_{0\,\mathrm{in}} < 2m_{\mathrm{SRS}}$, that portion of the beam in which the SRS threshold is exceeded ($m_{\mathrm{p\,in}} > m_{\mathrm{SRS}}$) is deflected by an angle $\overline{\varphi}_{\mathrm{SRS}} = 2\varphi_{\mathrm{SRS}}$, while the "subthreshold" portion ($m_{\mathrm{p\,in}} < m_{\mathrm{SRS}}$) does not change its direction. If $m_{0\,\mathrm{in}} > 2m_{\mathrm{SRS}}$, two deflected beams with the Stokes component are formed; the angle of deflection of one of them ($m_{\mathrm{p\,in}} > 2m_{\mathrm{SRS}}$) is φ_{SRS}, that of the other ($m_{\mathrm{SRS}} < m_{\mathrm{p\,in}} < 2m_{\mathrm{SRS}}$) is $\overline{\varphi}_{\mathrm{SRS}} = 2\varphi_{\mathrm{SRS}}$. Finally, the subthreshold portion of the beam propagates without change, as before. As the pumping incident intensity increases, the whole transformed power becomes concentrated in the first ray.

The experimental observation of SRS self-refraction in hydrogen can be found in [8.42]. The order of magnitude of the pumping field threshold intensity for SRS refraction is the same as that for SRS self-focussing.

8.6 Self-Action Effects at Nonlinear Interface

8.6.1 Nonlinear Properties of Interfaces

Over the past few years, a new direction in nonlinear optics and optical bistability [8.43, 58] emerged that deals with the self-action at nonlinear interfaces [8.44–52] (i.e., interfaces between linear and nonlinear transparent media). In the most studied configuration, light is incident from a linear medium (with susceptibility ε_0) onto a plane surface of the nonlinear medium (Fig. 8.7). The theory [8.44–49] predicted that for a plane wave incident on the interface at a small *grazing* angle the condition [8.44, 45] of proximity of susceptibilities ε_0 and ε_{NL} of both materials results in a strong dependence of the reflectivity of the interface on the intensity of the incident light. Under some conditions it should even become bistable.

The first experimental observation of the new phenomenon was made recently by the authors of the work [8.50–52] who demonstrated the hysteretic behavior of reflection with a "positive" nonlinear (Kerr-effect) medium such as CS_2 ($\varepsilon_2>0$). This experiment confirmed the theoretical prediction [8.44, 45] of an absolute threshold for the onset of a nonlinear switch from the total internal reflection (TIR) into the transmission regime. Under some conditions, a good agreement with the theory [8.44, 45] was also observed for reflectivity of the interface during the entire TIR-transmission-TIR cycle.

The effects of a nonlinear interface should appear in the case of "positive" nonlinearity ($\varepsilon_2>0$) as well as "negative" nonlinearity ($\varepsilon_2<0$). For $\varepsilon_2<0$, the plane-wave theory for incident waves predicts [8.46–48] an excitation of nonlinear waves of a new kind, (described in [8.46–48] as longitudinally inhomogeneous traveling waves). All effects are caused by the competition between linear component (i.e. difference between linear susceptibilities) and nonlinear component. If the difference is sufficiently small, this competition can result in a strong change of reflection even if the nonlinear component is small as well (as it usually is in the optical range). As the intensity of the incident field increases, the penetrated (or evanescent) field in the nonlinear medium changes the refractive index difference at the interface, therefore, at some critical intensity, an abrupt switch for TIR to partial transmission occurs if $\varepsilon_2>0$ and vice versa if $\varepsilon_2<0$.

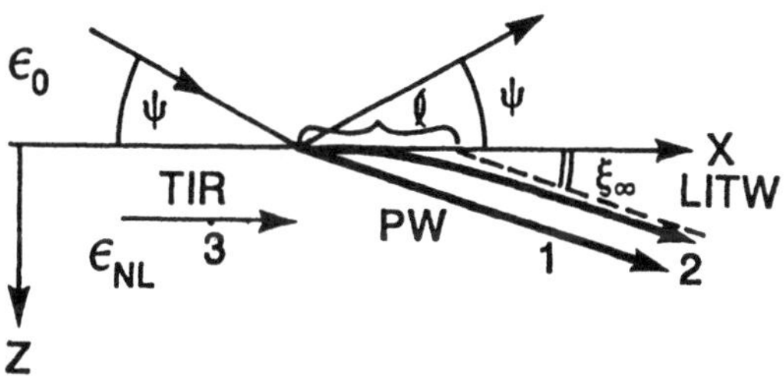

Fig. 8.7. Wave diagram of the nonlinear interface. Ray trajectors in a nonlinear medium: (*1*) traveling plane wave (PW), (*2*) longitudinally inhomogeneous traveling wave (LITW), (*3*) total internal reflection (TIR)

Effects due to nonlinear interface reflection predicted by the plane-wave theory include the following:

a) multistability and hysteresis jumps in the reflection coefficient [8.44–49, 53, 54];
b) change and scanning of the refraction angle and reflection coefficient by varying the intensity of incident light [8.44, 45, 47];
c) total bleaching of the interface by incident light with a certain intensity [8.44, 45, 47].
d) change of penetration depth of the field into a reflecting medium in the regime of total internal reflection (TIR) [8.45];
e) excitation of nonlinear waves of a new kind [8.45–49] (longitudinally inhomogeneous traveling waves-LITW) and effects related to LITW; namely,
f) strong nonlinear self-parallax of refracted rays along the interface [8.47] and
g) self-limitation of the energy flux of the refracted light [8.47, 54].

The conditions mentioned above (small grazing angle and proximity of susceptibilities [8.44–47]) are important for demonstrating the existence of all these effects in nonlinear optics. If these conditions are not met then the phenomenon does not occur. This may be the reason why such effects were not seen earlier e.g., in the work [8.55] discussing the generation of high harmonics due to reflection of light from nonlinear dielectrics or in the work [8.56] which tackled some aspects of nonlinear harmonics in evanescent field in TIR regime.

One of the most interesting opportunities suggested by the plane-wave theory [8.44, 45] and apparently demonstrated in the experiment [8.50–52] is a new kind of optical bistability. The concept of bistability (multistability, in general) implies that the system can support *one* of *two* (or more) stable stationary states under the same external conditions. Usually it correlates with hysteresis which occurs as some of the external parameters of the system vary (e.g., the intensity of incident radiation). This phenomenon is well known in physics, e.g., in the field of phase transitions, in the theory of nonlinear mechanical or electronic oscillators [8.57], etc. In the past years, optical bistability has became a very promising field in nonlinear optics [8.58]. The elements based on optical bistability can be utilized to perform an optically-controlled memory and logic operations in high-speed optical signal processing systems that should be constructed by all-optical devices.

The best known bistable optical devices (which were first proposed in [8.59] and [8.60]) are comprised of a Fabry-Perot interferometer filled with a nonlinear medium. Bistability in these systems was first observed in [8.61]. In these systems, bistability is due to the presence of a resonator (interferometer) which provides a feedback. The media used might have resonant saturated absorption [8.59–61] or Kerr-nonlinearity [8.62]. Use of Fabry-Perot resonators causes these devices to be strongly selective to the frequency of the incident light. In contrast to these devices, a nonlinear interface does not employ any resonators or interferometers. Therefore, no resonant tuning of the laser frequency is required; moreover, a broad spectrum of incident light might be used and a fast switching time is expected. A number of new effects at nonlinear interfaces, which are of considerable physical interest as well as of importance to applications, does not exist in resonators.

New optical switching devices related to nonlinear interfaces have been proposed in [8.49] and [8.52]. Work [8.77] suggested to use nonlinear interfaces as bistable mirrors for self-pulsing lasers. Interesting modifications of the nonlinear interface idea were proposed by the authors of work [8.78] who predicted optical bistability in the frustrated-total-reflection optical cavity and by the authors of work [8.79] who discussed in detail optical bistability of a dielectric cladded thin film near the TIR state. Exploiting surface plasmons at metallic surfaces the authors of [8.80] proposed an enhancement of nonlinear interface effects. A very interesting analogy to a bistable TIR state at nonlinear interfaces was found in [8.81] for bistability of the reflection from metallic diffraction grating (positioned at the nonlinear interface) in the vicinity of the Wood anomaly. One of the most recent trends in the theory and applications of nonlinear interfaces relates to the so called nonlinear waveguides, e.g., dielectric waveguides formed by a linear dielectric layer sandwiched between two nonlinear layers or vice versa (see e.g. [8.82–84] and references therein) which have promising characteristics.

The most comprehensive analytical theory of nonlinear interfaces has been developed for the case of incident plane waves [8.43–49] (which also implies a two-dimensional behavior of all waves in the system). For a Gaussian beam quite a few computer simulations done recently [8.51, 52, 63–65] demonstrated very contradictory results on the main issues of the problem. Some of them [8.63] (and, to some extent [8.65]) endorse the existence of bistability for the Gaussian beam case, while others [8.64] dispute it. On the other hand, [8.51, 65, 66] demonstrated the appearance of nonlinear surface waves; these findings should apparently be revised in the light of the later results [8.67, 68] showing that these surface waves are prohibited under conditions implied in [8.51, 63–65].

The incident-plane-wave theory, though, remains of considerable interest and significance, because it predicts the main features of the phenomenon observed in experiment, and is consistent with the absolute threshold for the onset of nonlinear behavior; it also is of general physical interest by itself, introducing new kinds of waves (such as LITW) and new problems into nonlinear electrodynamics (for instance, a "continuum problem" for LITW); it can be applied to other nonlinear wave problems, for example, in a plasma; it can suggest a new interesting direction of research for Gaussian beams. There are at least several possible experimental situations which can be described by the equations of plane-wave theory, for example, bistability at an electro-optically driven interface [8.69], which has recently been observed in experiment [8.70], and bistable reflection of light at the nonlinear-optical waveguide junction [8.49]. The main results of this theory are described below.

8.6.2 The Main Equations and Conditions

Let a plane wave with the amplitude E_{in} be incident from a linear medium with susceptibility ε_0 at the grazing angle ψ (Fig. 8.7) onto the boundary of a nonlinear medium whose susceptibility depends on the field amplitude E in the medium:

$$\varepsilon_{\text{NL}} = \varepsilon_0 + \Delta\varepsilon_{\text{L}} + \Delta\varepsilon_{\text{NL}}(|E|^2) \quad . \tag{8.6.1}$$

In the simplest case (for instance, Kerr-nonlinearity),

$$\Delta\varepsilon_{\text{NL}} = \varepsilon_2 |E|^2 \tag{8.6.2}$$

holds and $\Delta\varepsilon_{\text{L}}$ does not depend on the field (i.e., there is a linear detuning or mismatching between susceptibilities of both media). Obviously, the most interesting self-action effects might be expected only if the nonlinear component $\Delta\varepsilon_{\text{NL}}$ (even if small) can compete with linear detuning $\Delta\varepsilon_{\text{L}}$. Then the nonlinear component (caused by the field penetrating into the nonlinear medium) could dramatically change the total difference $\Delta\varepsilon = \Delta\varepsilon_{\text{L}} + \Delta\varepsilon_{\text{NL}}$ between the susceptibilities. Such a change, in turn, can result in an abrupt jump of the reflectivity of the interface at some critical incident intensity, e.g., due to the switch from the TIR state into the transmission state. For the same reason, the angle of the incident light has to be almost grazing, because the required magnitude of the grazing angle ψ should be of the order of the critical angle of TIR, which results in the above-mentioned conditions proposed in [8.44, 45]:

$$1 \gg |\Delta\varepsilon_{\text{L}}| \approx |\Delta\varepsilon_{\text{NL}}| \quad ; \quad 1 \gg \psi \approx \left(\frac{\Delta\varepsilon}{\varepsilon_0}\right)^{1/2} \tag{8.6.3}$$

which amounts to $\psi \approx 1°$ at $|\Delta\varepsilon_{\text{NL}}| \approx 10^{-4}$. These conditions also simplify the theory and make the phenomenon independent of the polarization of the incident light.

As usual, the field in a linear medium can be represented in the form of two homogeneous plane waves, one of which is the incident wave and the other is the reflected wave, determined by an unknown complex reflection coefficient r. The wave equation for the complex amplitude of the field E in a nonlinear medium in the two-dimensional case is written as

$$\frac{d^2E}{dz^2} + k_0^2 E\left[\frac{\varepsilon_{\text{NL}}(|E|^2)}{\varepsilon_0} - \cos^2\psi\right] = 0 \quad ; \quad k_0 = \frac{\omega\sqrt{\varepsilon_0}}{c} \quad , \tag{8.6.4}$$

where the z-axis is perpendicular to the boundary [the total field is $(1/2)E \exp(-i\omega t) + \text{c.c}$]. By comparing the tangential components of the fields on both sides of the boundary, the generalized boundary condition for the refracted wave E can be obtained [8.45, 47, 48]:

$$\mathrm{i}\frac{dE(0)}{dz} + k_0 \sin\psi[2E_{\mathrm{in}} - E(0)] = 0 \quad . \tag{8.6.5}$$

An expression for the reflection coefficient r is

$$r = \frac{E(0)}{E_{\mathrm{in}}} - 1 \tag{8.6.6}$$

which also follows from the continuity of the tangential components of the fields at the interface. The boundary conditions are a very important part of any problem related to the nonlinear interface. The failure to define these conditions adequately could result in missing the main effects. For example, inadequate conditions for the refracted plane wave in [8.71] made it impossible to determine for the case $\varepsilon_2{>}0$ refraction angle separately, so that the controversial statements in [8.71] about the existence of hysteresis are not based on the results of [8.71].

The field E in a nonlinear medium is written in the form [8.45–48]

$$E = u(z)\exp\left[\mathrm{i}k_0 \int_0^z \xi(z)dz + \mathrm{i}\phi + \mathrm{i}k_0 x \cos\psi\right] \quad ; \quad (\psi = \mathrm{const}) \quad , \tag{8.6.7}$$

where $u(z)$ and $\xi(z)$ are real quantities, ξ being the angle formed by the rays and the x-axis at a given point z. For plane waves (PW), u and ξ are constant, and for surface waves (SW), when TIR occurs, $\xi = 0$. In general, u and ξ are not constant.

In the theory of reflection from a semi-infinite medium, the condition at infinity must be defined. Since there are no sources inside the nonlinear medium, only traveling waves should propagate *away* from the interface in the nonlinear medium (or SW, which do not carry any energy along the z-axis at all). Therefore, in (8.67) for $z \to \infty$, the following conditions have to be satisfied [8.44–46]:

$$u \to \mathrm{const} \equiv u_\infty \geq 0 \quad , \quad \xi \to \mathrm{const} \equiv \xi \geq 0 \tag{8.6.8}$$

which is, essentially, the Sommerfeld radiation condition (the absence of the backward traveling wave). Substituting the field in the form of (8.67) into the wave equation (8.6.4), we obtain its first integral,

$$I = \xi u^2 = \mathrm{const} \equiv \xi_\infty u_\infty^2 \geq 0 \tag{8.6.9}$$

(which expresses the conservation of the energy flux) as well as the equation for the real amplitude u,

$$\frac{d^2u}{dz^2} + k_0^2 u\left(\frac{\varepsilon_{\mathrm{NL}}(u^2)}{\varepsilon_0} - \cos^2\psi - \frac{I^2}{u^4}\right) = 0 \quad . \tag{8.6.10}$$

The first integral of this equation, which satisfies the radiation condition (8.6.8), can be written in the form [8.46]

$$\left(\frac{du}{dz}\right)^2 = k_0^2 \int\limits_{u_\infty}^{u} [F(u) - F(u_\infty)] d\left(\frac{1}{u^2}\right) \quad , \tag{8.6.11}$$

where the nonlinear "characteristic" function $F(u)$ is introduced:

$$F(u) = u^4 \left[\frac{\varepsilon_{\mathrm{NL}}(u^2)}{\varepsilon_0} - \cos^2\psi\right] . \tag{8.6.12}$$

By integrating (8.6.11) and taking the boundary condition (8.6.5) into account, one can obtain all possible wave solutions for the problem.

8.6.3 Effects at "Positive" Nonlinearity

In the simplest case of "cubic" nonlinearity (8.6.2) with $\varepsilon_2 > 0$ (i.e., Kerr-nonlinearity), only two kinds of one-dimensional wave modes can exist:

a) the homogeneous plane wave (PW), $u(z) = \text{const}$, which corresponds to the transmission regime ($|r|<1$), and

b) the surface wave (SW), or evanescent wave, $\xi = 0$, which corresponds to TIR ($|r| = 1$).

Both of these waves are nonlinear analogs of corresponding linear waves, but differ now from those in their dependence on the incident light intensity. Using (8.6.4–7) with u and ξ constant for the transmission state under conditions (8.6.1), one readily obtains "Snell's nonlinear formula" for the transmission angle ξ [8.44, 45]:

$$\left(1 + \frac{\xi}{\psi}\right)^2 \left(\xi^2 - \psi^2 - \frac{\Delta\varepsilon_{\mathrm{L}}}{\varepsilon_0}\right) = 4\frac{\varepsilon_2}{\varepsilon_0} |E_{\mathrm{in}}^2| \tag{8.6.13}$$

or "Fresnel's nonlinear formula" for the reflection coefficient r,

$$4r\psi^2 + \frac{\Delta\varepsilon_{\mathrm{L}}}{\varepsilon_0}(1+r)^2 + \frac{\varepsilon_2}{\varepsilon_0}|E_{\mathrm{in}}^2|(1+r)^4 = 0 \tag{8.6.14}$$

(see Fig. 8.8). It was shown in [8.48] that in the general case of arbitrary magnitude of nonlinearity $\varepsilon_2|E_{\mathrm{in}}^2|$ and the incident angle ψ, (i.e., in the case when condition (8.6.3) is not satisfied any more), the "Snell's formula" for the angle ψ_{NL} of a refracted plane wave is

$$\left(\frac{\tan^2\psi_{\mathrm{NL}}}{\tan^2\psi} - 1 - \frac{\Delta\varepsilon_{\mathrm{L}}}{\varepsilon_0 \sin^2\psi}\right) = \frac{4\varepsilon_2|E_{\mathrm{in}}^2|}{\varepsilon_0 \sin^2\psi} \tag{8.6.15}$$

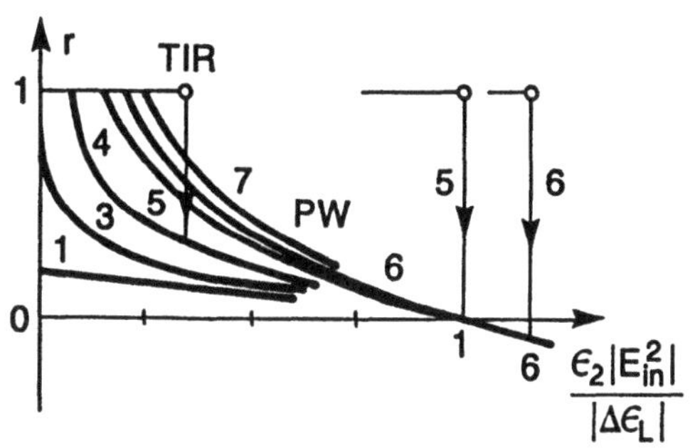

Fig. 8.8. Reflection coefficient r versus the incident light intensity $|E_{in}^2|$ at different glancing angles ψ in the case of negative linear detuning of susceptibilities ($\Delta\varepsilon_L<0$) and positive nonlinearity ($\varepsilon_2>0$). Curves (*1*) $\psi>\psi_{cr}$, (*2*) $\psi=\psi_{cr}$, (*3, 4*) $\psi/\psi_{cr}>1/2\sqrt{2}$, (*5*) $\psi=\psi_{cr}/2\sqrt{2}$ (nonlinear bleaching), (*6, 7*) $\psi<\psi_{cr}/2\sqrt{2}$, see [8.44]

(where now $\tan\psi_{NL}=\xi/\cos\xi$), while the "Fresnel's formula" for the amplitude u of reflected plane wave is

$$4\sin^2\psi\left(1-\frac{u}{|E_{in}|}\right)=\frac{u^2}{\varepsilon_0|E_{in}^2|}(\Delta\varepsilon_L+\varepsilon_2u^2)\quad . \tag{8.6.16}$$

The formula for the reflection coefficient r follows from (8.6.16) via (8.6.6) which again yields (8.6.14) where one has to replace ψ^2 by $\sin^2\psi$.

The nonlinear TIR state can be excited if linear detuning is negative ($\Delta\varepsilon_L<0$) and $\psi<\psi_{cr}$, where $\psi_{cr}=(|\Delta\varepsilon_L|/\varepsilon_0)^{1/2}$ is a critical angle of linear TIR. The profile of wave intensity now is not exponential; it is obtained from (8.6.11) with $u_\infty=0$. In the case of Kerr-nonlinearity (8.6.2), the general equation (8.6.11) under the condition of (8.6.3) can then be written in the form

$$\left(\frac{du}{dz}\right)^2=k_0^2u^2(\psi_{cr}^2-\psi^2-u^2/2)\quad . \tag{8.6.17}$$

Integration of this equation yields the amplitude profile $u(z)$ of a TIR evanescent field (surface wave) which has a "self-channel" shape which is well known from the soliton theory of two-dimensional self-focusing:

$$u=\left(\frac{2\varepsilon_0}{\varepsilon_2}\right)^{1/2}\frac{\gamma}{\cosh(k_0\gamma z+C)}\quad ;\quad \gamma=(\psi_{cr}^2-\psi^2)^{1/2}\quad , \tag{8.6.18}$$

where constant C is determined from the boundary conditions (8.6.5) and can take on one to four values (for different ψ and E_{in}) giving rise to multistability and hysteretic jumps (see Fig. 8.8). Indeed, from the complex boundary conditions (8.6.5), one obtains a condition for the real amplitude $u(0)$ of an evanescent field at the boundary:

$$\left[\frac{du(0)}{dz}\right]^2+k_0^2\psi^2[u^2(0)-4|E_{in}^2|]=0 \tag{8.6.19}$$

as well as a formula for the phases of an evanescent field ϕ and reflected wave ϕ_r [where ϕ_r is defined by the relationship $r=\exp(i\phi_r)$]:

$$\phi = \phi_r/2 = \mp \arccos|2u(0)/E_{in}|, \tag{8.6.20}$$

where the upper sign corresponds to $u'(0)<0$ and the lower to $u'(0)>0$. Now, eliminating $du(0)/dx$ from condition (8.6.20) and from (8.6.17) (taken at $z = 0$), one obtains an amplitude $u(0)$ [8.45]:

$$[u(0)]^2 = \frac{\varepsilon_0}{\varepsilon_2}\left[\psi_{cr}^2 \pm \left(\psi_{cr}^4 - 8\psi^2\frac{\varepsilon_2|E_{in}^2|}{\varepsilon_0}\right)^{1/2}\right] \quad . \tag{8.6.21}$$

Stipulating that the amplitude $u(0)$ (8.6.21) and the phase ϕ (8.6.20) be real quantities, one obtains the condition for the nonlinear TIR to occur:

$$0 \leq u^2(0) \leq 4|E_{in}^2| \quad ; \quad 8\psi^2\varepsilon_2|E_{in}^2| \geq \psi_{cr}^4\varepsilon_0 \quad . \tag{8.6.22}$$

Based on these conditions, one shows that the threshold grazing angle ψ_{th}, at which the switch from nonlinear TIR to partial transmission occurs, is given by [8.44, 45]

$$(\psi_{th}/\psi_{cr})^2 = \begin{cases} 1 - 2|\varepsilon_2 E_{in}^2/\Delta\varepsilon_L & ; \quad 4|\varepsilon_2 E_{in}^2| < |\Delta\varepsilon_L| \\ \frac{1}{8}\Delta\varepsilon_L/\varepsilon_2 E_{in}^2|; & \quad 4|\varepsilon_2 E_{in}^2| > |\Delta\varepsilon_L| . \end{cases} \tag{8.6.23}$$

Therefore, for a given grazing angle ψ, the threshold input intensity $|E_{th}^2 \ll$, at which such a switch occurs, is given by

$$\left|\frac{\varepsilon_2 E_{th}^2}{\Delta\varepsilon_L}\right|_1 = \begin{cases} \frac{1}{2}\left[\left(1 - \frac{\psi}{\psi_{cr}}\right)^2\right] & 1/\sqrt{2} \leq \psi/\psi_{cr} < 1 \\ \frac{1}{8}(\psi_{cr}/\psi)^2 \quad ; & 0 < \psi/\psi_{cr} \leq 1/\sqrt{2} \quad . \end{cases} \tag{8.6.24}$$

Thus, starting from TIR at the low input intensity and gradually increasing the intensity, the interface switches from the TIR state with $|r| = 1$ to the partial transmission state with $|r| \leq 1$, see Fig. 8.8, at the intensity given by (8.6.24). The reflection coefficient r at this point is defined by (8.6.14), where the value of $\varepsilon_2|E_{in}^2|$ is to be substituted from (8.6.24). In the reverse process (i.e., if the input intensity decreases), the switch from the partial transmission state to the TIR state occurs at the lower intensity, $|E_{th}^2|_2$, which is defined by (8.6.13) [or (8.6.14)] by assuming $\xi = 0$ [or $r = 1$]:

$$\left|\frac{\varepsilon_2 E_{th}^2}{\Delta\varepsilon_L}\right|_2 = \frac{1}{4}[1 - (\psi/\psi_{cr})^2] \quad , \tag{8.6.25}$$

i.e., $|E_{th}^2|_2 < |E_{th}^2|$, (see Fig. 8.8). The jump of the reflection coefficient only occurs when switching from TIR to transmission. For the backward direction, there is only a jump in the phase of reflection.

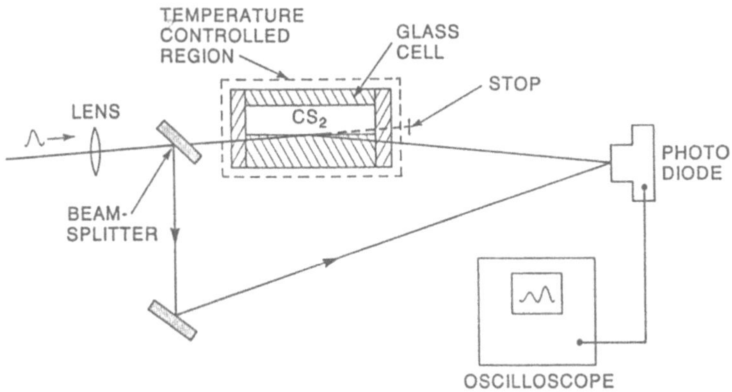

Fig. 8.9. Experimental setup, see [8.50]

One of the interesting features of nonlinear reflection is what can be called nonlinear bleaching (or transparentization) of the interface [8.44, 45]. It is seen from (8.6.14) that at $\Delta\varepsilon_L<0$, "bleaching" input intensity, $|E^2_{bl}|$ exists:

$$E^2_{bl} \equiv |\Delta\varepsilon_L/\varepsilon_2| \quad , \text{ i.e.,}$$
$$\Delta\varepsilon_L = -\Delta\varepsilon_{NL} \tag{8.6.26}$$

at which there is no reflection at all ($r = 0$) for any grazing angle ψ (see Fig. 8.8). This phenomenon cannot be regarded as a nonlinear analog of the Brewster bleaching (since it does not depend on either the angle or polarization of incident light). The nonlinear bleaching is attributed to the fact that the field equalizes the susceptibilities of the two media and, therefore, makes the boundary completely transparent. In this case, the jump from total reflection to total transmission occurs at $\psi = \psi_{cr}/2\sqrt{2}$ (see curve 5 in Fig. 8.8). For CS_2, where $n_2 \sim 10^{-11}$ cgs esu, we have $E_{bl} \sim 1.8 \times 10^6$ V/cm at $\Delta\varepsilon_L \sim -10^{-3}$, in which case, $\psi = 0.540$.

8.6.4 Experiments on a Nonlinear Interface

The first experiment, aimed to observe the effects predicted by theory [8.44, 45] has been done in work [8.50–52]. In this experiment, the liquid CS_2 was utilized as a nonlinear medium with "positive" nonlinearity ($\varepsilon_2>0$), and the glass was used as a linear medium. The temperature regulation of the cell served as a fine control of the difference $\Delta\varepsilon_L$ between both susceptibilities required by relationship (8.6.3). Following [8.50–52] the experiment is briefly described below.

Experiments were performed with the apparatus shown in Fig. 8.9. The input pulse was generated by a mode-locked traveling-wave ruby laser. Internal

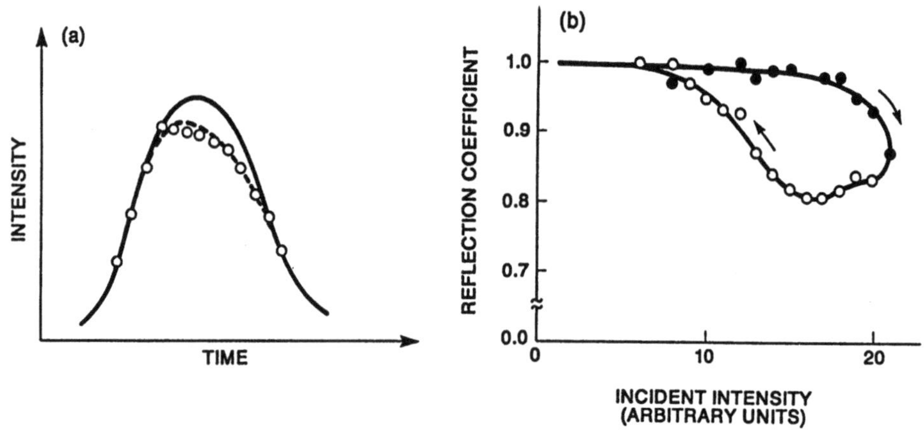

Fig.8.10a,b. Experimental measurements of incident and reflected pulse shapes for $\psi/\psi_c = 0.707$. **(a)** (—) is the incident pulse, and (•••) are the experimental measurements of the reflected pulse. (---) is the reflected pulse calculated from an incoherent plane-wave analysis for $I/I_0 = 1.125$ (see text). **(b)** The experimental data in **(a)** plotted in a way that demonstrates the optical hysteresis observed. See [8.50]

Fabry-Perot mode selection was employed so that the laser output consisted of a train of ≃1-ns pulses. The absence of internal structure in these pulses was verified with a streak camera. A single pulse was selected from this train and directed onto the set-up shown in Fig. 8.9. A portion of the beam was sampled and directed with an optical delay of ≃6 ns onto a fast photodiode. The remainder of the beam was focussed into the glass cell containing CS_2. The glass was chosen to have an index of refraction close to that of CS_2 at room temperature in order to satisfy condition (8.6.3). The entire cell was placed in a temperature-controlled holder and the temperature was adjusted so that for the 694.3 nm wavelength of the ruby laser the index difference takes on the value $\Delta \simeq 10^{-3}$. This corresponded to $\psi_{cr} \simeq 2.0°$ in the glass cell.

The reflected beam was monitored by the same photodiode used to monitor the incident pulse, and the diode output was displayed on a fast oscilloscope. The detector-oscilloscope combination had a measured response time of 320 ps. Figure 8.10a shows the measured incident and reflected pulse shapes for an incident intensity slightly above the threshold for hysteresis. The dashed curve is due to the incoherent plane-wave calculations based on the theory [8.44, 45] [see (8.6.14)] and was fitted to the data by adjusting the ratio of the peak intensity to the threshold intensity $I_0 = |E^2_{\rm th}|_1$ that corresponds to the switch from TIR into partial transmission. One can see that experimental and calculated curves agree almost ideally. The agreement is spoiled as the peak input intensity is increased (see Fig. 8.11), however, the plane-wave theory [8.45] still accurately predicts the threshold intensity $I_0 = |E^2_{\rm th}|_1$ [(8.6.24)] for the onset of the switch TIR → transmission.

From the published [8.72] value $n_2 = 3 \times 10^{-8} ({\rm MW/cm^2})^{-1}$ for CS_2, the theoretical value of $I_0 = 8.1 \times 10^9\,{\rm W/cm^2}$ was computed based on (8.6.24).

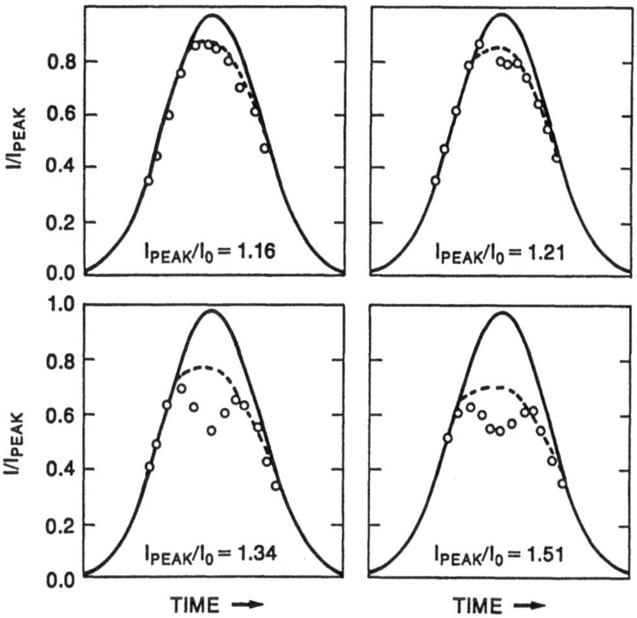

Fig. 8.11. Experimental measurements of incident and reflected pulse shapes for $\psi/\psi_{cr} = 0.707$. (—) are the incident pulse and (•••) are the experimental measurements of the reflected pulses. (- - -) are from the incoherent plane-wave analysis fitted to all four curves with a single value of the adjustable parameter, I_0. See [8.51]

The experimentally measured value was $I_0 = 7.5 \times 10^9\ \mathrm{W/cm^2}$, in a very good agreement with the theoretical one. The other important fact is that the backward switch from transmission to TIR always occurs at the "backward threshold" input intensity $I_2 = |E_{\mathrm{th}}^2|_2$ which is less than the threshold intensity I_0 at the onset of nonlinear behavior, i.e., $I_2 < I_0$ (see Figs. 8.10, 11) for any experimental situation. That confirms the hysteretic behavior of the system. In Fig. 8.10b, the same experimental data as in Fig. 8.10(a) are shown but now plotted in the form of reflectivity versus input intensity. This form emphasizes the optical hysteresis observed. Moreover, one can see from Figs. 8.10, 11 that always $I_2 \simeq \frac{1}{2} I_0$ which also confirms the prediction of plane-wave theory [see (8.6.25) in comparison with (8.6.24)]. Finally, the measured values of threshold intensity I_0 as a function of angle of incident ψ are shown in Fig. 8.12. We see that the results are again in a good agreement with the predictions of the plane-wave theory [8.45] [(8.6.24)].

Some of the experimental results show substantial disagreement with plane waves calculations. For instance, for sufficiently large peak input intensities, the measured reflected pulse shapes deviate significantly from those obtained from calculations. It was believed and seemingly backed by the computer simulations of [8.51, 52, 65] that at least some of these disagreements could be caused by the excitation of nonlinear surface waves propagating along the interface with no decay, beginning from the point of reflection of the Gaussian beam. Indeed,

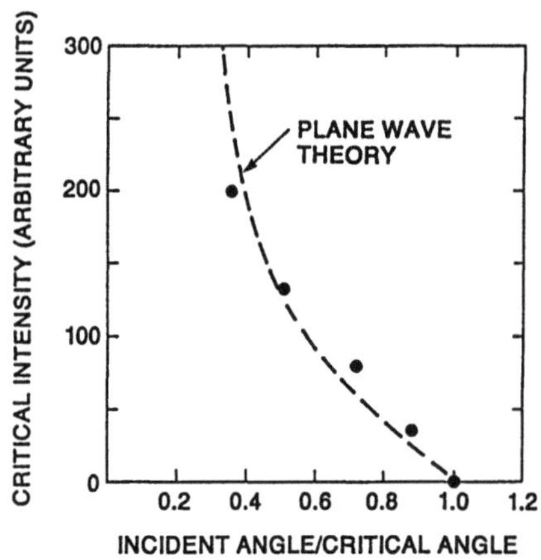

Fig. 8.12. Experimentally measured critical intensity, I_0, as a function of ψ/ψ_c. (—) shows the critical intensity predicted by the plane-wave theory. See [8.51, 52]

it was shown in [8.66] that a surface nonlinear wave in the form of a stationary channel can propagate along the interface, traveling from $x = -\infty$ to $x = +\infty$, in the absence of any wave incident from either linear or nonlinear media. (Similar waves were discussed previously in [8.85] and [8.86] in application to plasmas). However, the subsequent analytic [8.67] and computer [8.68] results did not confirm this suggestion in the case of incident two-dimensional Gaussian beams.

Using the general theory of the stability of solitons of a cubic-nonlinear Schrödinger equation, it was shown in [8.76] that if the surface nonlinear wave (in fact, a soliton solution) propagates in the direction $x \to \infty$ beyond the point where the incident Gaussian beam hits the interface, then the same soliton-like surface wave (albeit with a spatially shifted maximum) must exist even before this point, i.e. this wave can be observed even when $x \to -\infty$. This would be inconsistent with the formulation of the problem (indeed, it is presumed that at $x \to -\infty$, the entire energy of the field is concentrated in the linear semi-space); therefore a "semi-infinite" nonlinear surface wave [8.52, 65] is prohibited. This result was confirmed by the computer simulation [8.68]. Furthermore, it showed that when the intensity of a laser Gaussian beam exceeds a critical level, the TIR state is drastically disturbed. The laser beam, instead of being totally reflected, breaks through the interface and propagates in the nonlinear material *departing* from the interface. Furthermore, the intensity of the beam at this moment is sufficiently high in order for the beam to form a self-trapped channel. If the incident intensity increases further, a second channel is formed [8.87]. It is natural to expect formation of more and more channels as intensity increases. Formation of each self-trapped channel is accompanied by a downward jump in the reflection coefficient.

The computer results of [8.68] did not verify the existence of hysteresis. It must be noticed, however, that the computer simulation of [8.68] was not explicitly directed towards the search of multistability and hysteresis (which in general is not an easy task for nonanalytic methods). The problem with a steady-states hysteresis (or bistability) remained open. In the later experimental research, [8.88] an "artificial" nonlinear medium (dielectric spheres suspended in a liquid) was used in order to observe a switching at nonlinear interface in cw regime. These new results were very close to the computer results [8.68]; for the long observation time, hysteresis was not observed. However, the modes of nonlinear reflection described by the points at lower hysteretic branches of hysteretic curves at Fig. 8.8, which in general appear to be unstable in the experiment [8.88], occur to have a very long decay time (much longer than relaxation time of nonlinear medium). The nature of this effect remains unclear, although it suggests, at least, that for any practical application with a pulse mode of operation, the nonlinear interface should demonstrate hysteresis and bistability. Indeed, the most recent experimental work [8.89] done with essentially the same kind of media as in [8.88] using long pulse mode of operation, revealed very distinct hysteretic jumps.

The nonlinear switching can be accompanied by a large nonlinear Goos-Hänchen effect suggested in [8.45] and obtained in a computer simulation in [8.68]. The Goos-Hänchen effect consists in the displacement of a reflected beam along the interface with regard to the point of incidence (see e.g. [8.56]). It was shown in [8.68] that for intensities of the incident beam slightly lower than the critical intensity of formation of the first self-trapped channel, the point of reflection at the nonlinear interface is moving very far away from the point of incidence as the incident intensity increases.

8.6.5 Effects at "Negative" Nonlinearity Longitudinally Inhomogeneous Traveling Waves (LITW)

In a number of physical situations, the nonlinearity can be "negative" (i.e., $\varepsilon_2<0$), for instance, due to nonlinear resonant interactions (see Sects. 2.1, 8.1), or due to giant nonlinearities, recently observed in some semiconductor (e.g., in InSb [8.73]). In such a case, the nonlinear analogs of linear waves (i.e., PW and TIR) can exist as well. For plane waves (PW), Snell's (8.6.13, 15) as well as Fresnel's (8.6.14, 16) formulas remain valid (one has to take into consideration that now the sign of ε_2 is changed). Under this regime it is also possible to vary the values of the transmission angle ξ and reflection r by changing the incident light intensity (Fig. 8.13). Similarly to the "positive" nonlinearity, a nonlinear total bleaching of the interface also exists. The "bleaching" incident light intensity is given by the same expression (8.6.26) but now the value of the linear detuning should be positive (Fig. 8.13, curves 3 and 4).

In contrast to the "positive" nonlinear case, the surface waves at the TIR regime exist exactly in the same range of linear detunings and glancing angles [i.e., $\Delta\varepsilon_{\mathrm{L}}<0$ and $\psi<\psi_{\mathrm{cr}} = |\Delta\varepsilon_{\mathrm{L}}/\varepsilon_0|^{1/2}$], as in the linear case. However, now the amplitude profile of SW given by integration of (8.6.11) is [8.45]

$$u = \sqrt{2\frac{\varepsilon_0}{|\varepsilon_2|}}\frac{\gamma}{\sinh(k_0\gamma z + C)} \quad ; \quad \gamma = (\psi_{\mathrm{cr}}^2 - \psi^2)^{1/2} \quad , \qquad (8.6.27)$$

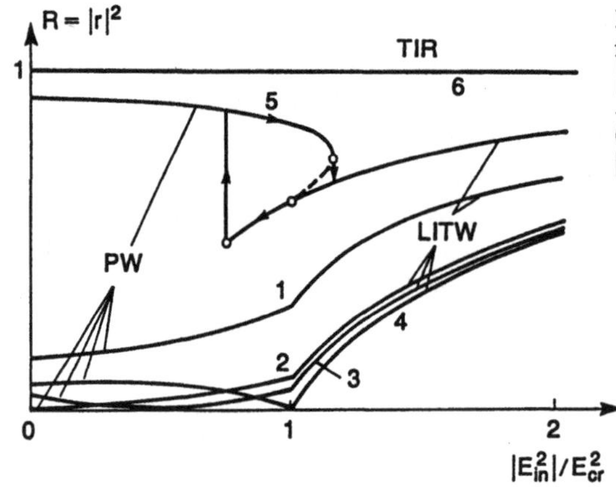

Fig. 8.13. Reflectivity $R = |r|^2$ as a function of the incident light intensity $|E_{\rm in}^2|$. Curves (*1*) $D<0$, $\psi^2>|D|$; (*2*) $D=0$; (*3*) $\psi^2>D/2>0$; (*4*) $\psi^2=D/2$; (*5*) $\psi^2<D/2$; (*6*) $D<0$, $\psi^2<|D|$. See [8.47]

where the constant C is determined by boundary conditions. The peculiarity of TIR in this case is that the depth of penetration of the field into the medium, which is defined by C, can be much smaller than the depth of linear penetration [8.45].

The amplitude of the field $u(0)$ at the boundary is given by (8.6.21), where one has to take the lowest sign in front of the radical. As $z \to \infty$, the evanescent field (8.6.27) decreases as $\sim\exp(-k_0\gamma z)$, just as in the linear case. But the depth of field penetration into the medium (at the half-intensity level), $L_{\rm NL}$, which is determined by the shape (8.6.27) and the value of $u(0) \equiv u_0$, decreases [8.45]:

$$L_{\rm NL} = L_{\rm L}\left[1 + \frac{2}{\ln 2}\ln\frac{1+(1+u_0^2/4\gamma^2)^{1/2}}{1+(1+u_0^2/2\gamma^2)^{1/2}}\right] \quad , \tag{8.6.28}$$

where $L_{\rm L} = (\ln 2)/2k_0\gamma$ is the depth of the linear penetration. In a strong field, as $u_0^2 \gg \gamma^2$, one has $L_{\rm NL} \simeq 0.6\gamma L_{\rm L}/u_0 \ll L_{\rm L}$.

The most interesting feature in the case of negative nonlinearity is the feasibility of excitation (under some special conditions) of waves of a new kind, so-called longitudinally inhomogeneous traveling waves (LITW), obtained analytically in [8.46–48, 74]. Being inhomogeneous and non-planar near the boundary, these waves transform into plane waves sufficiently far from the boundary. The concept of longitudinal inhomogeneity was introduced in [8.46] to distinguish this phenomenon from the self-actions due to transversely inhomogeneous waves, such as self-focussing [8.1–9] and self-bending [8.10] of the beams with limited cross-section.

The examination of [8.6, 11] shows that under the conditions

$$\frac{2}{3}q < \frac{|\varepsilon_2|}{\varepsilon_0}u_\infty^2 < q \quad ; \quad q \equiv \frac{\Delta\varepsilon_{\rm L}}{\varepsilon_0} + \sin^2\psi , \tag{8.6.29}$$

the excitation of LITW is possible. Integration of (8.6.11) yields two possible intensity profiles of LITW [8.46]:

$$u^2 = u_\infty^2 \pm 2B^2 \begin{cases} \sinh^{-2} \\ \cosh^{-2} \end{cases} (Bk_0 z + C) \quad ;$$

$$B \equiv \left[\frac{3}{2}\frac{|\varepsilon_2|}{\varepsilon_0} u_\infty^2 - q\right]^{1/2} \quad . \tag{8.6.30}$$

For the minimum possible energy density at infinity,

$$u_\infty^2 = u_m^2 = \frac{2}{3}\frac{\varepsilon_0}{|\varepsilon_2|} q \quad , \tag{8.6.31}$$

only one (limiting) type of LITW remains:

$$u_{\mathrm{lim}}^2 = 2\frac{\varepsilon_0}{|\varepsilon_2|}\left[\frac{q}{3} + \frac{1}{(k_0 z + C)^2}\right] \quad . \tag{8.6.32}$$

The unique "continuum" problem arises pertinent to the new solutions: we have only two conditions (which are equivalent to a single complex boundary condition (8.6.5)) to determine three unknown constants [v_∞ and C in (8.6.32), and ϕ in (8.6.7)]. The radiation conditions (8.6.8) have already been used for the construction of LITW, (8.6.11, 30). Therefore, there is a continuum of solutions even when boundary and radiation conditions are completely specified.

This differs drastically from the situation in a linear medium and a "positive" nonlinear medium. The energy criterion for selecting a unique kind of physically realized LITW was proposed [8.47, 74] which is based on the minimization of wave energy density (simultaneously, it results in a maximum energy flux). The only "surviving" kind of LITW occurs to be a limiting type (8.6.32), when u_∞^2 reduces to the minimum possible value for LITW equal to u_{m}^2 (8.6.31). This choice was verified analytically [8.47] by asymptotic method whereby a weak absorption was introduced and the asymptotic results were obtained when the absorption tended to zero. A small parameter of this limiting transition occurred to be $(k_0 L_{\mathrm{ab}})^{-2/5}$, where L_{ab} is characteristic length of the wave absorption.

Now, when the unique "surviving" *type* of LITW (but not unique *solution* yet), is found, one can obtain all characteristics of these waves, in particular, their amplitude at the boundary, u_0, [8.47, 48]. Based on this result and using the relationship for u_0 for PW (8.6.16), it is easy to find the domains of existence of all of the possible waves (i.e., PW, LITW and TIR) in the system [8.48]. These domains are shown in Fig. 8.14 in the space of generalized parameters P and Q:

$$P = \frac{\varepsilon_2 |E_{\mathrm{in}}^2|}{\varepsilon_0 \sin^2 \psi} \quad ; \quad Q = \frac{\Delta\varepsilon_{\mathrm{L}}}{\varepsilon_0 \sin^2 \psi} + 1. \tag{8.6.33}$$

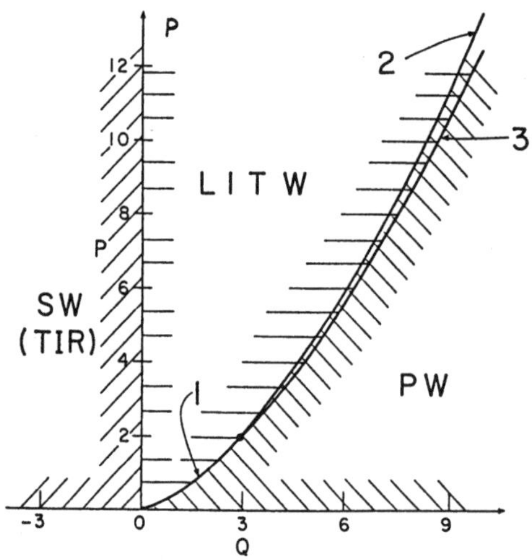

Fig. 8.14. Diagram of wave states in the space of parameters P and Q. Curve (*1*) is the boundary between PW and LITW states; curve (*2*) is the boundary of transition jump PW → LITW; curve (*3*) is the boundary of reverse transition jump LITW → PW. The space between curves *2* and *3* is the hysteresis region [8.48]

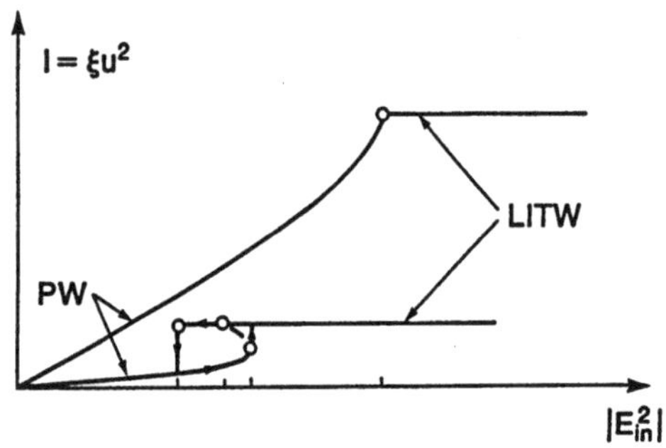

Fig. 8.15. Energy flux I of transmitted waves as a function of the incident light intensity $|E_{\rm in}^2|$ for hysteresis and nonhysteresis situations [8.47]

The LITW excitation gives rise to new physical effects. First of all, similarly to a "positive" nonlinear case, it leads to the hysteretic behavior of a system under certain conditions, but now the hysteresis jumps occur between states of two different traveling waves, PW and LITW (in contrast to "positive" nonlinearity, where they occur between PW and TIR). Hysteresis occurs only under the following conditions [8.47, 48, 74]:

$$|\varepsilon_2 E_{\rm in}^2| > \Delta\varepsilon_{\rm L} > 2\varepsilon_0 \sin^2 \psi \quad .$$

The behavior of reflectivity for different parameters of a system is shown in Fig. 8.13 (in particular, one can see hysteresis, curve 5, and self-bleaching, curves 3 and 4). In this figure, $E_{\rm cr}^2$ is the critical intensity of incident light required for excitation of LITW for each particular case, and $D \equiv \Delta\varepsilon_{\rm L}/\varepsilon_0$.

Another effect which might be of interest to applications, consists of the self-limitation of the energy flux of LITW penetrating into a nonlinear medium [8.47, 54]. It is a direct consequence of the selection of a unique value $u_\infty = u_{\rm m}$ (8.6.31). Behavior of the energy flux I is shown in Fig. 8.15 for hysteretic and non-hysteretic situations.

One of the most interesting effects attributed to the excitation of LITW is self-parallax [8.47, 74], i.e., displacement of the refracted rays along the interface. This effect is not a nonlinear analog of the Goss-Hänchen displacement of Gaussian beams under TIR since, in contrast to the Goos-Hänchen effect, self-parallax is now valid for plane waves and for the transmission regime, rather than for TIR. It corresponds to the initial "sticking" of the rays to the interface (Fig. 8.7, curve 2) which occurs because of conservation of the energy flux (8.6.9) (the angle ξ should increase as the intensity of u^2 decreases), and it is not related to the existence of hysteresis. The ray trajectory in the space is determined by the formula [8.47]

$$x = z(3/Q)^{1/2} + l(1 + z_l/t)^{-1} \quad , \tag{8.6.34}$$

where l is the parallax for $z \rightarrow \infty$ and z_l is the characteristic depth for half of the parallax; l and z_l are determined by E_{in}, ψ and $\Delta\varepsilon_{\text{L}}$. Let $k_0 = 10^5\,\text{cm}^{-1}$ $\Delta\varepsilon_{\text{L}} = -10^{-4}$, $\psi \sim 0.7°$ and $|\varepsilon_2/\varepsilon| = 10^{-10}$ cgs esu; then the critical value of E_{in} for excitation of LITW is $E_{\text{cr}} = 1.15 \times 10^5\,\text{V/cm}$, and if $E_{\text{in}} = 2E_{\text{cr}}$, one obtains $l = 1.2\,\text{cm}$ and $z_l = 0.01\,\text{mm}$; the parallax is very strong and occurs at a very small depth.

8.6.6 Theorems of LITW Existence for Arbitrary Kinds of Nonlinearity

The existence of LITW in the system with negative nonlinearity and its absence in a "positive" nonlinear system pose a number of questions related to the kind of nonlinearity of the medium, which may cause LITW, such as: is it possible to predict what kinds of nonlinearities and system characteristics give rise to LITW and what kinds of nonlinearities do not allow LITW? What are the parameters of possible LITW if allowed?

These questions have been addressed by the theorem formulated and proved in [8.46, 47]. This theorem relates the existence of different kinds of traveling waves to the behavior of a "characteristic" nonlinear function $F(u)$ (8.6.12). The main conclusion of the theorem is that the existence of LITW in a transparent medium is possible if, and only if:

a) A V exists such that for $u \in V$ one has $F(u) > 0$.
b) $F(u)$ falls at least somewhere in V (i.e., there is at least one interval $W \subset V$ where $F(u)$ falls monotonically).
c) If these conditions are satisfied, the LITW, if excited, can have a value of u_∞ which must belong only to this falling interval, i.e., always $u_\infty \in W$.

The proof of this theorem is based of the behavior of the integral in the rhs of (8.6.11).

One of the main consequences of this theorem is that LITW can be excited only in a nonlinear medium whose nonlinearity $\varepsilon_{\text{NL}}(u^2)$ has at least one interval in which it decreases. On the other hand, LITW are prohibited in the medium

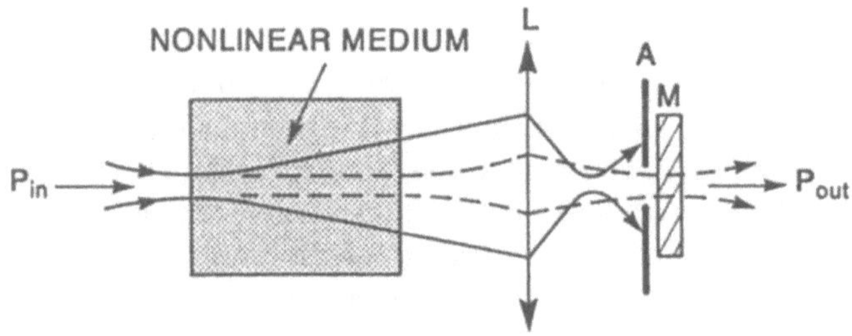

Fig. 8.16. Schematic diagram of the bistable optical device leading to self-trapping. The lens L images the optical field on the exit face of the nonlinear medium onto the partially transmitting mirror M. The aperture A causes a feedback to account for the laser beam spot size at the exit face of the medium. The situations corresponding to normal propagation and self-trapping are shown by the *solid* and *dashed lines*, respectively [8.19]

with increasing function $\varepsilon_{NL}(u^2)$. This is the reason why excitation of LITW is prohibited for "positive" nonlinearity.

The "continuum problem" arises for all kinds of LITW. Therefore, the special direction of the theory of two-dimensional nonlinear waves should be devoted to the theory of "survival" of nonlinear transmitted waves in the case of arbitrary nonlinearity. Several theorems related to this problem were formulated in [8.47]. In particular, the complete class of functions $F(u)$ was found which allowed for the "surviving" LITW, and it was proved that the principle of minimization of wave energy remains valid throughout this class.

8.7 Optical Bistability Based on Mutual Self-Action of Counterpropagating Light Beams

Very recently, a new effect was proposed in [8.19], [8.76], and observed experimentally [8.19]. It is based on the self-action of counterpropagating light beams in nonlinear media and gives rise to yet another kind of cavityless optical bistability. The first experiment [8.19] is based on the self-focussing and self-trapping of light due to the large nonlinearity caused by resonant excitation of atoms in sodium vapor. As a bistable device, the new system possesses the same feature as nonlinear interfaces: it does not have any cavities or Fabry-Perot resonators, as opposed to other intrinsic (i.e. all-optical) bistable systems [8.58–62]. This allows one to work with broadband sources of light, avoid resonant tuning, obtain a fast switching response, etc.

8.7.1 Experimental Observation of Bistability Based on Self-Trapping

The basic principles upon which this new class of devices operates can be understood from a simplified discussion based on the schematic diagram of Fig. 8.16. A laser beam of power P_{in}, assumed to have a Gaussian intensity profile, is focussed onto the interface of a medium having an intensity dependent index of refraction, with "positive" nonlinearity ($\varepsilon_2>0$). Suppose that P_{cr} is the critical power of self-focussing (see the beginning of Chap. 8). When $P_{in} = P_{cr}$, the input laser beam passes through the medium with no change in spot size, as illustrated by the dashed lines. This situation is referred to as "self-trapping." For $P_{in}<P_{cr}$ the laser beam diverges less rapidly than in the absence of the nonlinearity and for $P_{in}>P_{cr}$ the beam converges. For $P_{in}\ll P_{cr}$ there is essentially no self-focussing. The input focal spot size is chosen such that, for these conditions, the laser beam diverges appreciably when passing through the nonlinear medium, see the solid lines in Fig. 8.16.

The optical field at the output interface of the nonlinear medium is by the lens L transferred onto the partially transmitting mirror M which is aligned normal to the laser beam to provide optical feedback. Immediately in front of the mirror is an aperture of appropriately chosen size. The aperture size is adjusted to satisfy two criteria. First, it must be small enough that in the absence of self-focussing the fraction of the incident light fed back into the nonlinear medium by the mirror is small. Secondly, it has to be large enough that under self-trapping conditions essentially all the light passes through the aperture and is reflected back by the mirror. This strong feedback reinforces the self-focussing in the nonlinear medium and it allows self-trapping to be maintained even when the input power is subsequently reduced below P_{cr}. This is the mechanism which gives rise to the optical bistability (hysteresis).

The experimental demonstration [8.19] of these principles utilized atomic sodium vapor as the nonlinear medium, which provides an opportunity to get a large nonlinear index for frequencies near its resonance transition [8.12]. It was also of importance that the nonlinearity saturates at reasonable intensity levels so that catastrophic self-focussing does not occur. These factors have allowed steady-state self-focussing and self-trapping to be observed in sodium vapor using cw dye lasers [8.18].

Without feedback from the mirror, strong self-focussing and self-trapping were readily observed with approximately 150 mW of light tuned about 1.2 GHz above the resonant frequency of the $3S_{1/2}(F = 2) \rightarrow 3P_{1/2}$ sodium transition at 5896 BA. The sodium density was approximately $2 \times 10^{12}\,\mathrm{cm}^{-3}$. Figure 8.17 presents data obtained under conditions for which the device exhibited bistability; the laser was tuned about 1.2 GHz above the $3S_{1/2}(F = 2) \rightarrow 3P_{1/2}$ transition. As the input power was increased, upward switching from a low-transmission state to a high-transmission state occurred at an input power of about 130 mW.

The dashed lines show the calculated low and high transmission limits, based on the measurement of 45 mW incident onto the aperture for an input power of 140 mW and no feedback. The high transmission limit assumes 100 % transmission by the aperture.

The upward switch occurs at a power level roughly equal to the level at which self-trapping was observed without feedback. Visual observation of the resonance fluorescence induced by the laser beam shows dramatically that at

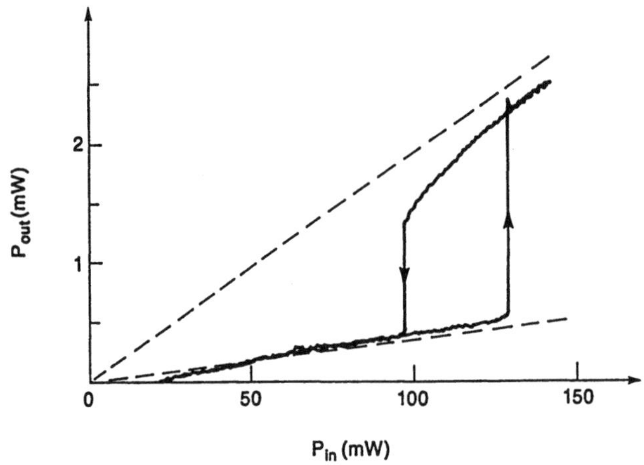

Fig. 8.17. Experimental curve of P_{out} vs. P_{in} with the parameters adjusted so that the device exhibits bistable operation. The aperture diameter was 150 μm and the arrows point into the switching directions [8.19]

switching the laser beam abruptly changes from diverging propagation to what appears to be self-trapping. With higher sodium density and a smaller input spot size, several switching levels and several hysteresis loops were obtained, perhaps corresponding to oscillations of the spot size as the laser beam propagated through the self-focusing medium [8.18]. The theory of this cross-self-trapping bistability poses quite a challenge (see also next section). The main problem seems to arise from the fact that the forward and backward beams overlap only partially and both of them oscillate in space which results in a very complicated structure of the refractive index profile at any cross-section of the beams. However, even the rough approximation used in [8.90] (assuming complete overlapping and no spatial oscillations) provided a fairly good qualitative description which agrees well with the experimental hysteretic loop [8.19]. Later cw experiments with cross-self-trapping bistability [8.91] using an artificial nonlinear medium (dielectric spheres suspended in a liquid) demonstrated qualitatively the same features of the effect and confirmed the validity of the approximate model [8.90].

8.7.2 Mutual Self-Action of Counterpropagating Beams in the General Case

The experiments described above utilized self-trapping and retroreflection of the incident beam. It was shown, however, [8.76] that in the general case the bistability could be obtained by employing mutual self-action (of any kind) of two counterpropagating beams (even without any mirrors). Such an optical bistable system could be based not only on self-trapping, but also on the mutual self-focusing (or self-defocusing, or self-bending) of counterpropagating light beams interacting with each other, and could work with no reflection feedback

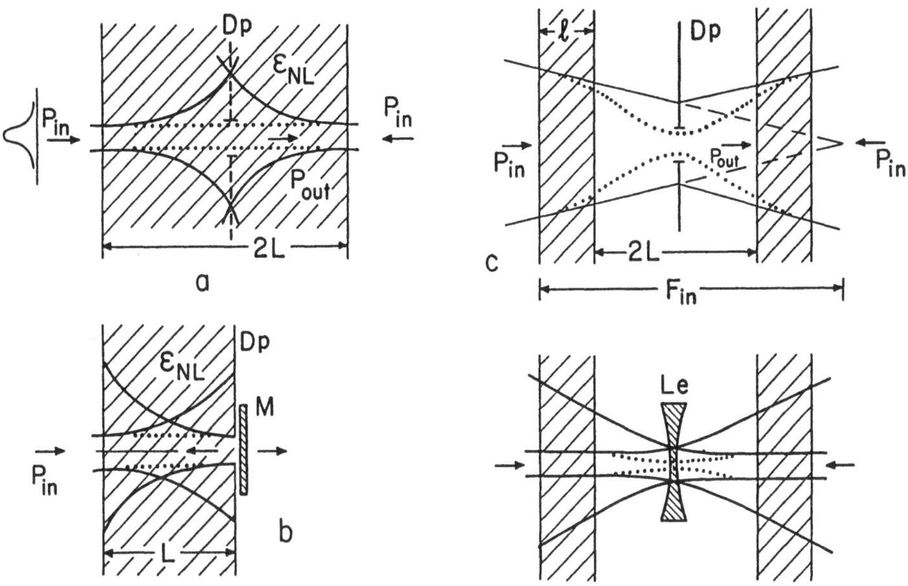

Fig. 8.18. Various configurations realizing optical bistability due to mutual self-focusing (or defocusing) of counterpropagating beams. M-mirror; Dp-diaphragm; Le-lens [8.76]

at all. In general, the system only requires a sample of nonlinear material (whose refractive index depends on the intensity of light) and two light beams incident from opposite sides (Fig. 8.18). This provides, probably, the simplest known opportunity to obtain optical bistability. The required feedback is caused by the mutual influence (or cross-interaction) of the beams, each of which affects the other by changing the average nonlinear component of the refractive index; the feedback is substantially enhanced by the diaphragm (Fig. 8.18) whose radius is to be of the order of the focal spot size or self-channel radius (Fig. 8.18a).

When using two incident beams, one of the beams can be of constant power, while the power of the other beam can be driven by an external source. This device can be used as an optical transistor, amplifying a small variation of the driving signal due to the large response in the output angle distribution of the forward beam with a constant input power. This system can be regarded as an *asymmetrical* one. The first experimental observation of such mirrorless optical bistability due to cross-trapping of counter-propagating beams in sodium vapor was done in [8.92], see also [8.58].

On the other hand, one can achieve bistability by simultaneous, equal driving power of both beams. A particular realization of this *symmetrical* configuration is a device using a mirror set in the central plane (see Fig. 8.18b), and a single input beam, while the backward beam is provided by the reflection from the mirror, e.g. like in the system [8.75] described above (see Sect. 8.7.2) which utilizes self-trapping in sodium vapor in cw regime. (Some other aspects of the same effect such as possible instability, first noticed in [8.75], and

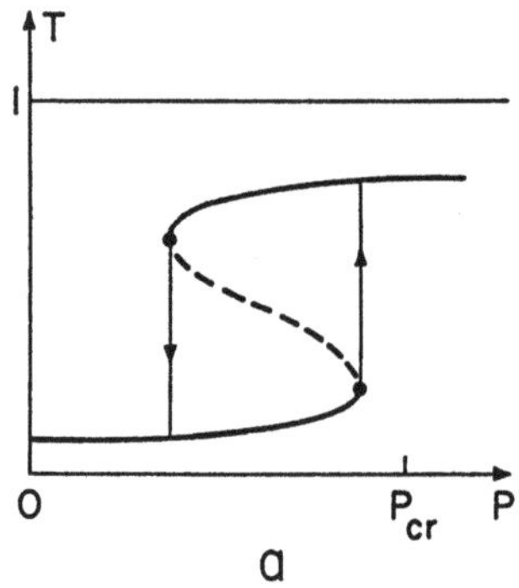

Fig. 8.19. Hysteresis of transmittivity $T = P_{out}/P_{in}$ in the case of external self-action [8.76]

chaos were explored in [8.93] and [8.92].) However, in the case of fast response nonlinear materials such as glasses or liquids (e.g. CS_2), one has to deal with an uncontrolled catastrophic collapse (i.e. non-trapping) of the light beam above a self-focusing threshold. Moreover, bistability due to self-trapping allows one to utilize only self-focusing materials. There exist other materials which have large self-defocusing nonlinearity (e.g. InSb [8.73]).

To solve these problems (i.e., to utilize self-defocusing and to avoid a catastrophic collapse of the beam inside the nonlinear material), it was proposed [8.76] to use the so-called "external" focusing (or defocusing) system [8.24] in which a focal spot of the beam is set outside of the nonlinear layer of thickness l. This layer is acting then as a nonlinear lens whose focal length is intensity dependent (see Fig. 8.18c,d). When using a self-focusing medium, the initial location $|F_{in}|$ of the focal spot of the incident beam should be set behind a diaphragm, i.e. $|F_{in}|>L + l/n$ (as it is shown in Fig. 8.18c) where L is half-distance between two layers, such that only a small portion of the incident power reaches the opposite layer (lower branch of Fig. 8.19). Under the action of strong light, the focal spot moves then toward the diaphragm, which creates a strong feedback, and at some critical intensity almost all the incident power passes through the diaphragm (if its diameter is of the same order as the focused spot size). In the case of a self-defocusing medium, the initial focus of the incident beam should be located in the front of the diaphragm ($|F_{in}|<L+l/n$). This system can perform either in a two-beam mode or in a single-incident beam mode. In the latter case, the mirror should be set immediately behind the diaphragm. The external cw cross-self-action bistability of such a kind was first experimentally observed and explored [8.94] utilizing self-focusing of light in nematic liquid crystal film. The observation and study of this effect in sodium vapor was reported in [8.92].

The "external" configuration can utilize linear lenses as well (see Fig. 8.18d). This provides an opportunity to work without diaphragms and to employ only narrow parallel beams. The switch appears when linear and nonlinear lenses almost cancel each other, so that one should use a defocusing lens for the focusing medium, and vice versa.

We consider now the basic features of the theory of the cross-self-action systems. Representing a total complex field E of the two beams, counterpropagating along axis z, as

$$E = E_1 \exp(ikz) + E_2 \exp(-ikz) \tag{8.7.1}$$

where envelopes E_j are slow complex functions of spatial coordinates, one can write a conventional parabolic equation for each beam;

$$\begin{aligned}&-2i(-1)^j k\partial E_j/\partial z + \Delta_\perp E_j + k^2 E_j \varepsilon_0^{-1} \Delta\varepsilon_j^{\rm NL}\\ &\quad \times (|E_1^2|, |E_2^2|) = 0 \quad ; \quad j = 1,2\end{aligned} \tag{8.7.2}$$

where in the case of a cylindrical beam $\Delta_\perp = \partial^2/\partial\varrho^2 + \varrho^{-1}\partial/\partial\varrho$ (ϱ is radial coordinate). If dealing with a cubic nonlinearity, (8.6.2), it turns out that the nonlinear factors in (8.7.1) have different magnitudes for opposite directions of propagation [8.76, 95–97]

$$\begin{aligned}&\Delta\varepsilon_1^{\rm NL} = \varepsilon_2(|E_1|^2 + 2|E_2|^2) \quad \text{and}\\ &\Delta\varepsilon_2^{\rm NL} = \varepsilon_2(|E_2|^2 + 2|E_1|^2) \quad .\end{aligned} \tag{8.7.3}$$

It is worth noting that the fact that $\Delta\varepsilon_1 \neq \Delta\varepsilon_2$ (if $|E_1|^2 \neq |E_2|^2$) is attributed to the formation of nonlinear index grating in a partially-standing EM wave (8.7.1). Although this fact is fairly well known (see e.g. [8.29]), the authors of work [8.95, 96] realized that it must result in a *light-induced nonreciprocity* which immediately suggests very interesting implications when this effect occurs simultaneously with linear nonreciprocal effects such as e.g. the Sagnac effect. They proposed a large nonlinear optical enhancement of optical gyroscopes [8.95] based on the Sagnac effect; they also found the so-called directional optical bistability in a nonlinear ring resonator. For a fiber-optic gyroscope both of these results have recently been generalized in [8.100]. The effect of light-induced nonreciprocity was recently observed experimentally [8.104, 105].

It was shown later [8.97] that the light-induced nonreciprocity depends on the relative polarizations of both of the beams. In particular, the so-called nonlinear eigen-polarizations of both of the beams were found. At these eigen-polarizations the beams counter-propagate without changing their polarizations. The interesting fact was that these eigenpolarizations have different light-induced nonreciprocities. In their most recent research [8.98], the general solution for the spatial dynamics of arbitrary polarized counter-propagated plane waves was found in a nonlinear medium with arbitrary anisotropic Kerr-effect. This analytical solution gives rise to exotic optical multistability in mirrorless systems with multiple isolated branches of solutions (the so-called isolas).

The development of the correct analytic theory of transverse (i.e. non-plane wave) effects, in particular for the self-trapping configuration, involves difficult problems. This is caused by a non-Gaussian distribution of the light intensity at the cross-section, and by the necessity to consider the highest terms of nonlinearity (e.g. saturation) to evaluate spatially-oscillating noncollapsing trapping above the threshold of self-focusing. The approximate analysis of cross-self-trapping bistability was developed in [8.90].

On the other hand, for the external configuration (Fig. 8.18c,d) one can obtain approximate analytical results in a straightforward way by using a Gaussian-beam approximation upon assumption that the radii of the beams (as well as their Gaussian profile) change very little while passing through each nonlinear layer. In particular, that is valid if the condition $l<L<L_d = kb_{\mathrm{in}}^2$ is satisfied. Using (8.7.2), and following the calculation of [8.24] which provides the diffractional theory of nonlinear lenses, one can obtain the formula for each nonlinear lens [8.76]

$$\left(F_{\mathrm{out}} + \frac{l}{n}\right)^{-1} + F_{\mathrm{in}}^{-1} = \left(F_0 + \frac{l}{n}\right)^{-1}\left(1 + \frac{l}{nF_{\mathrm{n}}}\right)^{-1} \tag{8.7.4}$$

where $n = \sqrt{\varepsilon_0}$, F_{in} is the radius of curvature of the phase front of the wave at the incident plane ($F_{\mathrm{in}}<0$ for a convergent beam), F_{out} is the radius of curvature at the back wall plane ($F_{\mathrm{out}}>0$ for a convergent beam), F_0 is the "nonlinear focal length" of the layer. In geometrical optics F_{in} and F_{out} correspond to the "object" and "image" positions. In the case of a single beam (8.7.4) is valid for an arbitrary thickness of layer [8.24]. Assuming $F_0 \sim L$ (which corresponds to the strongest feedback) and taking into consideration (8.7.3), one can obtain that under the above-mentioned conditions the value of F_0 with respect to the incident beam is

$$(F_0^{-1})_j \simeq \frac{4\varepsilon_2 l}{c\varepsilon_0}\left[\left(\frac{P_{\mathrm{in}}}{\varrho_{\mathrm{in}}^4}\right)_j + 2\left(\frac{P_{\mathrm{out}}}{\varrho_{\mathrm{out}}^4}\right)_{3-j}\right] \quad ; \quad j = 1,2 \tag{8.7.5}$$

where P_{in} is the total power of the incident beam, b_{in} is its radius [i.e. intensity is proportional to $\exp(-\varrho^2/b_{\mathrm{in}}^2)$], while P_{out} and b_{out} are the corresponding values for the backward beam transmitted through the diaphragm and exiting out from the system. Now, performing conventional calculations for the propagation of beams in the linear space between the layers (as well as their diffraction at the diaphragm [8.99]), considering the only case when both of the beams have the same parameters at their planes of incidence (and are driven simultaneously, which refers to the symmetrical configuration), one can obtain the required formula for the behavior of the system:

$$\frac{P_0}{P_{\mathrm{in}}} = \left\{1 + \frac{2T}{[(1+a^2)T + a^2]^2}\right\}\left[\frac{1}{A \pm \mu\sqrt{T^{-1} - 1 - a^2}} - \frac{l}{nL}\right] \tag{8.7.6}$$

where $T = P_{\mathrm{out}}/P_{\mathrm{in}}$ is a transmissivity of the system, and

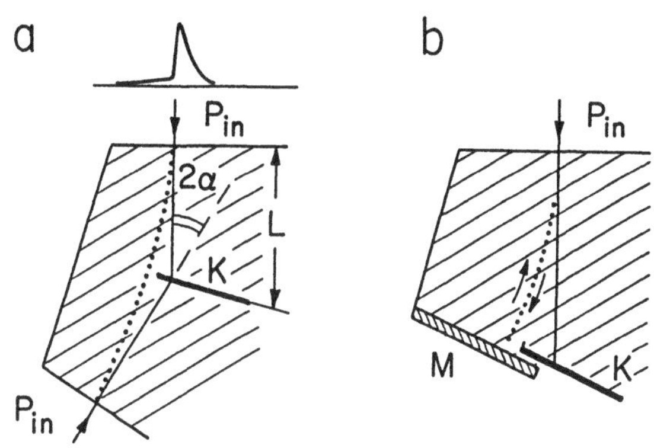

Fig.8.20. Configurations realizing bistability due to mutual self-bending of the light beams. M-mirror; K-knife-erge [8.76]

$$P_0 = \frac{cb_{in}^4 \varepsilon_0}{4lL\varepsilon_2} \quad ; \quad \mu = \frac{d}{b_{in}} \quad ; \quad a = \frac{L}{kdb_{in}} \quad ; \quad A = 1 + \frac{L + l/n}{F_{in}} \quad .$$

Here d is the radius of the diaphragm, and P_0 defines some scale power. Formula (8.7.6) is valid only if $b_{out}b \geq b_{in}$; this is always satisfied if $a \geq 1$. If $a = 1$, then the critical condition for hysteresis to appear, Fig. 8.19, is $|\mu/A| < 1/16$. In the case $b_{in} = 0.4$ mm and $F_{in} = \infty$, this gives $L < 10$ cm and $d < 0.025$ mm. For $\varepsilon_2 > 0$ (self-focusing), it is necessary to set $A > 0$, while for $\varepsilon_2 < 0$ (self-defocusing, i.e. $P_0 < 0$) it should be $A < 0$. The same formula (8.7.6) is valid for the reflection configuration, as well, if the mirror is located directly behind the diaphragm (the T then corresponds to the reflectivity of the system).

Optical bistability could also result [8.76] from the "self-bending" [8.10] of the light beam (see Sect. 8.3), which induces an angular deflection of the whole trajectory of a beam whose intensity profile has an asymmetrical shape. Using self-bending, optical bistability can be achieved either by utilizing two beams whose initial axes are slightly tilted (see Fig. 8.20a) or by employing one beam with a mirror located besides the initial axis (Fig. 8.20b). For both of these configurations the threshold power for the hysteresis P_{SB} can be estimated in the geometro-optical approximation [8.10] as

$$P_{SB} \sim 2\alpha n^3 cb_{in}^3 / \pi |\varepsilon_2| L \qquad (8.7.7)$$

where α is a half-angle of the initial tilt. This formula is valid under condition $2b_{in}/\alpha < L < kb_{in}^2$. If $|\varepsilon_2| \sim 10^{-2}$ esu [8.73], $L = 1$ cm, $B_{in} = 0.05$ mm and $\alpha = 2°$, then $P_{SB} \sim 1$ mW.

* *Acknowledgements (A.E.K.):* I discussed virtually all my new results with L.I. Gudzenko, a close friend of mine for about twenty years who deceased in 1978; a physicist with incredibly broad research interests, he contributed to a vast number of fields (for example, he was one of the first to predict the x-ray lasing scheme in 1965). His memory will always be dear to me. There are many other scientists and just friends whose moral support, professional expertise, and friendship were so indispensable and valuable to me in the country which I left in 1979

for the US. Although their names do not appear here, they know I remember them and miss them very much.

I am greatly indebted to P.L. Kelley and H. Schlossberg for their encouragement and substantial moral support of my recent research done in the US, some results of which are included in this edition. This research was sponsored by the U.S. Air Force Office of Scientific Research, whose support I gratefully acknowledge. I wish also to thank P.W. Smith, W.J. Tomlinson, J.E. Bjorkholm, J.P. Gordon, P. Meystre, and my colleagues of MIT, Purdue University, and Johns Hopkins University for their help, collaboration, and many exciting discussions.

References

Chapter 1

1.1 S.I. Vavilov, W.L. Levshin: Z. Phys. **35**, 932 (1926)
1.2 W. Kaiser, C.G.B. Garrett: Phys. Rev. Lett. **7**, 229 (1961)
1.3 R.W. Terhune, P.D. Maker, C.M. Savage: Phys. Rev. Lett. **14**, 681 (1965)
1.4 S. Singh, L.T. Bradley: Phys. Rev. Lett. **12**, 612 (1964)
1.5 N.N. Badalyan, V.A. Iradyan, M.E. Movsesyan: JETP Lett. **8**, 316 (1968)
1.6 V.M. Fain: *Photons and Nonlinear Media* (Sovietskoye radio, Moscow 1972) (in Russian)
1.7 N. Bloembergen: *Nonlinear Optics* (Benjamin, New York 1965)
1.8 S.A. Akhmanov, R.V. Khokhlov: *Problems of Non-Linear Optics* (USSR Acad. Sci., Moscow 1964) [English transl.: Gordon and Breach, New York 1972]
1.9 R. Karplus, J. Schwinger: Phys. Rev. **73**, 1020 (1948)
1.10 C.H. Townes, A.I. Schawlow: *Microwave Spectroscopy* (McGraw-Hill, New York 1955)
1.11 Yu.G. Khronopulo: Izv. vuzov. Radiofizika **7**, 674 (1964)
1.12 V.S. Butylkin, G.L. Gurevich, M.I. Kheifets, Yu.G. Khronopulo: Sov. Radiophys. **9**, 330 (1966)
1.13 V.S. Butylkin, Yu.G. Khronopulo: Sov. Radiophys. Quantum Electron. **10**, 106 (1967)
1.14 E.A. Manykin: In *Propagation of Radiation through Matter* (Atomizdat, Moscow 1968) p. 72 (in Russian)
1.15 B.P. Kirsanov: In *Nonlinear Optics*, Proc. Lebedev Phys. Inst. USSR Academy Sci., Vol 43 (Nauka, Moscow 1968) p. 187 (in Russian)
1.16 M. Takatsuji: Phys. Rev. A **11**, 619 (1975)
1.17 V.S. Butylkin, Yu.G. Khronopulo, E.I. Yakubovich: Sov. Phys.-JETP **44**, 897 (1976)
1.18 S.A. Akhmanov, A.S. Chirkin: *Statistical Phenomena in Nonlinear Optics* (Lomonossov Univ. Press, Moscow 1971) (in Russian)
1.19 A.M. Clogston: J. Phys. Chem. Solids **4**, 271 (1958); see also A.M. Clogston: *Quantum Paramagnetic Amplifiers* (Inostrannaya literatura, Moscow 1961), pp. 170–181 (in Russian)
1.20 N.N. Bogolyubov, Yu.A. Mitropol'sky: *Asymptotic Methods in the Theory of Nonlinear Oscillations* (Gordon and Breach, New York 1967)
1.21 Yu.A. Mitropol'sky: *The Averaging Method in Nonlinear Mechanics* (Naukova Dumka, Kiev 1971) (in Russian)
1.22 V.N. Fomin: *The Mathematical Theory of Parametric Resonance in Linear Distributed Systems* (Leningrad State Univ., Leningrad 1972) (in Russian)
1.23 L.D. Landau, E.M. Lifshitz: *Quantum Mechanics: Nonrelativistic Theory* (Pergamon, Oxford 1965)
1.24 I.I. Sobel'man: *Introduction into the Theory of Atomic Spectra* (Pergamon, Oxford 1972)
1.25 A.M. Bonch-Bruevich, V.A. Khodovoi: Sov. Phys.-Usp. **10**, 637 (1968)
1.26 W. Heitler: *The Quantum Theory of Radiation* (Clarendon Press, Oxford 1954)
1.27 A.M. Bonch-Bruevich, V.A. Khodovoi: Sov. Phys.-Usp. **8**, 1 (1965)
1.28 N. Bloembergen, Y.R. Shen: Phys. Rev. **133**, No. A 37 (1964)
1.29 V.S. Butylkin: Abstract of Ph.D. Thesis (USSR Acad. Sci. Inst. for Radioeng. and Electronics, Moscow 1971)

1.30 V.I. Arnol'd: Dokl. Akad. Nauk **161,** 9 (1965) (in Russian)
1.31 L. Allen: *Coherence and Quantum Opt. 5*. Proc. 5 Rochester Conf., June 13–15, 1983 (New York, London 1984) p. 373
1.32 M.V. Fedorov: J. Phys. B: Atom and Molec. Phys. **10**, 2573 (1977)
1.33 M.V. Fedorov: Izv. AN SSSR, Ser. fiz. **41**, 2569 (1977) (in Russian)
1.34 V.S. Butylkin, Yu.G. Khronopulo: *A Description of Resonant Multiphoton Interactions of Light and Matter by Means of the Generalized Two-Level System* (IRE Reprint USSR Acad. Sci., Moscow 1974) (in Russian); *The Generalized Two-Level System Populations Change under the Effect of Quasy-Monochromatic Fields* (IRE Reprint, USSR Acad. Sci., Moscow 1975) (in Russian)
1.35 V.S. Butylkin, Yu.G. Khronopulo: *The Generalized Two-Level System: Polarization and Susceptibility* (IRE Reprint, USSR Acad. Sci., Moscow 1976) (in Russian)

Chapter 2

2.1 N.D. Artamonova, V.T. Platonenko, R.V. Khokhlov: Sov. Phys.-JETP **31,** 1185 (1970)
2.2 L.V. Keldysh: Sov. Phys.-JETP **20,** 1307 (1965)
2.3 G.A. Delone, N.B. Delone, V.K. Zolotaryov, M.A. Tursunov: Brief Reports of Lebedev Phys. Inst., No. 2, 37 (1973) (in Russian)
2.4 G.S. Voronov: Sov. Phys.-JETP **24,** 1009 (1967)
2.5 V.S. Butylkin, O.V. Kudrevatova: Theor. Math. Phys. **5,** 1053 (1970)
2.6 O.V. Kudrevatova, L.V. Norinsky, V.A. Pryadein: Sov. Phys.-JETP **18,** 1482 (1974)
2.7 R. Karplus, J. Schwinger: Phys. Rev. **73,** 1020 (1948)
2.8 V.M. Fain: *Photons and Nonlinear Media* (Sovietskoye radio, Moscow 1972), p. 250 (in Russian)
2.9 V.S. Butylkin, A.E. Kaplan, Yu.G. Khronopulo: Sov. Phys.-JETP **32,** 501 (1971)
2.10 A. Javan, P. Kelley: IEEE J. Quant. El., **QE-2,** 470 (1966)
2.11 A.M. Bonch-Bruevich, V.A. Khodovoi: Sov. Phys.-Usp. **10,** 637 (1968)
2.12 I.I. Sobel'man: *Introduction into the Theory of Atomic Spectra* (Pergamon, Oxford 1972)
2.13 K.B. Eisenthal, K.E. Rieckhoff: J. Chem. Phys. **55,** 3317 (1971)
2.14 N. Bloembergen: *Nonlinear Optics* (Benjamin, New York 1965)
2.15 S.A. Akhmanov, A.P. Sukhorukov, R.V. Khokhlov: Sov. Phys.-Usp. **10,** 609 (1968)
2.16 N.G. Basov, P.G. Kryukov, Yu.V. Senatsky, S.V. Chekalin: Sov. Phys.-JETP **10,** 641 (1970)
2.17 V.V. Korobkin, A.J. Alcock: Phys. Rev. Lett. **21,** 1433 (1968)
2.18 B. Wilgelmi, E. Goiman: Zhurn. prikl. spektroskopii **19,** 550 (1973) (in Russian)
2.19 V.P. Protasov: Abstract of Ph.D. thesis (Moscow State Univ., Moscow 1974)
2.20 V.S. Butylkin, G.V. Venkin, L.L. Kulyuk, D.I. Maleev, V.P. Protasov, Yu.G. Khronopulo: JETP Letters **19,** 253 (1974)
2.21 V.R. Belan, V.V. Grigor'yants, M.E. Zhabotinsky: JETP Letters **6,** 200 (1967)
2.22 A.P. Aleksandrov, V.N. Genkin, M.I. Kheifets: Opt. Spectrosc. **20**, 69 (1966)
2.23 D.E. Burlankov, A.V. Rezvov: Opt. Spectrosc. **20,** 262 (1966)
2.24 V.R. Blok, G.M. Krochik, Yu.G. Khronopulo: Sov. Phys.-JETP **49**, 23 (1979)
2.25 V.V. Krasnikov, M.S. Pshenichnikov, V.S. Solomatin: JETP Lett. **43**, 148 (1986); Optika i spektroskopiya **62**, 10 (1987) (in Russian); Zhurn. exp. i teor. fiz. **92**, 1578 (1987) (in Russian)
2.26 V.R. Blok, G.M. Krochik, Yu.G. Khronopulo: Sov. J. Quant. Electron. **9**, 1444 (1979)
2.27 Cun-Kai Wu, Jun-Yin Fan: Uli syuebao. Acta Phys. Sinica **28**, 748 (1979)
2.28 M.A. Vasil'eva, Yu. Vishchakas, V. Gul'binas, V. Kabelka, A.V. Maslov, V. Syrus: Sov. J. Quant. Electron. **15**, 656 (1985)
2.29 G.B. Al'tshuler, S.A. Kozlov: Sov. J. Quant. Electron. **15**, 459 (1985)
2.30 A. Flusberg, D. Korff: J. Opt. Soc. Am. **B 3**, 1338 (1986)
2.31 A.Yu. Okulov, A.N. Oraevsky: Kvantovaya elektronika **15**, 369 (1988) (in Russian)
2.32 N.G. Basov, V.A. Danilychev, I.G. Rudoi, A.M. Soroka: Dokl. Akad. Nauk SSSR **284**, 1346 (1985) (in Russian)
2.33 A.I. Osipov, V.Ya. Panchenko, A.A. Fillippov: Sov. J. Quant. Electron. **15**, 465 (1985); Zhurn. Tekhn. Fiz. **55**, 1172 (1985) (in Russian)

2.34 E.N. Bazarov, G.A. Gerasimov, V.P. Gubin, N.I. Starostin, V.V. Fomin: Sov. J. Quant. Electron. **15**, 1036 (1985)
2.35 V.Yu. Baranov, L.A. Bol'shov, T.K. Kirichenko, S.M. Kozochkin, V.V. Likhansky, K.N. Makarov, D.D. Malyuta, Yu.A. Satov, L.K. Sokolova, A.P. Strel'tsov: Kvantovaya elektronika **14**, 707 (1987) (in Russian)
2.36 Y. Beaudoin, P. Galarneau, A. Normandin, S.L. Chin: Appl. Phys. B **42**, 225 (1987)
2.37 V.M. Arutyunyan, A.Zh. Muradyan, L.S. Petrosyan: Optika i Spektroskopiya **59**, 643 (1985) (in Russian)
2.38 G.G. Adonts, E.G. Kanetsyan, M.V. Slobodskoi: Optika i Spektroskopiya **64**, 258 (1988)
2.39 L.I. Burov, I.I. Gancharenok: Optika i Spektroskopiya **64**, 306 (1988) (in Russian)

Chapter 3

3.1 S.S. Alimpiev, I.V. Karlov: Sov. Phys.-JETP **39**, 260 (1974)
3.2 I.I. Sobelman: *Introduction to the Theory of Atomic Spectra* (Pergamon, Oxford 1972)
3.3 A. Yariv: *Quantum Electronics* (J. Wiley, New York 1975)
3.4 L.D. Landau, E.M. Lifshitz: *Quantum Mechanics: Non-Relativistic Theory* (Pergamon, Oxford 1965)
3.5 L. Allen, J.H. Eberly: *Optical Resonance and Two-Level Atoms* (Wiley, New York 1975)
3.6 I.I. Rabi: Phys. Rev. **51**, 652 (1937)
3.7 R. Karplus, J. Schwinger: Phys. Rev. **73**, 1020 (1948)
3.8 C.H. Townes, A.I. Schawlow: *Microwave Spectroscopy* (McGraw-Hill, New York 1955)
3.9 V.M. Fain: *Photons and Nonlinear Media* (Sovietskoye Radio, Moscow 1972) (in Russian)
3.10 G.P. Lyubimov, R.V. Khokhlov: Sov. Phys.-JETP **6**, 1074 (1958)
3.11 V.M. Fain: Izv. vuzov. Radiofiz. **2**, 167 (1959)
3.12 S.G. Rautian *Nonlinear Optics* (coll. pap.), Proc. Lebedev Phys. Inst., Vol. 43 (Nauka, Moscow 1968) p. 3 (in Russian)
3.13 Z.K. Yankauskas: Sov. Phys.-Tech. Phys. **13**, 1506 (1969)
3.14 A.E. Kaplan: Sov. Phys.-JETP **38**, 705 (1974)
3.15 T.A. Germogenova, S.G. Rautian: Sov. Phys.-JETP **17**, 507 (1964)
3.16 L.S. Turovtseva: USSR Comput. Math. Phys. **6**, 229 (1966)
3.17 S.G. Rautian, R.I. Sokolovsky: In *Nonlinear Processes in Optics* (coll. pap.) (Nauka, Novosibirsk 1970) p. 45 (in Russian)
3.18 S.G. Rautian, I.I. Sobelman: Sov. Phys.-JETP **14**, 328 (1962)
3.19 S.G. Rautian, I.I. Sobelman: Sov. Phys.-JETP **17**, 635 (1963)
3.20 W.F. Lamb: Phys. Rev. **134**, A1429 (1964)
3.21 F.V. Bunkin: Izv. vuzov. Radiofizika **5**, 687 (1962)
3.22 N. Bloembergen, Y.R. Shen: Phys. Rev. **133**, A37 (1964)
3.23 A.E. Kaplan: In Theses of the Sixth All-Union Symposium on Nonlinear Optics (Moscow State University, Moscow 1972) p. 25 (in Russian)
3.24 S.G. Przhibel'sky, V.A. Khodovoi: Opt. Spectrosc. **32**, 125 (1972)
3.25 B.A. Zon, B.G. Katsnel'son: Radiophys. Quantum Electron. **16**, 290 (1973)
3.26 A.I. Burshtein, Yu.I. Naberukhin: Sov. Phys.-JETP **25**, 799 (1967)
3.27 A.I. Burshtein: *Lectures on Quantum Kinetics* (Novosibirsk State University, Novosibirsk 1963) (in Russian)
3.28 L.D. Zusman, A.I. Burshtein: Sov. Phys.-JETP **34**, 520 (1972)
3.29 S.G. Przhibel'sky, V.A. Khodovoi: Opt. Spectrosc. **32**, 125 (1972)
3.30 T.I. Kuznetsova, S.G. Rautian: Sov. Phys.-JETP 22, 1098 (1966)
3.31 T.I. Kuznetsova: In *Nonlinear Optics* (coll. pap.), Proc. of Lebedev Phys. Inst., Vol. 43 (Nauka, Moscow 1968) p. 116 (in Russian)
3.32 V.M. Fain, Ya.I. Khanin: *Quantum Electronics* (MIT Press, Cambridge 1969)
3.33 B. Senitzky, G. Gould, S. Cutler: Phys. Rev. **130**, 1460 (1963)
3.34 D.N. Klyshko, Yu.S. Konstantinov, V.S. Tumanov: Sov. Radiophys. **8**, 368 (1965)
3.35 P.A. Apanasevich: Dokl. Akad. Nauk BSSR **12**, 878 (1968) (in Russian)

3.36 A.M. Bonch-Bruevich, N.N. Kostin, V.A. Khodovoi, V.V. Khromov: Sov. Phys.-JETP **29**, 82 (1969)
3.37 E.B. Aleksandrov, A.M. Bonch-Bruevich, V.A. Khodovoi, N.A. Chigir': JETP Lett. **18**, 58 (1973)
3.38 A.M. Bonch-Bruevich, V.A. Khodovoi: Sov. Phys.-Vsp. **10**, 637 (1968)
3.39 I.M. Beterov, R.I. Sokolovsky: Sov. Phys.-Vsp. **16**, 339 (1973)
3.40 A.M. Bonch-Bruevich, V.A. Khodovoi, V.V. Khromov: JETP Lett. **11**, 290 (1970)
3.41 A.E. Kaplan: Sov. Phys.-JETP **41**, 409 (1975)
3.42 A.E. Kaplan: Sov. J. Quantum Electron. **6**, 728 (1976)
3.43 S.L. McCall, E.L. Hahn: Phys. Rev. Lett. **18**, 908 (1967)
3.44 P.G. Krykov, V.S. Letokhov: Sov. Phys.-Vsp. **12**, 641 (1970)
3.45 I.A. Poluektov, Yu.M. Popov, V.S. Roitberg: Sov. Phys.-Vsp. **17**, 673 (1975); Sov. J. Quantum Electron. **4**, 424, 719 (1974)
3.46 S.M. Rytov: *Introduction to Statistical Radiophysics*, 2nd ed. (Nauka, Moscow 1976) (in Russian)
3.47 N.B. Delone, V.P. Krainov: *The Atom in a Strong Light Field* (Atomizdat, Moscow 1978) (in Russian)
3.48 L.P. Rapoport, B.A. Zon, N.L. Monakhov: *The Theory of Intraatomic Multiphoton Processes* (Atomizdat, Moscow 1978) (in Russian)
3.49 R. Boscaino, F.M. Gelurdi: J. Phys. **C11**, 1475 (1978); Proc. of the XXth Congress Ampere, Tallin, USSR (1978)
3.50 E. Kamke: *Differentialgleichungen, Lösungsmethoden und Lösungen* (Leipzig 1959)
3.51 H. Bateman, A. Erdélyi: *Higher Transcendental Functions*, Calif. Inst. Techn. (1953–1955)
3.52 L. Landau: Phys. Z. Sowjetunion **2**, 46 (1932);
C. Zener: Proc. R. Soc. London, **A137**, 696 (1932)

Chapter 4

4.1 G.A. Plachek: *Rayleigh Scattering and Raman Effect* (Ukraine State Publ. of Sci. and Tech. Literature, Kharkov 1935) (in Russian)
4.2 V.B. Berestetskii, E.M. Lifshitz, L.P. Pitaevskii: *Relativistic Quantum Theory* (Pergamon, Oxford 1974)
4.3 L.D. Landau, E.M. Lifshitz: *Quantum Mechanics: Nonrelativistic Theory* (Pergamon, Oxford 1965)
4.4 L.S. Vasilenko, V.P. Chebotaev, A.V. Shishaev: JETP Lett. **12**, 113 (1970)
4.5 E.I. Yakubovich: Sov. Phys.-JETP **30**, 1151 (1970)
4.6 R.F. Boikova, E.E. Fradkin: Sov. Radiophys. **8**, 780 (1965)
4.7 L.D. Landau, E.M. Lifshitz: *Electrodynamics of Continuous Media* (Pergamon, Oxford 1960)
4.8 E.I. Yakubovich: Sov. Phys.-JETP **29**, 370 (1969)
4.9 E.I. Yakubovich: Abstract of Ph.D. thesis (Gor'ky State University, Gor'ky 1971) (in Russian)

Chapter 5

5.1 L.A. Ostrovsky, E.I. Yakubovich: Sov. Radiophys. **8**, 63 (1966)
5.2 T.I. Kuznetsova, S.G. Rautian: Izv. vuzov. Radiofizika **7**, 682 (1964)
5.3 L.A. Ostrovsky, E.I. Yakubovich: Sov. Phys.-JETP **19**, 656 (1964)
5.4 A.G. Litvak, M.I. Petelin, E.I. Yakubovich: Sov. Phys.-Tech. Phys. **10**, 135 (1965)
5.5 V.M. Fain, Ya.I. Khanin: Sov. Phys.-JETP **14**, 1069 (1962)
5.6 A.P. Ivanov, B.M. Berkovsky, I.L. Katsev: Inzhen. fiz. zhurn. **5**, 58 (1962) (in Russian)

5.7 B.I. Stepanov, A.P. Ivanov, B.M. Berkovskii, I.L. Katsev: Opt. Spectrosc. **12**, 299 (1962)
5.8 E.S. Kovalenko: Radiophys. Quantum Electron. **10**, 102 (1967)
5.9 E.I. Yakubovich: Sov. Phys.-JETP **29**, 370 (1969)
5.10 I.L. Bershtein, V.A. Rogachev: Radiophys. Quantum Electron. **13**, 23 (1970)
5.11 C. Wieman, T.W. Hänsch: Phys. Rev. Lett. **36**, 1170 (1976)

Chapter 6

6.1 D.A. Kleinman: Phys. Rev. **128**, 1761 (1962)
6.2 B. Lax, J.G. Mavroides, D.F. Edwards: Phys. Rev. Lett. **8**, 166 (1962)
6.3 R. Loudon: Proc. Phys. Soc. **80**, 952 (1962)
6.4 A.M. Afanas'ev, E.A. Manykin: Sov. Phys.-JETP **21**, 323 (1965)
6.5 N.D. Artamonova, V.T. Platonenko, R.V. Khokhlov: Sov. Phys.-JETP **31**, 1185 (1970)
6.6 G.L. Gurevich, Yu.G. Khronopulo: Sov. Phys.-JETP **24**, 1012 (1967)
6.7 B.A. Akanaev, Yu.G. Khronopulo: In *Nonlinear Optics* (coll. pap.), Proc. of the 2nd All-Union Symposium on Nonlinear Optics (Nauka, Novosibirsk 1968) p. 260 (in Russian)
6.8 B.A. Akanaev, S.A. Akhmanov, Yu.G. Khronopulo: Sov. Phys.-JETP **28**, 656 (1969); Abstracts of Reports to the 3rd All-Union Symposium on Nonlinear Optics (Moscow State University, Moscow 1967) p. 52 (in Russian)
6.9 V.S. Butylkin, E.A. Lavrovsky, V.A. Kolesnikov: Abstracts of Reports to the 6th All-Union Conference on Nonlinear Optics, Minsk (1972) p. 202 (in Russian)
6.10 G.M. Krochik, Yu.G. Khronopulo: Sov. J. Quant. Electron. **4**, 1076 (1975)
6.11 A.I. Maimistov, L.R. Malov, E.A. Manykin: Sov. J. Quant. Electron. **5**, 375 (1975)
6.12 Yu.A. Il'insky, V.D. Taranukhin: Sov. J. Quant. Electron. **4**, 491 (1974)
6.13 Yu.A. Il'insky, V.D. Taranukhin: Sov. J. Quant. Electron. **4**, 828 (1975)
6.14 Yu.A. Il'insky, V.D. Taranukhin: Sov. J. Quant. Electron. **5**, 171 (1975)
6.15 D.C. Haueisen, H. Mahr: Phys. Rev. Lett. **26**, 838 (1971)
6.16 B.L. Davydov, V.V. Dunina, V.F. Zolin, L.G. Koreneva, M.A. Samokhina, E.P. Shliteris: Opt. Spectrosc. **32**, 119 (1972)
6.17 S.A. Akhmanov, R.V. Khokhlov: *Problems of Nonlinear Optics* (Gordon and Breach, New York 1972)
6.18 E. Kamke: *Differentialgleichungen, Lösungsmethoden und Lösungen*, Band I. Gewöhnliche Differentialgleichungen (Geest und Portig, Leipzig 1959)
6.19 G.A. Kasabov, V.V. Eliseev: *Spectroscopic Tables for Low Temperature Plasmas* (Atomizdat, Moscow 1973) (in Russian)
6.20 C.E. Moore: Atomic Energy Levels, Circ. No. 467 (Natl. Bur. Std., Washington 1949–58)
6.21 J.F. Ward, G.H. New: Phys. Rev. **185**, 57 (1969)
6.22 K. Washio, T. Yamane, S. Yoshikawa, K. Koizuma, Y. Iheda: IEEE J. Quant. El. **QE-9**, 807 (1973)
6.23 A.N. Nesmeyanov: *The Pressure of Vapor of Chemical Elements* (USSR Acad. of Sci., Moscow 1961) (in Russian)
6.24 A.H. Kung, J.F. Young, G.C. Bjorklund, S.E. Harris: Phys. Rev. Lett. **29**, 985 (1972)
6.25 P.N. Butcher, R. Loudon, T.P. McLean: Proc. Phys. Soc. **85**, 565 (1965)
6.26 R. Loudon: Proc. Phys. Soc. **82**, 393 (1963)
6.27 E. Garmire, E. Pandarese, C.H. Townes: Phys. Rev. Lett. **11**, 160 (1963)
6.28 F. De Martini: J. Appl. Phys. **37**, 4503 (1966)
6.29 V.S. Butylkin, D.Yu. Kozyarsky, E.N. Plyusnina, P.S. Fisher, Yu.G. Khronopulo: Sov. J. Quant. Electron. **5**, 1242 (1975)
6.30 M.G. Mayer: Compt. Rend. **266**B, 123 (1968)
6.31 V.S. Butylkin, P.S. Fisher, Yu.G. Khronopulo: Sov. Tech. Phys. Lett. **2**, 99 (1976)
6.32 F. Capasso, F. De Martini: Opt. Communs. **9**, 172 (1973)
6.33 V.T. Platonenko, R.V. Khokhlov: Sov. Phys.-JETP **19**, 378 (1964)
6.34 E.E. Hagenlocker, R.W. Minck, W.G. Rado: Phys. Rev. **154**, 226 (1967)

6.35 H.L. Buils, H.P. Gush: Canad. J. of Physics **49,** 2366 (1971)
6.36 R.W. Minck, W.G. Rado: In *The Effect of Laser Radiation* (Mir, Moscow 1968) p. 59 (in Russian)
6.37 J.M. Yarborough, S.S. Sussman, H.E. Puthoff, R.H. Pantell, B.C. Johnson: Appl. Phys. Lett. **15,** 102 (1969)
6.38 V.S. Butylkin, P.S. Fisher, M.F. Shalyaev: Sov. J. Quant. Electron. **10,** 202 (1980)
6.39 I. Meek, I. Craggs: *Electrical Breakdown of Gases* (Clarendon, Oxford 1953)
6.40 L.M. Blinov: *Elektro and Magneto-Optics of Liquid Crystals* (Nauka, Moscow 1978) (in Russian)
6.41 A. Flusberg, T. Mossberg, S. Hartmann: Phys. Rev. Lett. **38,** 59 (1977)
6.42 B.D. Sinclair, M.H. Dunn: Phys. Rev. A **34,** 3989 (1986)
6.43 V.S. Butylkin, M.F. Shalyaev: In *Nonlinear Resonant Conversion of the Laser Radiation Frequency. Theses of Reports* (Publ. House "Fan", Uzbek SSR, Tashkent 1979) p. 9 (in Russian)
6.44 B. Dick, R.M. Hochstasser: Chem. Phys. **91,** 1 (1984)
6.45 R.W. Boyd, D.J. Gouthier, J. Krasinski, M.S. Malcuit: In *Laser Techn. Extreme Ultraviolet. 2 Top. Meet., Boulder, Colo., 5–7 March, 1984* (New York 1984) p. 356

Chapter 7

7.1 R.W. Terhune: Bull. Am. Phys. Soc. II **8,** 359 (1963)
7.2 E. Garmire: "An Investigation of Stimulated Raman Emission", Submitted in Partial Fulfilment of the Requirements for Ph.D. at MIT, 1965
7.3 R.W. Minck, R.W. Terhune, W.G. Rado: Appl. Phys. Lett. **3,** 181 (1963)
7.4 A.I. Sokolovskaya, A.D. Kudryavtseva, M.M. Sushchinsky: In *Nonlinear Optics* (coll. pap.) (Proc. of the 2nd All-Union Symposium on Nonlinear Optics) (Nauka, Novosibirsk 1968) p. 277 (in Russian)
7.5 A.I. Sokolovskaya, E.A. Morozova, A.D. Kudryavtseva, M.M. Sushchinsky: Sov. J. Quant. Electr. **3,** 321 (1974)
7.6 K.A. Prokhorov, M.M. Sushchinsky: Kratkiye soobshcheniya po fizike, Brief Reports on Physics (Lebedev Physics Institute), No. 5, 48 (1970) (in Russian)
7.7 A.V. Bortkevich, Ya.S. Bobovich: Opt. Spectr. **29,** 475 (1970)
7.8 G.V. Venkin: Abstract of Ph.D. Thesis (Moscow State University, Moscow 1970) (in Russian)
7.9 V.T. Platonenko, R.V. Khokhlov: JETP Letters 2, 269 (1965)
7.10 E.A. Manykin, A.M. Afanas'ev: Sov. Phys.-JETP **21,** 619 (1965)
7.11 P.P. Sorokin, N.S. Shiren, J.R. Lankard, E.C. Hammond, T.G. Kazyka: Appl. Phys. Lett. **10,** 44 (1967); Bull. Am. Phys. Soc. **12,** 112 (1967)
7.12 M. Rokni, S. Yatsiv: Phys. Lett. **24A,** 277 (1967)
7.13 M. Rokni, S. Yatsiv: IEEE J. Quant. El. **QE-3,** 329 (1967)
7.14 O.J. Lumpkin: IEEE J. Quant. El. **QE-4,** 226 (1968)
7.15 A.M. Bonch-Bruevich, V.A. Khodovoi, V.V. Khromov: JETP Letters **14,** 333 (1971)
7.16 F.A. Korolyov, S.A. Bakhramov, V.I. Odintsov: Opt. Spectr. **30,** 425 (1971)
7.17 P.P. Sorokin, J.J. Wynne, J.R. Lankard: Appl. Phys. Lett. **22,** 342 (1973)
7.18 R.T. Hodgson, P.P. Sorokin, J.J. Wynne: Phys. Rev. Lett. **32,** 343 (1974)
7.19 D.M. Bloom, J.T. Yardley, J.F. Young, S.E. Harris: Appl. Phys. Lett. **24,** 427 (1974)
7.20 B.P. Kirsanov, A.S. Selivanenko: Opt. Spectr. **26,** 489 (1969)
7.21 S.E. Harris, D.M. Bloom: Appl. Phys. Lett. **24,** 229 (1974)
7.22 P.P. Sorokin, J.R. Lankard: IEEE J. Quant. El. **QE-9,** 227 (1973)
7.23 E.A. Stappaerts, S.E. Harris, J.F. Young: Appl. Phys. Lett. **29,** 669 (1976)
7.24 V.S. Butylkin, G.V. Venkin, V.P. Protasov, N.D. Smirnov, Yu.G. Khronopulo, M.F. Shalyaev: JETP Letters **17,** 285 (1973)
7.25 K. Shimoda: Jap. J. Appl. Phys. **5,** 86 (1966)
7.26 A.N. Arbatskaya, M.M. Sushchinsky: Sov. Phys.-JETP **39,** 981 (1974)
7.27 V.I. Barantsov, A.K. Popov, G.Kh. Tartakovsky: Sov. Phys.-JETP **40,** 449 (1975)
7.28 A.H. Kung, J.F. Young, S.E. Harris: Appl. Phys. Lett. **22,** 301 (1973)

7.29 A.H. Kung, J.F. Young, G.C. Bjorklund, S.E. Harris: Phys. Rev. Lett. **29**, 985 (1972)
7.30 N. Bloembergen: *Nonlinear Optics*, A lecture note (Benjamin, New York 1965)
7.31 V.S. Butylkin, G.M. Krochik, Yu.G. Khronopulo: Sov. Phys.-JETP **41**, 247 (1975)
7.32 G.V. Venkin, G.M. Krochik, L.L. Kulyuk, D.I. Maleev, Yu.G. Khronopulo: JETP Letters **21**, 105 (1975); Sov. Phys.-JETP **43**, 873 (1976)
7.33 J.A. Giordmaine, W. Kaiser: Phys. Rev. **144**, 676 (1966)
7.34 S.A. Akhmanov, N.I. Koroteev: Sov. Phys.-JETP **40**, 650 (1975)
7.35 J.A. Duardo, F.M. Johnson, L.J. Nugent: IEEE J. Quant. El. **QE-4**, 397 (1968)
7.36 V.N. Lugovoi: *Introduction into the Theory of Stimulated Raman Scattering* (Nauka, Msocow 1968) (in Russian)
7.37 V.N. Lugovoi, I.I. Sobelman: Sov. Phys.-JETP **31**, 690 (1970)
7.38 W.G. Rado: Appl. Phys. Lett. **11**, 123 (1967)
7.39 A.D. Kudryavtseva, A.I. Sokolovskaya, M.M. Sushchinsky: Sov. Phys.-JETP **32**, 849 (1971)
7.40 Yu.E. D'yakov: Kratkie soobshcheniya po fizike (Brief Report on Physics, Lebedev Physics Inst.) No. 12, 34 (1973) (in Russian)
7.41 D.N. Klyshko: Sov. Phys.-JETP **37**, 590 (1973)
7.42 V.S. Butylkin, G.V. Venkin, V.P. Protasov, P.S. Fisher, Yu.G. Khronopulo, M.F. Shalyaev: Sov. Phys.-JETP **43**, 430 (1976)
7.43 V.S. Butylkin, A.E. Kaplan, Yu.G. Khronopulo: Opt. Spectr. **31**, 120 (1971)
7.44 V.P. Protasov: Abstract of Ph.D. Thesis (Moscow State University, Moscow 1974) (in Russian)
7.45 A.G. Ford, J.C. Browne: Atomic Data **5**, 305 (1973)
7.46 P. Lallemand, P. Simova, G. Bret: Phys. Rev. Lett. **17**, 1239 (1966)
7.47 M. Born, E. Wolf: *Principles of Optics*; Electromagnetic Theory of Propagation, Interference and Diffraction of Light (Macmillan, New York 1964)
7.48 G.V. Venkin, G.M. Krochik, L.L. Kulyuk, D.I. Maleev, Yu.G. Khronopulo: Sov. J. Quant. Electr. **6**, 369 (1976)
7.49 V.B. Markov, T.Ya. Marusii, M.S. Soskin, A.I. Khizhnyak: Sov. Phys.-JETP **37**, 778 (1973)
7.50 C.E. Moore: Atomic Energy Levels, Natl. Bur. Stds (USA), Circ. No. 467, Washington, DC (1949–58)
7.51 A.N. Nesmeyanov: *The Pressure of Vapor of Chemical Elements* (USSR Acad. of Sci., Moscow 1961) (in Russian)
7.52 J.F. Ward, G.H.C. New: Phys. Rev. **185**, 57 (1969)
7.53 G.M. Krochik, Yu.G. Khronopulo: Sov. J. Quant. Electron. **5**, 917 (1975)
7.54 G.A. Kasabov, V.V. Eliseev: *Spectroscopic Tables for Low Temperature Plasmas* (Atomizdat, Moscow 1973) (in Russian)
7.55 A.S. Selivanenko: Opt. Spectr. **21**, 54 (1966)
7.56 P.P. Sorokin, N. Braslau: IBM J. Res. Develop. **8**, 182 (1964)
7.57 G.L. Gurevich, Yu.G. Khronopulo: Sov. Radiophys. **8**, 353 (1965)
7.58 V.S. Butylkin, G.L. Gurevich, M.I. Kheifets, Yu.G. Khronopulo: Sov. Radiophys. **9**, 330 (1966)
7.59 G.L. Gurevich, M.I. Kheifets: Radiophys. Quant. Electron. **10**, 22 (1967)
7.60 B.P. Kirsanov, A.S. Selivanenko: Opt. Spectr. **23**, 242 (1967)
7.61 A.J. De Maria, C.J. Ultee: Appl. Phys. Lett. **9**, 67 (1966)
7.62 J.R. Murray, S.K. Rhodes: Laurence-Livermore Lab. Report UCRL-51455 (1973)
7.63 J.R. Murray, H.T. Powell, S.K. Rhodes: IEEE J. Quant. El. **QE-10**, 781 (1974)
7.64 L.S. Vasilenko, V.P. Chebotaev, A.V. Shishaev: JETP Letters **12**, 113 (1970)
7.65 T.L. Andreeva, V.I. Malyshev, A.I. Maslov, I.I. Sobelman: JETP Letters **10**, 271 (1969)
7.66 T.L. Andreeva, S.S. Kuznetsova, A.I. Maslov, I.I. Sobelman, V.N. Sorokin: Khimiya vys. energii **6**, 418 (1972) (in Russian)
7.67 V.S. Zuev, V.A. Katulin, V.Yu. Nosach, O.Yu. Nosach: Sov. Phys.-JETP **35**, 870 (1972)
7.68 A.V. Vinogradov, E.A. Yukov: JETP Letters **16**, 447 (1972)
7.69 N.G. Basov, L.E. Golubev, V.S. Zuev, V.A. Katulin, V.N., Netemin, V.Yu. Nosach, O.Yu. Nosach, A.L. Petrov: Sov. J. Quant. Electr. **3**, 524 (1974)

7.70 G.M. Krochik, Yu.G. Khronopulo: JETP Letters **21**, 274 (1975)
7.71 W.R. Trutna, R.L. Byer: Appl. Optics **19**, 301 (1980)
7.72 V.S. Butylkin, G.V. Venkin, L.L. Kulyuk, D.I. Maleev, Yu.G. Khronopulo, M.F. Shalyaev: Sov. J. Quant. Electron. **7**, 867 (1977)
7.73 V.A. Gorbunov, K.Sh. Mustaev, V.A. Serebryakov: Pis'ma ZhTF **5**, 1244 (1979) (in Russian)
7.74 V.S. Butylkin, V.S. Grigor'yan, M.F. Shalyaev: In *Nonlinear Resonant Conversion of the Laser Radiation Frequency.* Theses of Reports (Fan, Tashkent 1979) (in Russian), p. 53
7.75 G.M. Krochik, Yu.G. Khronopulo: Sov. Phys.-JETP **51**, 245 (1980)
7.76 Ovadia Shlomo, Sargent III Murray, T. Hendow Sami: Opt. Commun. **49**, 447 (1984); Opt. Lett. **10**, 505 (1985)
7.77 N.P. Bashlakova, V.R. Blok, Yu.G. Khronopulo: Izv. vuzov. Radiofizika **28**, 1540 (1985) (in Russian)
7.78 T.R. Poyt, Chi H. Lee: Appl. Phys. Lett. **30**, 332 (1977)
7.79 I.A. Poluéktov: Sov. J. Quant. Electron. **7**, 364 (1977)
7.80 V.I. Anikin, K.N. Drabovich, A.N. Dubovik: Sov. Phys.-JETP **45**, 906 (1977)
7.81 J.N. Elgin, G.H.C. New: J. Phys. B **11**, 3439 (1978)
7.82 A.I. Maimistov, E.A. Manykin, L.B. Khodulev: Sov. Phys.-JETP **49**, 431 (1979)
7.83 V.S. Butylkin, V.S. Grigor'yan, M.E. Zhabotinskiĭ: Sov. J. Quant. Electron. **10**, 378 (1980); Sov. Phys.-JETP **51**, 832 (1980)
7.84 V.S. Butylkin, V.S. Grigor'yan, M.E. Zhabotinskiĭ: *Self-Induced and Parametric Transparency in Resonant Multi-Photon Interactions* (IRE Reprint, USSR Acad. Sci., Moscow 1980) (in Russian); Sov. Phys.-JETP **53**, 475 (1981)
7.85 D.G. Sarkisyan, S.O. Sapondzhyan: Sov. J. Quant. Electron. **14**, 1652 (1984)
7.86 V.S. Grigor'yan: Sov. Phys.-JETP **59**, 1166 (1984)
7.87 R. Beach, D. De Beer, S.R. Hartmann: Phys. Rev. A **32**, 3467 (1985)
7.88 G.V. Arutyunyan, G.P. Dzhotyan: Zhurn. prikladnoĭ spektroskopii **44**, 279 (1986) (in Russian)
7.89 V. Smith Arlee: Optics Lett. **10**, 341 (1985)
7.90 R.G. Caro, A. Costela, N.P. Smith, C.E. Webb: J. Phys. D **18**, 1291 (1985)
7.91 M.S. Dzhidzhoev, V.T. Platonenko, A.V. Chugunov: Sov. J. Quant. Electron. **15**, 1451 (1985)
7.92 V.F. Lukinykh, S.A. Myslivets, A.K. Popov, V.V. Slabko, L.V. Kirensky: Appl. Phys. B **38**, 143 (1985)
7.93 V.G. Bespalov, V.N. Krylov, D.I. Stasel'ko: Optika i spektroskopiya **59**, 486 (1985) (in Russian)
7.94 J. Burris, T.J. McIlrath: J. Opt. Soc. Am. B **2**, 1307 (1985); B **2**, 1313 (1985)
7.95 P.R. Herman, B.P. Stoicheff: Optics Lett. **10**, 502 (1985)
7.96 B.G. Sartakov: *Third Harmonic Generation under Resonant Radiation Interaction with Molecules* (Preprint of the Institute of General Physics of the USSR Acad. Sci., Moscow 1986) (in Russian)
7.97 S.G. Dinev, I.G. Koprinkov, I.L. Stefanov: Appl. Phys. B **39**, 65 (1986)
7.98 Yu. Oseledchik: Ukrainsky fizichesky zhurnal **30**, 1648 (1985) (in Russian)
7.99 M.S. Malcuit, D.J. Gauthier, R.W. Boyd: Phys. Rev. Lett. **55**, 1086 (1985)
7.100 A.V. Ivanova, G.G. Melikyan: J. Phys. B **18**, 557 (1985)
7.101 V.G. Arkhipkin, Yu.I. Heller, A.K. Popov, A.S. Provorov: Appl. Phys. B **37**, 93 (1985)
7.102 V.S. Butylkin, M.F. Shalyaev: Sov. J. Quant. Electron. **15**, 724 (1985); Kvantovaya elektronika **13**, 1685 (1986) (in Russian)
7.103 S.O. Sapondzhyan, D.G. Sarkisyan, G.A. Torosyan: Sov. J. Quant. Electron. **15**, 220 (1985)
7.104 N.A. Iskanderov, V.A. Kudryashov, N.E. Tuvaev: Sov. J. Quant. Electron. **15**, 561 (1985)
7.105 K.G.H. Baldwin, J.P. Marangos, D.D. Burgess, M.C. Gower: Opt. Commun. **52**, 351 (1985)
7.106 M. Bernardini, A. Palucci, A. Ferrario, S. Marchetti, M. Giorgi, S. Ribezzo: Nuovo cim. D **6**, 559 (1985)
7.107 V.G. Arkhipkin, A.L. Vysotin, Im Thek-de, O.P. Podavalova, A.K. Popov: Kvantovaya elektronika **13**, 1352 (1986) (in Russian)

7.108 V.F. Lukinykh, S.A. Myslivets, A.K. Popov, V.V. Slabko: Kvantovaya elektronika **13**, 1415 (1986) (in Russian)
7.109 L. Mannik, S.K. Brown: Opt. Commun. **57**, 360 (1986)
7.110 Kenzo Miyazaki, Hirofumi Sakai, Takuzo Sato: IEEE J. Quant. Electron. **QE-22**, 2266 (1986)
7.111 H.F. Döbele, M. Hörl, M. Röwekamp: Appl. Phys. B **42**, 67 (1987)
7.112 V. Vaîchaîtis, M. Ignatavichyus, V.A. Kudryashov, Yu.N. Pimenov, R. Yakute: Kvantovaya elektronika **14**, 762 (1987) (in Russian)
7.113 R.W. Boyd, M.S. Malcuit, D.J. Gauthier, K. Rzazewski: Phys. Rev. A **35**, 1648 (1987)
7.114 V.V. Lebedev, V.M. Plyasulya: Kvantovaya elektronika **15**, 127 (1988) (in Russian)
7.115 N.P. Makarov, A.K. Popov, V.P. Timofeev: Kvantovaya elektronika **15**, 757 (1988)
7.116 S.A. Bakhramov, G.Kh. Tartakovsky, P.K. Khabibullaev: *Nonlinear Resonant Processes and Frequency Conversion in Gases* (Publ. House "Fan", Uzbek SSR, Tashkent 1981) (in Russian)
7.117 R. Hilbig, G. Hilger, A. Lago, B. Wolf, R. Wallenstein: In *Short Wavelength Coher. Radiat.: Generat. and Appl.* Top. Meet. Monterey, Calif., March, 24–26, 1986 (New York, NY 1986) p. 382
7.118 A.K. Popov, V.P. Timofeev: *Frequency Addition in Nonlinear Gaseous Media under Resonant Two-photon Pumping (parts 1 and 2)* (Preprints of the Kirensky Phys. Inst., Krasnoyarsk 1984) (in Russian)
7.119 A.K. Popov: Infrared Phys. **25**, 21 (1985)
7.120 V.G. Arkhipkin, A.K. Popov: *Nonlinear Light Conversion in Gases* (Publ. House "Nauka", Novosibirsk 1987) (in Russian)

Chapter 8

8.1 G.A. Askar'yan: Sov. Phys.-JETP **15**, 1088 (1962)
8.2 V.I. Talanov: Izv. vuzov. Radiofizika **7**, 564 (1964); JETP Letters **2**, 138 (1965)
8.3 R. Chiao, E. Garmire, S. Townes: Phys. Rev. Lett. **13**, 479 (1964); P.L. Kelley: Phys. Rev. Lett. **15**, 1005 (1965)
8.4 S.A. Akhmanov, A.P. Sukhorukov, R.V. Khokhlov: Sov. Phys.-JETP **24**, 198 (1967)
8.5 V.N. Lugovoi, A.M. Prokhorov: JETP Letters **7**, 117 (1968)
8.6 V.E. Zakharov: Sov. Phys.-JETP **26**, 994 (1968)
8.7 S.A. Akhmanov, A.P. Sukhorukov, R.V. Khokhlov: Sov. Phys.-Usp. **10**, 609 (1968)
8.8 V.N. Lugovoi, A.M. Prokhorov: Sov. Phys.-Usp. **16**, 658 (1974)
8.9 G.A. Askar'yan: Sov. Phys.-Usp. **16**, 680 (1974)
8.10 A.E. Kaplan: JETP Letters **9**, 33 (1969); *Self-Refraction of Light in Nonlinear Media*, Abstracts of Rep. IV-th Nat. Symp. on Nonlinear Optics (Moscow State University, Moscow 1968) p. 100
8.11 G.A. Askar'yan: JETP Letters **4**, 270 (1966)
8.12 A. Javan, P.L. Kelley: IEEE J. Quant. El. **QE-2**, 470 (1966)
8.13 V.B. Krasovitsky, V.N. Kurilko: Sov. Phys.-Tech. Phys. **11**, 293 (1966)
8.14 A.E. Kaplan: Abstracts of Rep. V-th All-Union Conf. on Nonlinear Optics (Lomonosov Univ. Press, Moscow 1970) p. 69; V.S. Butylkin, A.E. Kaplan, Yu.G. Khronopulo: Sov. Phys.-JETP **34**, 276 (1972)
8.15 V.S. Butylkin, A.E. Kaplan, Yu.G. Khronopulo: Sov. Phys.-JETP **32**, 501 (1971)
8.16 A.M. Bonch-Bruevich, V.A. Khodovoi, V.V. Khromov: JETP Letters **11**, 290 (1970)
8.17 D. Grishkowsky: Phys. Rev. Lett. **24**, 866 (1970)
8.18 J.E. Bjorkholm, A. Ashkin: Phys. Rev. Lett. **32**, 129 (1974)
8.19 J.E. Bjorkholm, P.W. Smith, W.J. Tomlinson, A.E. Kaplan: Opt. Lett. **6**, 345 (1981)
8.20 G.A. Swartzlander, H. Yin, A.E. Kaplan: Opt. Lett. **13**, 1011 (1988)
8.21 M.G. Boshier, W.J. Sandle: Opt. Commun. **42**, 371 (1982); K. Tai et al.: Opt. Lett. **9**, 243 (1984)
8.22 Ch.C. Wang, G.W. Racette: Appl. Phys. Lett. **8**, 256 (1966)
8.23 W. Kaiser, A. Laubereau, M. Maier, J.A. Giordmaine: Phys. Lett. **22**, 60 (1966)
8.24 A.E. Kaplan: Radiophys. Quant. Electron. **12**, 692 (1969)
8.25 M.S. Brodin, A.M. Kamuz: JETP Letters **9**, 351 (1969)
8.26 A.A. Borshch, M.S. Brodin, V.I. Volkov, V.V. Ovchar: Sov. J. Quant. Electron. **5**, 340 (1975)

8.27 Yu.I. Kyzylasov: Abstracts of Rep. VIII-th All-Union Conf. on Coherent and Nonlinear Optics, Vol. 1 (Metsniereba, Tbilisi 1976) p. 290
8.28 S.A. Bakhramov, K.N. Drabovich, I.G. Kirin: Abstracts of Rep. VIII-th All-Union Conf. on Coherent and Nonlinear Optics, Vol. 1 (Metsniereba, Tbilisi 1976) p. 291
8.29 B.Ya. Zel'dovich: Kratkie soobshcheniya po fizike (Brief communications on Physics, Lebedev Physics Institute), No. 5, 20 (1970) (in Russian)
8.30 A.E. Kaplan: Abstracts of Rep. VIII-th All-Union Conf. on Coherent and Nonlinear Optics, Vol. 1 (Metsniereba, Tbilisi 1976) p. 292
8.31 V.S. Butylkin, A.E. Kaplan, Yu.G. Khronopulo: Opt. Spectr. **31**, 120 (1971)
8.32 R.Y. Chiao, M.A. Johnson, S. Krinsky, H.A. Smith, C.H. Townes, E. Garmire: IEEE J. Quant. El. **QE-2**, 467 (1966)
8.33 V.S. Butylkin, A.E. Kaplan, Yu.G. Khronopulo: Radiophys. Quant. Electron. **12**, 1395 (1969)
8.34 A.N. Sokolovskaya, A.D. Kudryavtseva, T.P. Zhbanova, M.I. Sushchinsky: Sov. Phys.-JETP **26**, 286 (1968)
8.35 Zh. Gizenzhel', A.D. Kudryavtseva, Zh. Rivua, A.I. Sokolovskaya: Sov. Phys.-JETP **44**, 917 (1976)
8.36 Zh. Gizenzhel', Zh. Rivua, A.D. Kudryavtseva, A.I. Sokolovskaya: Abstracts of Rep. VIII-th All-Union Conf. on Coherent and Nonlinear Optics, Vol. 1 (Metsniereba, Tbilisi 1976) p. 284
8.37 K. Deblan, Zh. Gizenzhel', Fu Suen Inguen, Zh. Rivua: Abstracts of Rep. VIII-th All-Union Conf. on Coherent and Nonlinear Optics, Vol. 2 (Metsniereba, Tbilisi 1976) p. 14
8.38 B.Kh. Bairamov, B.P. Zakharchenya, Z.M. Khashkhozhev: Sov. Phys.-Solid State **14**, 2357 (1970)
8.39 S.A. Bakhramov, U.G. Gulyamov, K.N. Drabovich, Ya.Z. Faizulaev: JETP Letters **21**, 102 (1975)
8.40 N. Bloembergen: *Nonlinear Optics.* A lecture note. (Benjamin, New York 1965)
8.41 G.L. Gurevich, Yu.A. Otmakhov: Sov. Radiophys. **9**, 673 (1966)
8.42 G.V. Venkin, L.L. Kulyuk, D.M. Maleev, V.P. Protasov: Sov. J. Quat. Electron. **41**, 1056 (1975)
8.43 P.W. Smith, W.J. Tomlinson: IEEE Spectrum **18**, 26 (1981)
8.44 A.E. Kaplan: JETP Lett. **24**, 114 (1976)
8.45 A.E. Kaplan: Sov. Phys. JETP **45**, 896 (1977)
8.46 A.E. Kaplan: Sov. J. Quantum Electronics **8**, 95 (1978)
8.47 A.E. Kaplan: Radiophys. Quant. Electron **22**, 229 (1979)
8.48 A.E. Kaplan: IEEE J. Quant. Electron **QE-17**, 336 (1981)
8.49 A.E. Kaplan: In *Optical Bistability,* ed. by C.W. Bowden, M. Ciftan, M.R. Rohl (Plenum, New York 1981) p. 447
8.50 P.W. Smith, T.-P. Herman, W.J. Tomlinson, P.L. Maloney: Appl. Phys. Lett. **35**, 846 (1979)
8.51 P.W. Smith, W.J. Tomlinson, P.J. Maloney, J.-P. Herman: IEEE J. Quant. Electron **QE-17**, 340 (1981)
8.52 P.W. Smith, W.J. Tomlinson: In *Optical Bistability,* ed. by C.W. Bowden, M. Ciftan, M.R. Robl (Plenum, New York 1981) p. 463
8.53 N.N. Rosanov: Sov. Tech. Phys. Lett. **3**, 583 (1977); **4**, 30 (1978)
8.54 V.A. Permyakov, O.V. Bagdasaryan: Radiophys. Quant. Electron **21**, 92 (1978)
8.55 N. Bloembergen, D.S. Pershan: Phys. Rev. **128**, 606 (1962);
N. Bloembergen, J. Ducuing: Phys. Lett. **6**, 5 (1963);
N. Bloembergen, C.H. Lee: Phys. Rev. Lett. **19**, 835 (1967)
8.56 H. Lotsch: Optik **32**, 776, 189, 299, 553 (1970/72)
8.57 J.J. Stoker: *Nonlinear Vibrations in Mechanical and Electrical Systems* (Interscience, 1950);
A.E. Kaplan, Yu.A. Kravtsov, V.A. Rylov: Parametric Oscillators and Frequency Dividers (Sov. Radio, Moscow 1966) (in Russian)
8.58 H.M. Gibbs: *Optical Bistability* (Academic, New York 1985)
8.59 H. Seidel: U.S. Patent No. 3,610,731 (1969)
8.60 A. Szöke, V. Daneu, T. Goldhar, N.A. Kurnit: Appl. Phys. Lett. **15**, 376 (1969)
8.61 H.M. Gibbs, S.L. McCall, T.N.C. Venkatesan: Phys. Rev. Lett. **36**, 1135 (1976)

8.62 F.S. Felber, J.H. Marburger: Appl. Phys. Lett. **28,** 731 (1976)
8.63 N.N. Rosanov: Opt. Spectr. **47,** 335 (1979)
8.64 A.A. Kolokolov, A.I. Sukov: Radiophys. Quant. Electron **21,** 1013 (1978)
8.65 D. Marcuse: Appl. Opt. **19,** 3130 (1980)
8.66 W.J. Tomlinson: Opt. Lett. **5,** 323 (1980)
8.67 A.E. Kaplan: J. Opt. Soc. Am. **71,** 1640 (1981)
8.68 W.J. Tomlinson, J.P. Gordon, P.W. Smith, A.E. Kaplan: Appl. Opt. **21,** 2041 (1982)
8.69 A.E. Kaplan: Appl. Phys. Lett. **38,** 67 (1981)
8.70 P.W. Smith, W.J. Tomlinson, P.J. Maloney, A.E. Kaplan: *Opt. Lett.* **7,** 57 (1982)
8.71 B.B. Boiko, I.Z. Dzhilavdari, N.S. Petrov: Appl. Spectr. **23,** 1511 (1975)
8.72 M.J. Moran, C.-Y. She, R.L. Carmen: IEEE J. Quant. Electr. **QE-11,** 259 (1975)
8.73 D.A.B. Miller, S.D. Smith, B.S. Wherett: Opt. Commun. **35,** 221 (1980)
8.74 A.E. Kaplan: Proc. 9th National Conf. on Coherent and Nonlinear Optics (Leningrad-Moscow, 1978) pp. 238, 241
8.75 I.C. Khoo, R.R. Michael, T.H. Liu, G. Finn, A.E. Kaplan: SPIE, Nonlinear Optics and Application **613,** 43 (1986);
I.C. Khoo, G.M. Finn, R.R. Michael, T.H. Liu: Optics Letters **11,** 227 (1986)
8.76 A.E. Kaplan: Opt. Lett. **6,** 360 (1981)
8.77 T.S. Dlodlo: Phys. Lett. **84A,** 107 (1981); Phys. Rev. **A27,** No. 2 (1983)
8.78 B. Bosacchi, L.M. Narducci: Opt. Lett. **8,** 324 (1983)
8.79 I.C. Khoo, J.Y. Hou: J. Opt. Soc. Am. **B2,** 761 (1985)
8.80 G.M. Vysin, H.J. Simon, R.T. Deck: Opt. Lett. **6,** 30 (1981)
8.81 G.Ya. Slepyan: Radiotekhnika i Electronika **8,** 1953 (1985) [Radio Eng. and Electr. Phys. **8** (1985)]
8.82 G.I. Stegeman: IEEE Trans. Micr. Theory and Applications **MTT-30,** 1598 (1982)
8.83 C.T. Seaton, J.D. Valera, R.L. Shoemaker, G.I. Stegeman, J.T. Chilwell, S.D. Smith: IEEE J. Quant. Electr. **QE-21,** 774 (1985)
8.84 H. Vach, C.T. Seaton, G.I. Stegeman, I.C. Khoo: Opt. Lett. **9,** 238 (1984)
8.85 A.G. Litvak, V.A. Mironov: Izv. Vyssh. Uchebn. Zaved. Radiofiz. **11,** 1911 (1968)
8.86 Yu. R. Alanakyan: Sov. Phys. Tech. Phys. **12,** 587 (1967)
8.87 Similar multisoliton emission from a nonlinear waveguide has recently been demonstrated in a computer simulation by E.M. Wright, G.E. Stegeman, C.T. Seaton, J.V. Moloney, and A.D. Boardman: to be published
8.88 P.W. Smith, W.J. Tomlinson: IEEE J. Quant. Electr. **QE-20,** 30 (1984)
8.89 G. Delfino, P. Mormile: Opt. Lett. **10,** 618 (1985)
8.90 J.E. Bjorkholm, P.W. Smith, W.J. Tomlinson: IEEE J. Quant. Electr. **QE-18,** 2016 (1982)
8.91 P.W. Smith, A. Ashkin, J.E. Bjorkholm, D.J. Eilenberger: Opt. Lett. **10,** 131 (1984)
8.92 M. LeBerre, E. Ressayre, A. Tallet, K. Tai, H.M. Gibbs: IEEE J. Quant. Electr. **QE-21,** 1404 (1985)
8.93 J.P. Moloney, F.A. Hopf, H.M. Gibbs: Phys. Rev. **A25,** 3442 (1982)
8.94 I.C. Khoo: Appl. Phys. Lett. **41,** 909 (1982);
I.C. Khoo, P.Y. Yan, T.H. Liu, S. Shepard, J.Y. Hou: Phys. Rev. **A29,** 2756 (1984)
8.95 A.E. Kaplan, P. Meystre: Opt. Lett. **6,** 590 (1981)
8.96 A.E. Kaplan, P. Meystre: Opt. Commun. **40,** 229 (1982)
8.97 A.E. Kaplan: Opt. Lett. **8,** 560 (1983)
8.98 A.E. Kaplan, C.T. Law: IEEE J. Quant. Electr. **QE-21,** 1529 (1985)
8.99 The diaphragm can be constructed as a "Gaussian" one, so that its transmissivity depends on ϱ as $\exp(-\varrho^2/d^2)$, which causes the beams to remain Gaussian.
8.100 E.M. Wright, P. Meystre, W.J. Firth, A.E. Kaplan: Phys. Rev. **A32,** 2857 (1985)
8.101 J.A. Hermann: Optics Comm. **62,** 367 (1987)
8.102 G.A. Swartzlander, A.E. Kaplan: J. Opt. Soc. Am. **B5,** 765 (1988)
8.103 I. Golub, Y. Beaudoin, S.L. Chin: Opt. Lett. **13,** 488 (1988);
JOSA **B5,** 2490 (1988)
8.104 R.A. Bergh, H.C. Lefevre, H.J. Shaw: Opt. Lett. **7,** 282 (1982)
8.105 S. Ezekiel, J.L. Davis, R.W. Hellwarth: Opt. Lett. **7,** 457 (1982)

Subject Index

MIX
Papier aus verantwortungsvollen Quellen
Paper from responsible sources
FSC® C105338

If you have any concerns about our products,
you can contact us on
ProductSafety@springernature.com

In case Publisher is established outside the EU,
the EU authorized representative is:
Springer Nature Customer Service Center GmbH
Europaplatz 3, 69115 Heidelberg, Germany

Printed by Libri Plureos GmbH
in Hamburg, Germany